# ENGINEERING MECHANICS

AF646349

# ENGINEERING MECHANICS

As per the Latest Syllabus of Rashtrasant Tukadoji Maharaj Nagpur University, Nagpur

**S S Bhavikatti**
Emeritus Professor
BVB College of Engineering & Technology
Hubli, Karnataka
(Formerly Principal, RYMEC, Bellary
Professor & Dean
SDMCET, Dharwad and NITK, Surathkal)

NEW AGE
RTM

*An Imprint of*

NEW AGE INTERNATIONAL (P) LIMITED, PUBLISHERS

New Delhi • Bangalore • Chennai • Cochin • Guwahati
Hyderabad • Kolkata • Lucknow • Mumbai
Visit us at **www.newagepublishers.com**

**Copyright © 2015, New Age International (P) Ltd., Publishers**
Published by New Age International (P) Ltd., Publishers
**First Edition:** 2015

---

**All rights reserved.**
No part of this book may be reproduced in any form, by photostat, microfilm, xerography, or any other means, or incorporated into any information retrieval system, electronic or mechanical, without the written permission of the publisher.

**BRANCHES**

- **Bangalore** 37/10, 8th Cross (Near Hanuman Temple), Azad Nagar, Chamarajpet, Bangalore-560 018
  Tel.: (080) 26756823, Telefax: 26756820, **E-mail: bangalore@newagepublishers.com**
- **Chennai** 26, Damodaran Street, T. Nagar, Chennai-600 017
  Tel.: (044) 24353401, Telefax: 24351463, **E-mail: chennai@newagepublishers.com**
- **Cochin** CC-39/1016, Carrier Station Road, Ernakulam South, Cochin-682 016
  Tel.: (0484) 2377004, Telefax: 4051303, **E-mail: cochin@newagepublishers.com**
- **Guwahati** Hemsen Complex, Mohd. Shah Road, Paltan Bazar, Near Starline Hotel, Guwahati-781 008
  Tel.: (0361) 2513881, Telefax: 2543669, **E-mail: guwahati@newagepublishers.com**
- **Hyderabad** 105, 1st Floor, Madhiray Kaveri Tower, 3-2-19, Azam Jahi Road, Near Kumar Theater Nimboliadda, Kachiguda, Hyderabad-500 027, Tel.: (040) 24652456, Telefax: 24652457
  **E-mail: hyderabad@newagepublishers.com**
- **Kolkata** RDB Chambers (Formerly Lotus Cinema) 106A, 1st Floor, S.N. Banerjee Road, Kolkata-700 014
  Tel.: (033) 22273773, Telefax: 22275247, **E-mail: kolkata@newagepublishers.com**
- **Lucknow** 16-A, Jopling Road, Lucknow-226 001
  Tel.: (0522) 2209578, 4045297, Telefax: 2204098, **E-mail:lucknow@newagepublishers.com**
- **Mumbai** 142C, Victor House, Ground Floor, N.M. Joshi Marg, Lower Parel, Mumbai-400 013
  Tel.: (022) 24927869, Telefax: 24915415, **E-mail: mumbai@newagepublishers.com**
- **New Delhi** 22, Golden House, Daryaganj, New Delhi-110 002
  Tel.: (011) 23262368, 23262370, Telefax: 43551305, **E-mail: sales@newagepublishers.com**

---

**ISBN: 978-81-224-3699-0**

**₹. 175.00**

**C-14-01-7317**

Printed in India at Print 'O' Pack, New Delhi.

Typeset at In-house, New Delhi.

**PUBLISHING GLOBALLY**
**NEW AGE INTERNATIONAL (P) LIMITED, PUBLISHERS**
7/30 A, Daryaganj, New Delhi-110002
Visit us at **www.newagepublishers.com**

# PREFACE

Engineering Mechanics is a basic subject taught to students of all branches. In this subject students learn how to apply laws of mechanics to actual engineering problems. By studying this subject students develop skill of splitting the larger practical problems into number of small problems and solve them easily without losing the continuity.

The author has clearly explained theories involved and illustrated by solving a number of problems. Neat free body diagrams have been drawn and solutions are presented. SI units and the standard notations suggested by Indian standard codes are used throughout. This has made this book to suit latest syllabus of RTM Nagpur University, Nagpur.

Author hopes the students and teachers of RTM Nagpur University, Nagpur will receive this book whole-heartedly. The suggestions and corrections, if any, are most welcome.

Author acknowledges the efforts of M/s New Age International Publishers for bringing out this book in nice form.

**S.S. Bhavikatti**

# Contents

*Preface* ............ (*v*)

**Chapter 1**

**Important Vector Quantities** ............ **1**

1.1 Vectorial Representation of Forces ............ 1
1.2 Addition of Vectors ............ 5
1.3 Subtraction of Vectors ............ 7
1.4 Product of a Vector with a Scalar Quantity ............ 7
1.5 Position Vector ............ 8
1.6 Displacement Vector ............ 10
1.7 Dot Product of Vectors ............ 11
1.8 Cross Product of Vectors ............ 14
1.9 Moment of a Force About a Point ............ 17
1.10 Components of Moments ............ 18
1.11 Moment of a Force About a given Axis ............ 19
1.12 Varignon's Theorem of Moments ............ 20
1.13 Couple Moment ............ 24
*Exercise* ............ 25
*Objective Type Questions* ............ 26

**Chapter 2**

**Equivalent Force System** ............ **28**

2.1 Resolution of Concurrent Coplanar Force System ............ 29
2.2 Resultant of Two Dimensional Concurrent Force System ............ 31
2.3 Transmissibility of a Force ............ 37
2.4 Resolution of a Force into a Force and a Couple ............ 37
2.5 Resultant of Non-concurrent Force System ............ 38
2.6 Resolution of Forces in Three Dimensional System ............ 45
2.7 Resultant of Concurrent Force System in Space ............ 49
2.8 Resultant of Three Dimensional General Force System ............ 53
2.9 Wrench Resultant ............ 55
*Exercise* ............ 59
*Objective Type Questions* ............ 64

## Chapter 3

**Equations of Equilibrium** ........ **66**

3.1 Types of Forces on a Body ........ 66
3.2 Free Body Diagram (FBD) ........ 71
3.3 Equilibrium of Concurrent Force System in a Plane ........ 72
3.4 Equilibrium of Concurrent Force System in Space ........ 85
3.5 Equilibrium of Co-planar Non-concurrent Systems ........ 90
3.6 Beams ........ 91
3.7 Equilibrium of General Spacial Force System ........ 111
*Questions* ........ 117
*Objective Type Questions* ........ 122

## Chapter 4

**Analysis of Pinjointed Frames** ........ **123**

4.1 Introduction ........ 123
4.2 Perfect, Deficient and Redundant Trusses ........ 123
4.3 Assumptions ........ 124
4.4 Nature of Forces in Members ........ 125
4.5 Methods of Analysis ........ 126
4.6 Method of Joints ........ 126
4.7 Method of Section ........ 138
*Important Formula* ........ 144
*Theory Questions* ........ 144
*Exercises* ........ 144
*Objective Type Questions* ........ 146

## Chapter 5

**Friction** ........ **147**

5.1 Introduction ........ 147
5.2 Laws of Coulomb Friction ........ 148
5.3 Equilibrium of Bodies Involving Dry Friction ........ 148
5.4 Belt Friction ........ 162
*Important Formulae* ........ 168
*Theory Questions* ........ 168
*Exercises* ........ 168
*Objective Type Questions* ........ 171

## CHAPTER 6

## CENTROID AND MOMENT OF INERTIA ........ 173

6.1 Determination of Areas and Volumes ........ 173
6.2 Centre of Gravity and Centroids ........ 177
6.3 Centroid of a Line ........ 179
6.4 First Moment of Area and Centroid ........ 183
6.5 Second Moments of Plane Area ........ 197
6.6 Moment of Inertia from First Principles ........ 200
6.7 Moment of Inertia of Composite Sections ........ 205
6.8 Product Moment of Plane Area ........ 218
6.9 Principal Moment of Inertia ........ 219
6.10 Mohr's Circle of Inertia ........ 224
6.11 Centre of Gravity of Solids ........ 226
6.12 Mass Moment of Inertia ........ 230
6.13 Determination of Mass Moment of Inertia From First Principles ........ 231
*Questions* ........ 239
*Exercises* ........ 239
*Objective Type Questions* ........ 245

## CHAPTER 7

## VIRTUAL WORK METHOD ........ 249

7.1 Work Done by Forces and Moments ........ 249
7.2 The Method ........ 251
7.3 Sign Convention ........ 251
*Exercises* ........ 267
*Question* ........ 268
*Objective Type Question* ........ 268

## CHAPTER 8

## D'ALEMBERT'S PRINCIPLE ........ 269

8.1 Newton's Second Law of Motion ........ 269
8.2 D'Alembert's Principle ........ 270
*Important Definition* ........ 282
*Questions* ........ 282
*Exercises* ........ 282

CHAPTER 9

WORK-ENERGY METHOD ........ 285

9.1 Work ........ 285
9.2 Work Done by a Varying Force ........ 286
9.3 Energy ........ 286
9.4 Power ........ 287
9.5 Work–Energy Equation for Translation ........ 288
9.6 Motion of Connected Bodies ........ 294
9.7 Work Done by a Spring ........ 296
*Important Definitions* ........ 299
*Important Formulae* ........ 300
*Questions* ........ 300
*Exercises* ........ 300
*Objective Type Question* ........ 302

CHAPTER 10

MOMENTUM METHOD ........ 303

10.1 Linear Impulse and Momentum ........ 303
10.2 Connected Bodies ........ 306
10.3 Force of Jet on a Vane ........ 312
10.4 Conservation of Momentum ........ 314
10.5 Pile and Pile Hammer ........ 318
*Important Definitions* ........ 321
*Important Equations* ........ 321
*Questions* ........ 321
*Exercises* ........ 321
*Objective Type Questions* ........ 323

CHAPTER 11

IMPACT OF ELASTIC BODIES ........ 324

11.1 Definitions ........ 324
11.2 Coefficient of Restitution ........ 324
11.3 Direct Central Impact ........ 327
11.4 Oblique Impact ........ 331
11.5 Loss of Kinetic Energy ........ 336
*Important Definitions* ........ 338
*Important Formulae* ........ 338
*Questions* ........ 338
*Exercises* ........ 338
*Objective Type Question* ........ 340

CHAPTER

# 1

# IMPORTANT VECTOR QUANTITIES

Various quantities used in engineering mechanics may be grouped into scalars and vectors. A quantity is said to be scalar if it is completely defined by its magnitude alone. Examples of scalars are time, length, area, mass, etc.

A quantity is said to be vector if it is completely defined only when its magnitudes as well as directions are specified. Examples of vectors are force, velocity, acceleration, momentum, etc. Vectors are represented by bold face letters in print and with bars above or below the letters when hand written. Thus, **A**, **B** or $\overline{A}, \overline{B}$ or $\underline{A}, \underline{B}$ are vectorial representations of vectors $A$ and $B$. The absolute magnitude of vector may be represented by ordinary letters like $A$, $B$ or $|A|$, $|B|$.

## 1.1 VECTORIAL REPRESENTATION OF FORCES

Vectorial representation of force **F** does not indicate the direction of the force. To make the direction clear, the vector is represented by its components in the cartesian coordinates. Referring to Fig. 1.1, the force **F** in a plane is written as

$$\mathbf{F} = \mathbf{F}_x + \mathbf{F}_y \qquad ...(1.1)$$

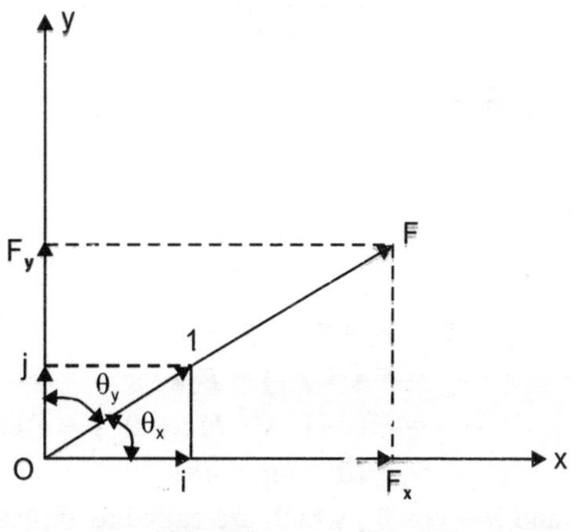


**Fig. 1.1**

More convenient way for handling vectors is to represent it in terms of unit vectors in the cartesian coordinate directions. If **i** and **j** are the unit vectors in the coordinate directions $x$ and $y$, then

$$\mathbf{F_x} = F_x\,\mathbf{i}$$
$$\mathbf{F_y} = F_y\,\mathbf{j}$$

where $F_x$ and $F_y$ represent only magnitudes.

Hence, $$\mathbf{F} = F_x\,\mathbf{i} + F_y\,\mathbf{j} \qquad \text{...(1.2)}$$

Obviously from Fig. 1.1,

$$F_x = F\cos\theta_x$$

and $$F_y = F\cos\theta_y$$

$\therefore$ $$\mathbf{F} = F\cos\theta_x\,\mathbf{i} + F\cos\theta_y\,\mathbf{j}$$
$$= F(l\mathbf{i} + m\mathbf{j}) \qquad \text{...(1.3)}$$

where $l = \cos\theta_x$ and $m = \cos\theta_y$ are the direction cosines of the force. In two dimensional problem,

$$\theta_y = 90 - \theta_x$$

$\therefore$ $$m = \cos\theta_y = \sin\theta_x$$

$\therefore$ Magnitude of the force

$$F = \sqrt{F_x^2 + F_y^2} = \sqrt{(Fl)^2 + (Fm)^2}$$
$$= F\sqrt{l^2 + m^2} \qquad \text{...(1.4)}$$

Similarly, a force in three dimensions (Ref. Fig. 1.2) may be written as

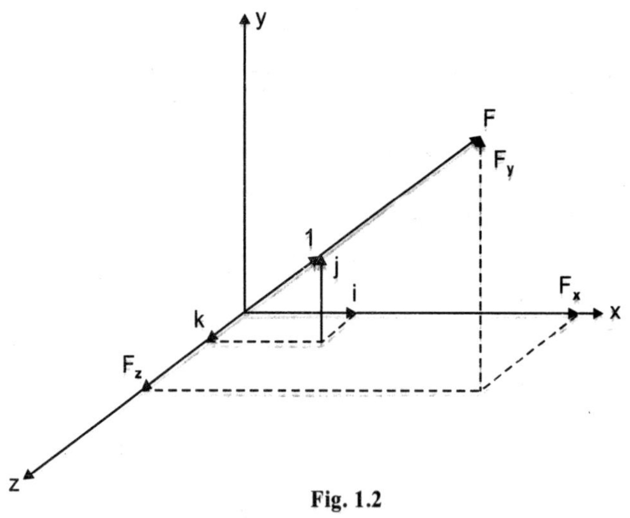


**Fig. 1.2**

$$\mathbf{F} = \mathbf{F_x} + \mathbf{F_y} + \mathbf{F_z}$$
$$= F_x\mathbf{i} + F_y\mathbf{j} + F_z\mathbf{k}$$
$$= F\cos\theta_x\,\mathbf{i} + F\cos\theta_y\,\mathbf{j} + F\cos\theta_z\,\mathbf{k} \qquad \text{...[1.5 (a)]}$$
$$= F(l\mathbf{i} + m\mathbf{j} + n\mathbf{k}) \qquad \text{...[1.5 (b)]}$$

where $l = \cos\theta_x$, $m = \cos\theta_y$ and $n = \cos\theta_z$, which are called as direction cosines and **i, j, k** are unit vectors in the directions $x$, $y$, $z$ respectively. The magnitude is given by

$$F = \sqrt{F_x^2 + F_y^2 + F_z^2} = \sqrt{(Fl)^2 + (Fm)^2 + (Fn)^2}$$

$$= F\sqrt{l^2 + m^2 + n^2} \qquad ...(1.6)$$

If **n** is the unit vector in the direction of the force, then

$$\mathbf{F} = F\mathbf{n} \qquad ...(1.7)$$

From Eqns. (1.5) and (1.7), we get

$$F(l\mathbf{i} + m\mathbf{j} + n\mathbf{k}) = F\mathbf{n}$$

or

$$\mathbf{n} = l\mathbf{i} + m\mathbf{j} + n\mathbf{k}$$

Thus a vector can be represented as,

$$\mathbf{F} = F\mathbf{n}$$
$$= F(l\mathbf{i} + m\mathbf{j} + n\mathbf{k})$$

where $F$ is the magnitude of the vector, **n** unit vector in the direction of vector and **i, j, k** are unit vectors in the coordinate directions *x, y, z*. **F** may be any vector like force, moment, velocity, etc. Usually in the literature, **F** is used to represent the force and **M** the moment.

**Example 1.1.** Determine the vectorial form of forces $\mathbf{F}_1$, $\mathbf{F}_2$ and $\mathbf{F}_3$, as shown in Fig. 1.3.

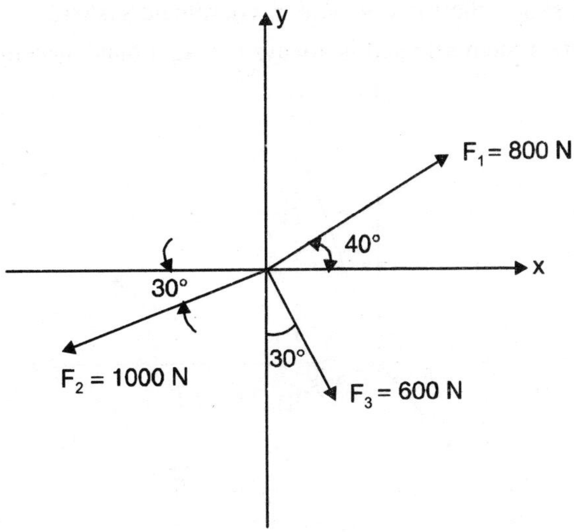


**Fig. 1.3** *Example 1.1*

**Solution.**

(*i*) Magnitude of force $\mathbf{F}_1 = 800$ N

$\theta_x = 40°$ $\quad\therefore \theta_y = 90° - 40° = 50°$

$\therefore \quad F_{1x} = 800 \cos 40° = 612.84$ N $\qquad F_{1y} = 800 \cos 50° = 514.23$ N

$\therefore \quad \mathbf{F}_1 = 612.84\,\mathbf{i} + 514.23\,\mathbf{j}$ **Ans.**

(*ii*) Magnitude of force $\mathbf{F}_2 = 1000$ N

$\theta_x = 180° + 30° = 210°$ and $\theta_y = 90° + 30° = 120°$

$\therefore \quad F_{2x} = 1000 \cos 210° = -866.03$ N

$F_{2y} = 1000 \cos 120° = -500$ N

$\therefore$ $\mathbf{F_2} = -866.03\ \mathbf{i} - 500\ \mathbf{j}$ **Ans.**

(*iii*) Magnitude of force $\mathbf{F_3} = 600$ N

$\theta_x = 270° + 30° = 300°$ $\theta_y = 180° + 30° = 210°$

$\therefore$ $F_{3x} = 600 \cos 300°$ $F_{3y} = 600 \cos 210°$

$= 300$ N $= -519.62$ N

$\therefore$ $\mathbf{F_3} = 300\ \mathbf{i} - 519.62\ \mathbf{j}$ **Ans.**

**Note:**

1. The component of forces may be found using acute angles with $x$ and $y$ axes, but carefully observing the signs of the components.
2. To get the direction of the force, move from tail to head of the vector in the desired directions.

**Example 1.2.** A force of magnitude 800 N makes 20° with $y$-axis and its projection on $x$-$z$ plane makes 30° with $x$-axis. Express the force in the vector form. Find the direction cosines and the direction of the force w.r.t. $x$, $y$ and $z$ directions.

**Solution.** Figure 1.4 shows the force in $x$, $y$, $z$, coordinate system.

[**Note:** The coordinate system selected is always the right hand system].

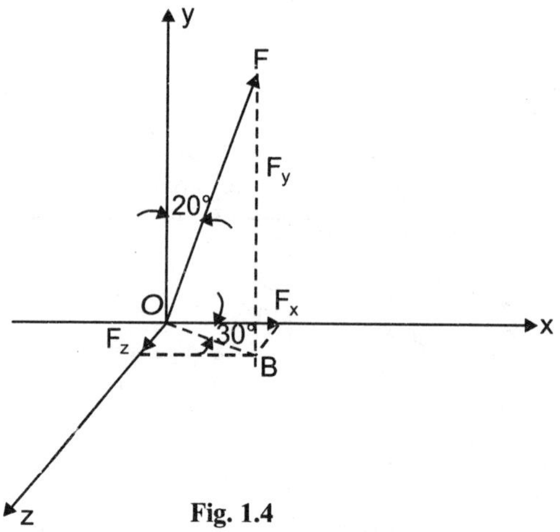


**Fig. 1.4**

$F = 800$ N

$F_y = 800 \cos 20° = 751.75$ N

$OB = 800 \cos (90° - 20°) = 273.62$ N

Resolving $OB$ in $x$-$z$ plane, we get

$F_x = OB \cos 30° = 273.62 \cos 30°$

$= 236.96$ N

$F_z = OB \cos (90° - 30°) = 273.62 \cos 60° = 136.81$ N

$\therefore$ $\mathbf{F} = 236.96\ \mathbf{i} + 751.75\ \mathbf{j} + 136.81\ \mathbf{k}$ **Ans.**

Let $l$, $m$, $n$ be direction cosines.

$F_x = F_l$

$$236.96 = 800\, l \quad \text{or} \quad l = 0.296$$

*i.e.,* $\cos \theta_x = 0.296 \quad \therefore \quad \theta_x = 72.77°$

Similarly, $F_y = F_m$

$$751.75 = 800\, m \quad \text{or} \quad m = 0.9396$$

*i.e.,* $\cos \theta_y = 0.9396 \quad \therefore \quad \theta_y = 20°$

and $F_z = F_n$

*i.e.,* $136.8 = 800\, n \text{ or } n = 0.171$

$\therefore \quad \theta_z = 80.15°$

Thus, the direction cosines are $l = 0.296$, $m = 0.9396$ and $n = 0.171$. Directions are $\theta_x = 72.77°$, $\theta_y = 20°$ and $\theta_z = 80.15°$. **Ans.**

## 1.2 ADDITION OF VECTORS

Addition of vectors is made as per the parallelogram law of vectors. This law is based on the experimental results and is an important law in mechanics. Though Stevinces employed it in 1586, the credit of presenting it as a law goes to Varignon and Newton (1687). Parallelogram law of vectors states that if two vectors acting simultaneously at a point are presented in magnitude and direction by two adjacent sides of a parallelogram, their resultant (addition) is represented in magnitude and direction by the diagonal of the parallelogram passing through that point.

In Fig. 1.5, **P** and **Q** are two vectors acting at point *O*. According to this law,

$$\mathbf{P} + \mathbf{Q} = \mathbf{R} \qquad ...(1.8)$$

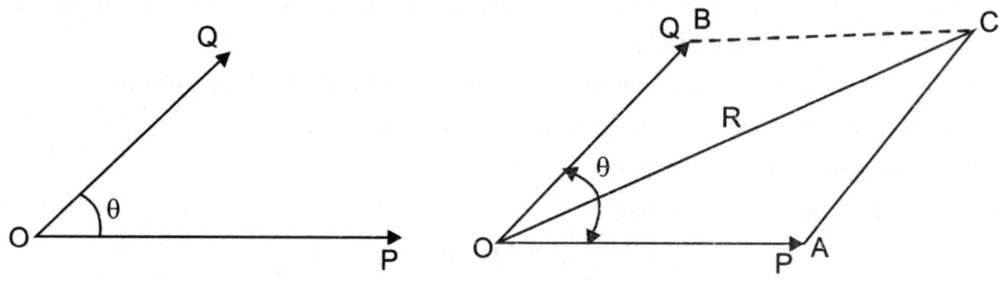


**Fig. 1.5** *Parelogram law of forces*

From parallelogram law of vectors, triangle law of vectors may be derived. Referring to Fig. 1.5, instead of drawing parallelogram *OACB* to get *R*, same result may be obtained by constructing the triangle *OAC*. In other words, to get vector addition of *P* and *Q*, construct triangle *OAC* to get *R*. Thus, by drawing vectors one after the other in tail to head fashion, triangle of vectors may be constructed. The closing line of the triangle, *i.e.*, *OC* represents the addition of **P** and **Q** as shown in Fig. 1.6.

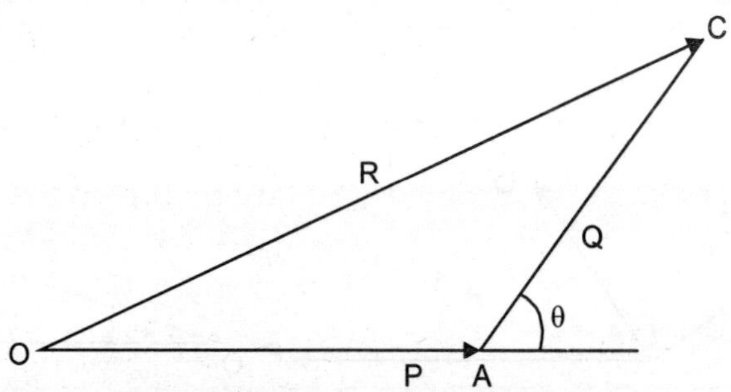


**Fig. 1.6** *Triangle law of forces*

Triangle law of forces may be used repeatedly to get addition of number of vectors. Let **P**, **Q**, **R** and **S** be the vectors to be added [Ref. Fig. 1.7].

From Fig. 1.7, using triangle law of vectors, we find

$$\mathbf{T} = \mathbf{P} + \mathbf{Q}$$

$$\mathbf{U} = \mathbf{T} + \mathbf{R} = \mathbf{P} + \mathbf{Q} + \mathbf{R}$$

$$\mathbf{V} = \mathbf{U} + \mathbf{S} = \mathbf{P} + \mathbf{Q} + \mathbf{R} + \mathbf{S} \quad \text{...(1.9)}$$

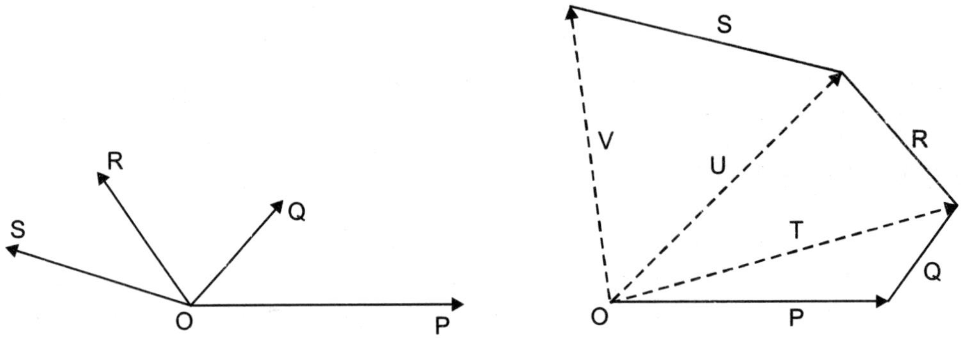


**Fig. 1.7** *Polygon law of vectors*

Instead of drawing triangles to add two vectors at a time, from Fig 1.7, it may be observed that the same result can be obtained if the vectors are added in tail to head fashion one after the other. Then the closing line of the polygon shows the addition in the direction from the first point to the last point. Hence, the following law can be concluded.

If vectors are drawn one after the other in tail to head fashion, the line joining the tail of the first vector to the head of the last vector represents the sum of the vectors.

This is called **Polygon Law of Vectors.**

From Fig. 1.5, it may be observed that

$$\mathbf{R} = \mathbf{P} + \mathbf{Q} = \mathbf{Q} + \mathbf{P} \quad \text{...(1.10)}$$

Thus, the **commutative law** of addition holds good.

Figure 1.8 shows addition of three vectors **P**, **Q**, **R** using polygon law of forces. In this, it may be seen that

$$(\mathbf{P} + \mathbf{Q}) + \mathbf{R} = \mathbf{P} + (\mathbf{Q} + \mathbf{R}) = \mathbf{T}$$

Since, $\mathbf{P} + \mathbf{Q} = \mathbf{T}_1$ and $\mathbf{Q} + \mathbf{R} = \mathbf{T}_2$

Thus, the **associative law** also holds good.

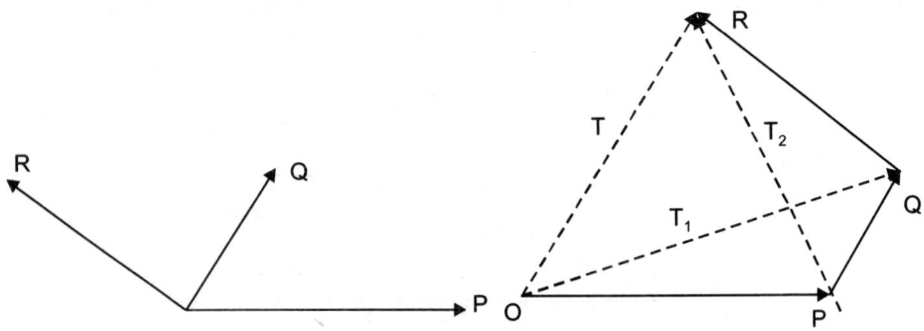


**Fig. 1.8**

All the above concepts have been developed using graphical method, since, parallelogram law of vector addition is the base for the vector addition. It is not convenient to use graphical method in the analysis, since, it needs drawing sheet and drawing accessories. **Analytical method** is more convenient for any design office. This method consists in expressing the quantities in the vectorial form as presented in Eqn. 1.5 (*b*). Then all the components in **i** direction can be added algebraically, since, they are collinear. Similarly, components in **j** and **k** directions may be added. Thus if

$$\mathbf{P} = 5\mathbf{i} + 6\mathbf{j} - 3\mathbf{k}$$
$$\mathbf{Q} = 7\mathbf{i} - 4\mathbf{j} + 5\mathbf{k}$$
$$\mathbf{R} = \mathbf{P} + \mathbf{Q} = (5 + 7)\mathbf{i} + (6 - 4)\mathbf{j} + (-3 + 5)\mathbf{k}$$
$$= 12\mathbf{i} + 2\mathbf{j} + 2\mathbf{k}$$

Commutative and associative laws hold good.

## 1.3 SUBTRACTION OF VECTORS

If **Q** is a vector, then '– **Q**' is a vector of the same magnitude and line of action, but opposite in the sense. Hence, to subtract **Q** from **P**, we can make use of law of addition to add **P** and '– **Q**'. Figure 1.9 shows the use of parallelogram and triangle laws for this purpose.

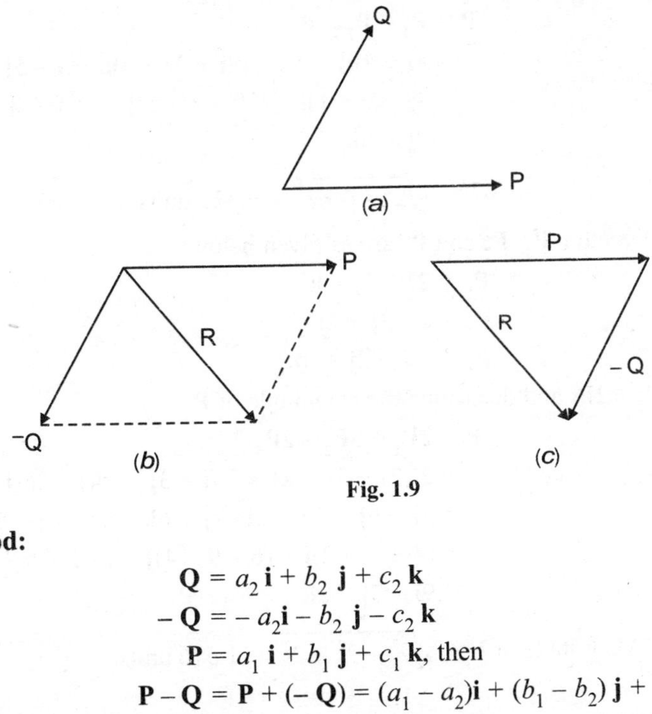


**Fig. 1.9**

**Analytic Method:**

If $$\mathbf{Q} = a_2\,\mathbf{i} + b_2\,\mathbf{j} + c_2\,\mathbf{k}$$

then $$-\mathbf{Q} = -a_2\mathbf{i} - b_2\,\mathbf{j} - c_2\,\mathbf{k}$$

If $$\mathbf{P} = a_1\,\mathbf{i} + b_1\,\mathbf{j} + c_1\,\mathbf{k}, \text{ then}$$

$$\mathbf{P} - \mathbf{Q} = \mathbf{P} + (-\mathbf{Q}) = (a_1 - a_2)\mathbf{i} + (b_1 - b_2)\,\mathbf{j} + (c_1 - c_2)\mathbf{k}$$

## 1.4 PRODUCT OF A VECTOR WITH A SCALAR QUANTITY

If we add **P** to itself, we get a vector of magnitude 2**P**. Similarly, if we add **P** to itself *m* times, we get vector of *m* times **P,** *i.e.,* it will be *m***P**. On the same lines, we can see that

$$(m+n)\mathbf{P} = m\mathbf{P} + n\mathbf{P}$$
$$m(\mathbf{P}+\mathbf{Q}) = m\mathbf{P} + m\mathbf{Q}$$
$$m(n\mathbf{P}) = mn\mathbf{P}$$

**Example 1.3.** $\mathbf{P}_1$, $\mathbf{P}_2$ and $\mathbf{P}_3$ are the vectors as given below:

$$\mathbf{P}_1 = 8\mathbf{i} + 10\mathbf{j} - 6\mathbf{k}$$
$$\mathbf{P}_2 = 9\mathbf{i} + 3\mathbf{j} + 4\mathbf{k}$$

and $\mathbf{P}_3 = \mathbf{i} - 5\mathbf{j} + 4\mathbf{k}$

Determine (*i*) $\mathbf{P} = \mathbf{P}_1 + \mathbf{P}_2 + \mathbf{P}_3$ and $P$

(*ii*) $\mathbf{P} = \mathbf{P}_1 - \mathbf{P}_2 + \mathbf{P}_3$ and $P$

**Solution.**

(*i*)

$$\begin{aligned}\mathbf{P} &= \mathbf{P}_1 + \mathbf{P}_2 + \mathbf{P}_3 \\ &= 8\mathbf{i} + 10\mathbf{j} - 6\mathbf{k} + 9\mathbf{i} + 3\mathbf{j} + 4\mathbf{k} + \mathbf{i} - 5\mathbf{j} + 4\mathbf{k} \\ &= (8+9+1)\mathbf{i} + (10+3-5)\mathbf{j} + (-6+4+4)\mathbf{k} \\ &= 18\mathbf{i} + 8\mathbf{j} + 2\mathbf{k}\end{aligned}$$

$$P = \sqrt{18^2 + 8^2 + 2^2} = 19.80 \text{ units}$$ **Ans.**

(*ii*)

$$\begin{aligned}\mathbf{P} &= \mathbf{P}_1 - \mathbf{P}_2 + \mathbf{P}_3 \\ &= 8\mathbf{i} + 10\mathbf{j} - 6\mathbf{k} - (9\mathbf{i} + 3\mathbf{j} + 4\mathbf{k}) + \mathbf{i} - 5\mathbf{j} + 4\mathbf{k} \\ &= (8-9+1)\mathbf{i} + (10-3-5)\mathbf{j} + (-6-4+4)\mathbf{k} \\ &= 2\mathbf{j} - 6\mathbf{k}\end{aligned}$$

$$P = \sqrt{2^2 + (-6)^2} = 6.325 \text{ units}$$ **Ans.**

**Example 1.4.** The vectors $\mathbf{P}_1$, $\mathbf{P}_2$ and $\mathbf{P}_3$ are as given below:

$$\mathbf{P}_1 = 2\mathbf{i} + 3\mathbf{j} - \mathbf{k}$$
$$\mathbf{P}_2 = \mathbf{i} - 3\mathbf{j} + 2\mathbf{k}$$
$$\mathbf{P}_3 = -\mathbf{i} + 2\mathbf{j} + 3\mathbf{k}$$

Find $\mathbf{P} = 2\mathbf{P}_1 + 3\mathbf{P}_2 - 2\mathbf{P}_3$ and determine the magnitude of $\mathbf{P}$.

**Solution.**

$$\begin{aligned}\mathbf{P} &= 2\mathbf{P}_1 + 3\mathbf{P}_2 - 2\mathbf{P}_3 \\ &= 2(2\mathbf{i} + 3\mathbf{j} - \mathbf{k}) + 3(\mathbf{i} - 3\mathbf{j} + 2\mathbf{k}) - 2(-\mathbf{i} + 2\mathbf{j} + 3\mathbf{k}) \\ &= 4\mathbf{i} + 6\mathbf{j} - 2\mathbf{k} + 3\mathbf{i} - 9\mathbf{j} + 6\mathbf{k} + 2\mathbf{i} - 4\mathbf{j} - 6\mathbf{k} \\ &= (4+3+2)\mathbf{i} + (6-9-4)\mathbf{j} + (-2+6-6)\mathbf{k} \\ &= 9\mathbf{i} - 7\mathbf{j} - 2\mathbf{k}\end{aligned}$$

Magnitude of $\mathbf{P} = \sqrt{9^2 + 7^2 + 2^2} = 11.576$ units. **Ans.**

## 1.5 POSITION VECTOR

*The position vector of a point in the space is defined as the vector represented by the line segment connecting the origin and the point.* The cartesian coordinates system used to locate a point in the space is always taken according to right hand rule. Figure 1.10 shows different orientations of cartesian system according to right hand rule.

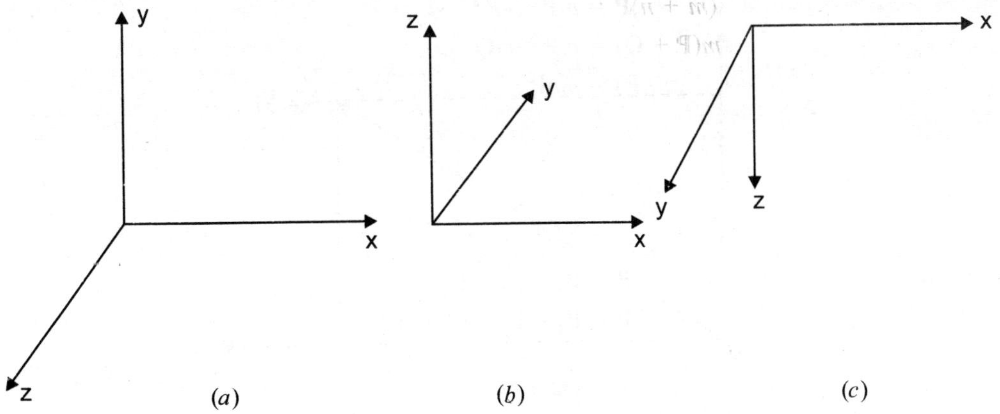


**Fig. 1.10** *Different orientations of right hand system of cartesian coordinates*

Let $A$ be a point in space and its coordinates be $x, y, z$ as shown in Fig. 1.11.

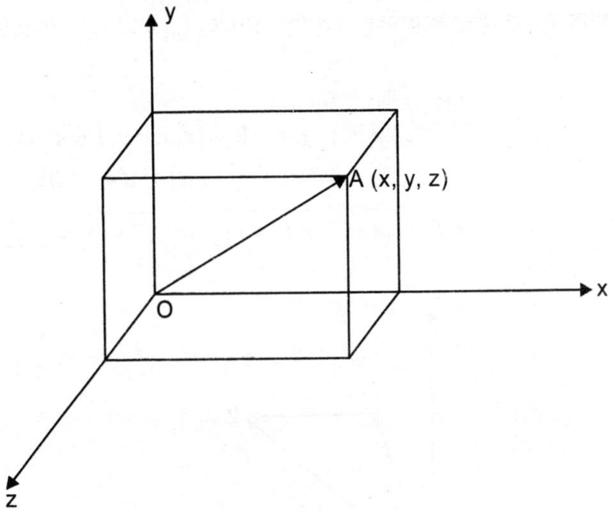


**Fig. 1.11** *Position vector*

Then the position vector of point $A$ is $OA$, represented as $r_{OA}$, and is given by

$$\mathbf{r_{OA}} = x\mathbf{i} + y\mathbf{j} + z\mathbf{k} \qquad ...[1.11(a)]$$

Magnitude of this vector is given by

$$r_{OA} = \sqrt{x^2 + y^2 + z^2} \qquad ...[1.11(b)]$$

In case of two dimensional problem, the position vector of $OA$ (Fig. 1.12) is

$$\mathbf{r_{OA}} = x\mathbf{i} + y\mathbf{j}$$

and $$r_{OA} = \sqrt{x^2 + y^2}$$

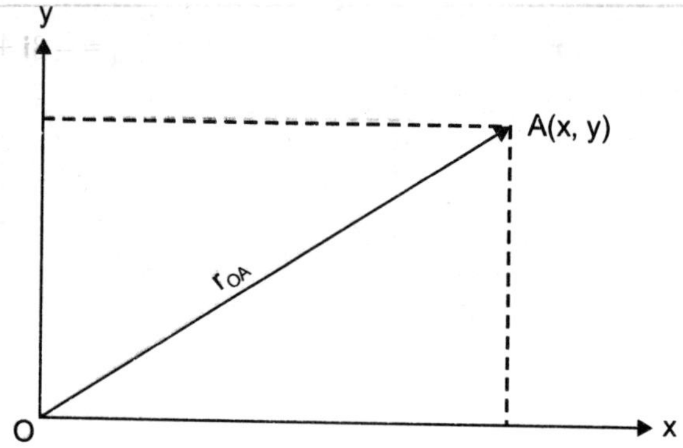


**Fig. 1.12**

## 1.6 DISPLACEMENT VECTOR

A vector in space with its tail at one point (say point 1) and the head at another point (say point 2) is termed as displacement vector. It represents the rectilinear distance between those two points.

In Fig. 1.13, the vector $\mathbf{r}_{12}$ is displacement vector. Since, $\mathbf{r}_{01}$ and $\mathbf{r}_{02}$ are referred as position vectors of points 1 and 2,

$$\mathbf{r}_{12} = \mathbf{r}_{02} - \mathbf{r}_{01}$$

$$= x_2\mathbf{i} + y_2\,\mathbf{j} + z_2\mathbf{k} - (x_1\,\mathbf{i} + y_1\,\mathbf{j} + z_1\mathbf{k})$$

$$= (x_2 - x_1)\mathbf{i} + (y_2 - y_1)\mathbf{j} + (z_2 - z_1)\mathbf{k} \qquad \text{...[1.12}(a)\text{]}$$

and hence

$$r_{12} = \sqrt{(x_2 - x_1)^2 + (y_2 - y_1)^2 + (z_2 - z_1)^2} \qquad \text{...[1.12}(b)\text{]}$$

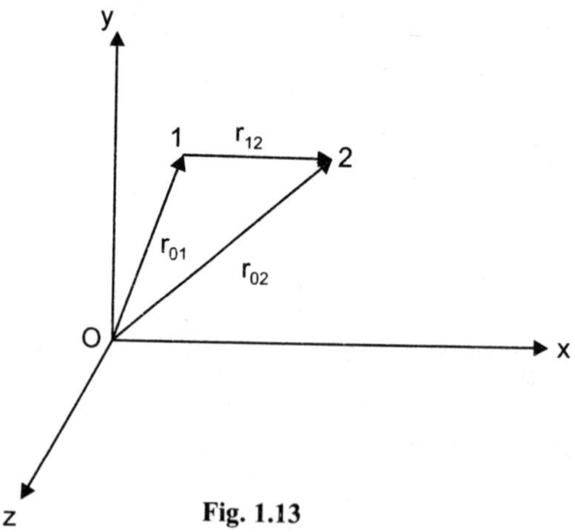


**Fig. 1.13**

**Example 1.5.** Determine the vector form of displacement vector *AB* connecting points *A* (15, –10) and *B* (–8, 20). Find its magnitude, direction cosines and directions also.

**Solution.**

$A\ (15, -10),\ \ B\ (-8, 20)$

$$\mathbf{r_{OA}} = 15\mathbf{i} - 10\mathbf{j} \quad \text{and} \quad \mathbf{r_{OB}} = -8\mathbf{i} + 20\mathbf{j}$$

$$\therefore \quad \mathbf{r_{AB}} = \mathbf{r_{OB}} - \mathbf{r_{OA}}$$

$$= -8\mathbf{i} + 20\mathbf{j} - (15\mathbf{i} - 10\mathbf{j})$$

$$= -23\mathbf{i} + 30\mathbf{j} \qquad \textbf{Ans.}$$

$$r_{AB} = \sqrt{(-23)^2 + 30^2} = 37.80 \text{ units} \qquad \textbf{Ans.}$$

$$r_{AB}\, l = -23 \qquad \therefore \quad l = -\frac{23}{37.80} = -0.608 \qquad \textbf{Ans.}$$

$$r_{AB}\, m = 30 \qquad \therefore \quad m = \frac{30}{37.80} = 0.794 \qquad \textbf{Ans.}$$

$$\cos \theta_x = l = -0.608 \qquad \therefore \quad \theta_x = 127.445° \qquad \textbf{Ans.}$$

$$\cos \theta_y = m = 0.794 \qquad \theta_y = 37.439° \qquad \textbf{Ans.}$$

**Example 1.6.** Determine the displacement vector *AB* when the coordinates of *A* and *B* in mm units are given by *A* (60, – 80, 20) and *B* (100, – 40, 80). Find its magnitude, direction cosines and directions.

**Solution.**

$$A\,(60, -80, 20) \qquad B\,(100, -40, 80)$$

$$\therefore \quad \mathbf{r_{OA}} = 60\mathbf{i} - 80\mathbf{j} + 20\mathbf{k} \qquad \mathbf{r_{OB}} = 100\mathbf{i} - 40\mathbf{j} + 80\mathbf{k}$$

$$\therefore \quad \mathbf{r_{AB}} = \mathbf{r_{OB}} - \mathbf{r_{OA}}$$

$$= 100\mathbf{i} - 40\mathbf{j} + 80\mathbf{k} - (60\mathbf{i} - 80\mathbf{j} + 20\mathbf{k})$$

$$= 40\mathbf{i} + 40\mathbf{j} + 60\mathbf{k} \qquad \textbf{Ans.}$$

$$\therefore \quad r_{AB} = \sqrt{40^2 + 40^2 + 60^2} = 82.46 \text{ mm} \qquad \textbf{Ans.}$$

$$r_{AB}\, l = 40 \qquad r_{AB}\, m = 40 \quad \text{and} \quad r_{AB}\, n = 60$$

$$\therefore \quad l = \frac{40}{82.46} \qquad m = \frac{40}{82.46} \quad \text{and} \quad n = \frac{60}{82.46}$$

$$l = 0.485 \qquad m = 0.485 \qquad n = 0.728 \qquad \textbf{Ans.}$$

$$\therefore \quad \theta_x = \cos^{-1}(0.485) \qquad \theta_y = \cos^{-1}(0.485) \quad \text{and} \quad \theta_z = \cos^{-1}(0.728)$$

$$\theta_x = 60.988° \qquad \theta_y = 60.988° \qquad \theta_z = 43.31° \qquad \textbf{Ans.}$$

## 1.7 DOT PRODUCT OF VECTORS

The dot product of two vectors **P** and **Q** is defined as

$$\mathbf{P} \cdot \mathbf{Q} = PQ \cos \theta \qquad \text{...(1.13)}$$

where θ is the acute angle between the two vectors. The left hand side of Eqn. (1.13) is read as *P* dot *Q*. Since, the right hand side of dot product is a scalar, this product is also known as *scalar product of vectors.*

The graphical representation of dot product is as shown in Fig. 1.14. The right hand side of Eqn. (1.13) may be looked as the magnitude of *P* multiplied by the projection of magnitude of *Q* on *P* [Fig. 1.14 (*b*)] or as magnitude of *Q* multiplied by the projection of magnitude of *P* on *Q* [as seen from Fig. 1.14. (*c*)].

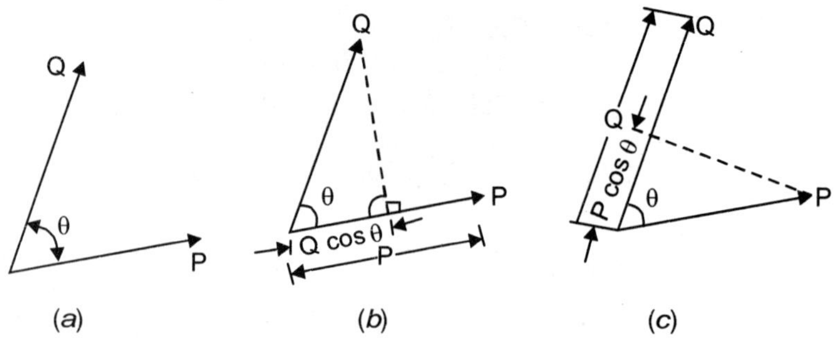


**Fig. 1.14**

Since, θ is the angle between the two vectors involved in the dot product, we get

$$\mathbf{i} \cdot \mathbf{i} = (1)\,(1) \cos 0 = 1$$
$$\mathbf{j} \cdot \mathbf{j} = (1)\,(1) \cos 0 = 1$$
$$\mathbf{k} \cdot \mathbf{k} = (1)\,(1) \cos 0 = 1$$
$$\mathbf{i} \cdot \mathbf{j} = (1)\,(1) \cos 90 = 0$$

Similarly, $\mathbf{j} \cdot \mathbf{k} = \mathbf{k} \cdot \mathbf{i} = \mathbf{i} \cdot \mathbf{k} = \mathbf{k} \cdot \mathbf{j} = \mathbf{j} \cdot \mathbf{i} = 0$

Thus in dot product, we have

$$\mathbf{i} \cdot \mathbf{i} = \mathbf{j} \cdot \mathbf{j} = \mathbf{k} \cdot \mathbf{k} = 1 \qquad \text{...[1.14(a)]}$$

and

$$\mathbf{i} \cdot \mathbf{j} = \mathbf{j} \cdot \mathbf{k} = \mathbf{k} \cdot \mathbf{i} = \mathbf{i} \cdot \mathbf{k} = \mathbf{k} \cdot \mathbf{j} = \mathbf{j} \cdot \mathbf{i} = 0 \qquad \text{...[1.14(b)]}$$

The following useful results from dot product may be noted:

(*i*) If $\mathbf{P} \cdot \mathbf{Q} = 0$

$$PQ \cos \theta = 0$$

$$\therefore \quad \cos \theta = 0 \qquad \text{...(1.15)}$$

*i.e.*, vectors **P** and **Q** are at right angles to each other.

(*ii*) Using Eqn. (1.11), we obtain

$$\mathbf{P} \cdot \mathbf{Q} = (P_x\mathbf{i} + P_y\mathbf{j} + P_z\mathbf{k}) \cdot (Q_x\mathbf{i} + Q_y\mathbf{j} + Q_z\mathbf{k})$$
$$= P_xQ_x + P_yQ_y + P_zQ_z \qquad \text{...[1.16(a)]}$$

and also

$$\mathbf{P} \cdot \mathbf{P} = \mathrm{P}_x^2 + \mathrm{P}_y^2 + \mathrm{P}_z^2 \qquad \text{...[1.16(b)]}$$

(*iii*) If **P** and **Q** are known, the angle between the two vectors is given by

$$\cos \theta = \frac{\mathbf{P.Q}}{PQ} = \frac{P_xQ_x + P_yQ_y + P_zQ_z}{PQ} \qquad \text{...(1.17)}$$

If $l_1$, $m_1$, $n_1$ are the direction cosines of **P** and $l_2$, $m_2$, $n_2$ are the direction cosines of **Q**, then

$$\cos \theta = \frac{Pl_1Ql_2 + Pm_1Qm_2 + Pn_1Qn_2}{PQ}$$

$$= l_1l_2 + m_1m_2 + n_1n_2 \qquad \text{...(1.18)}$$

(*iv*) The projection of a vector may be obtained in any desired direction, since, the projection of a vector **P** in direction θ to it is equal to $P \cos \theta$, $\cos \theta$ can be found using Eqn. (1.18).

**Example 1.7.** Determine the dot product and the angle between the vectors

$$\mathbf{F}_1 = 4\mathbf{i} + \mathbf{j} - \mathbf{k}$$

and

$$\mathbf{F}_2 = 3\mathbf{i} + 2\mathbf{j} + \mathbf{k}$$

**Solution.** We know that $\mathbf{i}\cdot\mathbf{i} = \mathbf{j}\cdot\mathbf{j} = \mathbf{k}\cdot\mathbf{k} = 1$ and $\mathbf{i}\cdot\mathbf{j} = \mathbf{j}\cdot\mathbf{k}$, etc. $= 0$

$$\mathbf{F}_1\cdot\mathbf{F}_2 = (4\mathbf{i} + \mathbf{j} - \mathbf{k})\cdot(3\mathbf{i} + 2\mathbf{j} + \mathbf{k})$$

$$= 4 \times 3 + 1 \times 2 - 1 \times 1$$

$$= 13 \qquad \textbf{Ans.}$$

$$F_1 = \sqrt{4^2 + 1^2 + (-1)^2} = 4.243$$

$$F_2 = \sqrt{3^2 + 2^2 + 1^2} = 3.742$$

From dot matrix rule, we know

$$\mathbf{F}_1\cdot\mathbf{F}_2 = F_1F_2 \cos\theta$$

$$\therefore \quad 13 = (4.243)\,(3.742)\cos\theta.$$

$$\therefore \quad \cos\theta = 0.8189 \text{ or } \theta = 35.037° \qquad \textbf{Ans.}$$

**Example 1.8.** If $\mathbf{F}_1 = 5\mathbf{i} + 4\mathbf{j} + 3\mathbf{k}$ and $\mathbf{F}_2 = 4\mathbf{i} - 2\mathbf{j} - 4\mathbf{k}$, show that $\mathbf{F}_1$ and $\mathbf{F}_2$ are at right angles to each other.

**Solution.**

$$F_1 = \sqrt{5^2 + 4^2 + 3^2} = 7.071$$

$$F_2 = \sqrt{4^2 + (-2)^2 + (-4)^2} = 6$$

$$\mathbf{F}_1\cdot\mathbf{F}_2 = (5\mathbf{i} + 4\mathbf{j} + 3\mathbf{k})\cdot(4\mathbf{i} - 2\mathbf{j} - 4\mathbf{k})$$

$$= 5 \times 4 - 4 \times 2 + 3\,(-4) = 0$$

*i.e.,* $F_1 F_2 \cos\theta = 0$ *i.e.,* $\cos\theta = 0$

$\therefore$ $F_1$ and $F_2$ are at right angles to each other. **Ans.**

**Example 1.9.** Determine the projection of force $\mathbf{F}_1 = 3\mathbf{i} + 2\mathbf{j} - \mathbf{k}$ on the line of action of the force $\mathbf{F}_2 = \mathbf{i} + 4\mathbf{j} + \mathbf{k}$.

**Solution.** $\mathbf{F}_1 = 3\mathbf{i} + 2\mathbf{j} - \mathbf{k}$ $\quad\therefore F_1 = \sqrt{3^2 + 2^2 + (-1)^2} = 3.742$

$\mathbf{F}_2 = \mathbf{i} + 4\mathbf{j} + \mathbf{k}$ $\quad\therefore F_2 = \sqrt{1^2 + 4^2 + 1^2} = 4.2426$

$$\therefore \quad \mathbf{F}_1\cdot\mathbf{F}_2 = (3\mathbf{i} + 2\mathbf{j} - \mathbf{k})\cdot(\mathbf{i} + 4\mathbf{j} + \mathbf{k})$$

$$= 3 + 2 \times 4 - 1 \times 1 = 10$$

*i.e.,* $F_1 F_2 \cos\theta = 10$

$\therefore$ Projection of $\mathbf{F}_1$ on the line of action of $\mathbf{F}_2$ is given as

$$F_1 \cos\theta = \frac{F_1 F_2 \cos\theta}{F_2} = \frac{10}{4.2426} = 2.357 \qquad \textbf{Ans.}$$

## 1.8 CROSS PRODUCT OF VECTORS

Cross product of vectors is the another type of product of vectors used in vector operation. This is also called *vector product of vectors*. The cross product of two vectors **P** and **Q** is defined by

$$\mathbf{P} \times \mathbf{Q} = PQ \sin \theta \, \mathbf{n} \quad ...(1.19)$$

The left hand side of Eqn. (1.19) is read as *P* cross *Q*. In the right hand side, 'θ' is the acute angle between *P* and *Q*, *i.e.*, sin θ is always a positive quantity and **n** is the unit vector in the direction perpendicular to the plane of *P* and *Q*. Its sense is given by the direction of advance of a right hand screw rotated from first to second vector. The vectors **P**, **Q** and their cross product vector **R** = **P** × **Q** are shown in Fig. 1.15. The following points of cross product may be carefully noted.

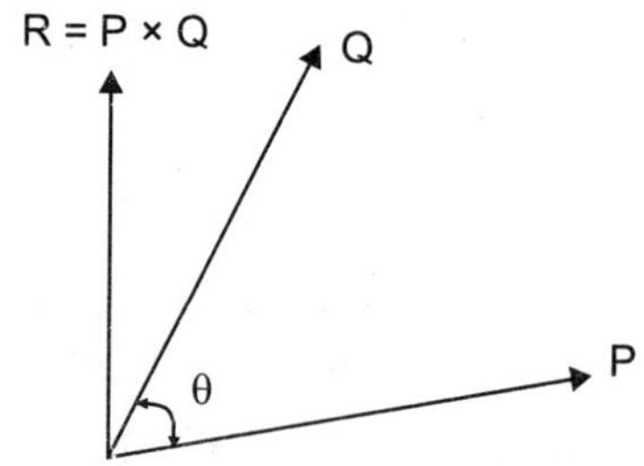


**Fig. 1.15**

(*i*) The magnitude is *PQ* sin θ.

(*ii*) It is a vector, with direction at right angles to the plane of *P* and *Q* in the sense advance of right hand screw from first to second vector is positive.

Figure 1.16 shows the cross product **Q** × **P**. Obviously its direction is opposite to that of **P** × **Q**, but the magnitude is same as that of **P** × **Q**.

Figure 1.17 shows the vectors **P** and **Q** drawn to the scale and the parallelogram with **P** and **Q** as sides completed.

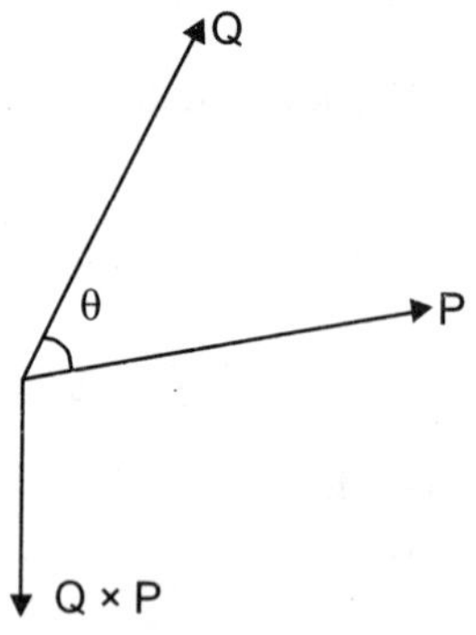


**Fig. 1.16**

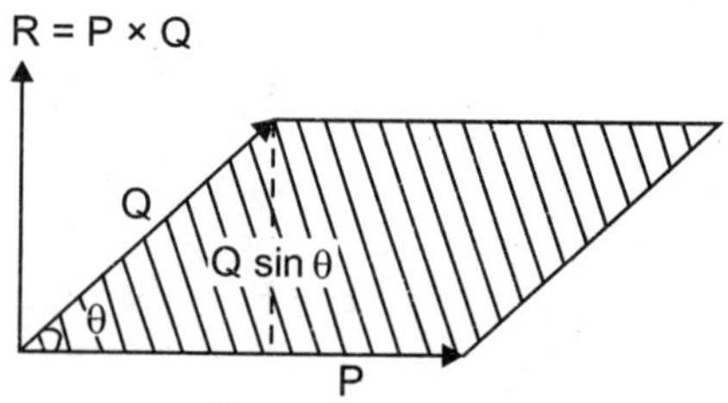


**Fig. 1.17**

Now, the magnitude of the product **P** × **Q** is

$$= PQ \sin \theta$$

$$= P (Q \sin \theta)$$

$$= \text{Area of the parallelogram} \quad ...(1.20)$$

Thus, the magnitude of **P** cross **Q** is equal to the area of parallelogram built with **P** and **Q** as sides.

From the cross product rule, we can note the following points:

(*i*) Since, **i**, **j** and **k** are unit vectors in three mutually perpendicular directions taken according to the right handed coordinate system, from the definition of cross product, we note

$$\mathbf{i}\times\mathbf{j} = \mathbf{k} \qquad \mathbf{j}\times\mathbf{k} = \mathbf{i} \qquad \text{and} \qquad \mathbf{k}\times\mathbf{i} = \mathbf{j} \qquad \text{...[1.21(a)]}$$

$$\mathbf{j}\times\mathbf{i} = -\mathbf{k} \qquad \mathbf{k}\times\mathbf{j} = -\mathbf{i} \qquad \text{and} \qquad \mathbf{i}\times\mathbf{k} = -\mathbf{j} \qquad \text{...[1.21(b)]}$$

(*ii*) The cross product of a collinear vector must vanish, since, in such case, $\theta = 0$.

*i.e.*,
$$\mathbf{i}\times\mathbf{i} = \mathbf{j}\times\mathbf{j} = \mathbf{k}\times\mathbf{k} = 0 \qquad \text{...(1.22)}$$

(*iii*) Now
$$\begin{aligned}\mathbf{P}\times\mathbf{Q} &= (P_x\mathbf{i} + P_y\mathbf{j} + P_z\mathbf{k})\times(Q_x\mathbf{i} + Q_y\mathbf{j} + Q_z\mathbf{k})\\ &= P_xQ_y\mathbf{k} + P_xQ_z(-\mathbf{j}) + P_yQ_x(-\mathbf{k}) + P_yQ_z\mathbf{i} + P_zQ_x\mathbf{j}\\ &\quad + P_zQ_y(-\mathbf{i})\\ &= (P_yQ_z - P_zQ_y)\mathbf{i} + (P_zQ_x - P_xQ_z)\mathbf{j} + (P_xQ_y - P_yQ_x)\mathbf{k}\\ &= \begin{vmatrix}\mathbf{i} & \mathbf{j} & \mathbf{k}\\ P_x & P_y & P_z\\ Q_x & Q_y & Q_z\end{vmatrix} \qquad \text{...(1.23)}\end{aligned}$$

(*iv*) The cross product may be used to find the angle between the two vectors. In this case,

$$\sin\theta = \frac{|P\times Q|}{PQ} \qquad \text{...(1.24)}$$

where $|P \times Q|$ is the absolute magnitude of **P** × **Q**.

(*v*) The commutative law does not hold good

*i.e.*,
$$\mathbf{P}\times\mathbf{Q} \neq \mathbf{Q}\times\mathbf{P} \qquad \text{...(1.25)}$$

(*vi*) The distributive law holds good, *i.e.*,

$$\mathbf{P}\times(\mathbf{Q}\times\mathbf{R}) = \mathbf{P}\times\mathbf{Q}\times\mathbf{R} \qquad \text{...[1.26(a)]}$$

and
$$n(\mathbf{P}\times\mathbf{Q}) = n\mathbf{P}\times\mathbf{Q} = \mathbf{P}\times n\mathbf{Q} \qquad \text{...[1.26(b)]}$$

**Example 1.10.** Using cross product, determine the angle between the vectors $\mathbf{F_1}$ and $\mathbf{F_2}$ given in Example 1.7. Also determine the magnitude and direction cosines of the resultant vector.

**Solution.** In this problem,

$$\mathbf{F_1} = 4\mathbf{i} + \mathbf{j} - \mathbf{k} \text{ and } \mathbf{F_2} = 3\mathbf{i} + 2\mathbf{j} + \mathbf{k}$$

$$\therefore \qquad \mathbf{F_1}\times\mathbf{F_2} = \begin{vmatrix}\mathbf{i} & \mathbf{j} & \mathbf{k}\\ 4 & 1 & -1\\ 3 & 2 & 1\end{vmatrix}$$

$$= \mathbf{i}(1 + 2) - \mathbf{j}(4 + 3) + \mathbf{k}(8 - 3)$$

$$= 3\mathbf{i} - 7\mathbf{j} + 5\mathbf{k}$$

$$\therefore \qquad |F_1 \times F_2| = \sqrt{3^2 + (-7)^2 + 5^2} = 9.110$$

$$F_1 = \sqrt{4^2 + 1^2 + (-1)^2} = 4.243$$

$$F_2 = \sqrt{3^2 + 2^2 + 1^2} = 3.742$$

$\therefore$ $$\sin\theta = \frac{|F_1 \times F_2|}{F_1 F_2} = \frac{9.110}{(4.243)(3.742)} = 0.5738$$

$\therefore$ $$\theta = 35.014°$$

Now $$\mathbf{R} = \mathbf{F}_1 \times \mathbf{F}_2 = 3\mathbf{i} - 7\mathbf{j} + 5\mathbf{k}$$

$\therefore$ $$R = \sqrt{3^2 + 7^2 + 5^2} = 9.110$$

Now $$R_x = Rl = 3$$

$$l = \frac{3}{R} = \frac{3}{9.11} = 0.329$$

$$m = \frac{-7}{9.11} = -0.768$$

$$n = \frac{5}{9.11} = 0.549$$

$\therefore$ The unit vector of $R$ is

$$\mathbf{n} = 0.329\ \mathbf{i} - 0.768\ \mathbf{j} + 0.549\ \mathbf{k}$$

**Example 1.11.** Determine the cross product of **P** and **Q**,

where $$\mathbf{P} = 2\mathbf{i} + 4\mathbf{j} - 3\mathbf{k}$$

and $$\mathbf{Q} = 3\mathbf{i} + 5\mathbf{j} + \mathbf{k}$$

and hence find angle θ between these two vectors.

**Solution.**

$$\mathbf{R} = \mathbf{P} \times \mathbf{Q} = \begin{vmatrix} \mathbf{i} & \mathbf{j} & \mathbf{k} \\ 2 & 4 & -3 \\ 3 & 5 & 1 \end{vmatrix}$$

$$= \mathbf{i}(4 + 15) - \mathbf{j}(2 + 9) + \mathbf{k}(10 - 12)$$
$$= 19\mathbf{i} - 11\mathbf{j} - 2\mathbf{k}$$

$\therefore$ $$R = \sqrt{19^2 + (-11)^2 + (-2)^2} = 22.045 \text{ units}$$ **Ans.**

$$P = \sqrt{2^2 + 4^2 + (-3)^2} = 5.385 \text{ units}$$

$$Q = \sqrt{3^2 + 5^2 + 1^2} = 5.916 \text{ units}$$

$\therefore$ $$\sin\theta = \frac{R}{PQ} = \frac{|P \times Q|}{PQ} = \frac{19.391}{5.385 \times 5.916}$$

$$= 0.60867$$

$\therefore$ $$\theta = 37.494°$$ **Ans.**

## 1.9 MOMENT OF A FORCE ABOUT A POINT

The *moment of a force* acting on a rigid body about a point in the body is defined as *the product of the magnitude of the force and the perpendicular distance of the line of action of the force from the point.*

In Fig. 1.18, $F$ is the force and $O$ is the point about which the moment is required. Taking the plane containing $F$ and $O$ as $x$-$z$ plane and $y$ as the axis at right angles to $x$-$z$ plane at $O$, the magnitude of the moment about $O$ is

$$M = Fd \qquad \text{...(1.27)}$$

where $d$ is the perpendicular distance of the line of action of $F$ from $O$.

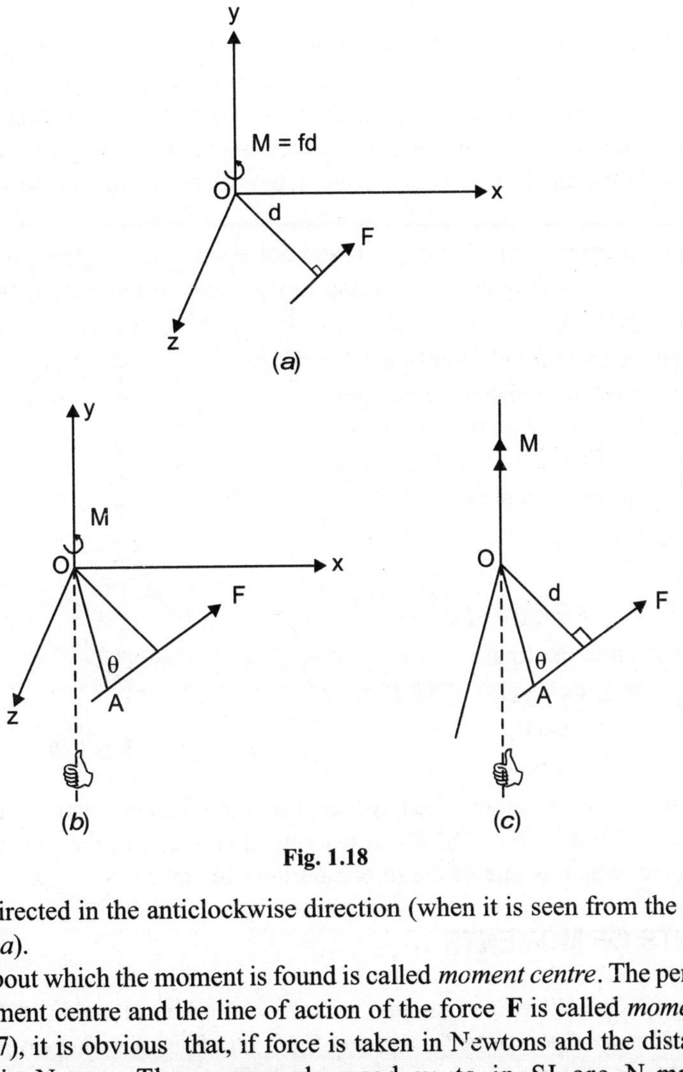


**Fig. 1.18**

Moment **M** is directed in the anticlockwise direction (when it is seen from the arrow head side) as shown in Fig. 1.18(*a*).

The point '*O*' about which the moment is found is called *moment centre*. The perpendicular distance '*d*' between the moment centre and the line of action of the force **F** is called *moment arm*.

From Eqn. (1.27), it is obvious that, if force is taken in Newtons and the distance '*d*' in mm, the unit of moment is N-mm. The commonly used units in SI are N-mm, kN-mm, N-m, kN-m.

Moment is a vector, since, it is associated with magnitude as well as direction (clockwise/anticlockwise). It can be proved that the moment of a force **F** about point *O* is

$$\mathbf{M} = \mathbf{r}_{OA} \times \mathbf{F} \qquad \text{...(1.28)}$$

where $A$ is a point on the line of action of **F**.

Referring to Fig. 1.18 (*b*).

$\mathbf{r_{OA}}$ is position vector of $A$, a point on the line of action of **F**.

$\theta$ is the acute angle between the position vector and the line of action of **F**.

From the cross product rule, we have

$$\begin{aligned}\mathbf{r_{OA}} \times \mathbf{F} &= r_{OA} F \sin\theta \\ &= F(r_{OA} \sin\theta) \\ &= Fd \\ &= \mathbf{M}.\end{aligned}$$

From the cross product rule, we also know that the product $r_{OA} \times F$ has direction at right angles to the plane of $r_{OA}$ and $F$. Hence, this product is directed as shown in Fig. 1.18(*b*). We look at this direction as the axis about which the moment is acting. If thumb of right hand is stretched and the other four fingers are closed, the thumb shows the axis about which moment is acting and the other four fingers indicate the direction of moment. Many times this moment is marked with double arrow heads as shown in Fig. 1.18 (*c*).

It is necessary to remember that $\mathbf{M} = \mathbf{r_{OA}} \times \mathbf{F}$ and not $\mathbf{F} \times \mathbf{r_{OA}}$, since, the cross product $\mathbf{r_{OA}} \times \mathbf{F}$ is exactly opposite of $\mathbf{F} \times \mathbf{r_{OA}}$ in direction. According to sign convention, anticlockwise moment is the positive moment and hence clockwise moment will be the negative moment.

It may be proved easily that any point on the line of action of force may be selected to get the moment about a point. Let 1 and 2 be two points on the line of action of force **F** as shown in Fig. 1.19. From the definition of moment about point $O$,

$$\mathbf{M} = \mathbf{r}_{01} \times \mathbf{F}$$

Its magnitude is

$$M = r_{01} F \sin\theta_1 = Fd$$

and the direction is shown in the figure.

If we consider point 2, the vector $\mathbf{r}_{02} \times \mathbf{F}$ has magnitude

$$\begin{aligned} &= r_{02} F \sin\theta_2 \\ &= Fd, \text{ as before}\end{aligned}$$

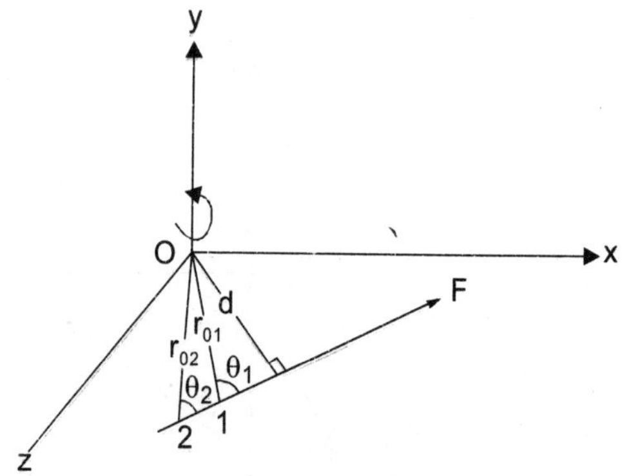


**Fig. 1.19**

and the direction is also same as before. Thus, we can take the position vector of any point on the line of action of the force to find its moment about a moment centre. In other words, it proves law of transmissibility of force, which is one of the important laws of statics.

## 1.10 COMPONENTS OF MOMENTS

Imagine the building of a parallelepiped on any point $B$ on the line of action of force **F** as shown in Fig. 1.20. Let the coordinate of point $B$ be $(x, y, z)$ and $F_x$, $F_y$ and $F_z$ be the components of force **F**. Noting that the force parallel to an axis cannot give any rotational effect (moment) about that axis, we get

$$M_x = yF_z - zF_y \qquad \text{...[1.29(a)]}$$

$$M_y = zF_x - xF_z \qquad \text{...[1.29(b)]}$$

and

$$M_z = xF_y - yF_x \qquad \text{...[1.29(c)]}$$

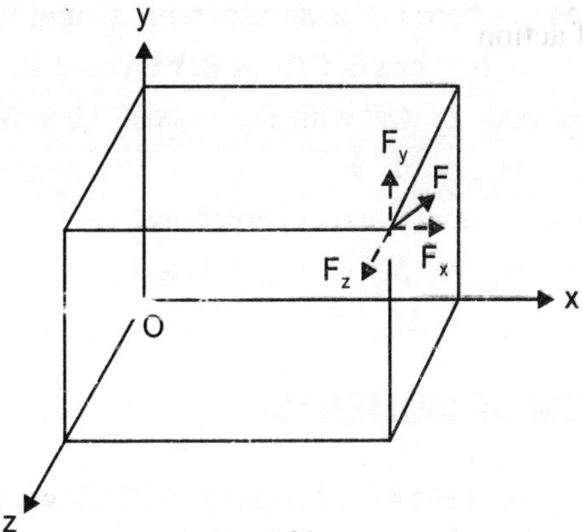


Fig. 1.20

From our earlier discussion, moment of force **F** about point $O$ is

$$\mathbf{M} = \mathbf{r}_{\mathbf{OB}} \times \mathbf{F}$$

$$= \begin{vmatrix} \mathbf{i} & \mathbf{j} & \mathbf{k} \\ x & y & z \\ F_x & F_y & F_z \end{vmatrix}$$

$$= (yF_z - zF_y)\mathbf{i} - \mathbf{j}(xF_z - zF_x) + \mathbf{k}(xF_y - yF_x)$$

$$= M_x\mathbf{i} + M_y\mathbf{j} + M_z\mathbf{k} \qquad ...(1.30)$$

Thus, the moment of a force about a point is the vector sum of the moment of the force about the orthogonal axes through that point. $M$, the magnitude of this vector, is equal to $\sqrt{M_x^2 + M_y^2 + M_z^2}$ and its unit vector is

$$\frac{M_x}{M}\mathrm{i} + \frac{M_y}{M}\mathrm{j} + \frac{M_z}{M}\mathrm{k}$$

## 1.11 MOMENT OF A FORCE ABOUT A GIVEN AXIS

If **F** is the force and $O$ is the moment centre, we know that it produces moment $\mathbf{M_0}$ about the moment centre at right angles to the plane containing the force and the moment centre (Ref. Fig. 1.21). In the form of components about $x$, $y$, $z$ plane, this moment is given by

$$\mathbf{M} = M_x\mathbf{i} + M_y\mathbf{j} + M_z\mathbf{k}$$

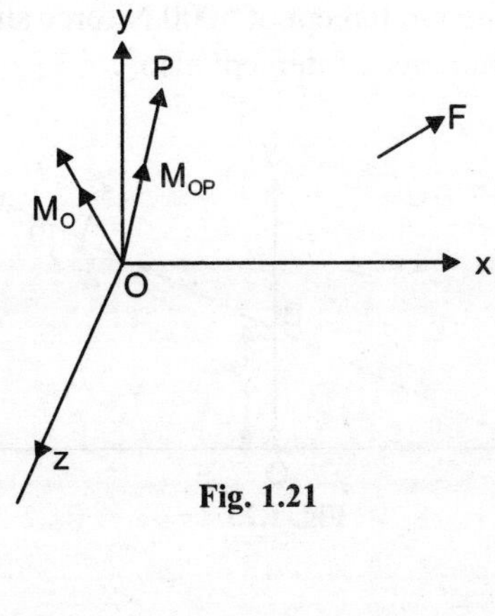


Fig. 1.21

Now our interest is to find $\mathbf{M}_{OP}$ where $OP$ is an axis with its unit vectors

$$\mathbf{n} = \cos\theta_x \mathbf{i} + \cos\theta_y \mathbf{j} + \cos\theta_z \mathbf{k}$$

where $\theta_x$, $\theta_y$ and $\theta_z$ are the angles made by $OP$ with $x$, $y$, $z$-axes. Then $\mathbf{M}_{OP}$ is given by

$$\mathbf{M}_{OP} = \mathbf{M} \cdot \mathbf{n} \qquad \text{...(1.31)}$$

For example, component of $M_{OP}$ about $x$-axis is given by

$$\mathbf{M}_{OP} = (M_x\mathbf{i} + M_y\mathbf{j} + M_z\mathbf{k})\cdot\mathbf{i}$$
$$= M_x$$

## 1.12 VARIGNON'S THEOREM OF MOMENTS

The French mathematician Varignon (1654–1722) gave the following theorem which is known as Varignon's theorem of moments.

"*The algebraic sum of the moments of a system of forces about a moment centre is equal to the moment of their resultant force about the same moment centre.*"

Actually, Varignon stated this theorem in the context of coplanar forces and proved it.

However, it can be generalised to two- as well as three-dimensional problems.

Let $\mathbf{F}_1$ and $\mathbf{F}_2$ be the two component forces acting at $A$ and $\mathbf{R}$ be the resultant force.

Then,
$$\mathbf{M}_O = \mathbf{r}_{OA} \times \mathbf{R} = \mathbf{r}_{OA} \times (\mathbf{F}_1 + \mathbf{F}_2)$$
$$= \mathbf{r}_{OA} \times \mathbf{F}_1 + \mathbf{r}_{OA} \times \mathbf{F}_2$$
$$= \mathbf{M}_1 + \mathbf{M}_2 \qquad \text{...(1.32)}$$

where $\mathbf{M}_1$ and $\mathbf{M}_2$ are the moments of component forces about $O$.

**Example 1.12.** Determine the moment of 400 N force acting at $B$ in $x$-$y$ plane about point $A$, as shown in Fig. 1.22.

**Solution.** Taking $A$ as origin, the position vector of $B$ is

$$\mathbf{r}_{AB} = 300\mathbf{i} + 400\mathbf{j}$$
$$\mathbf{F} = 400 \cos 30°\mathbf{i} + 400 \sin 30°\mathbf{j}$$
$$= 346.41\mathbf{i} + 200\mathbf{j}$$
$$\therefore \quad \mathbf{M}_A = \mathbf{r}_{AB} \times \mathbf{F}$$
$$= (300\mathbf{i} + 400\mathbf{j}) \times (346.41\mathbf{i} + 200\mathbf{j})$$
$$= 60000\mathbf{k} - 138564\mathbf{k}$$
$$= -78564\mathbf{k}$$
$$= 78564 \text{ N-mm, clockwise moment.}$$

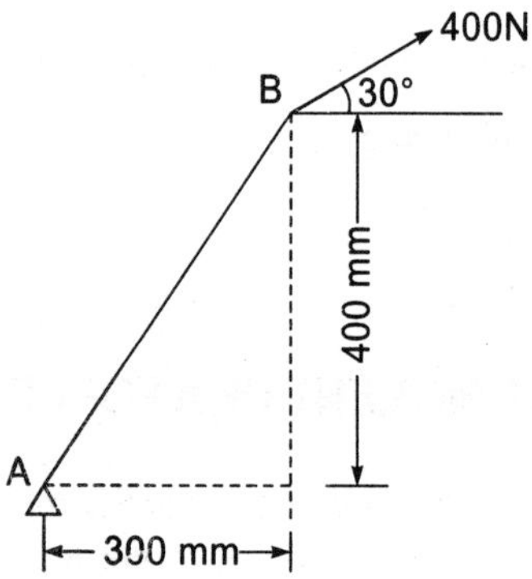

**Fig. 1.22**

**Example 1.13.** What will be the $y$-intercept of 5000 N force shown in Fig. 1.23, if its moment about $A$ is 8000 N-m clockwise? Determine its $x$ intercept also.

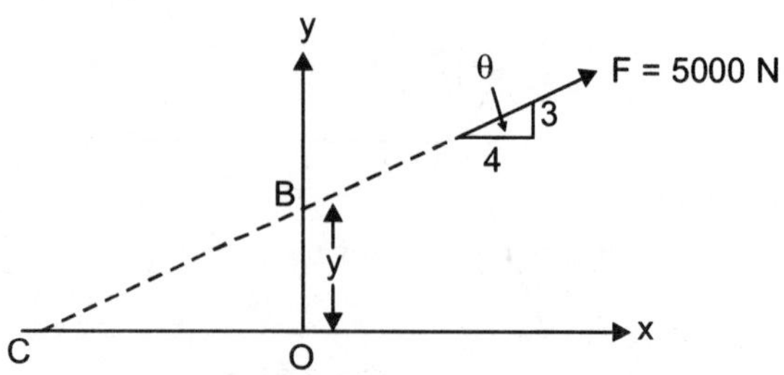

**Fig. 1.23**

**Solution.** Let 5000 N force intercept $y$-axis at $B$ and its intercept be $y$. Then,

$$\mathbf{r_{OB}} = 0\mathbf{i} + y\mathbf{j}$$

$$\theta = \tan^{-1} 3/4$$

$$\therefore \quad \theta = 36.87$$

$$F_x = 5000 \cos\theta = 4000 \text{ N}$$

$$F_y = 5000 \sin\theta = 3000 \text{ N}$$

$$\therefore \quad \mathbf{F} = 4000\mathbf{i} + 3000\mathbf{j}$$

$$\mathbf{M} = \mathbf{r_{OB}} \times \mathbf{F}$$

$$= (0\mathbf{i} + y\mathbf{j}) \times (4000\mathbf{i} + 3000\mathbf{j})$$

$$= 4000y\,(-\mathbf{k})$$

Now moment is given as 8000 N-m clockwise.

that is $-8000k$. Hence

$$-8000\,\mathbf{k} = -4000y\,\mathbf{k}$$

or $$y = 2 \text{ m}$$ **Ans.**

If $x$ is the intercept on $x$-axis, then the position vector of the point of intersection of force **F** with $x$-axis is

$$\mathbf{r_{OC}} = x\mathbf{i} + 0\mathbf{j}$$

$\therefore$ Using this position vector,

$$\mathbf{M_o} = (x\mathbf{i} + 0\mathbf{j}) \times (4000\mathbf{i} + 3000\mathbf{j})$$

$$= 3000x\,\mathbf{k}$$

Since, this moment is $-8000$ N-m, we get

$$-8000\mathbf{k} = 3000x\,\mathbf{k}$$

or $$x = -\frac{8}{3} = -2.667 \text{ m}$$ **Ans.**

**Example 1.14.** $A$ (50, 30, 40) and $B$(30, 20, 80) are the two points on the line of action of 400 N force which is acting in the direction $A$ to $B$. Find the moment of this force about the origin '$O$' and about the point $C$ (20, 20, 20). Millimetre units are used for the coordinate system in space.

**Solution.**

$$\mathbf{r_{OA}} = 50\mathbf{i} + 30\mathbf{j} + 40\mathbf{k} \text{ and } \mathbf{r_{OB}} = 30\mathbf{i} + 20\mathbf{j} + 80\mathbf{k}$$

$$\mathbf{r_{AB}} = (30 - 50)\mathbf{i} + (20 - 30)\mathbf{j} + (80 - 40)\mathbf{k} = -20\mathbf{i} - 10\mathbf{j} + 40\mathbf{k}$$

$$\therefore \quad r_{AB} = \sqrt{20^2 + 10^2 + 40^2} = 45.825 \text{ mm}$$

$$\therefore \quad \mathbf{n} = \frac{-20}{45.825}\mathbf{i} + \frac{(-10)}{45.825}\mathbf{j} + \frac{40}{45.825}\mathbf{k}$$

$$= -0.4364\mathbf{i} - 0.2182\mathbf{j} + 0.8728\mathbf{k}$$

$$\therefore \quad \mathbf{F} = F\mathbf{n} = 400\,(-0.4364\mathbf{i} - 0.2812\mathbf{j} + 0.8728\mathbf{k})$$

$$= -174.57\mathbf{i} - 87.28\mathbf{j} + 349.15\mathbf{k}$$

$\therefore$ Moment of this force about '$O$' is

$$\mathbf{M_O} = \mathbf{r_{OA}} \times \mathbf{F} \text{ (or } \mathbf{r_{OB}} \times \mathbf{F})$$

$$= (50\mathbf{i} + 30\mathbf{j} + 40\mathbf{k}) \times (-174.57\mathbf{i} - 87.28\mathbf{j} + 349.15\mathbf{k})$$

$$= \begin{vmatrix} \mathbf{i} & \mathbf{j} & \mathbf{k} \\ 50 & 30 & 40 \\ -174.57 & -87.28 & 349.15 \end{vmatrix}$$

$$= \mathbf{i}(30 \times 349.15 + 87.28 \times 40) - \mathbf{j}(50 \times 349.15 + 174.57 \times 40)$$
$$+ \mathbf{k}(-50 \times 87.28 + 174.57 \times 30)$$
$$= 13965.7\mathbf{i} - 24440.3\mathbf{j} + 873.1\mathbf{k}$$

*i.e.*, $M_x = 13965.7$ N-mm, $M_y = -24440.3$ N-mm **Ans.**

and $M_z = 873.1$ N-mm

Moment About C:

Now $\mathbf{r}_{CA} = (50 - 20)\mathbf{i} + (30 - 20)\mathbf{j} + (40 - 20)\mathbf{k}$

$= 30\mathbf{i} + 10\mathbf{j} + 20\mathbf{k}$

$\mathbf{M}_C = \mathbf{r}_{CA} \times \mathbf{F}$

$$\mathbf{M}_C = \begin{vmatrix} \mathbf{i} & \mathbf{j} & \mathbf{k} \\ 30 & 10 & 20 \\ -174.57 & -87.28 & 349.15 \end{vmatrix}$$

$$= \mathbf{i}(10 \times 349.15 + 20 \times 87.28) - \mathbf{j}(30 \times 349.15 + 20 \times 174.57)$$
$$+ \mathbf{k}(-30 \times 87.28 + 10 \times 174.57)$$
$$= 5237.1\mathbf{i} - 13965.9\mathbf{j} - 872.7\mathbf{k}$$

*i.e.*, $M_x = 5237.1$ N-mm, anticlockwise **Ans.**

$M_y = 13965.9$ N-mm, clockwise

and $M_z = 872.7$ N-mm, clockwise

**Example 1.15.** The coordinates of points *A* and *B* in metre units are *A*(1, 2, 3) and *B*(4, 3, 2). A force of magnitude 10 kN passes through these two points and is directed from *A* towards *B*. Find the moment of this force:

(*i*) about *C* (1, 1, 1)

(*ii*) about the axis passing through the points *D*(2.5, 1.5, 4) and *E* (4, 3, 5).

**Solution.** $\mathbf{r}_{AB} = (4 - 1)\mathbf{i} + (3 - 2)\mathbf{j} + (2 - 3)\mathbf{k}$

$= 3\mathbf{i} + \mathbf{j} - \mathbf{k}$

$\therefore$ $r_{AB} = \sqrt{3^2 + 1^2 + 1^2} = 3.3166$

$\therefore$ $\mathbf{n}_{AB} = \dfrac{3}{3.3166}\mathbf{i} + \dfrac{1}{3.3166}\mathbf{j} - \dfrac{1}{3.3166}\mathbf{k}$

$= 0.9045\mathbf{i} + 0.3015\mathbf{j} - 0.3015\mathbf{k}$

$\therefore$ $F = 10\,\mathbf{n}_{AB} = 9.045\mathbf{i} + 3.015\mathbf{j} - 3.015\mathbf{k}$

(*i*) Moment about *C* :

Displacement vector of *CA* is

$\mathbf{r}_{CA} = (1 - 1)\mathbf{i} + (2 - 1)\mathbf{j} + (3 - 1)\mathbf{k}$

$$= 0\mathbf{i} + \mathbf{j} + 2\mathbf{k}$$

$$\therefore \quad \mathbf{M_C} = \mathbf{r_{CA}} \times \mathbf{F}$$

$$= \begin{vmatrix} \mathbf{i} & \mathbf{j} & \mathbf{k} \\ 0 & 1 & 2 \\ 9.045 & 3.015 & -3.015 \end{vmatrix}$$

$$= \mathbf{i}(-3.015 - 6.030) - \mathbf{j}(-18.090) + \mathbf{k}(-9.045)$$

$$= -9.045\mathbf{i} + 18.090\mathbf{j} - 9.045\mathbf{k}$$

$M_x$ = 9.045 kN-m, clockwise

$M_y$ = 18.090 kN-m, anticlockwise

$M_z$ = 9.045 kN-m clockwise

$$M_C = \sqrt{(9.045)^2 + (18.090)^2 + (9.045)^2} = 22.16 \text{ kN-m}$$

(*ii*) Moment about the axis *DE*

$$\mathbf{r_{DE}} = (4 - 2.5)\mathbf{i} + (3 - 1.5)\mathbf{j} + (5.0 - 4)\mathbf{k}$$

$$= 1.5\mathbf{i} + 1.5\mathbf{j} + 1\mathbf{k}$$

$$r_{DE} = \sqrt{1.5^2 + 1.5^2 + 1^2} = 2.3452$$

$$\therefore \quad \mathbf{n}_{DE} = \frac{1.5}{2.3452}\mathbf{i} + \frac{1.5}{2.3452}\mathbf{j} + \frac{1}{2.3452}\mathbf{k}$$

$$= 0.6396\mathbf{i} + 0.6396\mathbf{j} + 0.4264\mathbf{k}$$

First we find moment about *D*

$$\mathbf{r_{DA}} = (1 - 2.5)\mathbf{i} + (2 - 1.5)\mathbf{j} + (3 - 4)\mathbf{k}$$

$$= -1.5\mathbf{i} + 0.5\mathbf{j} - \mathbf{k}$$

$$\mathbf{M_D} = \mathbf{r_{DA}} \times \mathbf{F}$$

$$= \begin{vmatrix} \mathbf{i} & \mathbf{j} & \mathbf{k} \\ -1.5 & 0.5 & -1 \\ 9.045 & 3.015 & -3.015 \end{vmatrix}$$

$$= \mathbf{i}(-0.5 \times 3.015 + 3.015) - \mathbf{j}(1.5 \times 3.015 + 9.045) + \mathbf{k}(-1.5 \times 3.015 - 0.5 \times 9.045)$$

$$= 0.1508\mathbf{i} - 13.5675\mathbf{j} - 9.045\mathbf{k}$$

$$\therefore \quad \mathbf{M_{DE}} = \mathbf{M_D} \cdot \mathbf{n_{DE}}$$

$$= (0.1508\mathbf{i} - 13.5675\mathbf{j} - 9.045\mathbf{k}) \cdot (0.6396\mathbf{i} + 0.6396\mathbf{j} + 0.4264\mathbf{k})$$

$$= 0.1508 \times 0.6396 - 13.5675 \times 0.6396 - 9.045 \times 0.4264$$

$$= -12.438 \text{ kN-m}$$

**Ans.**

## 1.13 COUPLE MOMENT

*Two parallel forces that are equal in magnitude but opposite in direction and separated by a definite distance are said to form a couple.* The rotational effect of couple is called '*Couple Moment*' and this moment is having interesting property.

Consider the couple as shown in Fig. 1.24. The forces $\mathbf{F}_1 = \mathbf{F}$ and $\mathbf{F}_2 = -\mathbf{F}$ form a couple. Let '$O$' be a moment centre and $A$ be an arbitrary point on $F_1$ and $B$ be an arbitrary point on $F_2$.

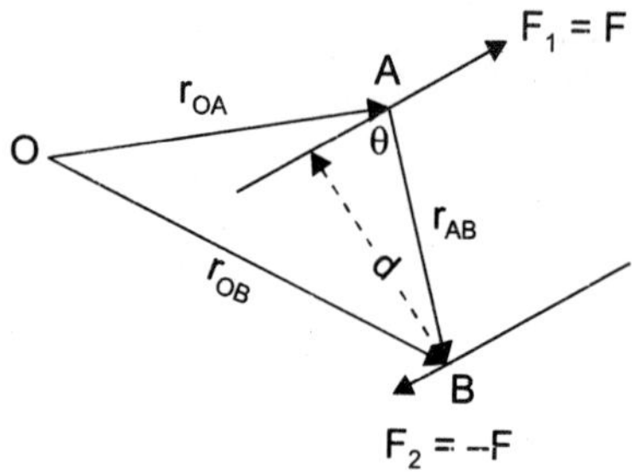


**Fig. 1.24**

$$
\begin{aligned}
\therefore \quad \mathbf{M}_O &= \mathbf{r}_{OA} \times \mathbf{F}_1 + \mathbf{r}_{OB} \times \mathbf{F}_2 \\
&= \mathbf{r}_{OA} \times \mathbf{F} + \mathbf{r}_{OB} \times (-\mathbf{F}) \\
&= (\mathbf{r}_{OA} - \mathbf{r}_{OB}) \times \mathbf{F} \\
&= -\mathbf{r}_{AB} \times \mathbf{F}
\end{aligned}
$$

Hence, magnitude of $M_O = r_{AB} F \sin\theta = Fd$, where $d$ is perpendicular distance between the two parallel forces, and it is clockwise moment. ...(1.33)

The position vector $\mathbf{r}_{AB}$ does not contain any reference to the moment centre. Hence, for any moment centre, the moment of system is the same. The direction of couple moment is normal to the plane containing the two forces forming the couple and its sense is established by the right hand screw rule.

The characteristics of the couple are listed below:

— A couple consists of a pair of equal and opposite parallel forces which are separated by a definite distance.

— The translatory effect of a couple on the body is zero.

— The rotational effect (moment) of a couple about any point is a constant and it is equal to the product of the displacement vector of any two points on the two forces forming the couple and the magnitude of the forces.

Since, the only effect of a couple is a moment and this moment is the same about any moment centre, the effect of couple is unchanged if

— the couple is rotated through any angle.

— the couple is shifted to any other position.

— the couple is replaced by any other pair of forces whose rotational effect is the same.

Hence, a couple moment is a free vector.

**Example 1.16.** Determine the couple moment and the distance between the line of action of the forces $\mathbf{F}_1 = 25\mathbf{i} + 20\mathbf{j} + 30\mathbf{k}$ acting through point $A$ (1.0, 1.2, 1.5) and $\mathbf{F}_2 = -25\mathbf{i} - 20\mathbf{j} - 30\mathbf{k}$ acting through point $B$ (1.5, 1.8, –1.0).

The force is in kN units and the coordinates are in m units.

**Solution.**

$$
\begin{aligned}
\mathbf{r}_{AB} &= (1.5 - 1.0)\mathbf{i} + (1.8 - 1.2)\mathbf{j} + (-1.0 - 1.5)\mathbf{k} \\
&= 0.5\mathbf{i} + 0.6\mathbf{j} - 2.5\mathbf{k}
\end{aligned}
$$

$$
\mathbf{M} = \mathbf{M}_A = \begin{vmatrix} \mathrm{i} & \mathrm{j} & \mathrm{k} \\ 0.5 & 0.6 & -2.5 \\ 25.0 & 20.0 & 30.0 \end{vmatrix}
$$

$$
= \mathbf{i}(18 + 50) - \mathbf{j}(15 + 62.5) + \mathbf{k}(10 - 15)
$$

$$= 68\mathbf{i} - 77.5\mathbf{j} - 5\mathbf{k}$$

$\therefore$ $$M = \sqrt{68^2 + 77.5^2 + 5^2} = 103.224 \text{ kN-m}$$ **Ans.**

$\therefore$ $$\mathbf{M} = 103.224\left(\frac{68}{103.224}\mathbf{i} - \frac{77.5}{103.224}\mathbf{j} - \frac{5}{103.224}\mathbf{k}\right)$$

*i.e.*, $$\mathbf{M} = 103.224\,(0.6588\mathbf{i} - 0.7508\mathbf{j} - 0.0484\mathbf{k})$$ **Ans.**

Now, $$\mathbf{F} = 25\mathbf{i} + 20\mathbf{j} + 30\mathbf{k}$$

$\therefore$ $$F = \sqrt{25^2 + 20^2 + 30^2} = 43.875 \text{ kN}$$

$$\mathbf{F} = 43.875\left(\frac{25}{43.875}\mathbf{i} + \frac{20}{43.875}\mathbf{j} + \frac{30}{43.875}\mathbf{k}\right)$$

$\therefore$ $$\mathbf{F} = 43.875\,(0.5698\mathbf{i} + 0.4558\mathbf{j} + 0.6838\mathbf{k})$$

Now from Eqn. (1.13),

$$M = Fd$$

$\therefore$ $$d = \frac{M}{F} = \frac{103.224}{43.875} = 2.353 \text{ m}$$ **Ans.**

## EXERCISE

**1.1** A force of magnitude 20 kN makes 30° with the $y$-axis and its projection on $x$-$z$ plane makes 45° with $x$-axis. Write the force in vector form. Determine its direction cosines and directions with respect to $x$, $y$, $z$-axes.

[**Ans.** $\mathbf{F} = 7.07\mathbf{i} + 17.321\mathbf{j} + 7.07\mathbf{k}$; $l = 0.3535$, $m = 0.866$, $n = 0.3535$
$\theta_x = 69.30°$, $\theta_y = 30°$, $\theta_z = 69.30°$]

**1.2** $\mathbf{F}_1$, $\mathbf{F}_2$ and $\mathbf{F}_3$ are the three forces as given below:

$$\mathbf{F}_1 = 4\mathbf{i} + 3\mathbf{j} + \mathbf{k}$$
$$\mathbf{F}_2 = 2\mathbf{i} - 2\mathbf{j} + 3\mathbf{k}$$
$$\mathbf{F}_3 = 5\mathbf{i} - \mathbf{j} + 2\mathbf{k}$$

Determine

(*i*) $\mathbf{F}_1 + \mathbf{F}_2 + \mathbf{F}_3$
(*ii*) $\mathbf{F}_1 - \mathbf{F}_2 - \mathbf{F}_3$
(*iii*) $2\mathbf{F}_1 + 3\mathbf{F}_2 - 2\mathbf{F}_3$

[**Ans.** (*i*) $11\mathbf{i} + 6\mathbf{k}$, (*ii*) $-3\mathbf{i} + 6\mathbf{j} - 4\mathbf{k}$, (*iii*) $4\mathbf{i} + 2\mathbf{j} + 7\mathbf{k}$]

**1.3** Determine the displacement vector $AB$, given that $A$ (2, 4, 5), $B$(3, 2, 1) in metre units. Determine the vector form of $AB$, its direction cosines and directions w.r.t. the $x$, $y$, $z$-axes.

[**Ans.** $\mathbf{r}_{AB} = \mathbf{i} - 2\mathbf{j} - 4\mathbf{k}$, $l = 0.218$, $m = -0.436$, $n = -0.872$,
$\theta_x = 77.4°$, $\theta_y = 115.88°$, $\theta_z = 190.794°$]

**1.4** The two vectors $\mathbf{P}_1$ and $\mathbf{P}_2$ are as given below:

$$\mathbf{P}_1 = \mathbf{i} + 2\mathbf{j} + 3\mathbf{k}$$
$$\mathbf{P}_2 = 3\mathbf{i} + 2\mathbf{j} + \mathbf{k}$$

Determine

(*i*) $\mathbf{P} \cdot \mathbf{P_2}$ and hence the acute angle between $\mathbf{P_1}$ and $\mathbf{P_2}$.

(*ii*) $\mathbf{P_1} \times \mathbf{P_2}$ and hence the acute angle between $\mathbf{P_1}$ and $\mathbf{P_2}$.

[**Ans.** (*i*) 10, $\theta_x = 44.415°$; (*ii*) $\theta_y = 9.798°$, $\theta_z = 44.415°$]

**1.5** Determine the projection of $\mathbf{P_1}$ on the line of action of $\mathbf{P_2}$, given

$$\mathbf{P_1} = \mathbf{i} + 2\mathbf{j} + 3\mathbf{k}$$

and $$\mathbf{P_2} = 3\mathbf{i} + 2\mathbf{j} + \mathbf{k}$$ [**Ans.** 2.676]

**1.6** If $\mathbf{P_1} = 4\mathbf{i} + 2\mathbf{j} + \mathbf{k}$ and $\mathbf{P_2} = 2\mathbf{i} - 3\mathbf{j} - 2\mathbf{k}$,
prove that they are at right angles to each other.

[ **Hint:** From dot product show that $\cos\theta = 0$, hence $\theta = 90°$ ]

**1.7** Explain the terms:

(*a*) Position vector

(*b*) Displacement vector.

**1.8** State and explain parallelogram law of vectors. How do you derive triangle law and polygonal law of vectors from parallelogram law?

**1.9** Explain dot product or scalar product of two vectors.

**1.10** Explain the properties of dot product of vectors.

**1.11** Explain cross product of two vectors.

**1.12** Explain the properties of cross product.

**1.13** Explain the term moment of a force about a point. Derive the vector representation of moment.

**1.14** State and prove Varignon's theorem of moments.

**1.15** Explain the term 'couple moment' and prove that its magnitude is same about any point in the plane. List its characters.

## OBJECTIVE TYPE QUESTIONS

**Select the correct option in the following:**

**1.1** What is the direction of unit vector of the point (4, 3, 12)?

(*a*) $\left(\frac{-4}{11}, \frac{-3}{11}, \frac{-12}{11}\right)$ (*b*) $\left(\frac{-4}{13}, \frac{-3}{13}, \frac{-12}{13}\right)$

(*c*) $\left(\frac{4}{11}, \frac{3}{11}, \frac{12}{11}\right)$ (*d*) $\left(\frac{4}{13}, \frac{3}{13}, \frac{12}{13}\right)$

**1.2** If *A* (6, 12) and *B* (10, 20), the displacement vector $r_{AB}$ is given by

(*a*) $-4\mathbf{i} - 8\mathbf{j}$ (*b*) $4\mathbf{i} + 8\mathbf{j}$

(*c*) $16\mathbf{i} + 32\mathbf{j}$ (*d*) None of the above

**1.3** If the resultant of two vectors $\overline{A}$ and $\overline{B}$ is perpendicular to $\overline{A}$, then the angle between $\overline{A}$ and $\overline{B}$ is (*a*) $\tan^{-1}\left(\frac{A}{B}\right)$ (*b*) $\tan^{-1}\left(\frac{-A}{B}\right)$

(*c*) $\cos^{-1}\left(\frac{A}{B}\right)$ (*d*) $\cos^{-1}\left(\frac{-A}{B}\right)$

**1.4** If $\bar{p} = 3\bar{i} - 4\bar{j}$, which one of the following is at right angle to it?

(*a*) $3\bar{i}$ (*b*) $4\bar{j}$

(*c*) $4\bar{i} - 3\bar{j}$ (*d*) $4\bar{i} + 3\bar{j}$

**1.5** If $P = \bar{i} + \bar{j}$ and $Q = \bar{j} + \bar{k}$, the angle between $P$ and $Q$ is

(*a*) 0° (*b*) 45°

(*c*) 60° (*d*) 90°

**1.6** If $\bar{P} = 3\bar{i} + 2\bar{j}$ and $\bar{Q} = \bar{i} + 4\bar{j}$, then $\bar{P} \times \bar{Q}$ is

(*a*) $4i + 6j$ (*b*) 11

(*c*) $10k$ (*d*) 10

**1.7** If $\bar{P} = 5\bar{i} + 4\bar{j} + 3\bar{k}$ and $Q = 4\bar{i} - 2\bar{j} - 4\bar{k}$, the angle between $\bar{P}$ and $\bar{Q}$ is

(*a*) $\tan^{-1}\frac{\sqrt{50}}{6}$ (*b*) $\cos^{-1}\frac{\sqrt{50}}{6}$

(*c*) 90° (*d*) 0°

**1.8** At what angle, the forces 2 N and $\sqrt{2}$ N act so that their resultant is $\sqrt{10}$?

(*a*) 0° (*b*) 30°

(*c*) 45° (*d*) 60°

**1.9** The rotational effect of a force is called

(*a*) shear force (*b*) moment

(*c*) centroid (*d*) couple

**1.10** Moment of a force

(*a*) varies directly with its distance from the moment centre

(*b*) varies inversely with its distance from the moment centre

(*c*) is independent of its distance from moment centre

(*d*) None of the above

**1.11** Which one of the following is wrong statement?

The effect of a couple is unchanged, if

(*a*) the couple is rotated through any angle

(*b*) the couple is shifted to any other position

(*c*) the magnitude of forces in a couple are doubled but the distance is halved

(*d*) the magnitude as well as distance are doubled

**1.12** The moment of a force about any point is numerically equal to '*n*' times the area of the triangle whose base is the line representing the force and the vertex is the point, where '*n*' is

(*a*) 0.5 (*b*) 1.0

(*c*) 2.0 (*d*) 4.0

CHAPTER

# 2

# EQUIVALENT FORCE SYSTEM

When several forces act simultaneously on a body, they constitute a *system of forces*. If all the forces in a system do not lie in a single plane, they constitute a *system of forces in space* which is also known as *three dimensional general force system*. If all the forces in a system lie in a single plane, it is called a *coplanar force system* which is also known as *two dimensional force system*. If the lines of action of all forces in a system pass through a single point, it is known as *concurrent force system*. If the lines of action of all forces in a system do not pass through a single point, it is known as *non-concurrent force system*.

A *particle* is a body which has mass but no size. Hence the systems of forces acting on a particle are concurrent forces. Actually there is no body which has no size. However, there are many engineering problems in which the shape and size of the body do not significantly affect the analysis. In all these problems, the system of forces can be considered as concurrent.

In many engineering problems, the size of the body may not be so small as to neglect it and treat the body as a particle. The system of forces acting on such bodies may not be concurrent. Hence the analysis for the system of non-concurrent forces is required. In engineering mechanics, such bodies are treated as rigid and analysed. A body is said to be *rigid*, if the relative positions of any two particles in it do not change under the action of forces. The forces acting on a body have not only the tendency to move the body but they have tendency to rotate also.

In this chapter, finding of equivalent simple system to a given concurrent and non-concurrent force systems in two dimensions and three dimensions is presented. Before going to explain equivalent system, resolution of concurrent forces is explained, which is basis for the analytical method for giving equivalent force.

## 2.1 RESOLUTION OF CONCURRENT COPLANAR FORCE SYSTEM

Resolution of the forces is the process of finding a number of component forces which will have the same effect on the body as the given single force. Thus the resolution of forces is exactly opposite to finding the resultant of two or more forces, for which we have seen parallelogram law of forces, triangle law of forces and polygon law of forces can be used.

In Fig. 2.1 (*a*), force **F** is resolved into two components making angles α and β with the force **F**. Use of parallelogram law and triangle law of forces in the reversed order is shown in the figure.

In Fig. 2.1 (*b*), force **F** is resolved into four components $\mathbf{F}_1$, $\mathbf{F}_2$, $\mathbf{F}_3$ and $\mathbf{F}_4$ is shown, in which polygon law of forces is used in the reverse order.

In Fig. 2.1 (*c*), force **F** is resolved into its rectangular components, $\mathbf{F}_x$ and $\mathbf{F}_y$ where *x* and *y* are cartesian coordinaters.

It may be noted that the component forces act at the same point *'O'* as the given force **F**. Resolution of a force into its rectangular components is more useful for the analysis. If $\mathbf{F}_x$ and $\mathbf{F}_y$ are its components in *x* and *y* coordinate directions and **i** and **j** are unit vectors in *x* and *y* coordinate directions, then

$$\mathbf{F} = \mathbf{F}_x + \mathbf{F}_y$$

$$= F_x\mathbf{i} + F_y\mathbf{j} \qquad ...(2.1)$$

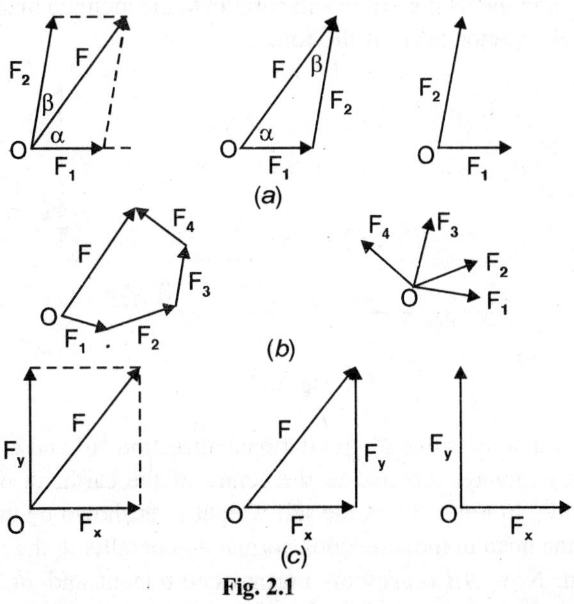


**Fig. 2.1**

**Example 2.1.** The guy wire of an electric pole, as shown in Fig. 2.2, makes 30° to the pole and is applying a force of 12 kN. Find the horizontal and vertical components of the force. Express it in the vector form taking horizontal direction as *x*-axis and vertical direction as *y*-axis.

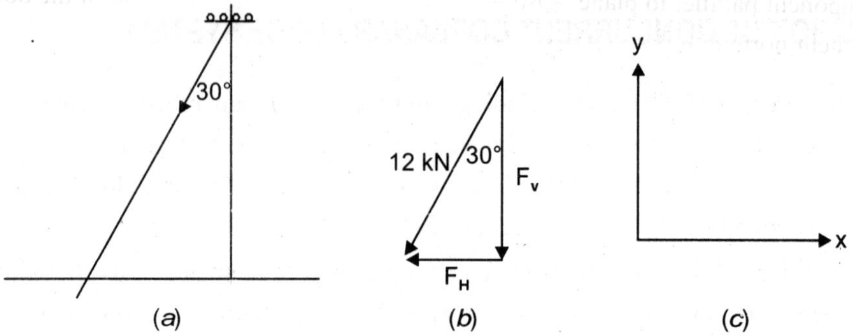


**Fig. 2.2**

**Solution.** To get the component of a force, assume the arrow marked represent the force to some scale. Move from tail to head of the arrow in the coordinate directions. Thus the vertical component

$$F_V = 12 \cos 30° = 10.392 \text{ kN (downward)} \quad \textbf{Ans.}$$

and Horizontal component $\quad F_H = 12 \sin 30° = 6 \text{ kN (towards left)}$ **Ans.**

Since the coordinates selected are as shown in Fig. 2.2 (*c*), taking **i** as unit vector in *x*-direction and **j** as unit vector in *y*-direction, we find

$$\mathbf{F}_x = -6\,\mathbf{i} \text{ and } \mathbf{F}_y = -10.392\,\mathbf{j}$$

Therefore, $\quad \mathbf{F} = -6\,\mathbf{i} - 10.392\,\mathbf{j}$ **Ans.**

**Example 2.2.** A block weighing $W = 12$ kN is resting on an inclined plane, as shown in Fig. 2.3 (*a*). Determine its components normal to and parallel to the inclined plane. Using *x* and *y* axes as shown in the figure, write the vector form of the force.

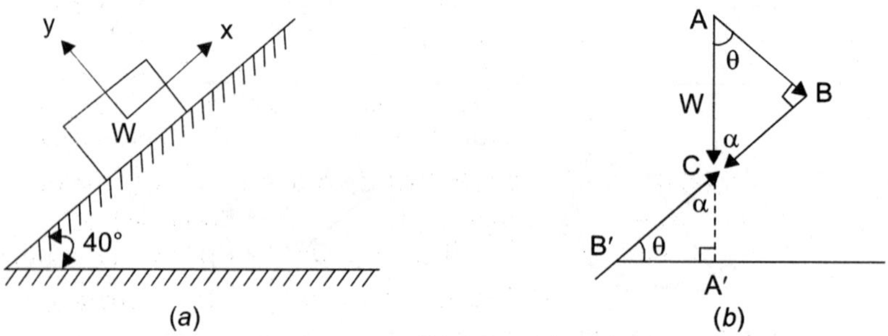


**Fig. 2.3**

**Solution.** Self weight of a body is due to gravitational attraction '*g*' and is equal to '*mg*', where *m* is the mass of the body. It is always directed to the centre of the earth; in other words, it is in the vertically downward direction. In Fig. 2.3 (*b*), the self weight *W* is shown by line *AC*. By moving from *A* (tail of the force line) to the head in the directions normal and parallel to the inclined plane direction, triangle *ABC* is completed. Now *AB* represents normal component and *BC* represents component parallel to *AB*. By extending *BC, AC* and drawing horizontal line *B′A′*, we find

$$\Delta ABC \parallel \Delta A'B'C, \text{ since } \angle ACB = \angle A'CB'$$

and $\quad \angle ABC = \angle CA'B' = 90°$

Hence $\quad \angle BAC = \angle A'B'C = \theta = 40°$

Thus normal to the plane makes angle θ with the vertical if the plane makes angle θ with the horizontal.

∴ Component parallel to plane = $BC = W \sin \theta$, where $W = mg$ is weight of the body.

and Component normal to plane = $AB = W \cos \theta$.

If $x$ and $y$ directions are selected as given in Fig. 2.3 ($a$).

$$F_x = -W \sin \theta = -12 \sin 40° = -7.713 \text{ kN} \qquad \textbf{Ans.}$$

$$F_y = -W \cos \theta = -12 \cos 40° = -9.193 \text{ kN} \qquad \textbf{Ans.}$$

Hence
$$\mathbf{F} = -7.713\,\mathbf{i} - 9.193\,\mathbf{j} \qquad \textbf{Ans.}$$

From the above two examples, the following points for finding component of forces and vector form may be noted:

(*i*) Imagine that arrow drawn to represent the force gives the magnitude of force to some scale.

(*ii*) Travel from the tail of the force line to arrow head in the direction of coordinates.

(*iii*) Then the direction of travel gives the direction of component forces.

(*iv*) From the triangle law of forces ($\Delta ABC$), the magnitude of the components can be found.

(*v*) Looking at the positive sense of the coordinate directions, the force may be expressed in the vector form.

## 2.2 RESULTANT OF TWO DIMENSIONAL CONCURRENT FORCE SYSTEM

It is possible to find a single force which will have the same effect as that of a number of forces acting on a particle. Such single force is called *Resultant force* and the process of finding out the resultant force is called *Composition of Forces.* Using parallelogram law, triangle law and polygon law, the resultant of forces can be found. These are the graphical methods, which need drawing sheets and drawing accessories.

In the analytical method, the components of each one of the forces in the coordinate directions are found and they are expressed in the form of vectors. Referring to Fig. 2.4, if $\mathbf{F}_1$, $\mathbf{F}_2$, $\mathbf{F}_3$ and $\mathbf{F}_4$ are the forces in the system, then we write

$$\left.\begin{aligned} \mathbf{F}_1 &= F_{1x}\,\mathbf{i} + F_{1y}\,\mathbf{j} \\ \mathbf{F}_2 &= F_{2x}\,\mathbf{i} + F_{2y}\,\mathbf{j} \\ \mathbf{F}_3 &= F_{3x}\,\mathbf{i} + F_{3y}\,\mathbf{j} \\ \mathbf{F}_4 &= F_{4x}\,\mathbf{i} + F_{4y}\,\mathbf{j} \end{aligned}\right\} \qquad \text{...(2.2)}$$

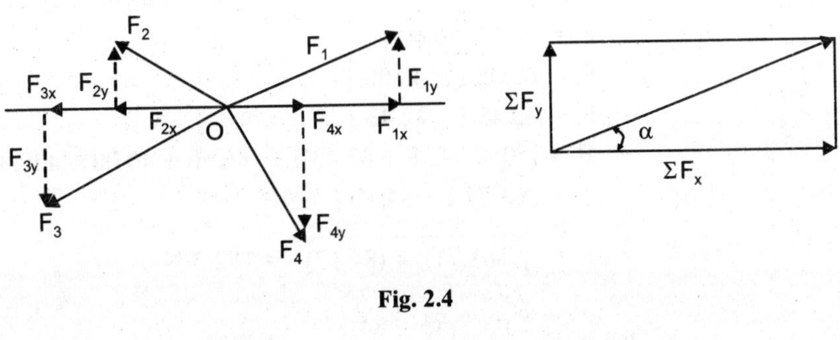


**Fig. 2.4**

**Note:** In the above expression, $F_{2x}$, $F_{3x}$, $F_{3y}$ and $F_{4y}$ have negative values.

Then the resultant **R** is the vector sum of $\mathbf{F}_1$, $\mathbf{F}_2$, $\mathbf{F}_3$ and $\mathbf{F}_4$. *i.e.*,

$$\begin{aligned}\mathbf{R} &= \mathbf{F}_1 + \mathbf{F}_2 + \mathbf{F}_3 + \mathbf{F}_4 \\ &= F_{1x}\,\mathbf{i} + F_{1y}\,\mathbf{j} + F_{2x}\,\mathbf{i} + F_{2y}\,\mathbf{j} + F_{3x}\,\mathbf{i} + F_{3y}\,\mathbf{j} + F_{4x}\,\mathbf{i} + F_{4y}\,\mathbf{j} \\ &= (F_{1x} + F_{2x} + F_{3x} + F_{4x})\,\mathbf{i} + (F_{1y} + F_{2y} + F_{3y} + F_{4y})\,\mathbf{j} \\ &= R_x\,\mathbf{i} + R_y\,\mathbf{j}\end{aligned}$$

where $R_x = \Sigma F_x$ and $R_y = \Sigma F_y$.

$$\therefore \quad \mathbf{R} = \Sigma F_x\,\mathbf{i} + \Sigma F_y\,\mathbf{j} \qquad \text{...(2.3)}$$

$$\therefore \quad R = \sqrt{(\Sigma F_x)^2 + (\Sigma F_y)^2} \qquad \text{...(2.4)}$$

If $\alpha$ is the angle between $x$-axis and the force, then

$$\tan\alpha = \frac{\Sigma F_y}{\Sigma F_x} \qquad \text{...(2.5)}$$

**Example 2.3.** Determine the resultant of the three forces acting on a hook as shown in Fig. 2.5 (*a*).

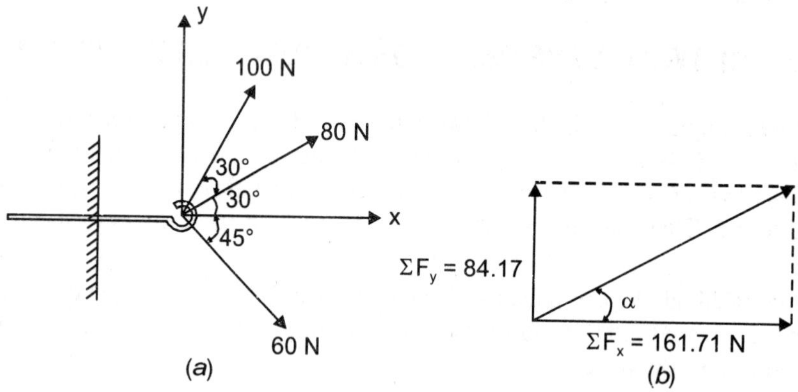


**Fig. 2.5**

**Solution.**

| Force | $x$-Component | $y$-Component |
|---|---|---|
| $F_1 = 100$ N | $100\cos 60° = 50$ N | $100\sin 60° = 86.60$ N |
| $F_2 = 80$ N | $80\cos 30° = 69.28$ N | $80\sin 30° = 40.00$ N |
| $F_3 = 60$ N | $60\cos 45° = 42.43$ N | $-60\sin 45° = -42.43$ N |

$$\therefore \quad \begin{aligned}\mathbf{F}_1 &= 50\,\mathbf{i} + 86.60\,\mathbf{j} \\ \mathbf{F}_2 &= 69.28\,\mathbf{i} + 40.00\,\mathbf{j} \\ \mathbf{F}_3 &= 42.43\,\mathbf{i} - 42.43\,\mathbf{j}\end{aligned}$$

$$\therefore \quad \begin{aligned}\mathbf{R} &= (50 + 69.28 + 42.43)\mathbf{i} + (86.60 + 40.00 - 42.43)\mathbf{j} \\ &= 161.71\,\mathbf{i} + 84.17\,\mathbf{j}\end{aligned}$$ **Ans.**

$$\therefore \quad R = \sqrt{(161.71)^2 + (84.17)^2} = 182.3 \text{ N}$$ **Ans.**

$$\tan\alpha = \frac{\Sigma F_y}{\Sigma F_x} = \frac{R_y}{R_x} = \frac{84.17}{161.71} = 0.5205$$

$$\alpha = 27.5°$$ **Ans.**

**Example 2.4.** A system of four forces acting on a body is shown in Fig. 2.6(*a*). Determine the resultant.

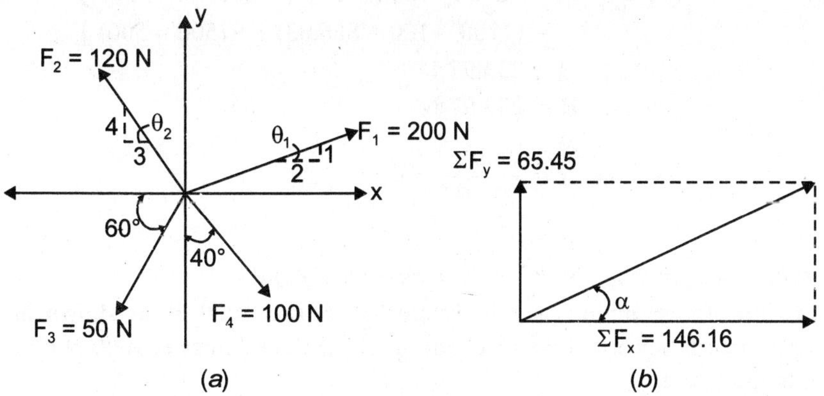


**Fig. 2.6**

**Solution.** If $\theta_1$ is the inclination of 200 N force to *x*-axis, then

$$\tan \theta_1 = \frac{1}{2} \text{ or } \theta_1 = 26.565°$$

Similarly if $\theta_2$ is the inclination of 120 N force to *x*-axis, as shown in figure, then

$$\tan \theta_2 = \frac{4}{3} \text{ or } \theta_2 = 53.13°$$

$$\mathbf{F}_1 = 200 (\cos \theta_1\, \mathbf{i} + \sin \theta_1\, \mathbf{j}) = 178.88\, \mathbf{i} + 89.44\, \mathbf{j}$$

$$\mathbf{F}_2 = -120 \cos \theta_2\, \mathbf{i} + 120 \sin \theta_2\, \mathbf{j} = -72\, \mathbf{i} + 96\, \mathbf{j}$$

$$\mathbf{F}_3 = -50 \cos 60°\, \mathbf{i} - 50 \sin 60°\, \mathbf{j} = -25\, \mathbf{i} - 43.30\, \mathbf{j}$$

$$\mathbf{F}_4 = 100 \sin 40°\, \mathbf{i} - 100 \cos 40°\, \mathbf{j} = 64.28\, \mathbf{i} - 76.60\, \mathbf{j}$$

$$\therefore \quad \mathbf{R} = \mathbf{F}_1 + \mathbf{F}_2 + \mathbf{F}_3 + \mathbf{F}_4$$

$$= (178.88 - 72 - 25 + 64.28)\mathbf{i} + (89.44 + 96 - 43.30 - 76.60)\mathbf{j}$$

$$= 146.16\, \mathbf{i} + 65.54\, \mathbf{j}$$

*i.e.*, $R_x = 146.16$ N and $R_y = 65.54$ N.

$$\therefore \quad R = \sqrt{(146.16)^2 + (65.54)^2} = 160.18 \text{ N}$$ **Ans.**

$$\alpha = \tan^{-1} \frac{R_y}{R_x} = \tan^{-1} \frac{65.54}{146.16} = 24.15°$$ **Ans.**

**Example 2.5.** A system of forces acting on a body and resting on an inclined plane is shown in Fig. 2.7. Determine the resultant force if θ = 60°, **W** = 1000 N, **N** = 500 N, **F** = 100 N and **T** = 1200 N.

**Solution.** In this problem, it is convenient to select the coordinate system as parallel to and perpendicular to the plane. The vector forms of various forces are given below:

$$\mathbf{T} = 1200\, \mathbf{i}$$

$$\mathbf{F} = -100\, \mathbf{i}$$

$$\mathbf{N} = 500\, \mathbf{j}$$

$$\mathbf{W} = -1000 \sin 60\, \mathbf{i} - 1000 \cos 60\, \mathbf{j}$$

$$= -866.03\, \mathbf{i} - 500\, \mathbf{j}$$

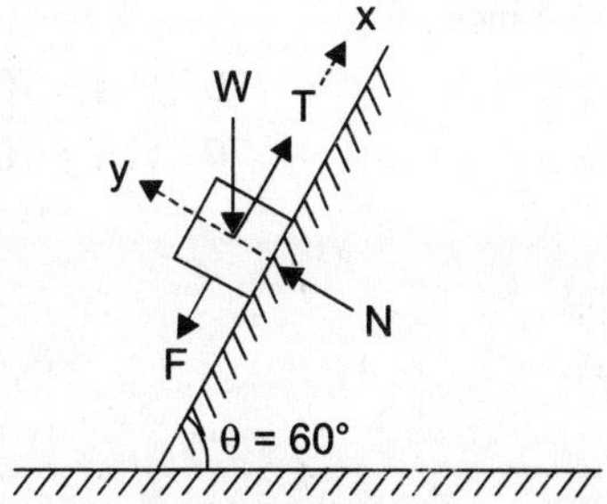


**Fig. 2.7**

$$\therefore \quad \mathbf{R} = \mathbf{T} + \mathbf{F} + \mathbf{N} + \mathbf{W}$$
$$= 1200\,\mathbf{i} - 100\,\mathbf{i} + 500\,\mathbf{j} - 866.03\,\mathbf{i} - 500\,\mathbf{j}$$
$$= (1200 - 100 - 866.03)\,\mathbf{i} + (500 - 500)\,\mathbf{j}$$
$$= 233.97\,\mathbf{i} \qquad \textbf{Ans.}$$
$$R = 233.97 \text{ N} \qquad \textbf{Ans.}$$
$$\tan\alpha = \frac{0}{233.97} = 0 \qquad \textbf{Ans.}$$
$$\alpha = 0° \qquad \textbf{Ans.}$$

Thus the *resultant is a force of 233.97 N directed up the plane.* **Ans.**

**Example 2.6.** Two forces acting on a body are 500 N and 1000 N, as shown in Fig. 2.8 (*a*). Determine the third force $\mathbf{F}_3$ such that the resultant of all the three forces is 1000 N directed at 45° to *x*-axis, as shown in the figure.

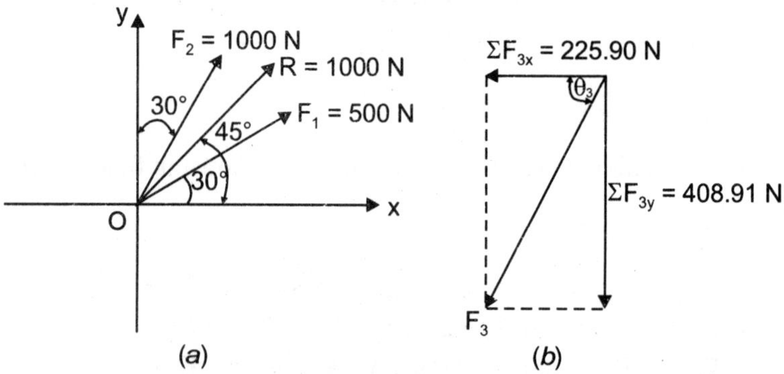


**Fig. 2.8**

**Solution.** Let the third force $\mathbf{F}_3$ makes an angle $\theta_3$ with *x*-axis.

Now
$$\mathbf{F}_1 = 500\cos 30°\,\mathbf{i} + 500\sin 30°\,\mathbf{j}$$
$$= 433.01\,\mathbf{i} + 250\,\mathbf{j}$$
$$\mathbf{F}_2 = 1000\sin 30°\,\mathbf{i} + 1000\cos 30°\,\mathbf{j}$$
$$= 500\,\mathbf{i} + 866.02\,\mathbf{j}$$
$$\mathbf{F}_3 = F_{3x}\,\mathbf{i} + F_{3y}\,\mathbf{j}$$
$$= F_3\cos\theta_3\,\mathbf{i} + F_3\sin\theta_3\,\mathbf{j}$$
$$\mathbf{R} = 1000\cos 45°\,\mathbf{i} + 1000\sin 45°\,\mathbf{j}$$
$$= 707.11\,\mathbf{i} + 707.11\,\mathbf{j}$$

Since $\mathbf{R} = \mathbf{F}_1 + \mathbf{F}_2 + \mathbf{F}_3$, we get

$$707.11\,\mathbf{i} + 707.11\,\mathbf{j} = (433.01 + 500 + F_3\cos\theta_3)\mathbf{i} + (250 + 866.02 + F_3\sin\theta_3)\mathbf{j}$$

*i.e.*,
$$707.11\,\mathbf{i} + 707.11\,\mathbf{j} = (933.01 + F_3\cos\theta_3)\mathbf{i} + (1116.02 + F_3\sin\theta_3)\mathbf{j}$$
$$\therefore \quad 707.11 = 933.01 + F_3\cos\theta_3$$

and
$$707.11 = 1116.02 + F_3\sin\theta_3$$

*i.e.*,
$$F_3\cos\theta_3 = -225.90 \text{ and } F_3\sin\theta_3 = -408.91$$
$$\therefore \quad \mathbf{F}_3 = -225.90\,\mathbf{i} - 408.91\,\mathbf{j} \qquad \textbf{Ans.}$$
$$\therefore \quad F_3 = \sqrt{(-225.90)^2 + (-408.91)^2} = 467.16 \text{ N} \qquad \textbf{Ans.}$$

$$\tan \theta_3 = \frac{408.91}{225.9} = 1.810$$

$$\theta_3 = 61.08° \text{ as shown in Fig. 2.8 } (b)$$ **Ans.**

**Example 2.7.** Three forces acting at a point are as shown in Fig. 2.9. The direction of 300 N forces may vary, but the angle between them is always 40°. Determine the value of θ for which the resultant of the three forces is directed to *x*-axis.

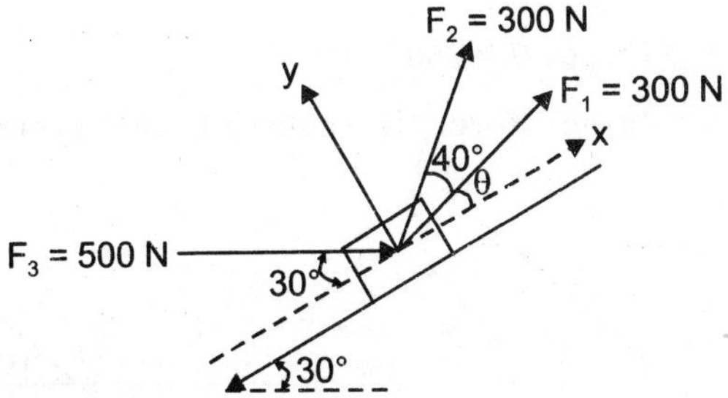


**Fig. 2.9**

**Solution.**

$$\mathbf{F}_1 = 300 \cos \theta\mathbf{i} + 300 \sin \theta\mathbf{j}$$
$$\mathbf{F}_2 = 300 \cos (40° + \theta)\mathbf{i} + 300 \sin (40° + \theta)\mathbf{j}$$
$$\mathbf{F}_3 = 500 \cos 30°\mathbf{i} - 500 \sin 30°\mathbf{j}$$
$$= 433.01\mathbf{i} - 250\mathbf{j}$$

$$\therefore \quad \mathbf{R} = \mathbf{F}_1 + \mathbf{F}_2 + \mathbf{F}_3$$
$$= [(300 \cos \theta + 300 \cos (40 + \theta) + 433.01)\ \mathbf{i} + (300 \sin \theta + 300 \sin (40 + \theta) - 250)\mathbf{j})]$$

According to condition specified in the problem, component of resultant in *y*-direction is zero.

Hence $$300 \sin \theta + 300 \sin (40° + \theta) - 250 = 0$$

$$\sin \theta + \sin (40° + \theta) = \frac{250}{300} = 0.8333$$

*i.e.,* $$2 \sin \left(\frac{40° + \theta + \theta}{2}\right) \cos \left(\frac{40° + \theta - \theta}{2}\right) = 0.8333$$

*i.e.,* $$2 \sin (20° + \theta) \cos 20° = 0.8333$$

$$\sin (20° + \theta) = 0.4434$$

$$\therefore \quad 20° + \theta = 26.32$$

or $$\theta = 6.32°$$ **Ans.**

**Example 2.8.** The resultant of two forces, one of which is double the other, is 260 N. If the direction of the larger force is reversed and the other remains unaltered, the resultant reduces to 180 N. Determine the magnitude of the forces and the angle between the forces.

**Solution.** Let $\mathbf{F}_1$ and $\mathbf{F}_2$ be the forces and the angle between them be θ. Taking the direction of the force $\mathbf{F}_1$ as *x*-axis and normal to it as *y*-axis [Ref. Fig. 2.10 (*a*)], we have

$$\mathbf{F}_1 = F_1\ \mathbf{i}$$
$$\mathbf{F}_2 = F_2 \cos \theta\ \mathbf{i} + F_2 \sin \theta\ \mathbf{j}$$

$$= 2F_1 \cos\theta\, \mathbf{i} + 2F_1 \sin\theta\, \mathbf{j}$$

$\therefore$ $$\mathbf{R} = (F_1 + 2F_1 \cos\theta)\, \mathbf{i} + 2F_1 \sin\theta\, \mathbf{j}$$

$\therefore$ $$R = \sqrt{(F_1 + 2F_1 \cos\theta)^2 + (2F_1 \sin\theta)^2} = 260$$

*i.e.,* $$F_1^2 + 4F_1F_1 \cos\theta + 4F_1^2 \cos^2\theta + 4F_1^2 \sin^2\theta = 260^2$$

or $$F_1^2 + 4F_1^2 \cos\theta + 4F_1^2 = 260^2, \quad [\text{Since } \cos^2\theta + \sin^2\theta = 1]$$

*i.e.,* $$5F_1^2 + 4F_1^2 \cos\theta = 260^2 \qquad ...(i)$$

When the direction of larger force is reversed, the system of forces is shown in Fig. 2.10 (*b*).

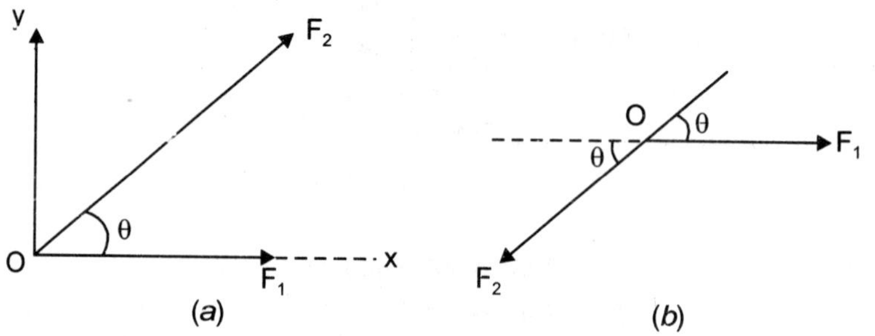


**Fig. 2.10**

Now
$$\mathbf{F}_1 = F_1\, \mathbf{i}$$
$$\mathbf{F}_2 = -F_2 \cos\theta\, \mathbf{i} - F_2 \sin\theta\, \mathbf{j}$$
$$= -2F_1 \cos\theta\, \mathbf{i} - 2F_1 \sin\theta\, \mathbf{j}$$
$$\mathbf{R} = (F_1 - 2F_1 \cos\theta)\, \mathbf{i} - 2F_1 \sin\theta\, \mathbf{j}$$
$$= F_1 (1 - 2\cos\theta)\, \mathbf{i} - 2F_1 \sin\theta\, \mathbf{j}$$

$\therefore$ $$R = \sqrt{F_1^2 (1 - 2\cos\theta)^2 + (2F_1 \sin\theta)^2} = 180$$

$\therefore$ $$F_1^2 (1 - 4\cos\theta + 4\cos^2\theta) + 4F_1^2 \sin^2\theta = 180^2$$

*i.e.,* $$5F_1^2 - 4F_1^2 \cos\theta = 180^2 \qquad ...(ii)$$

Adding Eqns. (*i*) and (*ii*), we get

$$10F_1^2 = 260^2 + 180^2$$

$\therefore$ $$F_1 = 100 \text{ N}$$ **Ans.**

Hence $$F_2 = 200 \text{ N}$$ **Ans.**

Substituting this value in (*i*), we get

$$5(100)^2 + 4 \times 100^2 \cos\theta = 260^2$$

$\therefore$ $$\cos\theta = \frac{260^2 - 5 \times 100^2}{4 \times 100^2} = 0.44$$

Therefore, $$\theta = 63.896°$$ **Ans.**

## 2.3 TRANSMISSIBILITY OF A FORCE

Before finding resultant of non-concurrent force system, law of transmissibility of force is explained. It is frequently used in the engineering mechanics.

According to law of transmissibility of force, *the state of rest or motion of the rigid body is unaltered if a force acting on the body is replaced by another force of the same magnitude and direction but acting anywhere on the body along the line of action of the replaced force.*

Let **F** be the force acting on a rigid body at point **A**, as shown in Fig. 2.11. According to the law of transmissibility of force, this force has the same effect on the state of body as force **F** applied at point *B*.

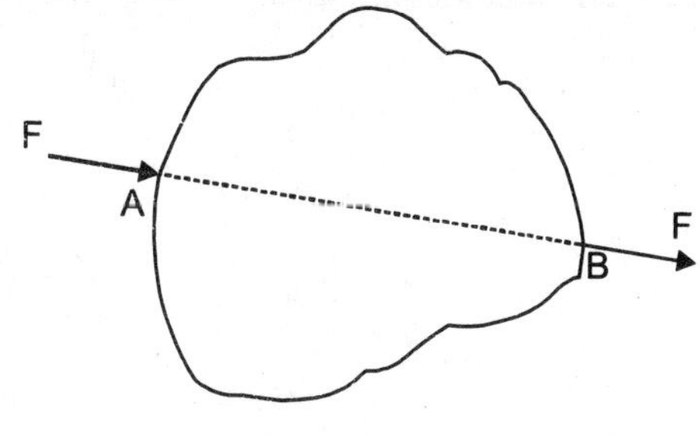


**Fig. 2.11**

In using law of transmissibility of force, it should be carefully noted that it is applicable only if the body can be treated as rigid. In this text, the engineering mechanics is restricted to the study of state of rigid bodies, and hence, this law is frequently used. Same thing cannot be done in the subject 'solid mechanics', where the bodies are treated as deformable and internal forces in the body are studied.

The law of transmissibility of force can be proved using *the law of superposition,* which can be stated as *the action of a given system of forces on a rigid body is not changed by adding or subtracting another system of forces in equilibrium.*

Consider the rigid body, as shown in Fig. 2.12(*a*). It is subjected by a force **F** at *A*. *B* is another point on the line of action of the force. From the law of superposition, it is obvious that if two equal and opposite forces of magnitude *F* are applied at *B* along the line of action of given force **F** [Ref. Fig. 2.12(*b*)], the effect of given force on the body is not altered. Force **F** at *A* and opposite force **F** at *B* form a system of forces in equilibrium. If these two forces are subtracted from the system, the resulting system is as shown in Fig. 2.12(*c*). Looking at the system of forces in Figs. 2.12(*a*) and 2.12(*c*), we can conclude that the law of transmissibility of forces is proved.

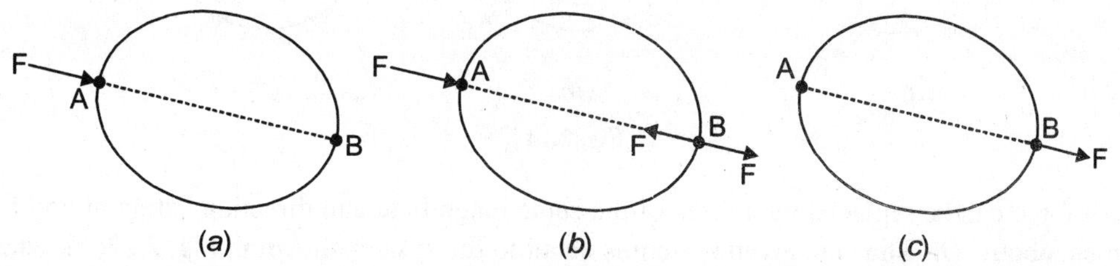


**Fig. 2.12**

## 2.4 RESOLUTION OF A FORCE INTO A FORCE AND A COUPLE

Similar to finding components of forces in case of concurrent force system, for non-concurrent force system, resolution of a force into a force and a couple is a useful concept.

In many engineering mechanics problems, it will be advantageous to resolve a force acting on a body at a particular point into a force acting at some other suitable point and a couple moment. Let **F** be a force acting at point *A* on a body, as shown in Fig. 2.13(*a*). Imagine, we apply two equal and opposite forces **F** at *B*. By doing so, the system of forces is not disturbed, *i.e.*, the system of forces in

Fig. 2.13 (*b*) is the same as the system shown in Fig. 2.13(*a*). Now, the given force **F** at *A* and the opposite force **F** at *B* form a couple of magnitude $r_{AB} \times F$ (or *Fd*). Hence the system shown in Fig. 2.13(*b*) is the same as that shown in Fig. 2.13(*c*). Thus the given force **F** at *A* is replaced by a force **F** at *B* and a couple-moment of magnitude $r_{AB} \times F$ (or *Fd*).

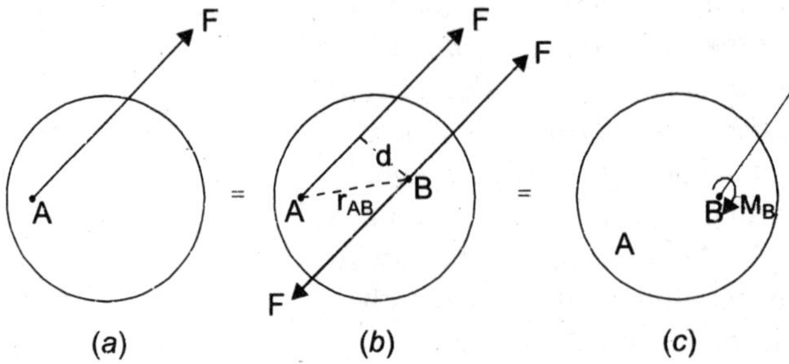


**Fig. 2.13**

## 2.5 RESULTANT OF NON-CONCURRENT FORCE SYSTEM

The resultant of a force system is the one which will have the same rotational and translatory effect as the given system of forces. It may be a single force, a pure moment or a force and a moment.

Consider the system of forces shown in Fig. 2.14(*a*).

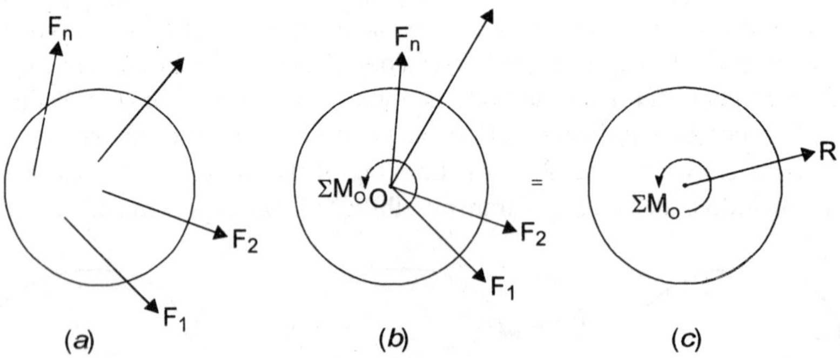


**Fig. 2.14**

Each force can be replaced by a force of the same magnitude and direction acting at point '*O*' and its moment about '*O*'. Thus the given system is equal to the system shown in Fig. 2.22(*b*), where $\Sigma \mathbf{M_o}$ is the algebraic sum of the moments of the given forces about *O*.

At *O*, the concurrent forces $\mathbf{F_1}, \mathbf{F_2}, ..., \mathbf{F_n}$ can be combined as usual to get the resultant force $\mathbf{R} = \mathbf{F} + \mathbf{F_2} + ... + \mathbf{F_n}$. Now the resultant of the given system is equal to a force **R** at *O* and a moment $\Sigma \mathbf{M_O}$, as shown in Fig. 2.22(*c*). Thus the resultant of the non-concurrent forces is

$$\mathbf{R} = \sum_{i=1}^{n} \mathbf{F_i} \text{ and } \mathbf{M_o} = \sum_{i=1}^{n} \mathbf{M_{oi}} \qquad ...(2.6)$$

The above equations are equivalent to the following scalar expressions

$$R_x = \Sigma F_{xi};\ R_y = \Sigma F_{yi};\ R_z = \Sigma F_{zi} \qquad ...(2.7)$$

$$M_x = \Sigma M_{xi};\ M_y = \Sigma M_{yi};\ M_z = \Sigma M_{zi} \qquad ...(2.8)$$

$$R = \sqrt{(\Sigma F_{xi})^2 + (\Sigma F_{yi})^2 + (\Sigma F_{zi})^2} \quad ...(2.9)$$

$$M = \sqrt{(\Sigma M_{xi})^2 + (\Sigma M_{yi})^2 + (\Sigma M_{zi})^2} \quad ...(2.10)$$

***x* and *y* Intercepts of the Resultant**

In some two dimensional problems, we may be interested in finding the intersection point of resultant $R$ along $x$ and $y$ axes, *i.e.*, $x$ and $y$ intercepts. Let **R** be the resultant intersecting $x$, $y$ coordinates through point $O$ at $A$ and $B$ as shown in Fig. 2.15. Let moment about the moment centre $O$ be $\mathbf{M}_O$. The position vector of $A$ is $x\mathbf{i}$ and the position vector of $B$ is $y\mathbf{j}$. Hence

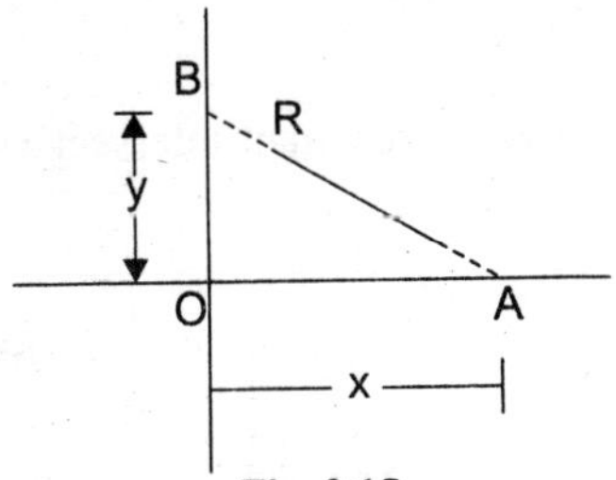

**Fig. 2.15**

$$\mathbf{M}_O = x_i \times \mathbf{F} = y_i \times \mathbf{F} \quad ...(2.11)$$

**Example 2.9.** The system of loads acting on a beam is as shown in Fig. 2.16. Determine the resultant of the loads.

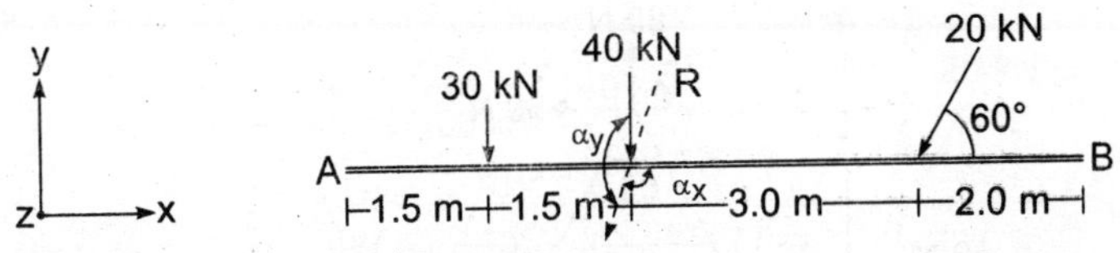

**Fig. 2.16**

**Solution.** Selecting the coordinate directions as shown in Fig. 2.16,

30 kN downward load $\mathbf{F}_1 = -30\mathbf{j}$

40 kN downward load $\mathbf{F}_2 = -40\mathbf{j}$

and 20 kN downward load, $\mathbf{F}_3 = (-20 \cos 60°\mathbf{i} - 20 \sin 60°\mathbf{j})$

$$= -10\mathbf{i} - 17.321\mathbf{j}$$

Taking $A$ as origin, the position vectors of the intersection points of the forces with beam are

$$\mathbf{r}_{A1} = 1.5\mathbf{i},\ \mathbf{r}_{A2} = 3.0\mathbf{i} \text{ and } \mathbf{r}_{A3} = 6\mathbf{i}$$

$$\therefore \quad \mathbf{M}_A = \mathbf{r}_{A1} \times \mathbf{F}_1 + \mathbf{r}_{A2} \times \mathbf{F}_2 + \mathbf{r}_{A3} \times \mathbf{F}_3$$

$$= -1.5\mathbf{i} \times (-30\mathbf{j}) + 3.0\mathbf{i} \times (-40\mathbf{j}) + 6\mathbf{i} \times (-10\mathbf{i} - 17.321\mathbf{j})$$

$$= -45\mathbf{k} - 120\mathbf{k} - 103.926\mathbf{k}$$

$$= -268.926\mathbf{k} \quad \textbf{Ans.}$$

*i.e.*, $M_A = 268.926$ kN-m, clockwise

$$\mathbf{R} = \mathbf{F}_1 + \mathbf{F}_2 + \mathbf{F}_3$$

$$\mathbf{M}_O = \mathbf{r}_{OA} \times \mathbf{F} = \mathbf{r}_{OB} \times \mathbf{F}$$

$$= x\mathbf{i} \times \mathbf{F} = y\mathbf{j} \times \mathbf{F}$$

Hence $x$ and $y$ intercepts can be found.

$$\mathbf{R} = -30\mathbf{j} - 40\mathbf{j} - 10\mathbf{i} - 17.321\mathbf{j}$$

$$= -10\mathbf{i} - 87.321\mathbf{j} \quad \textbf{Ans.}$$

$$\therefore \quad R = \sqrt{10^2 + 87.321^2} = 87.892 \text{ kN} \quad \textbf{Ans.}$$

Let $\alpha$ be the indication to the $x$-axis. Then

$$\cos \alpha_x = \frac{-10}{87.892} = -0.113776$$

or

$$\alpha_x = 96.533° \text{ (as shown in figure)}$$

$$\alpha_y = \cos^{-1} \frac{-87.321}{87.892} = 173.467°$$

Let the resultant intersect $x$-axis at $D$. Let $AD$ be $x$ metres. Then

$$\mathbf{r}_{AD} = x\mathbf{i}$$

$$\therefore \quad \mathbf{M} = \mathbf{r}_{AD} \times \mathbf{F}$$

$$-268.926\mathbf{k} = x\mathbf{i} \times (-10\mathbf{i} - 87.321\mathbf{j})$$

$$= 87.321k$$

$$\therefore \quad x = 3.08 \text{ m}$$

**Ans.**

**Example 2.10.** Find the resultant of the force system, as shown in Fig. 2.17, acting on a lamina of equilateral triangle of sides 120 mm.

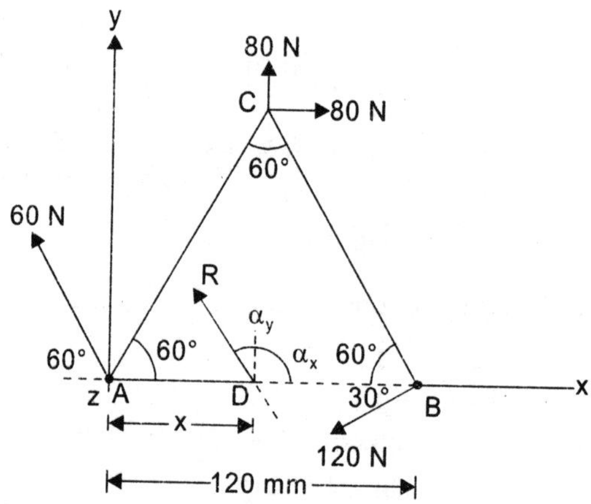


**Fig. 2.17**

**Solution.** Taking $x$, $y$ axes in the plane of lamina and $z$-axis normal to it,

60 N force:

$\mathbf{F}_1 = -60 \cos 60°\mathbf{i} + 60 \sin 60°\mathbf{j} = -30\mathbf{i} + 51.96\mathbf{j}$

80 N vertical force : $\mathbf{F}_2 = 80\mathbf{j}$

80 N horizontal force: $\mathbf{F}_3 = 80\mathbf{i}$

120 N force:

$$\mathbf{F}_4 = -120 \cos 30°\mathbf{i} - 120 \sin 30°\mathbf{j} = -103.923\mathbf{i} - 60\mathbf{j}$$

$$\therefore \quad \mathbf{R} = \mathbf{F}_1 + \mathbf{F}_2 + \mathbf{F}_3 + \mathbf{F}_4$$

$$= (-30 + 80 - 103.923)\mathbf{i} + (51.96 + 80 - 60)\mathbf{j}$$

$$= -53.923\mathbf{i} + 71.96\mathbf{j}$$

**Ans.**

$$\therefore \quad R = \sqrt{(53.923)^2 + (71.96)^2} = 89.922 \text{ N}$$

$$\cos \alpha_x = \frac{-53.923}{89.922} = -0.5997$$

$$\alpha_x = 126.845° \text{ as shown in Fig. 2.17.}$$

$$\alpha_y = \frac{71.96}{89.922} = 36.846°$$

Taking point $A$ as origin, the position vector of forces $\mathbf{F}_1$ to $\mathbf{F}_4$ are

$$\mathbf{r}_{A1} = 0 \quad \mathbf{r}_{A2} = 60\mathbf{i} + 120 \sin 60°\mathbf{j}$$

$$\mathbf{r}_{AB} = 60\mathbf{i} + 120 \sin 60°\mathbf{j} \text{ and } \mathbf{r}_{A4} = 120\mathbf{i}$$

$\therefore$

$$\mathbf{M}_4 = 0 + (60\mathbf{i} + 120 \sin 60°\mathbf{j}) \times (80\mathbf{j})$$
$$+ (60\mathbf{i} + 120 \sin 60°\mathbf{j}) \times 80\mathbf{i} + 120\mathbf{i} \times (-103.923\mathbf{i} - 60\mathbf{j})$$
$$= 4800\mathbf{k} + 8313.8\,(-\mathbf{k}) + 7200\mathbf{k} = 3686.16\mathbf{k}$$

**Note:**

$\mathbf{i} \times \mathbf{i} = \mathbf{j} \times \mathbf{j} = 0; \mathbf{i} \times \mathbf{j} = \mathbf{k}; \mathbf{j} \times \mathbf{i} = -\mathbf{k}$

Let the resultant interest $x$-axis at $D$.

Then $\mathbf{r}_{AD} = x\mathbf{i}$

$\therefore$ $\mathbf{r}_{AD} \times \mathbf{R} = \mathbf{M}_A$

$$x\mathbf{i} \times (-53.923\mathbf{i} + 71.96\mathbf{j}) = 3686.16\mathbf{k}$$

$$x\,(71.96\mathbf{k}) = 3686.16\mathbf{k}$$

or $x = 51.225$ mm as shown in figure. **Ans.**

**Example 2.11.** Find the resultant of a set of coplanar forces acting on a lamina as shown in Fig. 2.18. Each square has a side of 10 mm.

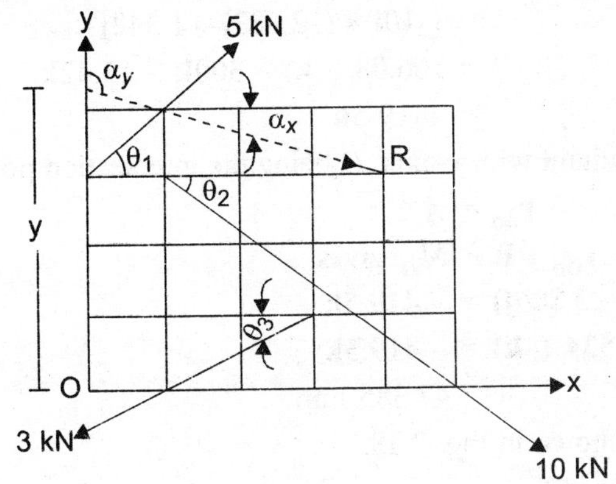


**Fig. 2.18**

**Solution.** Let $\theta_1$, $\theta_2$ and $\theta_3$ be the inclinations of the forces $\mathbf{F}_1 = 5$ kN, $\mathbf{F}_2 = 10$ kN and $\mathbf{F}_3 = 3$ kN. Then,

$$\theta_1 = \tan^{-1}\frac{10}{10} = 45°$$

$$\theta_2 = \tan^{-1}\frac{30}{40} = 36.87°$$

and

$$\theta_3 = \tan^{-1}\frac{10}{20} = 26.565°$$

The vector form of the given forces are

$$\mathbf{F}_1 = 5 \cos \theta_1\mathbf{i} + 5 \sin \theta_1\mathbf{j} = 3.536\mathbf{i} + 3.536\mathbf{j}$$

$$\mathbf{F}_2 = 10 \cos \theta_2\mathbf{i} - 10 \sin \theta_2\mathbf{j}$$
$$= 8\mathbf{i} - 6\mathbf{j}$$

$$\mathbf{F}_3 = -3 \cos \theta_3 \mathbf{i} - 3 \sin \theta_3 \mathbf{j}$$
$$= -2.683\mathbf{i} - 1.342\mathbf{j}$$

$\therefore$
$$\mathbf{R} = \mathbf{F}_1 + \mathbf{F}_2 + \mathbf{F}_3$$
$$= (3.536 + 8 - 2.683)\mathbf{i} + (3.536 - 6 - 1.342)\mathbf{j}$$
$$= 8.853\mathbf{i} - 3.806\mathbf{j}$$ **Ans.**

$$R = \sqrt{(8.853)^2 + (3.806)^2} = 9.636 \text{ kN}$$

$$\cos \alpha_x = \frac{8.853}{9.636}$$

$\therefore$
$$\alpha_x = 23.263°$$

$$\cos \alpha_y = \frac{-3.806}{9.636}$$

$\therefore$
$$\alpha_y = 113.263°$$

It is as shown in Fig. 2.18.

Now,
$$\mathbf{r}_{01} = 0\mathbf{i} + 30\mathbf{j}$$
$$\mathbf{r}_{02} = 50\mathbf{i} + 0\mathbf{j}$$
$$\mathbf{r}_{03} = 10\mathbf{i} + 0\mathbf{j}$$

$\therefore$
$$\mathbf{M}_O = \mathbf{r}_{01} \times \mathbf{F}_1 + \mathbf{r}_{02} \times \mathbf{F}_2 + \mathbf{r}_{03} \times \mathbf{F}_3$$
$$= 30\mathbf{j} \times (3.536\mathbf{i} + 3.536\mathbf{j}) + 50\mathbf{i} \times (8\mathbf{i} - 6\mathbf{j})$$
$$+ 10\mathbf{i} \times (-2.683\mathbf{i} - 1.342\mathbf{j})$$
$$= 106.08\,(-\mathbf{k}) - 300\mathbf{k} - 13.42\mathbf{k}$$
$$= -419.5\mathbf{k}$$

If $y$ is the intercept of resultant with $y$-axis, $D$ being the intersection point, then

$$\mathbf{r}_{OD} = y\mathbf{j}$$

$\therefore$
$$\mathbf{r}_{OD} \times \mathbf{R} = \mathbf{M}_O, \text{ gives}$$
$$y\mathbf{j} \times (8.853\mathbf{i} - 3.806\mathbf{j}) = -419.5\mathbf{k}$$

*i.e.*,
$$8.8534\,(-\mathbf{k}) = -419.5\mathbf{k}$$

or
$$y = 47.385 \text{ mm}$$ **Ans.**

Hence the resultant is as shown in Fig. 2.18.

**Example 2.12.** The system of forces acting on a bell crank is as shown in Fig. 2.19. Determine the magnitude, direction and the point of application of the resultant.

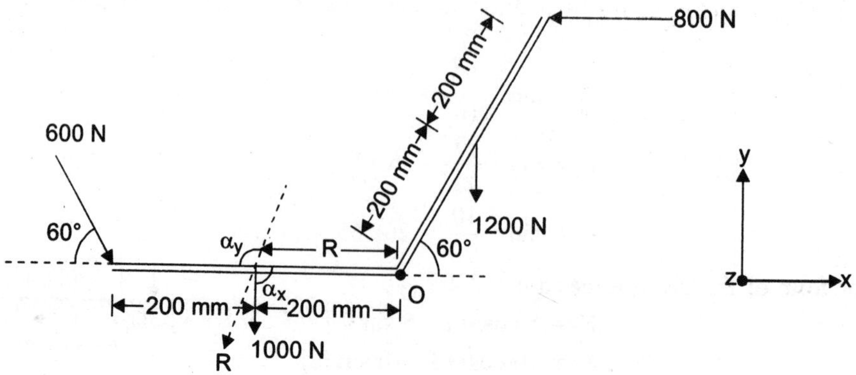


**Fig. 2.19**

**Solution.** The vector forms of the forces are given as

600 N force : $\mathbf{F}_1 = 600 \cos 60°\mathbf{i} - 600 \sin 60°\mathbf{j}$

$= 300\mathbf{i} - 519.615\mathbf{j}$

1000 N force : $\mathbf{F}_2 = -1000\mathbf{j}$

1200 N force : $\mathbf{F}_3 = -1200\mathbf{j}$

800 N force : $\mathbf{F}_4 = -800\mathbf{i}$

Taking '*O*' as the origin, the position vectors of these forces are given as

$$\mathbf{r}_{01} = -400\mathbf{i},\ r_{02} = -200\mathbf{i}$$

$$\mathbf{r}_{03} = 200 \cos 60°\mathbf{i} + 200 \sin 60°\mathbf{j} = 100\mathbf{i} + 173.205\mathbf{j}$$

$$\mathbf{r}_{04} = 400 \cos 60°\mathbf{i} + 400 \sin 60°\mathbf{j} = 200\mathbf{i} + 346.41\mathbf{j}$$

$$\mathbf{R} = \mathbf{F}_1 + \mathbf{F}_2 + \mathbf{F}_3 + \mathbf{F}_4$$

$$= (300 - 800)\mathbf{i} + (-519.615 - 1000 - 1200)\mathbf{j}$$

$$= -500\mathbf{i} - 2719.615\mathbf{j}$$

$\therefore$ $$R = \sqrt{(500)^2 + (2719.615)^2} = 2765.195 \text{ N}$$ **Ans.**

$$\alpha_x = \cos^{-1} \frac{-500}{2765.195} = 100.42°$$

$$\alpha_y = \cos^{-1} \frac{-2719.615}{2765.195} = 169.58°$$ **Ans.**

$$\mathbf{M_O} = -400\mathbf{i} \times (300\mathbf{i} - 519.615\mathbf{j}) - 200\mathbf{i} \times (-1000\mathbf{j})$$

$$+ (100\mathbf{i} + 173.205\mathbf{j}) \times (-1200\mathbf{j})$$

$$+ (200\mathbf{i} + 346.41\mathbf{j}) \times (-800\mathbf{i})$$

$$= 207846\mathbf{k} + 200000\mathbf{k} - 120000\mathbf{k} - 277128\ (-\mathbf{k})$$

$$= 564974\mathbf{k}$$

Let the point of application '*D*' of the resultant be at a distance *x* along *x*-axis. Then

$$\mathbf{r_{OD}} = x\mathbf{i}$$

$\therefore$ $$\mathbf{r_{OD}} \times \mathbf{R} = \mathbf{M_O}$$

$$x\mathbf{i} \times (-500\mathbf{i} - 2719.615\mathbf{j}) = 564974\mathbf{k}$$

$$-2719.615x\ \mathbf{k} = 564994\mathbf{k}$$

$\therefore$ $$x = -\ 207.74 \text{ mm}$$ **Ans.**

Hence the resultant acts as shown in Fig. 2.19.

**Example 2.13.** Various forces to be considered for the stability analysis of a dam are as shown in Fig. 2.20. The dam is safe if the resultant force passes through middle third of the base. Verify whether the dam is safe.

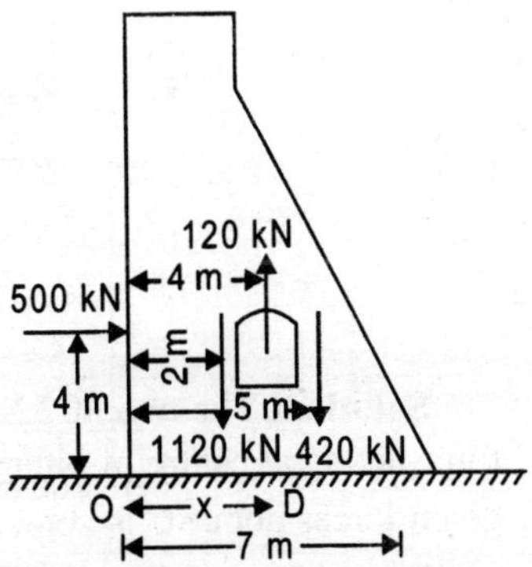

**Fig. 2.20**

**Solution.** The vector forms of the forces are as given below

500 kN force : $\mathbf{F}_1 = 500\mathbf{i}$

1120 kN force : $\mathbf{F}_2 = -1120\mathbf{j}$

120 kN force : $\mathbf{F}_3 = 120\mathbf{j}$

420 kN force : $\mathbf{F}_4 = -420\mathbf{j}$

$$\therefore \quad \mathbf{R} = \mathbf{F}_1 + \mathbf{F}_2 + \mathbf{F}_3 + \mathbf{F}_4$$
$$= 500\mathbf{i} + (-1120 + 120 - 420)\mathbf{j}$$
$$= 500\mathbf{i} - 1420\mathbf{j}$$

$$R = \sqrt{(500)^2 + (1420)^2} = 1505.457 \text{ kN}$$

Let the resultant pass through point $D$, which is at a distance $x$ from point $O$. Then

$$\mathbf{r}_{OD} = x\mathbf{i}$$

Now $\quad \mathbf{r}_{01} = 4\mathbf{j}, \ \mathbf{r}_{02} = 2\mathbf{i}, \mathbf{r}_{03} = 4\mathbf{i} \text{ and } \mathbf{r}_{04} = 5\mathbf{i}$

Then $\quad \mathbf{r}_{OD} \times \mathbf{R} = \Sigma \mathbf{M}_O$

$$x\mathbf{i} \times (500\mathbf{i} - 1420\mathbf{j}) = 4\mathbf{j} \times 500\mathbf{i} + 2\mathbf{i}\,(-1120\mathbf{j}) + 4\mathbf{i} \times (120\mathbf{j}) + 5\mathbf{i} \times (-420\mathbf{j})$$
$$-1420x\mathbf{k} = 2000\,(-\mathbf{k}) - 2240\mathbf{k} + 480\mathbf{k} - 2100\mathbf{k}$$
$$= -5860\mathbf{k}$$

$$\therefore \quad x = \frac{-5860}{-1420} = 4.126 \text{ m}$$

Thus $\frac{1}{3} \times 7 < x < \frac{2}{3} \times 7$

Hence the resultant passes through point $D$ which is within the middle third $\left(\frac{1}{3} \times 7 \text{ to } \frac{2}{3} \times 7\right)$ of the base. Therefore, the *dam is safe*. **Ans.**

**Example 2.14.** Determine the magnitude, direction and line of action of the equilibriant of the given set of coplanar forces acting on a truess as shown in Fig. 2.21.

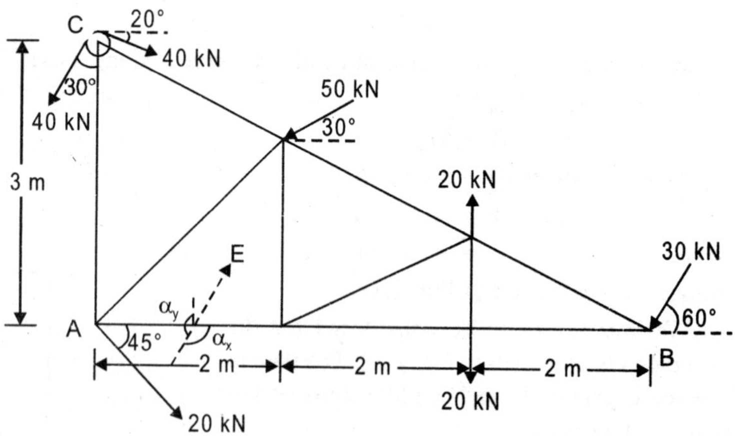


**Fig. 2.21**

**Solution.** The two 40 kN forces acting on the smooth pulley may be replaced by a pair of 40 kN force acting at centre of pulley $C$ and parallel to the given forces, since the sum of moment of the two given forces about $C$ is zero.

Vector forms of the forces are given as

40 kN force at 30° to vertical:

$$\mathbf{F}_1 = -40 \sin 30°\mathbf{i} - 40 \cos 30°\mathbf{j} = -20\mathbf{i} - 34.641\mathbf{j}$$

40 kN force at 20° to horizontal:

$$\mathbf{F}_2 = 40 \cos 20°\mathbf{i} - 40 \sin 20°\mathbf{j} = 37.588\mathbf{i} - 13.681\mathbf{j}$$

50 kN force:

$$\mathbf{F}_3 = -50 \cos 30°\mathbf{i} - 50 \sin 30°\mathbf{j} = -43.301\mathbf{i} - 25\mathbf{j}$$

20 kN upward load:

$$\mathbf{F}_4 = 20\mathbf{j}$$

30 kN laod:

$$\mathbf{F}_5 = -30 \cos 60°\mathbf{i} - 30 \cos 60°\mathbf{j} = -15\mathbf{i} - 25.981\mathbf{j}$$

20 kN downward load:

$$\mathbf{F}_6 = -20\mathbf{j}$$

20 kN inclined load:

$$\mathbf{F}_7 = 20 \cos 45°\mathbf{i} - 20 \sin 45°\mathbf{j} = 14.142\mathbf{i} - 14.142\mathbf{j}$$

Taking $A$ as origin,

$$\mathbf{r}_{A1} = 3\mathbf{j}, \quad \mathbf{r}_{A2} = 3\mathbf{j}, \quad \mathbf{r}_{A3} = 2\mathbf{i} + 2\mathbf{j}$$
$$\mathbf{r}_{A4} = 4\mathbf{i} + \mathbf{j}, \quad \mathbf{r}_{A5} = 6\mathbf{i}, \quad \mathbf{r}_{A6} = 4\mathbf{i}, \mathbf{r}_{A7} = 0$$

$$\therefore \quad \mathbf{R} = \mathbf{F}_1 + \mathbf{F}_2 + \mathbf{F}_3 + \mathbf{F}_4 + \mathbf{F}_5 + \mathbf{F}_6 + \mathbf{F}_7$$
$$= (-20 + 37.588 - 43.301 - 15 + 14.142)\mathbf{i}$$
$$+ (-34.641 - 13.681 - 25 + 20 - 25.981 - 20 - 14.142)\mathbf{j}$$
$$= -26.571\mathbf{i} - 113.445\mathbf{j}$$

$$R = \sqrt{(26.571)^2 + (113.445)^2} = 116.515 \text{ kN}$$

$$\alpha_x = \cos^{-1} \frac{-26.571}{116.515} = 103.182°$$

$$\alpha_y = \cos^{-1} \frac{-113.445}{116.515} = 166.818°$$

Let **R** intersect $AB$ at a distance $x$ from $A$.

Then $\mathbf{r}_{AD} = x\mathbf{i}$

$$\therefore \quad \mathbf{r}_{AD} \times \mathbf{R} = \mathbf{r}_{A1} \times \mathbf{F}_1 + \mathbf{r}_{A2} \times \mathbf{F}_2 + \mathbf{r}_{A3} \times \mathbf{F}_3 + \mathbf{r}_{A4} \times \mathbf{F}_4$$
$$+ \mathbf{r}_{A5} \times \mathbf{F}_5 + \mathbf{r}_{A6} \times \mathbf{F}_6 + \mathbf{r}_{A7} \times \mathbf{F}_7$$

$$x\mathbf{i} \times (-26.571\mathbf{i} - 113.445\mathbf{j}) = 3\mathbf{j} \times (-20\mathbf{i} - 34.641\mathbf{j}) + 3\mathbf{j} \times (37.588\mathbf{i} - 13.681\mathbf{j})$$
$$+ (2\mathbf{i} + 2\mathbf{j}) \times (-43.301\mathbf{i} - 25\mathbf{j}) + (4\mathbf{i} + \mathbf{j}) \times 20\mathbf{j}$$
$$+ 6\mathbf{i} \times (-15\mathbf{i} - 25.981\mathbf{j}) + (4\mathbf{i}) \times (-20\mathbf{j}) + 0$$

*i.e.*,

$$-113.445x\,\mathbf{k} = -60(-\mathbf{k}) + 112.764(-\mathbf{k}) - 50\mathbf{k} - 86.602(-\mathbf{k})$$
$$+ 80\mathbf{k} - 155.886\mathbf{k} - 80\mathbf{k} = -172.048\mathbf{k}$$

$$x = 1.516 \text{ m}$$

Equilibriant is opposite the resultant. Hence it is as shown in Fig. 2.21.

## 2.6 RESOLUTION OF FORCES IN THREE DIMENSIONAL SYSTEM

To represent a force in space, we need three mutually perpendicular directions say $x$, $y$ and $z$. In literature, coordinate directions $x, y, z$ are selected as per right hand rule. According to this, when thumb, index finger and middle finger are stretched to show three mutually perpendicular directions, thumb indicates $x$-coordinate direction, index finger shows $y$-coordinate direction and middle finger indicates $z$-coordinate direction. As shown in Fig. 2.22, different orientation of coordinates are possible and in this book orientation as shown in Fig. 2.22($a$) is used. The general equations derived in any one orientation hold good for any other orientation, provided right hand rule is followed.

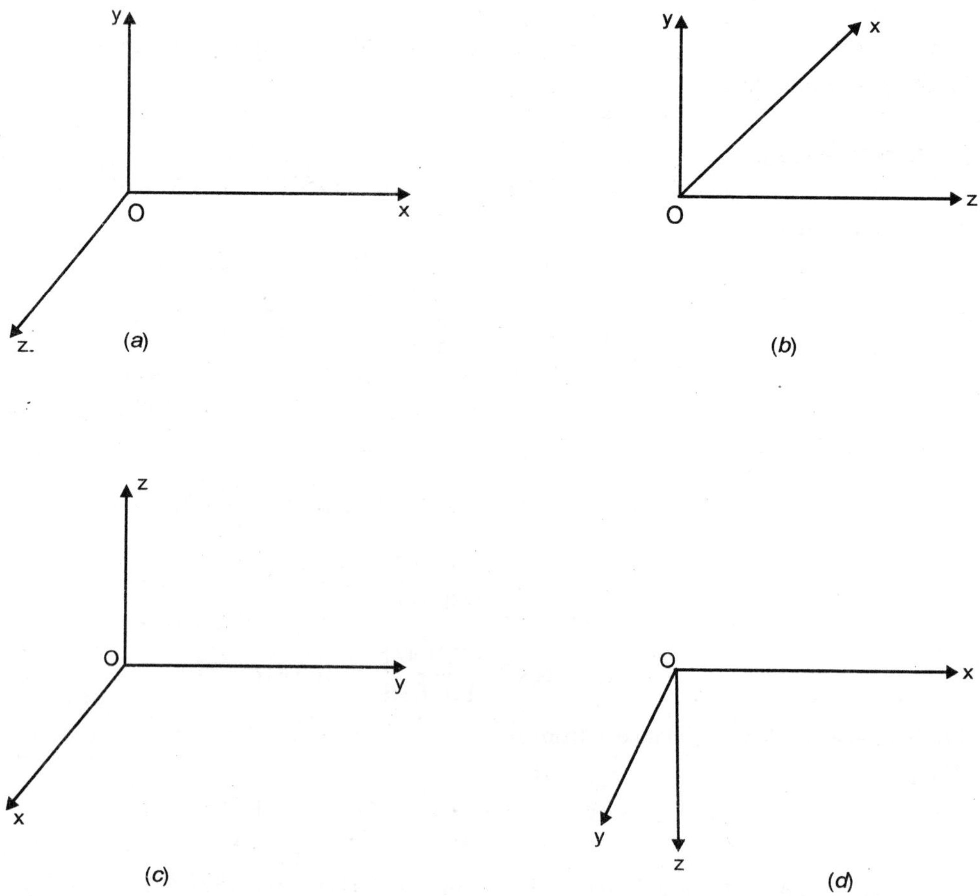


**Fig. 2.22** *Different orientation of coordinates with right hand rule*

In Fig. 2.23, $OA$ represents a force **F**. The projection of $OA$ in the plane of $x$-$z$ is $OB$. $OC$ and $OD$ are the projections of $OB$ on $z$ and $x$ coordinates. Then,

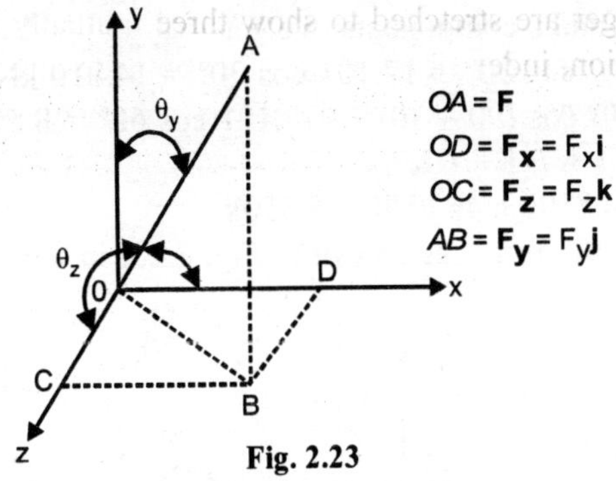


Fig. 2.23

$$F = OA = \sqrt{OB^2 + AB^2}$$
$$= \sqrt{OC^2 + OD^2 + AB^2}$$
$$= \sqrt{F_z^2 + F_x^2 + F_y^2}$$
$$= \sqrt{F_x^2 + F_y^2 + F_z^2} \quad \ldots(2.12)$$

If we consider vector summation, we get

$$\mathbf{F} = \mathbf{OA} = \mathbf{OB} + \mathbf{BA}$$
$$= \mathbf{OD} + \mathbf{OC} + \mathbf{BA}$$
$$= \mathbf{F}_x + \mathbf{F}_z + \mathbf{F}_y$$
$$= F_x\mathbf{i} + F_y\mathbf{j} + F_z\mathbf{k} \quad \ldots(2.13)$$

where **i, j** and **k** are unit vectors in the coordinate directions *x, y* and *z* respectively. $F_x$, $F_y$ and $F_z$ are the components of *F* in *x, y* and *z* directions. If $\theta_x$, $\theta_y$ and $\theta_z$ are the angles made by vector with *x, y* and *z* directions, then

$$\mathbf{F} = F \cos \theta_x \mathbf{i} + F \cos \theta_y \mathbf{j} + F \cos \theta_z \mathbf{k}$$
$$= F(l\mathbf{i} + m\mathbf{j} + n\mathbf{k})$$

If **n** is the direction of unit vector of *F*, then

$$\mathbf{F} = F\mathbf{n} = F(l\mathbf{i} + m\mathbf{j} + n\mathbf{k})$$

Hence,
$$\mathbf{n} = l\mathbf{i} + m\mathbf{j} + n\mathbf{k} \quad \ldots(2.14)$$

**Example 2.15.** Determine the components of 50 kN force as shown in Fig. 2.24 and express the force in the vector form.

**Solution.** Resolving the force in the plane containing *y*-axis and the force, we get

$$F_y = 50 \cos 20° = 46.985 \text{ kN}.$$
$$F_{OB} = 50 \cos (90 - 20)° = 17.101 \text{ kN}$$

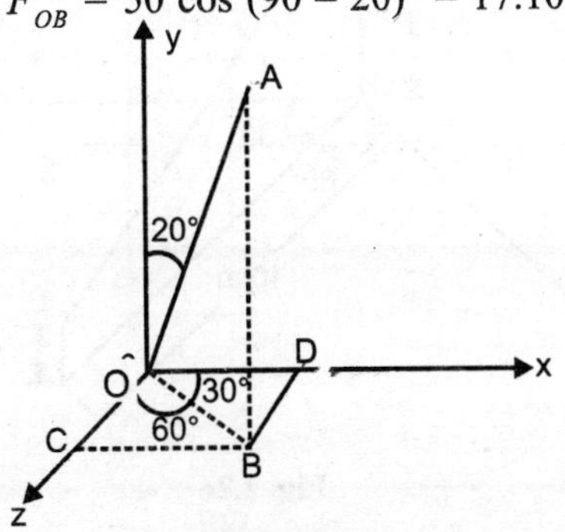


Fig. 2.24

Resolving $OB$ into its $x$ and $z$ components, we get

$$F_x = OB \cos 30° = 17.101 \cos 30° = 14.810 \text{ kN}$$
$$F_z = OB \cos (90 - 30)° = 17.101 \cos 60° = 8.550 \text{ kN}$$

Hence,
$$\mathbf{F} = F_x\mathbf{i} + F_y\mathbf{j} + F_z\mathbf{k}$$
$$= 14.810\mathbf{i} + 46.985\mathbf{j} + 8.550\mathbf{k}$$

**Example 2.16.** Express the force, as shown in Fig. 2.25, in the vector form.

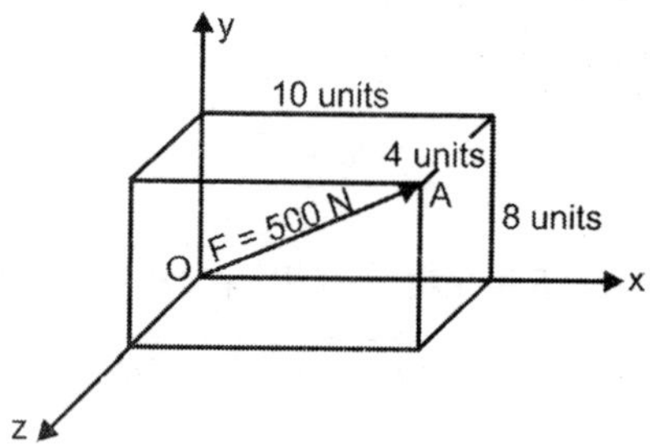


**Fig. 2.25**

**Solution.** $OA = \sqrt{10^2 + 8^2 + 4^2} = 13.416$

$$\therefore \quad l = \frac{10}{13.416} = 0.745\,; \quad m = \frac{8}{13.416} = 0.596\,; \quad n = \frac{4}{13.416} = 0.298$$

$\therefore$ $F_x = Fl = 500 \times 0.745$, $F_y = Fm = 500 \times 0.596$;

$F_z = Fn = 500 \times 0.298 = 372.5 \text{ N} = 298.14 \text{ N} = 149.07 \text{ N}$

Hence, $F = 372.5\mathbf{i} + 298.14\mathbf{j} + 149.07\mathbf{k}$

**Example 2.17.** A rope is tied to a pole at a height of 5 m and is pulled with a force of 500 N by a person standing at a distanc of 10 m in $S$ 60°E (South 60° East) direction. His hands are 1 m above the base of the pole. Determine the component of the pull exerted in vertical, horizontal east and horizontal south directions. Express the force in vector form.

**Solution.** Let $A$ be the point on the pole, where the rope is tied and $B$ be the point on the rope where force is applied (Ref Fig. 2.26). Taking the coordinate system as shown in the figure, the coordinates of the points $A$ and $B$ are $A(0, 5, 0)$ and $B$ (10 cos 30°, 1.0, 10 cos 60°), *i.e.*, $B$ (8.66, 1.0, 5.0)

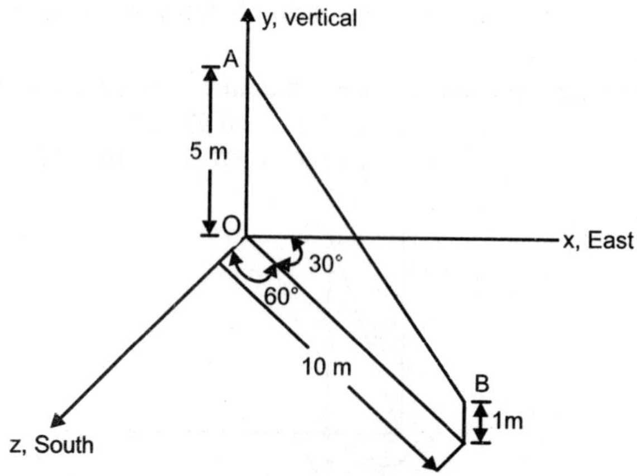


**Fig. 2.26**

∴ The position vector of *AB* is

$$\mathbf{r}_{AB} = \mathbf{r}_{OB} - \mathbf{r}_{OA} = (8.66 - 0)\mathbf{i} + (1.0 - 5)\mathbf{j} + (5 - 0)\mathbf{k}$$
$$= 8.66\mathbf{i} - 4\mathbf{j} + 5\mathbf{k}$$

$$\therefore \quad r_{AB} = \sqrt{8.66^2 + (-4)^2 + 5^2} = 10.77 \text{ m}$$

∴ The direction cosines are

$$l = \frac{8.66}{10.77} = 0.804; \quad m = \frac{-4}{10.77} = -0.371; \quad n = \frac{5}{10.77} = 0.64$$

Since 500 N force is in the direction *AB*, its direction cosines are the same as that of *AB*. Hence

$$F_x = 500 \times 0.804 = 402 \text{ N}$$
$$F_y = -0.37 \times 500 = -185.5 \text{ N} \quad \textbf{Ans.}$$
$$F_z = 0.464 \times 500 = 232.12 \text{ N} \quad \textbf{Ans.}$$

Hence, $\mathbf{F} = 402\mathbf{i} - 185.5\mathbf{j} + 232.12\mathbf{k}$ **Ans.**

## 2.7 RESULTANT OF CONCURRENT FORCE SYSTEM IN SPACE

The system of forces in space acting on a particle constitutes concurrent force system in space. The best way of finding their resultant force is to express each force in vector form and then use vector addition. Thus, if $\mathbf{F}_1$, $\mathbf{F}_2$ and $\mathbf{F}_3$ are the three forces acting on a particle in space, then their resultant is given by

$$\mathbf{R} = \mathbf{F}_1 + \mathbf{F}_2 + \mathbf{F}_3$$
$$= F_{1x}\mathbf{i} + F_{1y}\mathbf{j} + F_{1z}\mathbf{k} + F_{2x}\mathbf{i} + F_{2y}\mathbf{j} + F_{2z}\mathbf{k} + F_{3x}\mathbf{i} + F_{3y}\mathbf{j} + F_{3z}\mathbf{k}$$
$$= (F_{1x} + F_{2x} + F_{3x})\mathbf{i} + (F_{1y} + F_{2y} + F_{3y})\mathbf{j} + (F_{1z} + F_{2z} + F_{3z})\mathbf{k}$$
$$\mathbf{F}_1 = \Sigma F_x\mathbf{i} + \Sigma F_y\mathbf{j} + \Sigma F_z\mathbf{k}$$

**Example 2.18.** The directions of the three forces acting at point *O* are shown in Fig. 2.27 by drawing a parallelepiped of size 6, 4 and 2. Determine the resultant of the three forces.

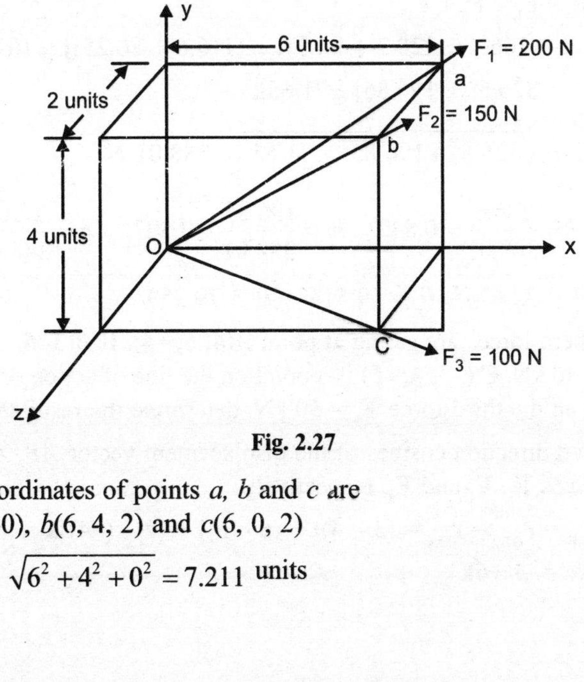


**Fig. 2.27**

**Solution.** The coordinates of points *a, b* and *c* are

*a*(6, 4, 0), *b*(6, 4, 2) and *c*(6, 0, 2)

$$\therefore \quad r_{OA} = \sqrt{6^2 + 4^2 + 0^2} = 7.211 \text{ units}$$

$$r_{OB} = \sqrt{6^2 + 4^2 + 2^2} = 7.483 \text{ units}$$

$$r_{OC} = \sqrt{6^2 + 0^2 + 2^2} = 6.325 \text{ units}$$

and

$$\mathbf{r_{OA}} = 6\mathbf{i} + 4\mathbf{j} + 0\mathbf{k} = r_{OA}(l_1\mathbf{i} + m_1\mathbf{j} + n_1\mathbf{k})$$
$$= 7.211\ (l_1\mathbf{i} + m_1\mathbf{j} + n_1\mathbf{k})$$
$$\mathbf{r_{OB}} = 6\mathbf{i} + 4\mathbf{j} + 2\mathbf{k} = 7.483(l_2\mathbf{i} + m_2\mathbf{j} + n_2\mathbf{k})$$
$$\mathbf{r_{OC}} = 6\mathbf{i} + 0\mathbf{j} + 2\mathbf{k} = 6.325(l_3\mathbf{i} + m_3\mathbf{j} + n_3\mathbf{k})$$

where $l_i$, $m_i$ and $n_i$ are the drection cosines of $i^{th}$ force.

$$\therefore \quad l_1 = \frac{6}{7.211} = 0.832, \quad m_1 = \frac{4}{7.211} = 0.558, \quad n_1 = 0$$

$$l_2 = \frac{6}{7.483} = 0.802, \quad m_2 = \frac{4}{7.483} = 0.535, \quad n_2 = \frac{2}{7.483} = 0.267$$

$$l_3 = \frac{6}{6.325} = 0.947, \quad m_3 = 0, \quad n_3 = \frac{2}{6.325} = 0.316$$

$$\mathbf{F_{OA}} = \mathbf{F}_1 = \mathrm{F}_1\mathbf{n}_1 = 200(l_1\mathbf{i} + m_1\mathbf{j} + n_1\mathbf{k})$$
$$= 200(0.832\mathbf{i} + 0.558\mathbf{j}) = 166.4\mathbf{i} + 111.6\mathbf{j}$$
$$\mathbf{F_{OB}} = F_2\mathbf{n}_2 = 150\ (0.832\mathbf{i} + 0.558\mathbf{j} + 0.267\mathbf{k})$$
$$= 120.3\mathbf{i} + 80.25\mathbf{j} + 40.05\mathbf{k}$$
$$\mathbf{F_{OC}} = F_3\mathbf{n}_3 = 100\ (0.947\mathbf{i} + 0.316\mathbf{k})$$
$$= 94.7\mathbf{i} + 31.6\mathbf{k}$$

$\therefore$ The resultant force **R** is given by

$$\mathbf{R} = \mathbf{F}_1 + \mathbf{F}_2 + \mathbf{F}_3$$
$$= (166.4 + 120.3 + 94.7)\mathbf{i} + (116.6 + 80.25)\mathbf{j} + (0 + 40.05 + 31.6)\mathbf{k}$$
$$= 326.6\mathbf{i} + 196.85\mathbf{j} + 71.65\mathbf{k}$$

Therefore, $$R = \sqrt{326.6^2 + 196.85^2 + 71.65^2} = 388.01 \text{ N}$$

$$l = \frac{326.6}{388.01} = 0.842, \quad m = \frac{196.85}{388.01} = 0.507, \quad n = \frac{17.65}{388.01} = 0.185$$ **Ans.**

$$\theta_x = 32.676°, \ \theta_y = 59.514°, \ \theta_z = 79.359°$$ **Ans.**

**Example 2.19.** There forces are acting at point $A(4, 6, -4)$. If $B(3, 6, 2)$ is a point on the line of action of the force $\mathbf{F}_1 = 40$ kN, C $(-3, 4, -5)$ is a point on the line of action of the force $\mathbf{F}_2 = 50$ kN and $D$ $(-5, -6, 8)$ is a point on the third force $\mathbf{F}_3 = 60$ kN, determine the resultant of the three forces.

**Solution.** We need direction cosines of the displacement vector $AB$, $AC$ and $AD$, which will be same as those of the forces $\mathbf{F}_1$, $\mathbf{F}_2$ and $\mathbf{F}_3$, respectively

$$\mathbf{r_{AB}} = \mathbf{r_{OB}} - \mathbf{r_{OA}} = (3 - 4)\mathbf{i} + (6 - 6)\mathbf{j} + [2 - (-4)]\mathbf{k}$$
$$= -\mathbf{i} + 6\mathbf{k}$$

Similarly, $\mathbf{r}_{AC} = (-3 - 4)\mathbf{i} + (4 - 6)\mathbf{j} + (-5 + 4)\mathbf{k} = -7\mathbf{i} - 2\mathbf{j} - \mathbf{k}$

$\mathbf{r}_{AD} = (-5 - 4)\mathbf{i} + (-6 - 6)\mathbf{j} + (8 + 4)\mathbf{k} = -9\mathbf{i} - 12\mathbf{j} + 12\mathbf{k}$

$$r_{AB} = \sqrt{(-1)^2 + 6^2} = 6.083$$

$$r_{AC} = \sqrt{(-7)^2 + (-2)^2 + (-1)^2} = 7.348$$

$$r_{AD} = \sqrt{(-9)^2 + (-12)^2 + (12)^2} = 19.209$$

$$\therefore \quad l_1 = \frac{-1}{6.083} = -0.164; \quad m_1 = 0; \quad n_1 = \frac{6}{6.083} = 0.986$$

$$l_2 = \frac{-7}{7.348} = -0.953; \quad m_2 = \frac{-2}{7.348} = -0.272; \quad n_2 = \frac{-1}{7.348} = -0.136$$

$$l_3 = \frac{-9}{19.209} = -0.469; \quad m_3 = \frac{-12}{19.203} = -0.625; \quad n_3 = \frac{12}{19.209} = 0.625$$

$\therefore$ $\mathbf{F}_1 = F_1(l_1\mathbf{i} + m_1\mathbf{j} + n_1\mathbf{k})$

$= 40(-0.164\mathbf{i} + 0.982\mathbf{k}) = -6.56\mathbf{i} + 39.454\mathbf{k}$

Similarly, $\mathbf{F}_2 = 50\ (0.953\mathbf{i} + 0.272\mathbf{j} - 0.13\mathbf{k})$

$= -\ 47.65\mathbf{i} - 13.6\mathbf{j} - 6.8\mathbf{k}$

and $\mathbf{F}_3 = 60(-\ 0.469\mathbf{i} - 0.625\mathbf{j} + 0.625\mathbf{k})$

$= -\ 28.14\mathbf{i} - 37.5\mathbf{j} + 37.5\mathbf{k}$

$\therefore$ The resultant force **R** is given by

$\mathbf{R} = \mathbf{F}_1 + \mathbf{F}_2 + \mathbf{F}_3$

$= (-\ 6.56 - 47.65 - 28.14)\mathbf{i} + (0 - 13.6 - 37.5)\mathbf{j} + (39.454 - 6.8 + 37.5)\mathbf{k}$

$= -\ 82.35\mathbf{i} - 51.1\mathbf{j} + 70.154\mathbf{k}$ **Ans.**

$$R = \sqrt{82.35^2 + 51.1^2 + 70.154^2} = 119.64 \text{ kN}$$ **Ans.**

$$l = \frac{-82.35}{119.64} = 0.688, \quad m = \frac{-51.1}{119.64} = -0.427, \quad n = \frac{70.154}{119.64} = 0.586$$ **Ans.**

$\theta_x = 133.47°, \ \theta_y = 115.277°, \ \theta_z = 54.1°$ **Ans.**

**Example 2.20.** After making a cut at ground level, a tree is to be pulled down with the help of two ropes tied at 8 m high above the ground level. Rope *OA* is in S 60° E direction at a horizontal distance of 10 m and here a group of labourers are applying a force of 6 kN at a height of 1 m (Ref. Fig. 2.28). Rope *O'B* is in N 45° E at a distance of 8 m and another group of labourers is applying a force 1 m above the ground level. What should be the magnitude of the force in rope *OB* so that the tree falls in eastern direction?

**Solution.** Coordinate system is selected with *O* as origin, east direction as *x*-axis, south direction as *z*-axis and vertical upward direction as *y*-direction as shown in the figure. Hence the coordinates of *O'*, *A* and *B* are

$O'(0, 8, 0)$, $A\ (10 \cos 30°, 1.0, 10 \cos 60°)$, $B\ (8 \cos 45°, 1.0, -\ 8 \cos 45°)$

*i.e.*, $O'(0, 8, 0)$, $A\ (8.660, 1.0, 5.0)$, $B\ (5.657, 1.0, -\ 5.657)$

$\therefore$ $\mathbf{r}_{O'A} = \mathbf{r}_{OA} - \mathbf{r}_{OO'}$

$= 8.660\mathbf{i} + 1.0\mathbf{j} + 5\mathbf{k} - (0\mathbf{i} + 8.0\mathbf{j} + 0\mathbf{k})$

$= 8.66\mathbf{i} - 7.0\mathbf{j} + 5\mathbf{k}$

$$r_{O'A} = \sqrt{8.66^2 + 7.0^2 + 5^2} = 12.207 \text{ m.}$$

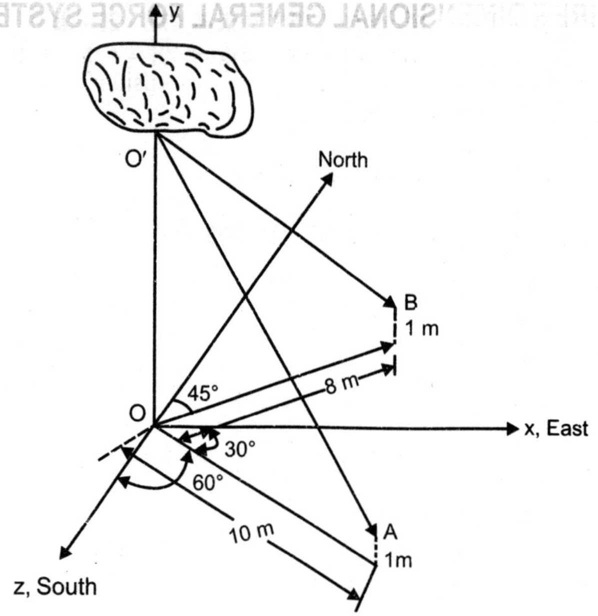


**Fig. 2.28**

III$^{ly}$, $\mathbf{r}_{O'B} = \mathbf{r}_{O'B} - \mathbf{r}_{OO'} = 5.657\mathbf{i} + 1.0\mathbf{j} - 5.657\mathbf{k} - (0\mathbf{i} + 8.0\mathbf{j} + 0\mathbf{k})$

$= 5.657\mathbf{i} - 7.0\mathbf{j} - 5.657\mathbf{k}$

$\therefore \quad r_{O'B} = \sqrt{5.6557^2 + 7^2 + 5.657^2} = 10.535$ m

$\therefore$ Direction cosines of $O'A$ are

$$l_1 = \frac{8.66}{12.207} = 0.709, \quad m_1 = \frac{-7.0}{12.207} = -0.573, \quad n_1 = \frac{5}{12.207} = 0.410$$

Direction cosines of $O'B$ are

$$l_2 = \frac{5.657}{10.535} = 0.537, \quad m_2 = \frac{-7}{10.535} = -0.664, \quad n_2 = \frac{5.657}{10.535} = -0.537$$

$\therefore$ Force in rope $O'A$ is

$$\mathbf{F}_1 = F_1(l_1\mathbf{i} + m_1\mathbf{j} + n_1\mathbf{k})$$
$$= 6(0.709\mathbf{i} - 0.573\mathbf{j} + 0.410\mathbf{k})$$
$$= 4.254\mathbf{i} - 3.438\mathbf{j} + 2.458\mathbf{k}$$

Force in rope $O'B$ is

$$\mathbf{F}_2 = F_2(0.537\mathbf{i} - 0.664\mathbf{j} - 0.537\mathbf{k})$$

The resultant force is

$$\mathbf{R} = \mathbf{F}_1 + \mathbf{F}_2$$
$$= (4.254 + 0.537\,F_2)\mathbf{i} + (-3.438 - 0.664\,F_2)\mathbf{j} + (2.458 - 0.537\,F_2)\mathbf{k}$$

If the tree has to fall in $x$-direction, the component of the resultant in $z$-direction should be zero, *i.e.*,

$$2.458 - 0.537\,F_2 = 0$$

$\therefore \quad F_2 = 4.577$ kN **Ans.**

## 2.8 RESULTANT OF THREE DIMENSIONAL GENERAL FORCE SYSTEM

The necessary equations to find the resultant in case of three dimensional general force system have been already presented in Art. 2.5. The equations are numbered as 2.6 to 2.11. Their application to three dimension general force system is illustrated with examples in this article.

**Example 2.21.** The loads transferred by four columns to concrete mat foundation are as shown in Fig. 2.29. Determine the resultant and the point of application.

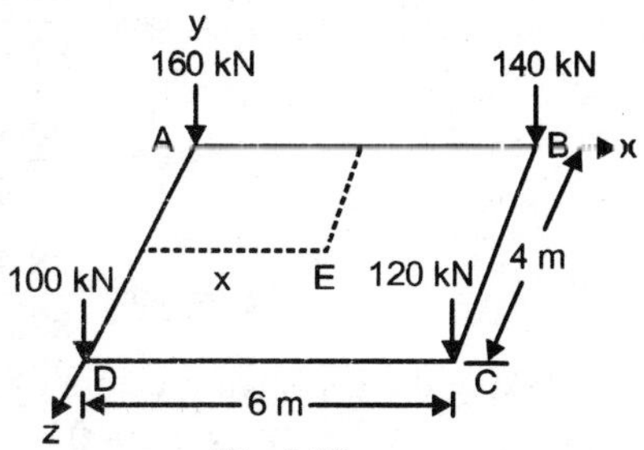


Fig. 2.29

**Solution.** Taking $A$ as the origin and the coordinate system as shown in Fig 2.29, we have

$$\mathbf{F}_A = -160\mathbf{j},\ \mathbf{F}_B = -140\mathbf{j};$$
$$\mathbf{F}_C = -120\mathbf{j} \text{ and } \mathbf{F}_D = -100\mathbf{j}$$
$$A\,(0, 0, 0);\ B\,(6, 0, 0),$$
$$C\,(6, 0, 4);\ D\,(0, 0, 4),$$
$$\mathbf{r}_{AA} = 0;\ \mathbf{r}_{AB} = 6\mathbf{i};\ \mathbf{r}_{AC} = 6\mathbf{i} + 4\mathbf{k};\ \mathbf{r}_{AD} = 4\mathbf{k}$$
$$\therefore \quad \mathbf{R} = \mathbf{F}_1 + \mathbf{F}_2 + \mathbf{F}_3 + \mathbf{F}_4$$
$$= (-160 - 140 - 120 - 100)\mathbf{j} = -520\mathbf{j} \qquad \textbf{Ans.}$$
$$\mathbf{M}_A = 0 \times (-160\mathbf{j}) + 6\mathbf{i} \times (-140\mathbf{j}) + (6\mathbf{i} + 4\mathbf{k})\,(-120\mathbf{j}) + 4\mathbf{k} \times (-100\mathbf{j})$$
$$= -840\mathbf{k} - 720\mathbf{k} - 480\,(-\mathbf{i}) - 400\,(-\mathbf{i}) = -1560\mathbf{k} + 880\mathbf{i}$$

Let the resultant pass through point $E$ on the slab $E(x, 0, z)$. Then

$$\mathbf{r}_{AE} = x\mathbf{i} + z\mathbf{k}$$
$$\therefore \quad \mathbf{M} = \mathbf{r}_{AE} \times \mathbf{R}$$

*i.e.,*
$$-1560\mathbf{k} + 880\mathbf{i} = (x\mathbf{i} + z\mathbf{k})\,(-520\mathbf{j})$$
$$= -520x\,\mathbf{k} - 520z\,(-\mathbf{i})$$
$$= -520x\,\mathbf{k} + 520z\,\mathbf{i}$$

*i.e.,*
$$520x = 1560 \text{ and } 520z = 880$$
$$\therefore \quad x = 3.0 \text{ m and } z = 1.692 \text{ m} \qquad \textbf{Ans.}$$

**Example 2.22.** *a b c d a′ b′ c′ d′* is a parallelepiped as shown in Fig. 2.30. A force of 30 kN acts in the direction of *ac′* and a force of 60 kN acts in the direction of *db′*. A force of 40 kN acts along *a′ d′*. Determine the resultant force-couple.

**Solution.** Taking '*a*' as origin and the coordinate system as shown in Fig. 2.30.

$$a\,(0, 0, 0);\ b'(6, 4, 0);\ c'(6, 4, 3);\ d\,(0, 0, 3),$$
$$\therefore \quad \mathbf{r}_{ac'} = 6\mathbf{i} + 4\mathbf{j} + 3\mathbf{k}$$
$$r_{ac'} = \sqrt{6^2 + 4^2 + 3^2} = 7.810 \text{ m}$$
$$\mathbf{n}_{ac} = 0.768\mathbf{i} + 0.512\mathbf{j} + 0.384\mathbf{k}$$
$$\therefore \quad \mathbf{F}_1 = 30(0.768\mathbf{i} + 0.512\mathbf{j} + 0.384\mathbf{k})$$
$$= 23.047\mathbf{i} + 15.36\mathbf{j} + 11.523\mathbf{k}$$
$$\mathbf{r}_{db'} = 6\mathbf{i} + 4\mathbf{j} - 3\mathbf{k}$$
$$r_{db'} = \sqrt{6^2 + 4^2 + 3^2} = 7.810 \text{ m}$$
$$\mathbf{n}_{db'} = 0.768\mathbf{i} + 0.512\mathbf{j} - 0.384\mathbf{k}$$
$$\therefore \quad \mathbf{F}_2 = 60(0.768\mathbf{i} + 0.512\mathbf{j} + 0.384\mathbf{k})$$
$$= 46.08\mathbf{i} + 30.72\mathbf{j} - 23.04\mathbf{k}$$
$$\mathbf{F}_3 = 40\mathbf{k}$$
$$\therefore \quad \mathbf{R} = \mathbf{F}_1 + \mathbf{F}_2 + \mathbf{F}_3$$

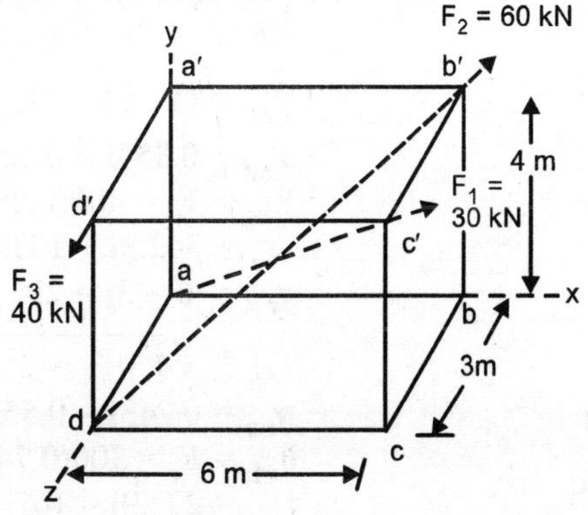


Fig. 2.30

$= (23.047 + 46.08)\mathbf{i} + (15.36 + 30.72)\mathbf{j} + (11.523 - 23.04 + 40)\mathbf{k}$

$= 69.127\mathbf{i} + 46.08\mathbf{j} + 28.483\mathbf{k}$ **Ans.**

$$R = \sqrt{(69.127)^2 + (46.08)^2 + (28.483)^2} = 87.825 \text{ kN}$$

$$\alpha_x = \cos^{-1}\frac{69.127}{87.825} = 38.08° \quad \textbf{Ans.}$$

$$\alpha_y = \cos^{-1}\frac{46.08}{87.825} = 58.353°$$

$$\alpha_y = \cos^{-1}\frac{28.483}{87.825} = 70.84°$$

Taking displacement vectors of convenient points on $F_1$, $F_2$, $F_3$

$\mathbf{r}_{a1} = \mathbf{r}_{\mathbf{ac}} = 0$

$\mathbf{r}_{a2} = \mathbf{r}_{\mathbf{ad}} = 3\mathbf{k}$

$\mathbf{r}_{a3} = \mathbf{r}_{\mathbf{aa'}} = 4\mathbf{j}$

$\mathbf{M}_{\mathbf{a}} = \mathbf{r}_{a1} \times \mathbf{F}_1 \times \mathbf{r}_{a2} \times \mathbf{F}_2 \times \mathbf{r}_{a3} \times \mathbf{F}_3$

$= 0 + 3\mathbf{k} \times (46.08\mathbf{i} + 30.72\mathbf{j} - 23.04\mathbf{k}) + 4\mathbf{j} \times (40\mathbf{k})$

$= 138.24\ (\mathbf{j}) + 92.16\ (-\mathbf{i}) + 160\mathbf{i}$

$= 67.84\mathbf{i} + 138.24\mathbf{j}$

Thus the resultant is a force $\mathbf{R} = 67.127\mathbf{i} + 46.08\mathbf{j} + 28.483\mathbf{k}$ and a couple moment $67.84\mathbf{i} + 138.24\mathbf{j}$.

**Example 2.23.** A rectangular block of size 6 m × 4 m × 3 m is divided into 6 × 4 × 3 cubes of side 1 m as shown in Fig. 2.31. It is subjected to forces as given below:

$\mathbf{F}_1 = 400$ N from *A* to *B*

$\mathbf{F}_2 = 300$ N from *C* to *B*

and $\mathbf{F}_3 = 200$ N from *E* to *F*.

Find the resultant force–couple system acting at the origin *O* and at *G*.

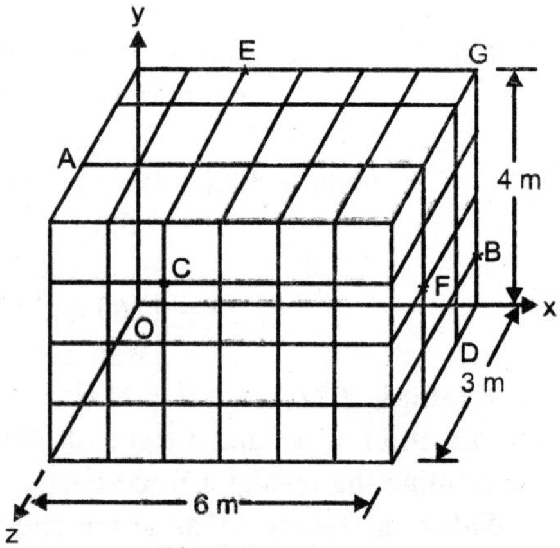


**Fig. 2.31**

**Solution.** The coordinates of various points are

*O* (0, 0, 0); *A* (0, 4, 2);

*B* (6, 1, 0); *C* (2, 3, 3), *D* (6, 0, 1)

*E* (2, 4, 0), *F* (6, 2, 2) and *G* (6, 4, 0),

$\therefore \quad \mathbf{r}_{\mathbf{AB}} = 6\mathbf{i} - 3\mathbf{j} - 2\mathbf{k} \qquad \mathbf{r}_{\mathbf{AB}} = \sqrt{6^2 + 3^2 + 2^2} = 7 \text{ m}$

$\mathbf{r}_{\mathbf{AB}} = 0.857\mathbf{i} - 0.428\mathbf{j} - 0.286\mathbf{k}$

$\mathbf{F}_{\mathbf{AB}} = \mathbf{F}_1 = 400(0.857\mathbf{i} - 0.428\mathbf{j} - 0.286\mathbf{k})$

$= 342.8\mathbf{i} + 171.2\mathbf{j} - 114.4\mathbf{k}$

$\mathbf{r}_{\mathbf{CD}} = 4\mathbf{i} - 3\mathbf{j} - 2\mathbf{k}$

$\therefore \quad r_{CD} = \sqrt{4^2 + 3^2 + 2^2} = 5.385$

$\mathbf{n}_{\mathbf{CD}} = 0.743\mathbf{i} - 0.557\mathbf{j} - 0.371\mathbf{k}$

$\therefore \quad \mathbf{F}_{\mathbf{CD}} = \mathbf{F}_2 = 300(0.743\mathbf{i} - 0.557\mathbf{j} - 0.371\mathbf{k})$

$= 222.9\mathbf{i} - 167.1\mathbf{j} - 111.3\mathbf{k}$

$\mathbf{r}_{\mathbf{EF}} = 4\mathbf{i} - 2\mathbf{j} + 2\mathbf{k}$

$r_{EF} = \sqrt{4^2 + 2^2 + 2^2} = 4.899$

$$\mathbf{n}_{EF} = 0.816\mathbf{i} - 0.408\mathbf{j} + 0.408\mathbf{k}$$

$$\mathbf{F}_{EF} = \mathbf{F}_3 = 200(0.816\mathbf{i} - 0.408\mathbf{j} + 0.408\mathbf{k})$$

$$= 163.30\mathbf{i} - 81.65\mathbf{j} + 81.65\mathbf{k}$$

$$\therefore \quad \mathbf{R} = \mathbf{F}_1 + \mathbf{F}_2 + \mathbf{F}_3$$

$$\mathbf{R} = (342.8 + 222.9 + 163.30)\mathbf{i} + (-171.2 - 167.7 - 81.65)\mathbf{j} + (-114.4 - 111.3 + 81.65)\mathbf{k}$$

$$= 729\mathbf{i} - 419.95\mathbf{j} - 144.05\mathbf{k}$$

**Ans.**

Now $\mathbf{r}_{OA} = 4\mathbf{j} + 2\mathbf{k}$

$$\mathbf{r}_{OC} = 2\mathbf{i} + 3\mathbf{j} + 3\mathbf{k}$$

$$\mathbf{r}_{OE} = 2\mathbf{i} + 4\mathbf{j}$$

$\therefore$ Moment of all the forces about $O$

$$\mathbf{M}_O = \begin{vmatrix} \mathbf{i} & \mathbf{j} & \mathbf{k} \\ 0 & 4 & 2 \\ 342.8 & -171.2 & -114.4 \end{vmatrix} + \begin{vmatrix} \mathbf{i} & \mathbf{j} & \mathbf{k} \\ 2 & 3 & 3 \\ 222.9 & -167.1 & -111.3 \end{vmatrix}$$

$$+ \begin{vmatrix} \mathbf{i} & \mathbf{j} & \mathbf{k} \\ 2 & 4 & 0 \\ 163.3 & -81.65 & 81.65 \end{vmatrix}$$

$$= \mathbf{i}(-457.6 + 342.4) - \mathbf{j}(0 - 685.6) + \mathbf{k}(0 - 1371.2)$$

$$+ \mathbf{i}(-333.9 + 501.3) - \mathbf{j}(-222.6 - 668.7) + \mathbf{k}(-334.2 - 668.7)$$

$$+ \mathbf{i}(326.6 - 0) - \mathbf{j}(163.3 - 0) + \mathbf{k}(-163.3 - 653.2)$$

$$= 378.8\mathbf{i} + 354.3\mathbf{j} - 2856\mathbf{k}$$

Hence force couple system at $O$ is

$$\mathbf{R} = 729.0\mathbf{i} - 419.95\mathbf{j} - 144.05\mathbf{k}$$

$$\mathbf{M}_O = 378.8\mathbf{i} + 354.3\mathbf{j} - 2856.0\mathbf{k}$$

Force couple system at $G$ is given by

$$\mathbf{M}_G = \mathbf{M}_O + \mathbf{r}_{GO} \times \mathbf{R}$$

$$\mathbf{r}_{GO} = -6\mathbf{i} - 4\mathbf{j}$$

$$\mathbf{M}_G = 378.8\mathbf{i} + 354.3\mathbf{j} - 2856.0\mathbf{k} + \begin{vmatrix} \mathbf{i} & \mathbf{j} & \mathbf{k} \\ -6 & -4 & 0 \\ 729.0 & -419.95 & -114.05 \end{vmatrix}$$

$$= 378.8\mathbf{i} + 354.3\mathbf{j} - 2856.0\mathbf{k} + \mathbf{i}(576.2) - \mathbf{j}(864.3) + \mathbf{k}(2519.7 + 2916)$$

$$= 955\mathbf{i} - 510\mathbf{j} + 2579.7\mathbf{k}$$

**Ans.**

## 2.9 WRENCH RESULTANT

In a two dimensional problem, the resultant $R$ and the moment $M$ at any point are at right angles to each other, the force being in the plane ($x$-$y$ plane) and the moment at right angles to the plane (about $z$-axis). Hence it was possible to replace the resultant force-couple system into a single force acting at a distance $d = \frac{M}{F}$ or having $x$ or $y$ intercepts given by $x\mathbf{i} \times \mathbf{F} = \mathbf{M}_O$ and $y\mathbf{j} \times \mathbf{F} = \mathbf{M}_O$. However in case of a three dimensional problem, it is not possible to replace the resultant force-couple system by a single force. Hence we presented the resultant in the form of a force and moment about a specified point.

The simplest equivalent form of the resultant system is a force and a moment about the axis of the force as shown in Fig. 2.32. This form of the resultant system is called a wrench. The word wrench is used because the resulting combination is a push and twist similar to that exerted by a wrench. A wrench is said to be positive if the resultant force and moment have the same sense of direction [Fig. 2.32(*a*)], and is said to be negative if the two vectors have the opposite directions [Fig. 2.32(*b*)].

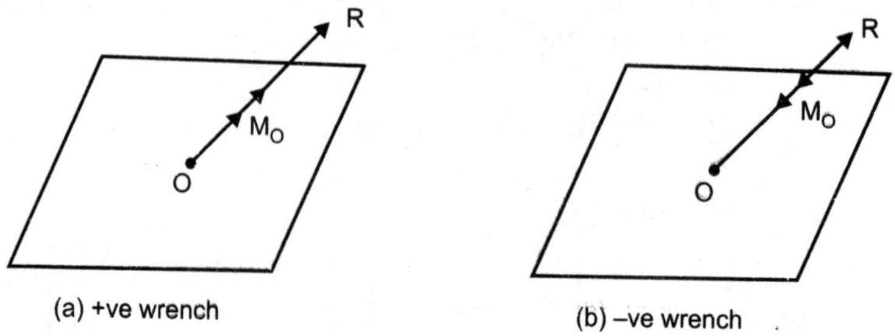


**Fig. 2.32** *Wrench system of resultant*

In Examples 2.22 and 2.23, we have seen how a given force system can be reduced to a resultant force **R** and moment $\mathbf{M}_0$, **R** and $\mathbf{M}_0$ may be in different directions as shown in Fig. 2.33(*a*). Let *N* be the plane normal to the plane **R** and $\mathbf{M}_0$. Let $\mathbf{M}_1$ and $\mathbf{M}_2$ be the component of $\mathbf{M}_0$ in the direction of **R** and normal to **R** respectively.

Now the resultant force **R** may be shifted to point *O*′ in the plane such that $Fd = \mathbf{M}_2$ as shown in Fig. 2.33 (*c*). Thus the resultant **R** at *O*′ replaces the resultant force **R** and moment $\mathbf{M}_2$ at point *O*. Hence the resultant force-moment system at *O*′ is a resultant force **R** and moment $\mathbf{M}_1$ about the axis of the force, *i.e.*, system of resultant is a wrench.

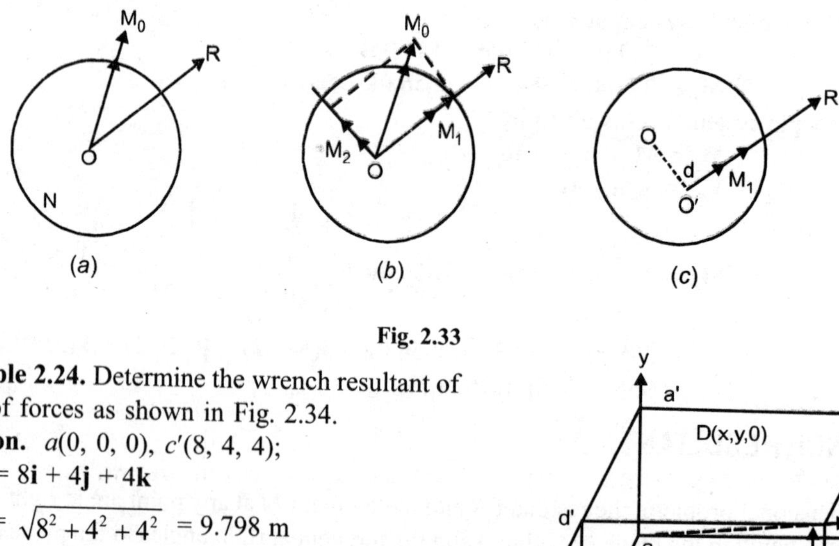


**Fig. 2.33**

**Example 2.24.** Determine the wrench resultant of the system of forces as shown in Fig. 2.34.

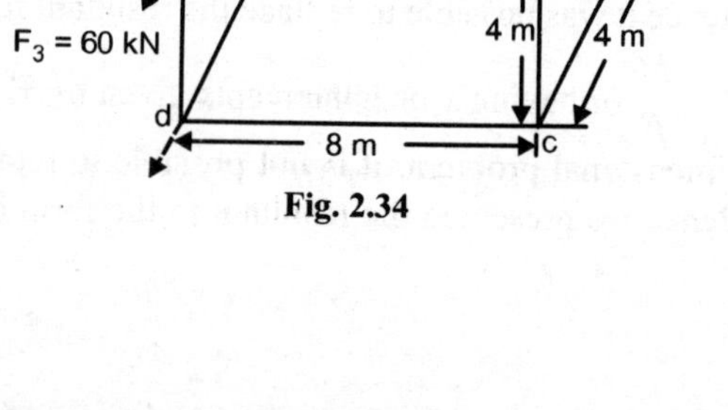


**Fig. 2.34**

**Solution.** $a(0, 0, 0)$, $c'(8, 4, 4)$;

$$\therefore \mathbf{r}_{ac'} = 8\mathbf{i} + 4\mathbf{j} + 4\mathbf{k}$$

$$r_{ac'} = \sqrt{8^2 + 4^2 + 4^2} = 9.798 \text{ m}$$

$$\mathbf{n}_{ac'} = \frac{8}{9.798}\mathbf{i} + \frac{4}{9.798}\mathbf{j} + \frac{4}{9.798}\mathbf{k}$$

$$= 0.816\mathbf{i} + 0.408\mathbf{j} + 0.408\mathbf{k}$$

$$\mathbf{F}_1 = 40(0.816\mathbf{i} + 0.408\mathbf{j} + 0.408\mathbf{k})$$

$$= 32.64\mathbf{i} + 16.32\mathbf{j} + 16.32\mathbf{k}$$

$$\mathbf{F}_2 = 20\mathbf{j} \text{ and } \mathbf{F}_3 = 60\mathbf{k}$$

$$\therefore \quad \mathbf{R} = \mathbf{F}_1 + \mathbf{F}_2 + \mathbf{F}_3$$

$$= 32.46\mathbf{i} + 36.32\mathbf{j} + 76.32\,\mathbf{k}$$

$$R = \sqrt{(32.64)^2 + (36.32)^2 + (76.32)^2} = 90.605 \text{ kN } \textbf{Ans.}$$

$$\mathbf{n} = \frac{32.64}{90.605}\mathbf{i} + \frac{36.32}{90.605}\mathbf{j} + \frac{76.32}{90.605}\mathbf{k}$$

$$= 0.360\mathbf{i} + 0.401\mathbf{j} + 0.842\mathbf{k}$$

Let $D$ $(x, y, 0)$ be a point on $x$-$y$ plane through which wrench resultant **F** passes.

$a(0, 0, 0)$, $b(8, 0, 0)$ and $a'(0, 4, 0)$

$$\therefore \quad \mathbf{r}_{\mathbf{Da}} = -x\mathbf{i} - y\mathbf{j},\ \mathbf{r}_{\mathbf{Db}} = (8-x)\mathbf{i} - y\mathbf{j} \text{ and } \mathbf{r}_{\mathbf{Da'}} = -x\mathbf{i} + (4-y)\mathbf{j}$$

$$\mathbf{M}_\mathbf{D} = \mathbf{r}_{\mathbf{DA}} \times \mathbf{F}_1 + \mathbf{r}_{\mathbf{Db}} \times \mathbf{F}_2 + \mathbf{r}_{\mathbf{Da'}} \times \mathbf{F}_3$$

$$= \begin{vmatrix} \mathbf{i} & \mathbf{j} & \mathbf{k} \\ -x & -y & 0 \\ 32.64 & 16.32 & 16.32 \end{vmatrix} + [(8-x)\mathbf{i} - y\mathbf{j}] \times 10\mathbf{j} + [-x\mathbf{i} + (4-y)\mathbf{j}] \times 30\mathbf{k}$$

$$= \mathbf{i}(-16.32y) - \mathbf{j}(-16.32x) + \mathbf{k}(-16.32x + 32.64y) + 10(8-x)\mathbf{k} - 30x(-\mathbf{j}) + 30(4-y)\mathbf{i}$$

$$= (-16.32y + 120 - 30y)\mathbf{i} + \mathbf{j}(16.32x + 30x) + \mathbf{k}(-16.32y + 32.64y + 80 - 10x)$$

$$= (120 - 46.32y)\mathbf{i} + 46.32x\mathbf{j} + (80 - 26.32x + 32.64y)\mathbf{k}$$

$$\mathbf{n}' = \frac{120 - 46.32y}{M}\mathbf{i} + \frac{46.32x}{M}\mathbf{j} + \frac{80 - 26.32x + 32.64y}{M}\mathbf{k},$$

where **M** is the moment. Then the direction cosine of **M** and **F** are the same in a wrench system.

$$\therefore \quad \frac{120 - 46.32y}{M} = 0.360 \qquad \ldots(i)$$

$$\frac{46.32x}{M} = 0.401 \qquad \ldots(ii)$$

and

$$\frac{80 - 26.32x + 32.64y}{M} = 0.84 \qquad \ldots(iii)$$

From Eqn. (*i*),

$$y = \frac{120 - 0.36M}{46.32}$$

From Eqn. (*ii*),

$$x = \frac{0.401}{46.32}M$$

$\therefore$ From Eqn. (*iii*),

$$80 - 26.32\frac{0.401}{46.32}M + \frac{32.64(120 - 0.36M)}{46.32} = 0.84\,M$$

*i.e.*, $\quad (0.84 + 0.2279 + 0.2537)M = 80 + 84.56$

$$M = 124.515 \text{ kN-m.}$$

$$x = \frac{0.401}{46.32} M = 1.078 \text{ m}$$ **Ans.**

$$y = \frac{120 - 0.36\,M}{46.32} = 1.623 \text{ m}$$

**Example 2.25.** A system of forces has resultant force couple as

$$\mathbf{R} = 700\mathbf{i} - 400\mathbf{j} - 150\mathbf{k} \text{ and}$$

$$\mathbf{M}_0 = 400\mathbf{i} + 350\mathbf{j} - 2500\mathbf{k}$$

Find the wrench system of resultant, indicating its intersection with $x$, $y$ plane.

**Solution.**

$$\mathbf{R} = 700\mathbf{i} - 400\mathbf{j} - 150\mathbf{k}$$

$\therefore$

$$R = \sqrt{700^2 + 400^2 + 150^2} = 820.06$$

$$\mathbf{nR} = \frac{700}{820.06}\mathbf{i} - \frac{400}{820.06}\mathbf{j} - \frac{150}{820.06}\mathbf{k}$$

$$= 0.854\mathbf{i} - 0.488\mathbf{j} - 0.183\mathbf{k}$$

$$\mathbf{M}_0 = 400\mathbf{i} + 350\mathbf{j} - 2500\mathbf{k}$$

Let $O'$ be the point of intersection of wrench resultant with $x$, $y$ plane. Its coordinates be $(x, y, 0)$.

Then, $\mathbf{M}_{o'} = \mathbf{M}_o + \mathbf{r}_{o'o} \times \mathbf{R}$

$$= 400\mathbf{i} + 350\mathbf{j} - 2500\mathbf{k} + \begin{vmatrix} \mathbf{i} & \mathbf{j} & \mathbf{k} \\ x & y & 0 \\ 700 & -400 & -150 \end{vmatrix}$$

$$= 400\mathbf{i} + 350\mathbf{j} - 2500\mathbf{k} + \mathbf{i}(-150y) - \mathbf{j}(-150x) + \mathbf{k}(-400x - 700y)$$

$$= (400 - 150y)\mathbf{i} + (350 + 150x)\mathbf{j} + (-2500 - 400x - 700y)\mathbf{k}$$

If $M'$ is the magnitude of the wrench moment, then

$$\mathbf{M}_{0'} = \frac{400 - 150y}{M'}\mathbf{i} + \frac{350 + 150x}{M'}\mathbf{j} - \frac{2500 + 400x + 700y}{M'}\mathbf{k}$$

Since $\mathbf{M}_0'$ has to be the same direction cosines as $\mathbf{R}$, we get

$$\frac{400 - 150y}{M'} = 0.854 \qquad \text{...}(i)$$

$$\frac{350 + 150x}{M'} = -0.488 \qquad \text{...}(ii)$$

$$-\frac{2500 + 400x + 700y}{M'} = -0.183 \qquad \text{...}(iii)$$

From Eqn. (*i*),

$$y = \frac{400 - 0.854\,M'}{150}$$

From Eqn. (*ii*),

$$x = \frac{-0.488\,M' - 350}{150}$$

Hence from Eqn. (*iii*),

$$2500 + 400\left(\frac{-0.488M' - 350}{150}\right) + \frac{700(400 - 0.854M')}{150} = 0.183\ M'$$

$$2500 - 1.301\ M' - 933.33 + 1886.67 - 3.985\ M' = 0.183\ M'$$

$\therefore$ $(0.183 + 1.301 + 3.985)\ M' = 2500 - 933.33 + 1886.67$

$\therefore$ $M' = 627.71$ units

$\therefore$ $$x = \frac{-0.488(627.71) - 350}{150} = -4.375$$

$$y = \frac{400 - 0.854(627.71)}{150} = -0.907$$ **Ans.**

$z = 0$

$\mathbf{R} = 700\mathbf{i} - 400\mathbf{j} - 150\mathbf{k}$

## EXERCISE

**2.1** Determine the resultant of the five coplanar forces acting on a particle as shown in Fig. 2.35.

[**Ans. R** = 105.359**i** – 41.962**j**, **R** = 113.408 N, α = – 21.716°]

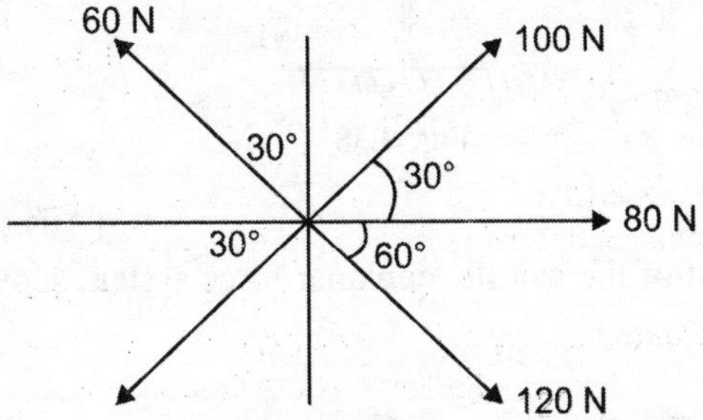


**Fig. 2.35**

**2.2** A body is subjected to the system of three forces as shown in Fig. 2.36. If possible, determine the direction of the force **F** so that the resultant is in *x*-direction, when

(*a*) **F** = 10 kN (*b*) **F** = 6 kN

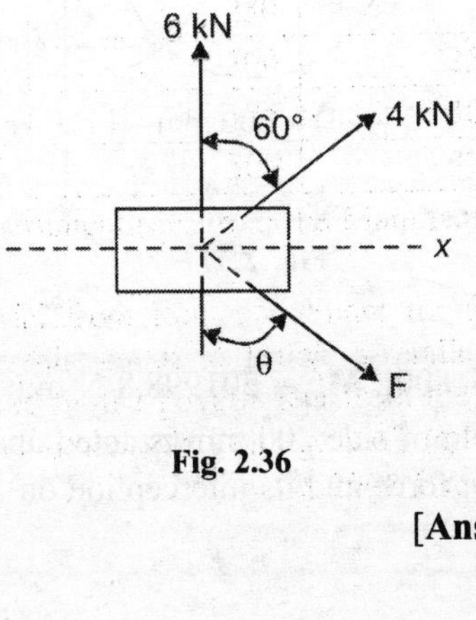


**Fig. 2.36**

[**Ans.** (*a*) θ = 36.87°, (*b*) Not possible]

**2.3** A disabled car is pulled along a road as shown in Fig. 2.37. Find the force $\mathbf{F_1}$ and the resultant **R** along the direction of the road.

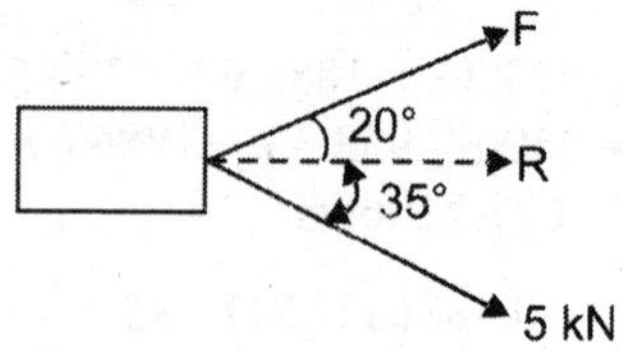


**Fig. 2.37**

[**Ans.** $\mathbf{F_1}$ = 8.385 kN, **R** = 11.975 kN]

**2.4** Determine the force **F** as shown in Fig. 2.38 such that the resultant of the three forces is 12 kN, vertically downward on the pole.

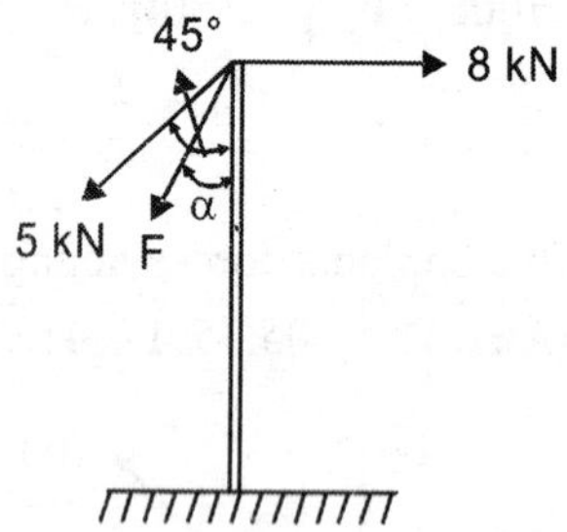


**Fig. 2.38**

[**Ans.** α = 27.809°, **F** = 9.570 kN]

**2.5** Determine the resultant of the parallel coplanar force system shown in Fig. 2.39.

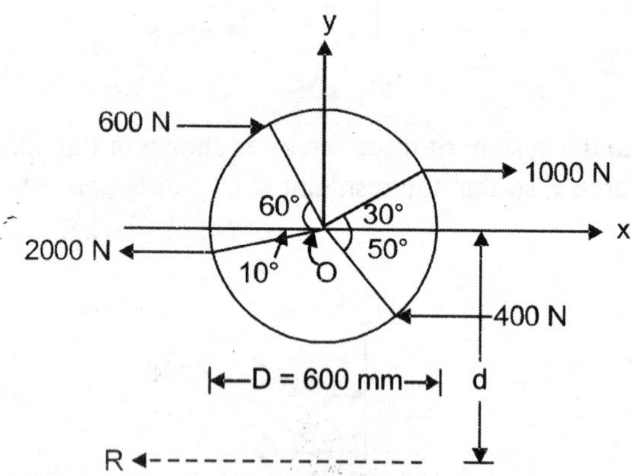


**Fig. 2.39**

[**Ans.** **R** = – 800**i**; $\mathbf{M_O}$ = 501998.8 N-mm *i.e.*, $d$ = 627.50 mm below $x$-axis]

**2.6** An equilateral triangular plate of side 200 mm is acted upon by four forces as shown in Fig. 2.40. Determine the resultant force and its interception on $x$-axis.

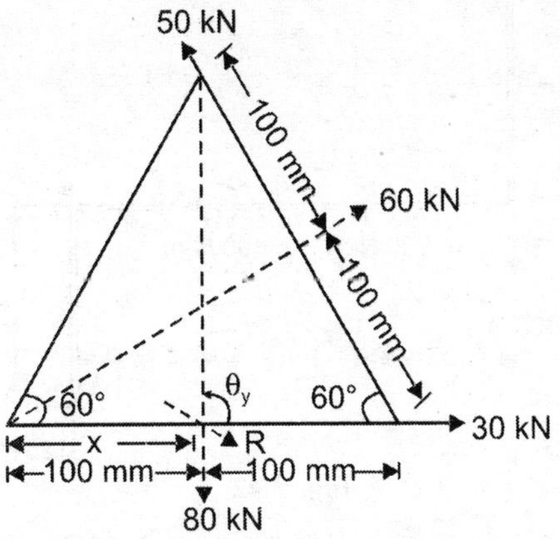


**Fig. 2.40**

[**Ans.** **R** = 56.961**i** – 6.699**j**; $\theta_x$ = 6.711°, $\theta_y$ = 96.707°; $x$ from $A$ = 98.56 mm]

**2.7** Determine the resultant of the four forces acting on a rigid frame as shown in Fig. 2.41.

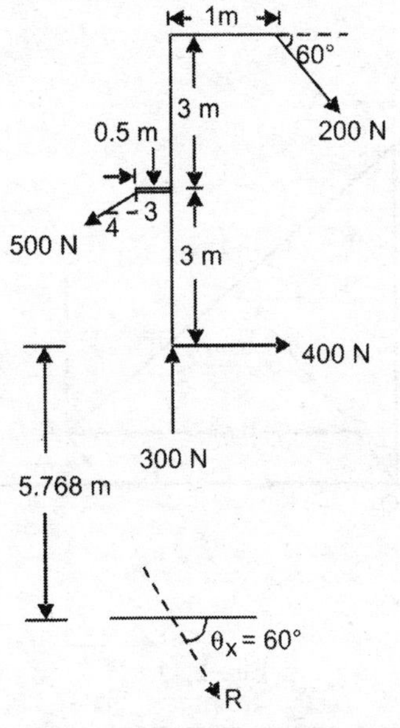


**Fig. 2.41**

[**Ans.** **R** = 100**i** – 173.205**j**,
*i.e.*, **R** = 200 kN, $\theta_x$ = 60°,
$y$ = 5.768 m below 0]

**2.8** A bracket is subjected to the system of forces and couples-moments as shown in Fig. 2.42. Determine the resultant of the system and the point of intersection of the resultant with *BC*.

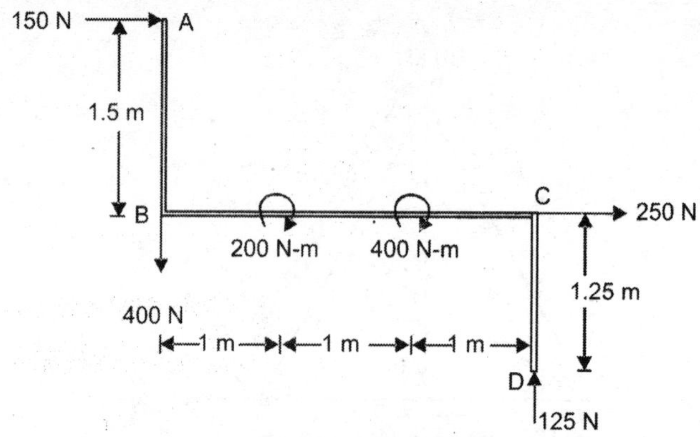


**Fig. 2.42**

[**Ans. R** = 400**i** – 175**j**; **R** = 485.51 N, $\theta_x = 34.5°$, $\theta_y = 124.5°$. Its line of action intercepts *BC* at 1.636 m from *B*]

**2.9** Distinguish between:

(*a*) Resolution and composition of forces (*b*) Resultant and equilibriant of forces.

**2.10** Prove that a force acting at point *A* may be resolved into a force and a couple at another point *B* as shown in Fig. 2.43.

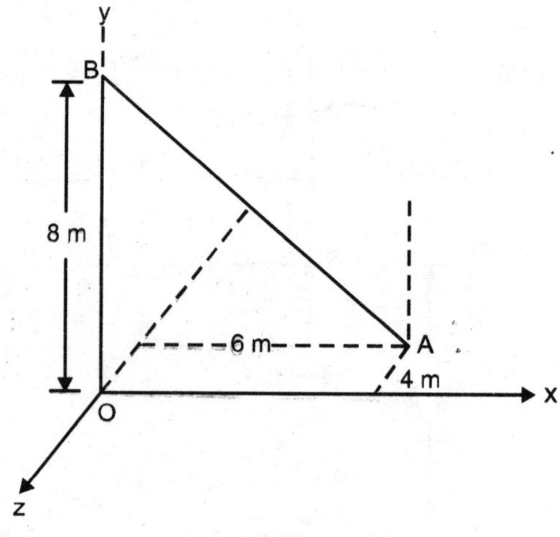


**Fig. 2.43**

**2.11** Explain how a system of force acting on a rigid body can be resolved into a force and a moment.

**2.12** *ab* is a guy wire as shown in Fig. 2.44 and is carrying a force of 40 kN. Determine its vector form, direction cosines and the angles with the coordinate directions *x*, *y* and *z*.

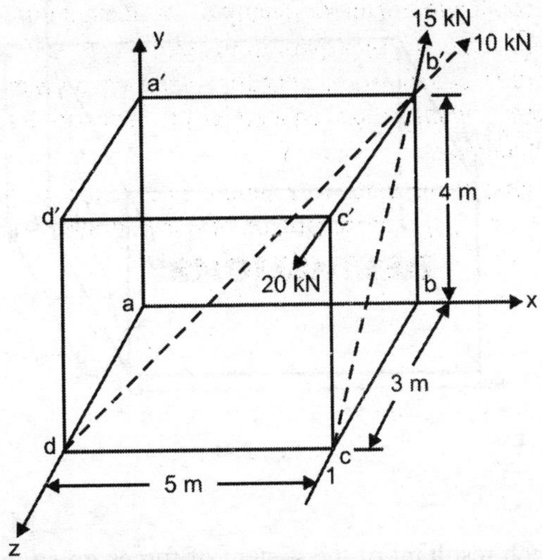


Fig. 2.44

[**Ans.** – 22.28**i** + 14.84**j** + 29.74**k**, $l$ = – 0.557, m = 0.371,
$n$ = 0.742, $\theta_x$ = 12385°, $\theta_y$ = 68.22°, $\theta_z$ = 42.01°]

**2.13** Three concurrent forces acting on a parallelepiped at $b'$ are as shown in Fig. 2.45. Determine the resultant force.

[**Ans. R** = 7.07**i** + 17.67**j** + 6.67**k**, **R** = 20.197 kN,
$\theta_x$ = 69.51, $\theta_y$ = 28.57°, $\theta_z$ = 70.45°]

**2.14** Determine the moment of 10 kN force as shown in Fig. 2.45 about point $O$ and about the axis $ac'$.

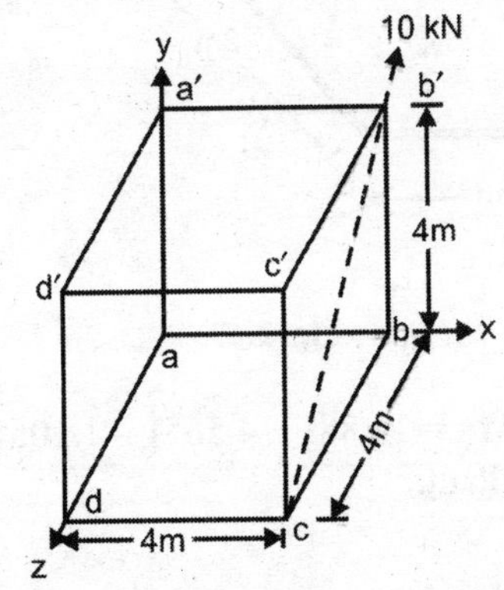


Fig. 2.45

[**Ans.** $\mathbf{M_a}$ = – 28.28**i** + 28.28**j** + 28.28**k**
$\mathbf{M_{ac'}}$ = – 16.318**i** + 16.318**j** + 16.318**k**]

**2.15** The forces $\mathbf{F_1}$ = 10 kN, $\mathbf{F_2}$ = 8 kN and $\mathbf{F_3}$ = 6 kN are acting on the parallelepiped of size 6 m × 3 m × 3 m as shown in Fig. 2.46. Determine the resultant of the system of forces at ‘ $a$’.

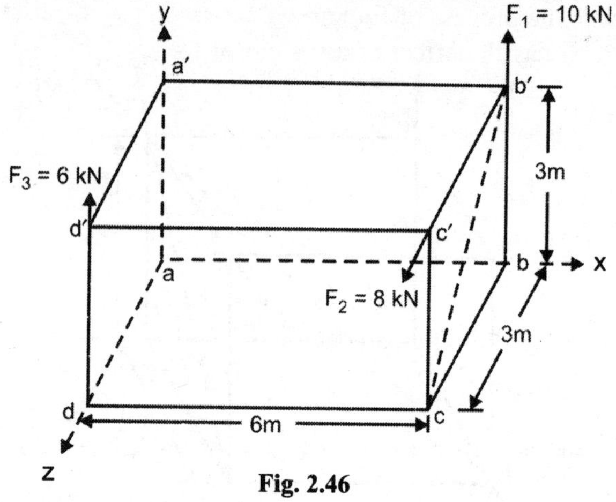


Fig. 2.46

[**Ans. R** = 13.07**j** + 0.93**k**
$\mathbf{M}_a$ = – 15.07**i** – 5.58**j** + 42.48**k**]

**2.16** Determine the wrench resultant of the system of forces given in exercise Problem 6.
[**Ans. R** = 13.07**j** + 0.93**k** acting on *x*-*y* plane
at (3.233, – 1710); **M** = 2.35 kN-m]

**2.17** The bent *ABC*, as shown in Fig. 2.47, is in the vertical plane *x*-*y*. This is subjected to a 5 kN force along *Ca′* as shown in the figure. Determine the resultant force-moment at *A* and *B*.

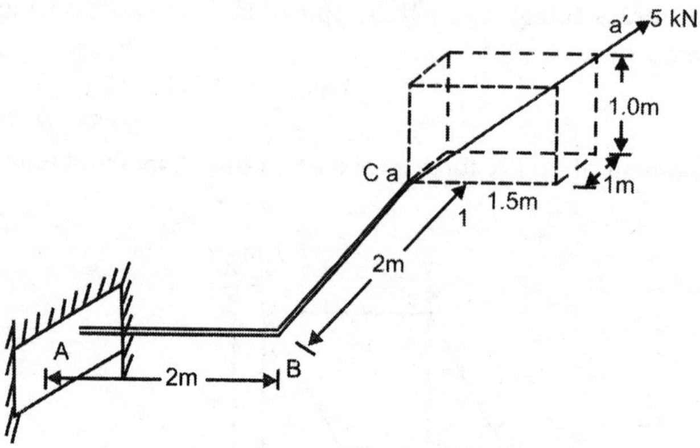


Fig. 2.47

[**Ans. F** = 4.01**i** + 2.66**j** – 1.335**k**,
$\mathbf{M}_A$ = 1.888**i** + 4.565**j** – 1.105**k**; $\mathbf{M}_B$ = 1.888**i** + 1.888**j** – 1.888**k**]

**2.18** Explain the term wrench resultant.

## OBJECTIVE TYPE QUESTIONS

**Select the correct options in the following:**

**2.1** A force and a moment acting at a point on a body can be replaced by suitable
(*a*) single force (*b*) a moment
(*c*) a set of concurrent forces (*d*) None of these

**2.2** Resolution of a force is the process of finding
(*a*) single force which has the effect of a system of force
(*b*) a set of forces having the same effect as a given force
(*c*) the equilibrium condition
(*d*) None of the above

**2.3** A force applied at the free end of a rigid bar creates maximum moment at the hinge point when the line of action and the bar make :
(*a*) 0° (*b*) 45°
(*c*) 60° (*d*) 90°

**2.4** Which one of the following is the fundamental law of mechanics?
(*a*) Lami's theorem (*b*) triangle law
(*c*) parallelogram law (*d*) law of polygon

**2.5** $\mathbf{F_1}$ and $\mathbf{F_2}$ are two forces acting at a point on a body. The angle between them is θ. Their resultant is given by
(*a*) $\sqrt{F_1^2 + F_2^2 + 2F_1F_2 \sin\theta}$ (*b*) $\sqrt{F_1^2 + F_2^2 + 2F_1F_2 \cos\theta}$
(*c*) $F_1 + F_2 \cos\theta$ (*d*) $F_1 + F_2 \sin\theta$

**2.6** $\mathbf{F_1}$ and $\mathbf{F_2}$ are the two concurrent forces. The angle between them is θ. Then the direction of the resultant with respect to line of action of $\mathbf{F_1}$ is
(*a*) $\tan^{-1}\dfrac{F_1 \cos\theta}{F_1 + F_2 \cos\theta}$ (*b*) $\tan^{-1}\dfrac{F_1 \sin\theta}{F_1 + F_2 \cos\theta}$
(*c*) $\tan^{-1}\dfrac{F_2 \sin\theta}{F_1 + F_2 \cos\theta}$ (*d*) $\tan^{-1}\dfrac{F_2 \cos\theta}{F_1 + F_2 \cos\theta}$

**2.7** The angles between two forces to make their resultant a minimum and a maximum are respectively
(*a*) 0° and 90° (*b*) 180° and 90°
(*c*) 90° and 180° (*d*) 180° and 0°

**2.8** Two forces each equal to $p/2$ act at right angles to each other. Their resultant is having magnitude
(*a*) $2p$ (*b*) $\sqrt{2}\,p$
(*c*) $\dfrac{p}{2}$ (*d*) $\dfrac{p}{\sqrt{2}}$

CHAPTER

# 3

# EQUATIONS OF EQUILIBRIUM

A body is said to be in equilibrium when it is at rest or continues to be in steady linear motion. According to Newton's second law of motion, the resultant of the force system acting on the body is zero. Analytically it means

$$\mathbf{R} = 0 \text{ and } \mathbf{M} = 0 \qquad ...(3.1)$$

Equations 3.1 are called equations of equilibrium.

While applying equilibrium conditions to any body in equilibrium, it is necessary that all forces acting on the body are considered and marked in proper directions on the diagram of the body, freed from all its contact surfaces. Hence first various forces to be considered are explained and the method of representing them on the free body is discussed. After this using equations of equilibrium is illustrated with a number of solved example.

## 3.1 TYPES OF FORCES ON A BODY

The various forces acting on a body may be grouped as

1. Applied Forces.
2. Non-applied Forces.

### 1. Applied Forces

Applied forces are the forces applied externally on the body under consideration. If a person stands on a ladder, his weight is an applied force on the ladder. If a car is pulled with a rope, the force in the rope is applied force on the car. In case of concurrent force system, the applied forces are always point force which may be referred as concentrated loads. In case of rigid bodies, the following type of applied forces may be found.

### Point Force

If a load is acting on the body over a very small area compared to the sizes of the body, without much loss of accuracy in the analysis, it may be approximated as a force acting at a point. Such load is also known as concentrated load.

## Linear Force

A force acting along a line of the body is known as the linear force. Figure 3.1 shows a linear loading on a beam. In this the ordinate represents, the intensity of the force and the abscissa represents the position of the force on the beam. The force **dF** on an elemental length $dx$ is given by

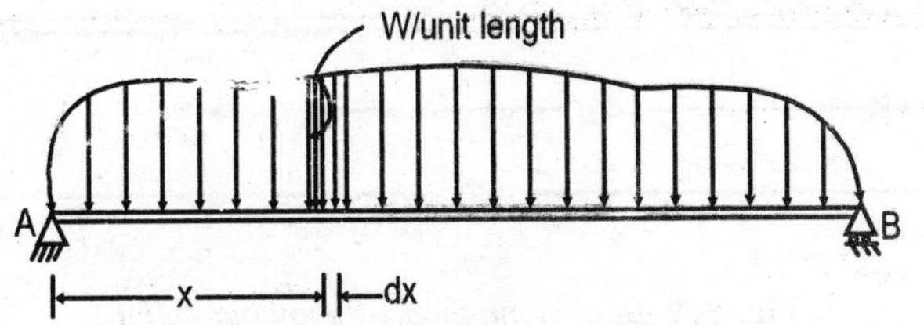


**Fig. 3.1** *Linear force*

$$\mathbf{dF} = w\,dx \qquad ...(3.2)$$

where $w$ is the intensity of the force.

Uniformly distributed loads (Fig. 3.2) and uniformly varying loads (Fig. 3.3) are the particular cases of linear forces.

From Eqn. 3.2, it is obvious that area of load diagram gives total load and centroid of this load is the centroid of the total load. Thus, in Fig. 3.2.

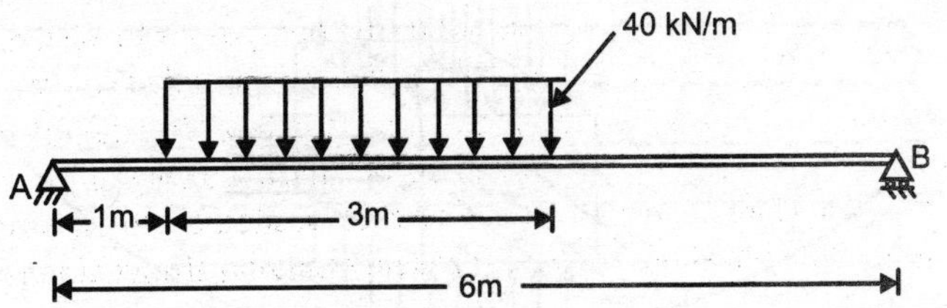


**Fig. 3.2** *Uniformly Distributed Load (UDL)*

Total load $\quad \mathbf{W} = 40 \times 3 = 120$ kN

and its centroid is at = $1 + \frac{3}{2} = 2.5$ m from $A$ and in Fig. 3.2, Total load = $\frac{1}{2} \times 80 \times 3 = 120$ kN

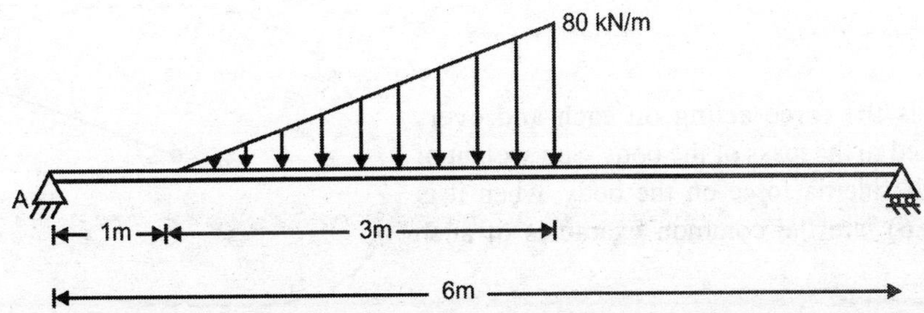


**Fig. 3.3** *Uniformly Varying Loads (UVL)*

and its centroid is at $1 + \frac{2}{3} \times 3 = 3$ m from $A$.

Many times, for simplicity, uniformly distributed load is indicated as shown in Fig. 3.4.

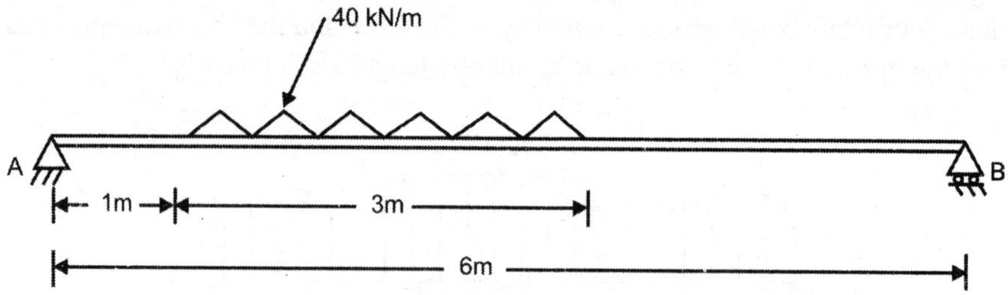


**Fig. 3.4** *Another method of showing (UDL)*

### Surface Force

A force acting on the surface of a body is termed as surface force. Common example of such a force is the hydrostatic pressure acting on a body immerced in water (Ref. Fig. 3.5). If '$p$' is the intensity of force acting on an elemental area $dA$, then the force acting on the element is

$$\mathbf{dF} = p.\ dA \qquad \ldots(3.3)$$

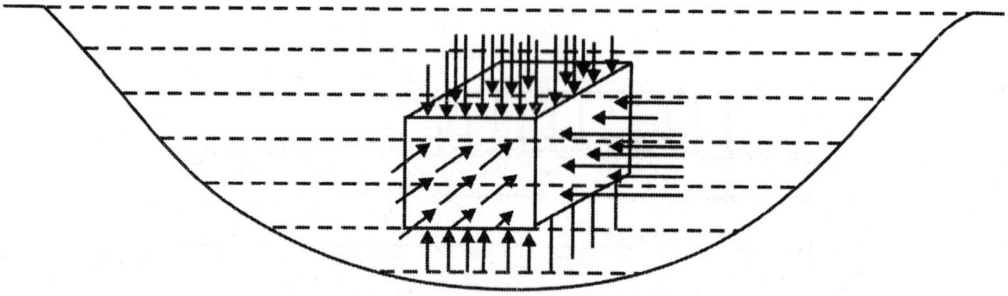

**Fig. 3.5** *Hydrostatic force (a surface force)*

### Body Force

A body force is the force acting on each and every particle contained in the mass of the body. Self weight of the body and the inertia force on the body, when it is rotated (Fig. 3.6), are the common examples of such forces.

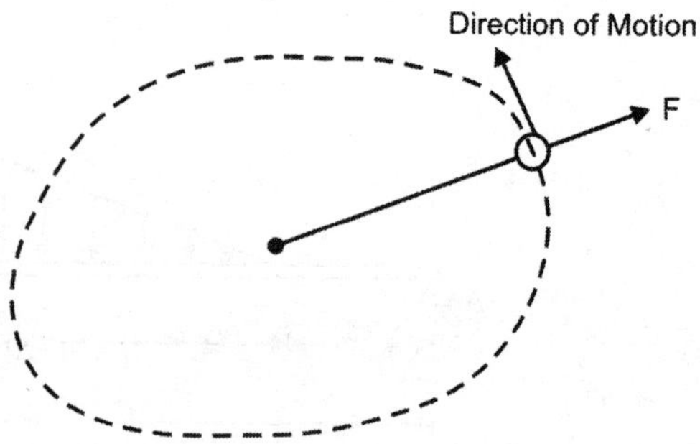


**Fig. 3.6** *Inertia force (a body force)*

### External Moment

Sometimes a couple moment may act on a body as shown in Fig. 3.7 ($a$). It is equivalent to an external moment $\mathbf{M} = \mathbf{F} \times \mathbf{a}$ as shown in Fig. 3.7 ($b$).

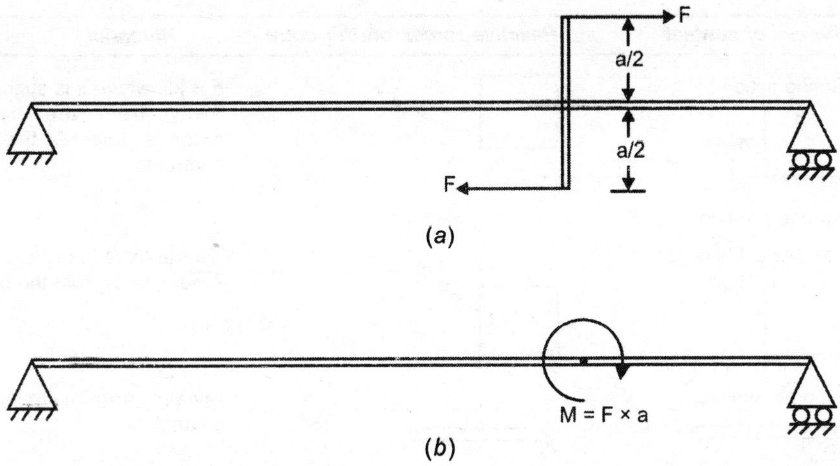


**Fig. 3.7** *External moment*

## 2. Non-Applied Forces

There are two types of non-applied forces:

1. Self weight
2. Reactions from other bodies in contact.

### Self Weight

Every body is subjected to gravitational attraction and hence has got self weight:

$$W = mg \qquad ...(3.4)$$

where $m$ is mass of the body and $g$ is acceleration due to gravity. The value of $g$ is 9.81 $m/s^2$ near the earth surface.

Since self weight is due to gravitational attraction, it is always directed towards the centre of the earth, *i.e.*, in vertically downward direction. In the analysis, self weight is treated as acting through the centre of gravity of the body. If self weight is small compared to the applied forces, in some engineering problems, it may be neglected.

### Reactions

Reactions are the self adjusting forces exerted by the other bodies which are in contact with the body under consideration. According to Newton's third law of motion, the reactions are equal and opposite to the actions. For example, if a block weighing 2 kN is kept on the table, action of the block is 2 kN load acting vertically downward on the table while the reaction from the table is 2 kN vertically upward acting on the block. The reactions adjust themselves to bring the body to equilibrium.

If the surface of contact is smooth, the direction of reaction is normal to the surface of contact. If the surface of contact is not smooth, there will be frictional reaction also. In case of beams, reactions depend upon the type of supports provided. Tables 3.1 and 3.2 show the different types of contacts between the body and other bodies and also the reactive forces on the body.

**Table 3.1:** *Types of Reactive Forces in Two-Dimensional Problems*

| Sl no. | Types of contact | Reactive forces on the body | Remarks |
|---|---|---|---|
| 1. | Spring action 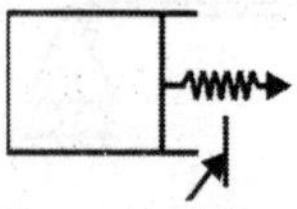  | 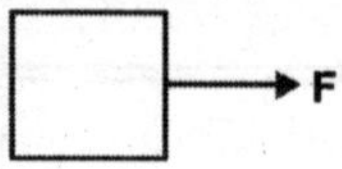  | **F** = **k**δ, where *k* is spring constant. Away from the body if spring extends, towards the body if spring contracts. |
| 2. | Flexible cable or rope 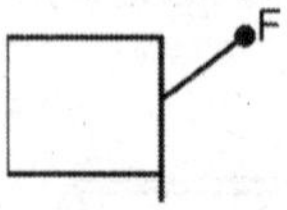  | 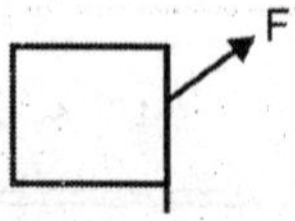  | Tensile force in cable. Always away from the body. |
| 3. | Smooth surface  |   | Always normal to the surface of contact. |
| 4. | Rough surface  |   | **N**-Normal to the surface. Frictional force **F** acting so as to oppose the motion of the body. |
| 5. | Roller support/simple support 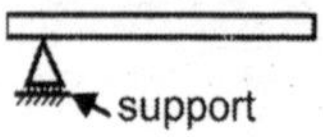  |   | Always normal to support. |
| 6. | Pinned/hinged support  | 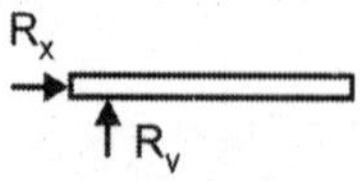  | No moment but reactive force in any direction, which may be conveniently represented by its components in two mutually perpendicular directions. |
| 7. | Fixed or built-in support  | 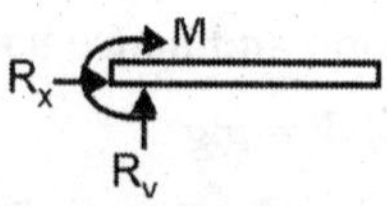  | Moment and force in any direction |

**Table 3.2:** *Types of Reaction Force in Three Dimensional Problems*

| Sl no. | Type of contact | Reactive forces on the body | Remarks |
|---|---|---|---|
| 1. | Smooth surface 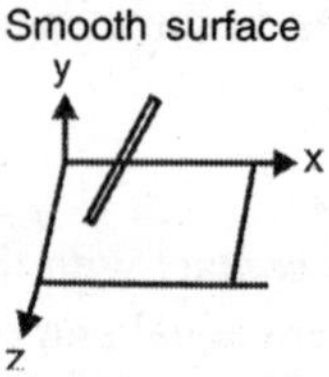  |   | Force normal to surface of contact. |
| 2. | Rough surface 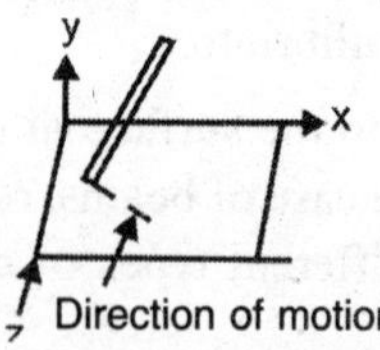  | 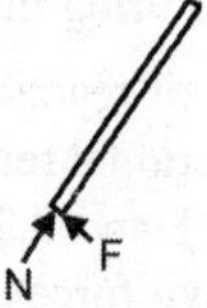  | **N**-Normal to surface, **F**-Frictional force opposite to the direction of motion. |

*(Contd.)*

| | | |
|---|---|---|
| 3. Roller, with lateral constraint 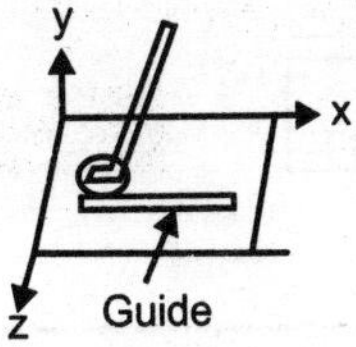  | 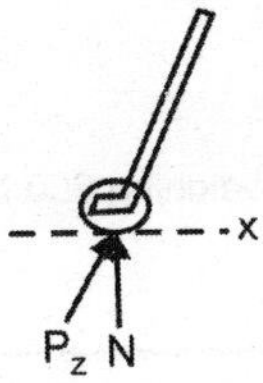  | **N**-Normal to surface on which it rolls<br>**P**-Lateral force exerted by guide |
| 4. Ball and socket joint 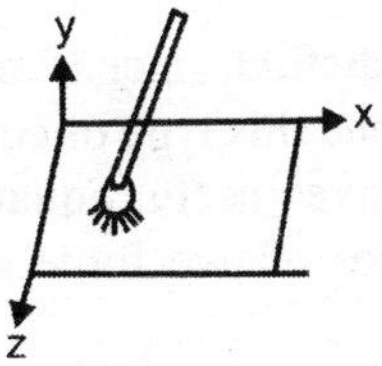  | 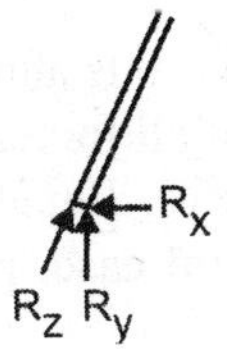  | Force in any direction, which may be split into its components in three mutually perpendicular directions and represented by $\mathbf{R_x}$, $\mathbf{R_y}$ and $\mathbf{R_z}$. |
| 5. Fixed joint 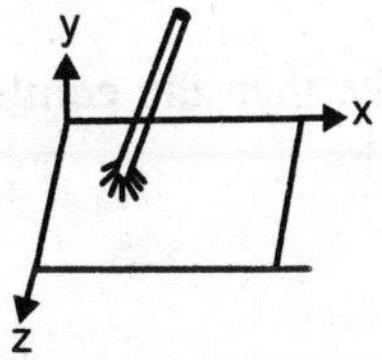  | 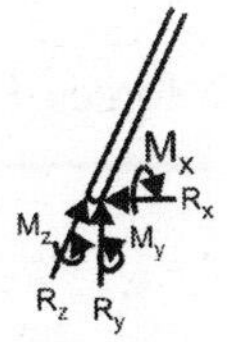  | Resultant force and moment in any direction, which may be represented by reaction components, $\mathbf{R_x}$, $\mathbf{R_y}$, $\mathbf{R_z}$, $\mathbf{M_x}$, $\mathbf{M_y}$ and $\mathbf{M_z}$ |

## 3.2 FREE BODY DIAGRAM (FBD)

In the analysis, it is necessary to iso late the body under consideration from the other bodies in contact and draw all the forces acting on the body. For this, first the body is drawn and then applied forces, self weight and the reactions are drawn. Such a diagram of the body in which the body under consideration is freed from all the contact surfaces and all the forces acting on it (including self weight and reactions) are drawn, is termed as Free Body Diagram (FBD). Free body diagrams are shown for typical cases in Table 3.3.

**Table 3.3:** *Free Body Diagrams (FBD) for a Few Typical Cases*

| *Reacting bodies* | *FBD required for* | *FBD* |
|---|---|---|
| 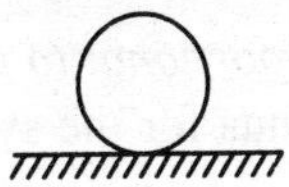 | Ball | 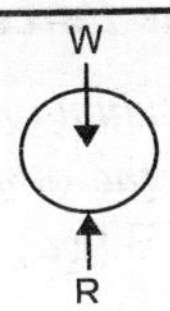  |
| 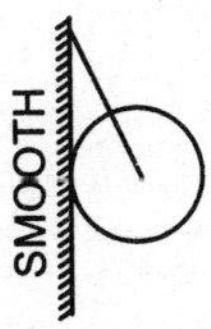  | Ball | 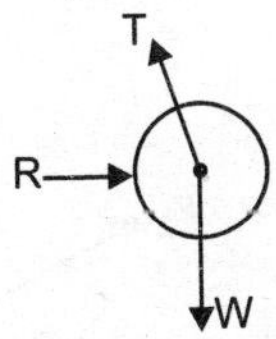  |
| 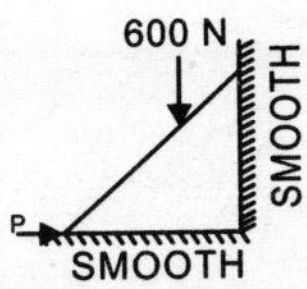  | Ladder | 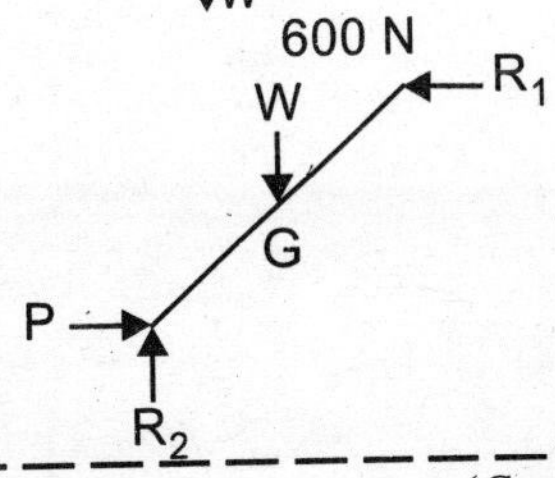  |

*(Contd.)*

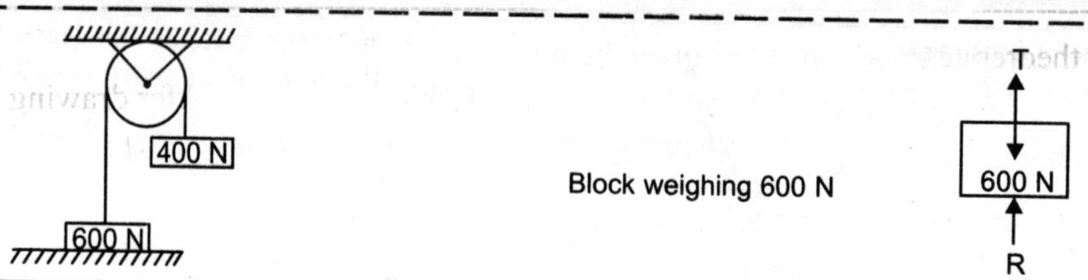


## 3.3 EQUILIBRIUM OF CONCURRENT FORCE SYSTEM IN A PLANE

In concurrent force system, the equilibrium equation **M** = 0 is automatically satisfied, since if moment about the point of concurrency is considered, moment of all forces is zero. Hence this type of equation do not yield any useful equation to find unknown forces. Equation **R** = 0 gives useful equation to determine unknown forces. Before taking up the general case, particular cases of two force system and three force system are presented in this article.

### *Equilibrium of Two Force System*

When a body is in equilibrium under the action of two forces $\mathbf{F_1}$ and $\mathbf{F_2}$ only, then the equilibrium condition is

$$\mathbf{R} = 0$$

$$\mathbf{F_1} + \mathbf{F_2} = 0$$

*i.e.,*

$$\mathbf{F_1} = -\mathbf{F_2}$$

In other words, the two forces must have same magnitude and direction but opposite in sense as shown in Fig. 3.8.

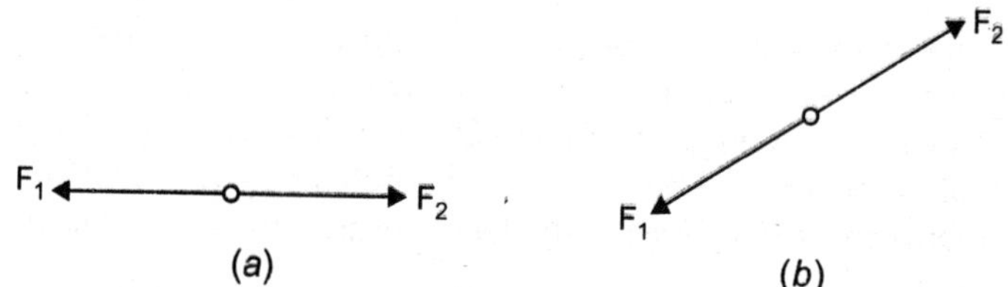


**Fig. 3.8**

### *Equilibrium of Three Force System*

*If a body is in equilibrium under the action of only three forces, each force is proportional to the sine of the angle between the other two forces*. This is known as **Lami's theorem**. Thus for the system of forces shown in Fig. 3.9(*a*), we have

$$\frac{F_1}{\sin\alpha} = \frac{F_2}{\sin\beta} = \frac{F_3}{\sin\gamma} \qquad \text{...(3.5)}$$

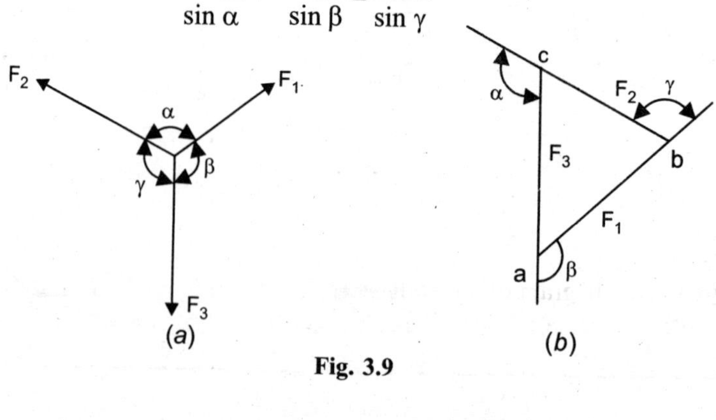


**Fig. 3.9**

The above theorem can be proved as given below:

Draw the three forces $\mathbf{F}_1$, $\mathbf{F}_2$ and $\mathbf{F}_3$ from tail to head fashion as explained for drawing polygon of forces. If '$a$' is the starting point of polygon of forces, the last point of the polygon must also be $a$, since the resultant is zero.

Thus it results into triangle of forces $abc$ as shown in Fig. 3.9($b$). Now the external angles at $a$, $b$ and $c$ are equal to $\beta$, $\gamma$ and $\alpha$, since

$ab$ is $\parallel$ to $F_1$

$bc$ is $\parallel$ to $F_2$

and $ca$ is $\parallel$ to $F_3$

In the triangle of forces $abc$,

$$ab = F_1$$

$$bc = F_2$$

and
$$ca = F_3$$

Applying sine rule to triangle $abc$, we get

$$\frac{ab}{\sin(180-\alpha)} = \frac{bc}{\sin(180-\beta)} = \frac{ca}{\sin(180-\gamma)}$$

i.e.,
$$\frac{F_1}{\sin\alpha} = \frac{F_2}{\sin\beta} = \frac{F_3}{\sin\gamma}$$

***General Approach***

Whatever may be number of forces in the concurrent system, it has to satisfy

$$\mathbf{R} = 0$$

i.e.,
$$\Sigma F_x \mathbf{i} + \Sigma F_y \mathbf{j} = 0$$

i.e.,
$$\Sigma F_x = 0 \qquad \text{...[3.6(a)]}$$

and
$$\Sigma F_y = 0 \qquad \text{...[3.6(b)]}$$

Use of equilibrium equations in two dimensional concurrent forces is illustrated by solving examples below:

**Example 3.1.** A sphere weighing 120 N is tied to a smooth wall by a string as shown in Fig. 3.10($a$). Find the tension **T** in the string and reaction **R** of the wall.

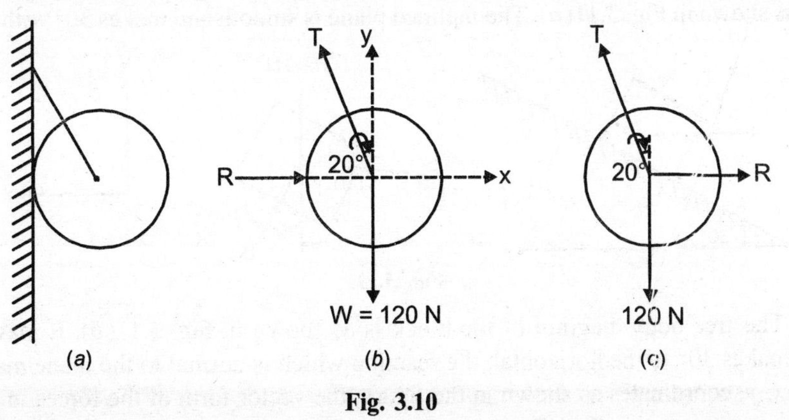


**Fig. 3.10**

**Solution.** Free body diagram of the body is as shown in Fig. 3.10($b$). Let **T** be the tension in the string and **R** be the reaction from the wall. Selecting the coordinate directions $x$, $y$ as shown in figure, the vector form of various forces in the system are

$$\mathbf{F}_1 = -T \sin 20\mathbf{i} + T \cos 20\,\mathbf{j}$$
$$\mathbf{F}_2 = R_\mathbf{i}$$
$$\mathbf{F}_3 = -120\mathbf{j}$$

Since the body is in equilibrium, the resultant = 0

$$\mathbf{F}_1 + \mathbf{F}_2 + \mathbf{F}_3 = 0$$
$$-T \sin 20°\mathbf{i} + T \cos 20°\mathbf{j} + R\mathbf{i} - 120\mathbf{j} = 0.$$
$$(-T \sin 20° + R)\mathbf{i} + (T \cos 20° - 120)\mathbf{j} = 0$$

*i.e.*, $-T \sin 20° + R = 0$ ...(*i*)

and $T \cos 20° - 120 = 0$ ...(*ii*)

From Eqn. (*ii*), $T = \dfrac{120}{\cos 20°} = 127.70$ N **Ans.**

From (*i*), $R = T \sin 20° = 127.7 \sin 20° = 43.68$ N **Ans.**

Alternately, the above problem may be solved using Lami's theorem. Using law of transmissibility of forces, $R$ is shifted as shown in Fig. 3.10(*c*), so that all the forces emerge out from the centre of gravity of the ball. Then applying Lami's theorem, we get

$$\frac{T}{\sin 90°} = \frac{R}{\sin(180-20)°} = \frac{120}{\sin(90+20)°}$$

∴ $T = 127.70$ N **Ans.**

and $R = 43.68$ N **Ans.**

**Note:**

1. A flexible cable like string can pull a body, but cannot push. Hence the force in the string is always directed away from the body.
2. The reaction from the wall on the ball is a push but not a pull. Hence this reaction is directed towards the body at contact surface. The vertical wall being smooth, the reaction from it is horizontal.
3. However, if the sense of the reactions are taken opposite, the values of the reactions obtained will be negative. While interpreting the results, one can take case of the negative signs.

**Example 3.2.** Determine the horizontal force **P** to be applied to a block weighing 1500 N to hold it in the position as shown in Fig. 3.11(*a*). The inclined plane is smooth and makes 30° with the horizontal.

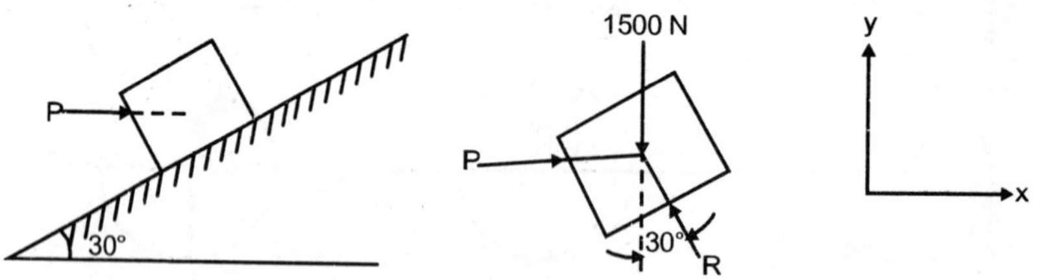


**Fig. 3.11**

**Solution.** The free body diagram of the block is as shown in Fig. 3.11(*b*). It may be noted that since the plane makes 30° to the horizontal, the reaction which is normal to the plane makes 30° to the vertical. Taking $x$, $y$, coordinates as shown in the figure, the vector form of the forces in the system are

$$\mathbf{F}_1 = \mathbf{P}_\mathbf{i}$$
$$\mathbf{F}_2 = -R \sin 30\mathbf{i} + R \cos 30\mathbf{j}$$
$$\mathbf{F}_3 = -1900\mathbf{j}$$

Since the body is in equilibrium, the resultant is zero.

*i.e.,* $$\mathbf{F}_1 + \mathbf{F}_2 + \mathbf{F}_3 = 0$$

$$P\mathbf{i} - R \sin 30°\mathbf{i} + R \cos 30°\mathbf{j} - 1500\mathbf{j} = 0.$$

$$(P - R \sin 30°)\mathbf{i} + (R \cos 30° - 1500)\mathbf{j} = 0$$

*i.e.,* $$P - R \sin 30° = 0 \qquad ...(i)$$

and $$R \cos 30° - 1500 = 0 \qquad ...(ii)$$

From (*ii*), $$R = \frac{1500}{\cos 30°} = 1732.05 \text{ N}$$ **Ans.**

Hence from (*i*), $P = R \sin 30° = 1732.05 \sin 30° = 866.03$ N. **Ans.**

**Example 3.3.** A roller weighing 10 kN rests on a smooth horizontal floor. It is connected to the floor using bar *AC* as shown in Fig. 3.12(*a*). Determine the force in the bar *AC* and reaction from the floor, if the roller is subjected to a horizontal force of 5 kN and an inclined force of 7 kN as shown in the figure.

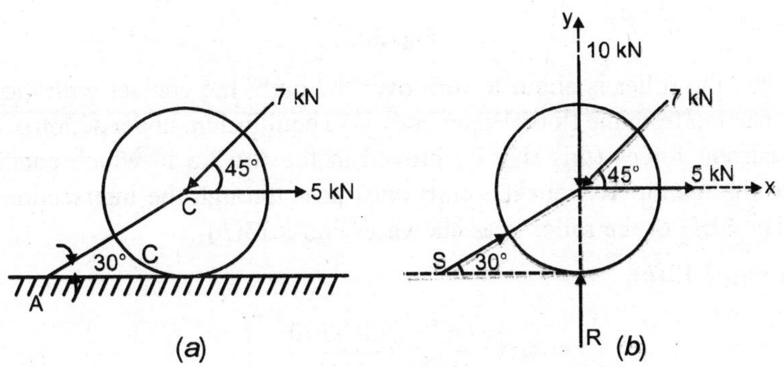


**Fig. 3.12**

**Solution.** A bar can develop tensile force or compressive force. Let the force developed be a compressive force 'S' (push on the cylinder). Free body diagram of the cylinder is as shown in Fig. 3.12(*b*).

Since there are more than three forces in the system. Lami's equation cannot be applied

$$\mathbf{R} = \mathbf{F}_1 + \mathbf{F}_2 + \mathbf{F}_3 + \mathbf{F}_4 = 0$$

*i.e.,* $$5\mathbf{i} - 7 \cos 45°\mathbf{i} - 7 \sin 45°\mathbf{j} - 10\mathbf{j} + S \cos 30°\mathbf{i} + S \sin 30°\mathbf{i} + R\mathbf{j} = 0$$

$$(5 - 7 \cos 45° + S \cos 30°)\mathbf{i} + (- 7 \sin 45° - 10 + S \sin 30° R)\mathbf{j} = 0$$

*i.e.,* $$5 - 7 \cos 45° + S \cos 30° = 0 \qquad ...(i)$$

and $$-7 \sin 45° - 10 + S \sin 30° + R = 0 \qquad ...(ii)$$

From Eqn. (*i*), $$S = \frac{7 \cos 45° - 5}{\cos 30°} = -0.058 \text{ kN}$$ **Ans.**

From Eqn. (*ii*), $$-7 \sin 45° - 10 + (-0.058 \sin 30°) + R = 0$$

or $$R = 14.979 \text{ kN}$$ **Ans.**

*Since the magnitude of S is negative the force exerted by the bar is not a push but a pull, i.e., the bar is under tension.*

**Example 3.4.** A roller of radius $r = 300$ mm and weighing 2000 N is to be pulled over a curb of height 150 mm [Fig. 3.13(*a*)] by horizontal force **F** applied to the end of a string wound tightly around the circumference of the roller. Find the magnitude of **F** required to start the roller move over the curb. What is the least **F** through the centre of the roller to just turn it over the curb?

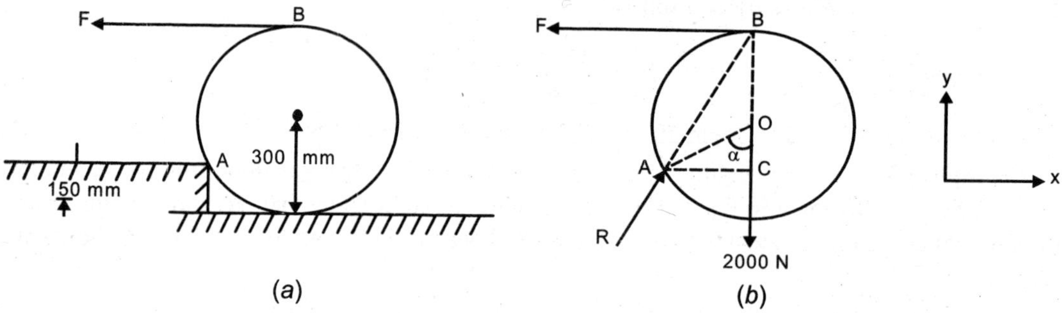


**Fig. 3.13**

**Solution.** When the roller is about to turn over the curb, the contact with the floor is lost, and hence, there is no reaction from the floor. When body is in equilibrium under action of only three forces, they must be concurrent forces (this will be proved in the Art 3.5 in which concept of moment is developed). Hence the reation **R** from the curb must pass through the intersection of the other two forces **F** and **W**. The FBD of the roller is as shown in Fig. 3.13(*b*).

Referring to Fig. 3.13(*b*),

$$\cos\alpha = \frac{OC}{OA} = \frac{300-150}{300} = \frac{1}{2}$$

$$\alpha = 60°$$

In $\Delta AOB$, $\angle OAB = \angle OBA$, since $OA = OB = r$

But $\angle OAB + \angle OBA = \alpha = 60°$

$\therefore$ $2\angle OBA = 60°$

or $\angle OBA = 30°$, *i.e.*, reaction makes 30° with vertical.

From the equilibrium condition, resultant = 0

$$R \sin 30°\mathbf{i} + R\cos 30°\mathbf{j} - F\mathbf{i} - 2000\mathbf{j} = 0$$

$$(R\sin 30° - F)\mathbf{i} + (R\cos 30° - 2000)\mathbf{j} = 0$$

*i.e.*,

$$R\sin 30° - F = 0 \qquad \ldots(i)$$

$$R\cos 30° - 2000 = 0 \qquad \ldots(ii)$$

From Eqn. (*ii*), $R = \dfrac{2000}{\cos 30} = 2309.4$ N

Hence from Eqn. (*i*), $F = R\sin 30° = 2309.4 \sin 30°$

$= 1154.70$ N. **Ans.**

## Least Force Through the Centre of Roller

Now the reaction from the curb must pass through the centre of the wheel since the other two forces pass through that point. Its inclination to vertical is 60°, [Ref. Fig. 3.14(*b*)].

Let the direction of the least force **F** make an angle θ with the horizontal as shown in Fig. 3.14.

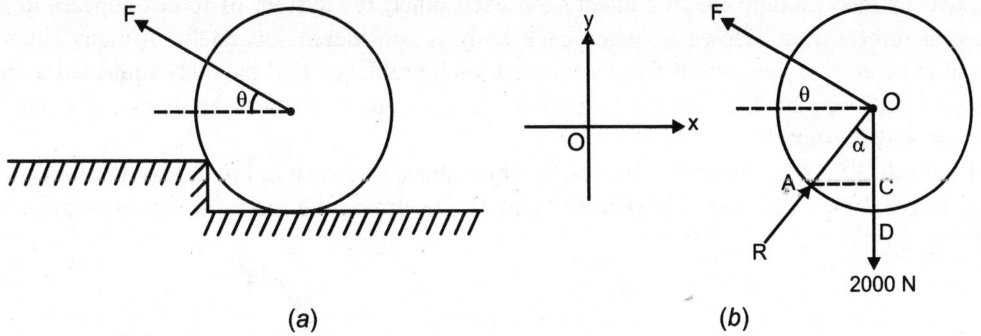


**Fig. 3.14**

The free body diagram of the roller when it is on the verge of rolling over the curb is as shown in Fig. 3.14(*b*). Since the body is in equilibrium,

Resultant force = 0

*i.e.,* $R \sin 60°\mathbf{i} + R \cos 60°\mathbf{j} - F \cos \theta\mathbf{i} + F \sin \theta\mathbf{j} - 2000\mathbf{j} = 0$

*i.e.,* $(R \sin 60° - F \cos \theta)\mathbf{i} + (R \cos 60° + F \sin \theta - 2000)\mathbf{j} = 0$

$\therefore$ $R \sin 60° - F \cos \theta = 0$ ...(*i*)

and $R \cos 60° + F \sin \theta - 2000 = 0$ ...(*ii*)

From (*i*), $R = \dfrac{F \cos\theta}{\sin 60°}$

Hence from (*ii*),

$$\frac{F\cos\theta}{\sin 60}\cos 60° + F\sin\theta = 2000$$

$\therefore$ $F(\cos\theta \cot 60° + \sin\theta) = 2000$

or $$F = \frac{2000}{\cos\theta\cot 60° + \sin\theta}$$

*F* is least when $\dfrac{dF}{d\theta} = 0$

*i.e.,* $$2000\frac{-\sin\theta.\cot 60° + \cos\theta}{(\cos\theta\cot 60° + \sin\theta)^2} = 0$$

*i.e.,* $\cos\theta = \sin\theta \cot 60°$

or $\cot\theta = \cot 60°$

*i.e.,* $\theta = 60°$

*i.e.,* **F** is at right angle to reaction **R**.

$\therefore$ $$\mathbf{F}_{least} = \frac{2000}{\cos 60 \cot 60 + \sin 60°} = 1732.05 \text{ N}$$ **Ans.**

## Equilibrium of Connected Bodies

When two or more bodies are in contact with each other, the system of forces appears to be non-concurrent force system. However when each body is considered separately, in many situations, it turns out to be a set of concurrent force system. In such problems, first the body subjected to only two unknown forces may be analysed (in case of planar problems) and then the others. This type of examples are solved below:

**Example 3.5.** A system of connected flexible cables, as shown in Fig. 3.15(*a*), is supporting two vertical forces 300 N and 400 N at points *B* and *D*. Determine the forces in various segments of the cable.

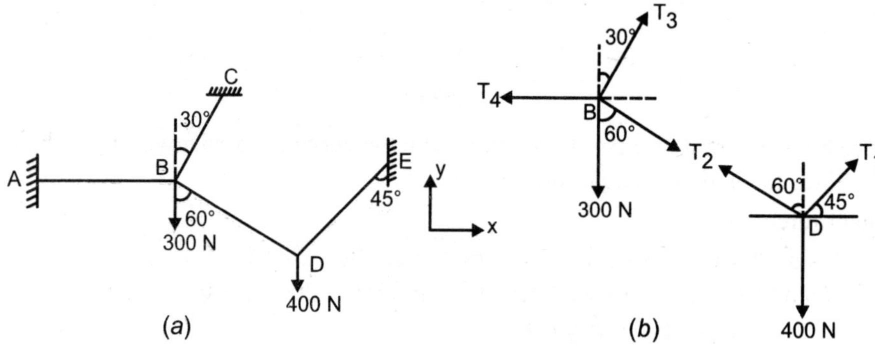


**Fig. 3.15**

**Solution.** The free body diagrams of points *B* and *D* are as shown in Fig. 3.15(*b*). Let the tensile forces in the various segments of the flexible cables be $\mathbf{T}_1$, $\mathbf{T}_2$, $\mathbf{T}_3$ and $\mathbf{T}_4$ as shown in the figure.

From the equilibrium condition of point *D*,

$$T_1 \cos 45°\mathbf{i} + T_1 \sin 45°\mathbf{j} - T_2 \sin 60°\mathbf{i} + T_2 \cos 60°\mathbf{j} - 400\mathbf{j} = 0$$

$$(T_1 \cos 45° - T_2 \sin 60°)\mathbf{i} + (T_1 \sin 45° + T_2 \cos 60° - 400)\mathbf{j} = 0$$

*i.e.*,
$$T_1 \cos 45° - T_2 \sin 60° = 0 \quad \text{...(i)}$$
and
$$T_1 \sin 45° + T_2 \cos 60° - 400 = 0 \quad \text{...(ii)}$$

From Eqn. (*i*),
$$T_1 = \frac{\sin 60°}{\cos 45°} T_2 = 1.2247\ T_2$$

Substituting it in Eqn. (*ii*), we get

$$1.2247\ T_2 \sin 45° + T_2 \cos 60° - 400 = 0$$

or
$$T_2\ [1.2247 \sin 45° + \cos 60°] = 400$$

$\therefore$
$$T_2 = 292.82 \text{ N} \qquad \textbf{Ans.}$$

Hence
$$T_1 = 1.2247 \times 292.82 = 358.62 \text{ N} \qquad \textbf{Ans.}$$

Considering the equilibrium of point *B*, we get

$$T_2 \sin 60°\mathbf{i} - T_2 \cos 60°\mathbf{j} + T_3 \sin 30°\mathbf{i} + T_3 \cos 30°\mathbf{j} - T_4\mathbf{i} - 300\mathbf{j} = 0$$

*i.e.*,
$$(T_2 \sin 60° + T_3 \sin 30° - T_4) = 0 \quad \text{...(iii)}$$
and
$$-T_2 \cos 60° + T_3 \cos 30° - 300 = 0 \quad \text{...(iv)}$$

Substituting the value of $T_2$, from Eqn. (*iv*), we get

$$T_3 \cos 30° = 292.82 \cos 60° + 300$$

$\therefore$
$$T_3 = 515.47 \text{ N}$$

and then from Eqn. (*iii*), we get

$$292.82 \sin 60° + 515.47 \sin 30° - T_4 = 0$$

$\therefore$
$$T_4 = 511.32 \text{ N} \qquad \textbf{Ans.}$$

**Example 3.6.** A rope *AB*, 4.5 m long is connected at two points *A* and *B* at the same level 4 m apart. A load of 1500 N is suspended from a point *C* on the rope 1.5 m from *A* as shown in Fig. 3.16(*a*). What load connected at point *D* on the rope, 1 m from *B* will be required to keep the position *CD* horizontal?

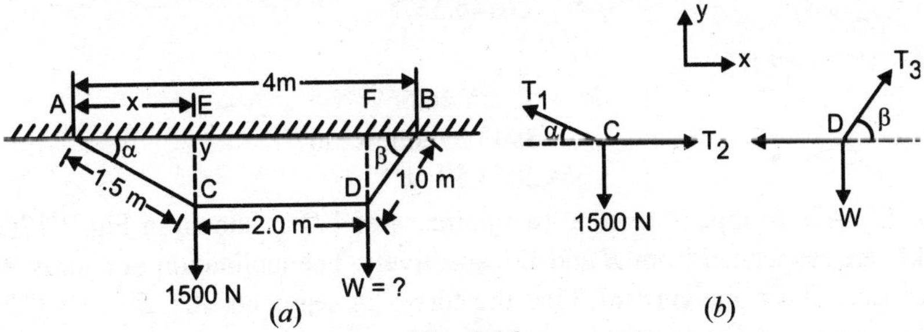


**Fig. 3.16**

**Solution.** Drop perpendiculars *CE* and *DF* on *AB*.

Let $CE = y$ and $AE = x$ ...(*i*)

From $\Delta AEC, x^2 + y^2 = 1.5^2 = 2.25$

It is given that

$$AC + CD + BD = 4.5 \text{ m}$$

$\therefore$ $CD = 4.5 - 1.5 - 1.0 = 2.0$ m

$\therefore$ $BF = AB - AE - EF$

$= 4.0 - x - 2.0 = 2 - x$ ....(*ii*)

From triangle *BFD*, we have

$$BF^2 + DF^2 = 1^2$$

$$(2 - x)^2 + y^2 = 1 \quad ...(iii)$$

Subtracting Eqn. (*iii*) from Eqn. (*i*), we get

$$x^2 - (2 - x)^2 = 1.25$$

*i.e.*, $x^2 - 4 + 4x - x^2 = 1.25$

$\therefore$ $x = 1.3125$ m

$$\alpha = \cos^{-1}\frac{1.3125}{1.5} = 28.955°$$

$$\beta = \cos^{-1}\frac{2 - 1.3125}{1} = 46.567°$$

Considering the equilibrium of point *C*, and selecting cartesian coordinates *x*, *y* as shown in Fig. 3.16(*b*), we obtain

$$- T_1 \cos 28.955°\mathbf{i} + T_1 \sin 29.955°\mathbf{j} + T_2\mathbf{i} - 1500\mathbf{j} = 0$$

*i.e.*, $(-T_1 \cos 28.955° + T_2)\mathbf{i} + (T_1 \sin 29.955° - 1500)\mathbf{j} = 0$

*i.e.*, $-T_1 \cos 28.955° + T_2 = 0$ ...(*iv*)

and $T_1 \sin 28.955° - 1500 = 0$ ...(*v*)

From Eqn. (*v*),

$$T_1 = 3098.39 \text{ N}$$

and hence from Eqn. (*v*), we get

$$T_2 = T_1 \cos 28.955° = 27111.09 \text{ N}$$

Considering the equilibrium of point *D*, we get

$$- T_2\mathbf{i} + T_3 \cos 46.567°\mathbf{i} + T_3 \sin 46.567°\mathbf{j} - W\mathbf{j} = 0$$

*i.e.*, $(- T_2 + T_3 \cos 46.567°)\mathbf{i} + (T_3 \sin 46.567° - W)\mathbf{j} = 0$

*i.e.*, $- 2711.09 + T_3 \cos 46.567° = 0$ ...(*vi*)

and $$T_3 \sin 46.567° - W = 0 \quad ...(vii)$$

From Eqn. (*vi*), we get

$$T_3 = \frac{2711.09}{\cos 46.567°} = 3943.37 \text{ N}$$

From Eqn. (*vii*), we get

$$W = T_3 \sin 46.567°$$
$$= 3943.37 \sin 46.567°$$
$$= 2863.59 \text{ N}$$

**Ans.**

**Example 3.7.** A wire rope is fixed at two points *A* and *D* as shown in Fig. 3.17(*a*). Two weights 20 kN and 30 kN are suspended from *B* and *C* respectively. The inclination of chords *AB* and *BC* are at 30° and 50° respectively to the vertical. Find the forces in segments *AB, BC* and *CD*. Determine the inclination of the segment *CD* to vertical.

**Solution.** Figure 3.17(*b*) shows free body diagrams of the points *B* and *C*. The coordintes *x, y* are taken as indicated in the figure. From the equilibrium condition of point *B*:

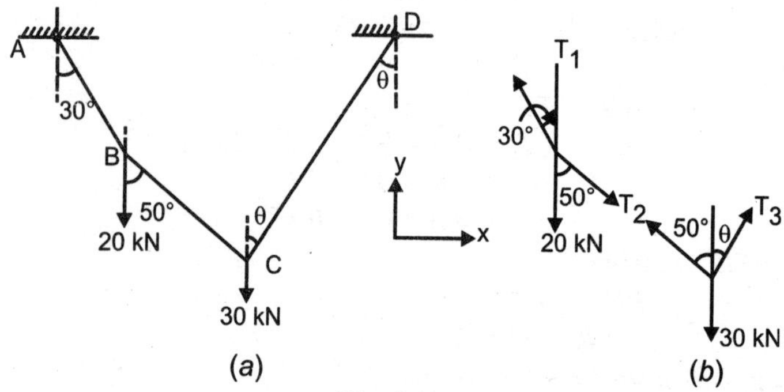


**Fig. 3.17**

$$-T_1 \sin 30°\mathbf{i} + T_1 \cos 30°\mathbf{j} + T_2 \sin 50°\mathbf{i} - T_2 \cos 50°\mathbf{j} - 20\mathbf{j} = 0$$

*i.e.*, $$(-T_1 \sin 30° + T_2 \sin 50°)\mathbf{i} + (T_1 \cos 30° - T_2 \cos 50° - 20)\mathbf{j} = 0$$

*i.e.*, $$-T_1 \sin 30° + T_2 \sin 50° = 0 \quad ...(i)$$

and $$T_1 \cos 30° - T_2 \cos 50° - 20 = 0 \quad ...(ii)$$

From Eqn. (*i*),

$$T_2 = T_1 \frac{\sin 30°}{\sin 50°} = 0.6527\, T_1$$

Substituting it in Eqn. (*ii*),

$$T_1 \cos 30° - 0.6527\, T_1 \cos 50° = 20$$

$$\therefore \quad T_1 = \frac{20}{\cos 30° - 0.6527 \cos 50°} = 44.80 \text{ kN}$$ **Ans.**

$$\therefore \quad T_2 = 0.6527\, T_1 = 29.24 \text{ kN}$$ **Ans.**

From the equilibrium condition of point *C*, we get

$$-T_2 \sin 50°\mathbf{i} + T_2 \cos 50°\mathbf{j} + T_3 \sin \theta\mathbf{i} + T_3 \cos \theta\mathbf{j} - 30\mathbf{j} = 0$$

*i.e.*, $$(-T_2 \sin 50° + T_3 \sin \theta)\mathbf{i} + (T_2 \cos 50° + T_3 \cos \theta - 30)\mathbf{j} = 0$$

*i.e.*, $$-T_2 \sin 50° + T_3 \sin \theta = 0 \quad ...(iii)$$

and $$T_2 \cos 50° + T_3 \cos \theta = 30 \quad ...(iv)$$

From Eqn. (*iii*), $$T_3 \sin \theta = T_2 \sin 50° = 29.24 \sin 50° = 22.4 \quad ...(v)$$

From Eqn. (*iv*),
$$T_3 \cos\theta = 30 - T_2 \cos 50°$$
$$= 30 - 29.24 \cos 50°$$
$$= 11.21 \qquad \ldots(vi)$$

From Eqns. (*v*) and (*vi*), we get

$$\tan\theta = \frac{22.4}{11.21}$$

$\therefore$ $$\theta = 63.41° \qquad \textbf{Ans.}$$

and hence from Eqn. (*v*), $$T_3 = \frac{22.4}{\sin 63.41°} = 25.05 \text{ kN} \qquad \textbf{Ans.}$$

**Example 3.8.** A wire is fixed at two points *A* and *D* as shown in Fig. 3.18(*a*). Two weights 20 kN and 25 kN are suspended at *B* and *C* respectively. When equilibrium is reached, it is found that inclination of *AB* is 30° and that of *CD* is 60° to the vertical. Determine the tension in the segment *AB*, *BC* and *CD* of the rope and also the inclination of *BC* to the vertical.

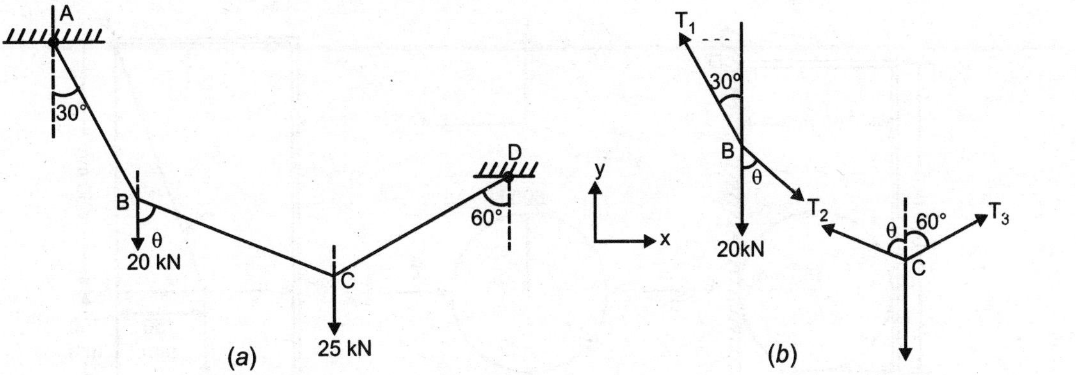


**Fig. 3.18**

**Solution.** Let $\mathbf{T}_1$, $\mathbf{T}_2$ and $\mathbf{T}_3$ be the forces in the segments *AB*, *BC* and *CD*. Let θ be the inclination of *BC* to vertical. Figure 3.18(*b*) shows FBD for points *B* and *C*.

From the equilibrium condition at *B*, we have

$$-T_1 \sin 30°\mathbf{i} + T_1 \cos 30°\mathbf{j} + T_2 \sin\theta\mathbf{i} - T_2 \cos\theta\mathbf{j} - 20\mathbf{j} = 0$$

*i.e.*, $$(-T_1 \sin 30° + T_2 \sin\theta)\mathbf{i} + (T_1 \cos 30° - T_2 \cos\theta - 20)\mathbf{j} = 0$$

*i.e.*, $$-T_1 \sin 30° + T_2 \sin\theta = 0 \qquad \ldots(i)$$

and $$T_1 \cos 30° - T_2 \cos\theta - 20 = 0 \qquad \ldots(ii)$$

From the equilibrium conditions at *C*, we get

$$-T_2 \sin\theta\mathbf{i} + T_2 \cos\theta\mathbf{j} + T_3 \sin 60°\mathbf{i} + T_3 \cos 60°\mathbf{j} - 25\mathbf{j} = 0$$

*i.e.*, $$(-T_2 \sin\theta + T_3 \sin 60°)\mathbf{i} + (T_2 \cos\theta + T_3 \cos 60° - 25)\mathbf{j} = 0$$

or $$-T_2 \sin\theta + T_3 \sin 60° = 0 \qquad \ldots(iii)$$

and $$T_2 \cos\theta + T_3 \cos 60° - 25 = 0 \qquad \ldots(iv)$$

From Eqns. (*i*) and (*iii*), we get

$$-T_1 \sin 30° + T_3 \sin 60° = 0$$

*i.e.*, $$T_1 = \frac{\sin 60°}{\sin 30°} T_3 = 1.732\, T_3 \qquad \ldots(v)$$

From Eqns. (*ii*) and (*iv*)

$$T_1 \cos 30° + T_3 \cos 60° - 45 = 0$$

Subsituting the value of $T_1$ from (*v*), we get

$$1.732\ T_3 \cos 30° + T_3 \cos 60° = 45$$

$$T_3\ [1.732 \cos 30° + \cos 60°] = 45$$

or $T_3 = 22.5$ kN **Ans.**

From (*v*), $T_1 = 1.732\ T_3 = 38.97$ kN **Ans.**

and from (*i*), $T_2 \sin \theta = T_1 \sin 30° = 19.485$ ...(*vi*)

and from Eqn. (*iv*), $T_2 \cos \theta = 25 - T_3 \cos 60°$

$$= 25 - 22.5 \cos 60° = 13.75 \quad ...(vii)$$

From Eqns. (*vi*) and (*vii*), we get

$$\tan \theta = \frac{19.485}{13.75} \quad \therefore \quad \theta = 54.78° \quad \textbf{(Ans.)}$$

From Eqn. (*vi*), $T_2 = 23.85$ kN **Ans.**

**Example 3.9.** A 600 N cylinder is supported by the frame *BCD* as shown in Fig. 3.19(*a*). The frame is hinged at *D*. Determine the reactions at *A*, *B*, *C*, and *D*.

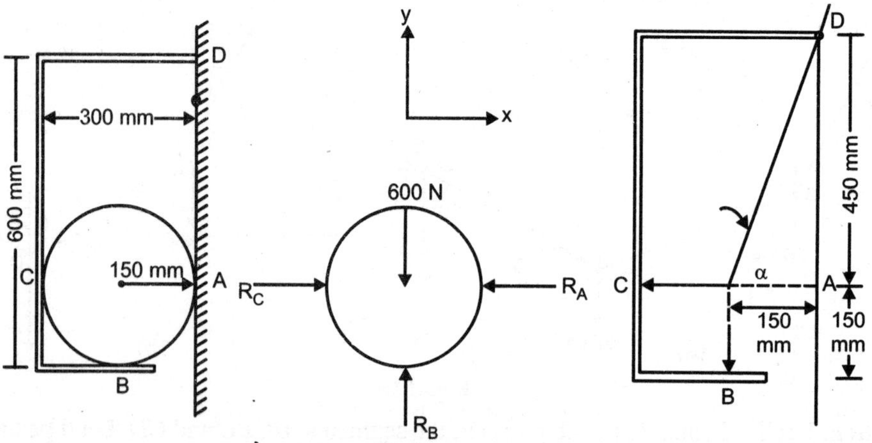


**Fig. 3.19**

**Solution.** Free body diagrams of sphere and the frame are shown in Fig. 3.19(*b*). The coordinates selected are also indicated.

From the equilibrium condition of the sphere, we have

$$-R_A\mathbf{i} + R_B\mathbf{j} + R_C\mathbf{i} - 600\mathbf{j} = 0$$

*i.e.*, $(-R_A + R_C)\mathbf{i} + (R_B - 600)\mathbf{j} = 0$

*i.e.*, $-R_A + R_C = 0$ and $R_B - 600 = 0$

$\therefore$ $R_A = R_C$ ...(*i*)

and $R_B = 600$ N **Ans.**

Consider the FBD of the frame. Since the frame is in equilibrium under the action of three forces only, they must be concurrent. In other words, the reaction at *D* is in the direction *OD*. Hence its inclination to the horizontal is given by,

$$\tan \alpha = \frac{450}{150} = 3 \quad \text{or} \quad \alpha = 71.565°$$

$\therefore$ The equation of equilibrium is,

$$R_D \cos 71.565°\mathbf{i} + R_D \sin 71.565°\mathbf{j} - R_B\mathbf{j} - R_C\mathbf{i} = 0 \quad ...(ii)$$

i.e., $(R_D \cos 71.565° - R_C)\mathbf{i} + (R_D \sin 71.565° - R_B)\mathbf{j} = 0$

i.e., $R_D \cos 71.565° - R_C = 0$ ...(iii)

and $R_D \sin 71.569° - R_B = 0$ ...(iv)

From Eqn. (iv), $R_D = \dfrac{R_B}{\sin 71.565°} = \dfrac{600}{\sin 71.565°} = 632.46 \text{ N}$ **Ans.**

∴ From Eqn. (iii), $R_C = R_D \cos 71.565° = 632.46 \cos 71.565°$

$= 200 \text{ N}$ **Ans.**

**Example 3.10.** Two identical rollers, each weighing 100 N, are supported by an inclined plane and a vertical wall as shown in Fig. 3.20(*a*). Assuming all contact surfaces are smooth, find the reactions developed at the contact surfaces *A*, *B*, *C*, and *D*.

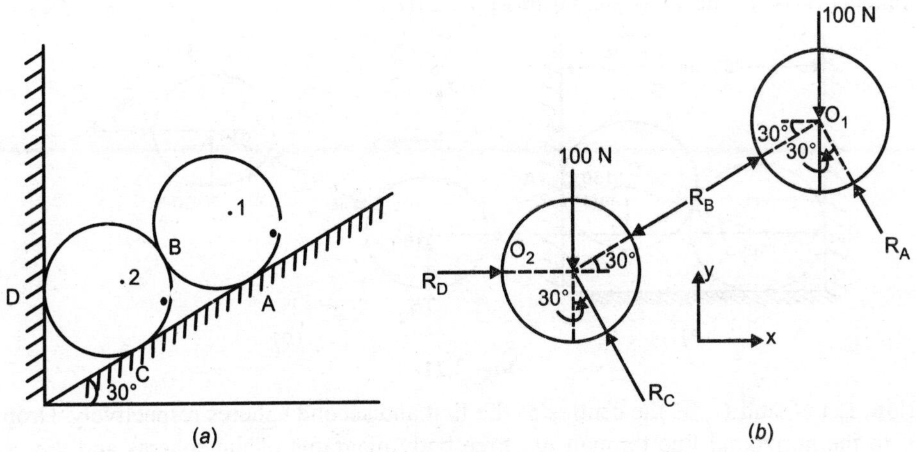


**Fig. 3.20**

**Solution.** Free body diagrams of the two rollers and cartesian coordinate system used for the analysis are shown in Fig. 3.20(*b*). It may be noted that $\mathbf{R_A}$ is normal to the plane and hence at 30° to vertical; $\mathbf{R_B}$ is parallel to the plane. From the equilibrium condition of cylinder 1, we have

$$R_B \cos 30°\mathbf{i} + R_B \sin 30°\mathbf{j} - R_A \sin 30°\mathbf{i} + R_A \cos 30°\mathbf{j} - 100\mathbf{j} = 0$$

$$(R_B \cos 30° - R_A \sin 30°)\mathbf{i} + (R_B \sin 30° + R_A \cos 30° - 100)\mathbf{j} = 0$$

i.e., $R_B \cos 30° - R_A \sin 30° = 0$ ...(i)

and $R_B \sin 30° + R_A \cos 30° - 100 = 0$ ...(ii)

From (i), $R_A = R_B \cot 30°$

Substituting it in (ii), we get

$$R_B \sin 30° + R_B \cot 30° \cos 30° = 100$$

$$R_B \frac{\sin^2 30° + \cos^2 30°}{\sin 30°} = 100$$

$R_B = 50 \text{ N}$ **Ans.**

∴ $R_A = R_B \cot 30° = 86.67 \text{ N}$ **Ans.**

From the equilibrium condition of cylinder 2,

$$-R_B \cos 30°\mathbf{i} - R_B \sin 30°\mathbf{j} - 100\mathbf{j} + R_D\mathbf{i} - R_C \sin 30°\mathbf{i} + R_C \cos 30°\mathbf{j} = 0$$

i.e., $(-R_B \cos 30° + R_D - R_C \sin 30°)\mathbf{i} + (-R_B \sin 30° - 100 + R_C \cos 30°)\mathbf{j} = 0$

i.e., $-R_B \cos 30° + R_D - R_C \sin 30° = 0$ ...(iii)

and $$-R_B \sin 30° + R_C \cos 30° - 100 = 0 \quad \ldots(iv)$$

Substituting the value of $R_B$ in Eqn. (*iv*), we get

$$-50 \sin 30° + R_C \cos 30° = 100$$

$$\therefore \quad R_C = \frac{100 + 50\sin 30°}{\cos 30°} = 144.34 \text{ N} \quad \textbf{Ans.}$$

Substituting it in Eqn. (*iii*), we get

$$-50 \cos 30° + R_D - 144.34 \sin 30° = 0$$

$$R_D = 144.34 \sin 30° + 50 \cos 30°$$
$$= 115.47 \text{ N} \quad \textbf{Ans.}$$

**Example 3.11.** Two smooth spheres each of radius 100 mm and weight 100 N rest in a horizontal channel having vertical walls, the distance between the walls being 360 mm. Find the reactions at the points of contacts *A*, *B*, *C* and *D* as shown in Fig. 3.21(*a*).

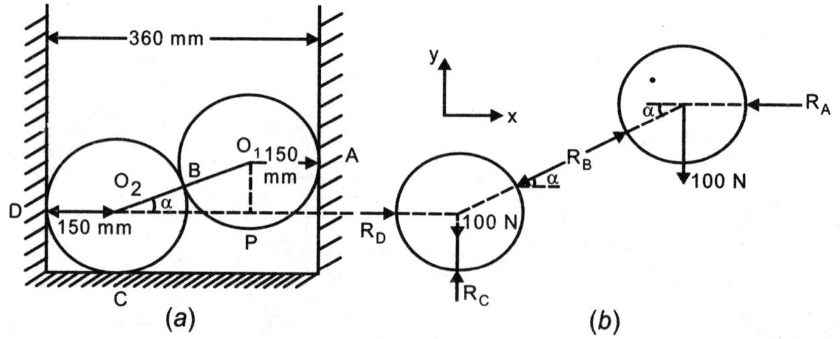


**Fig. 3.21**

**Solution.** Let $O_1$ and $O_2$ be the centres of the first and second spheres respectively. Drop perpendicular $O_1P$ to the horizontal line through $O_2$. Free body diagrams of the spheres and the coordinate directions are shown in Fig. 3.21(*b*). Since the surfaces of contact are smooth, reaction at *B* is at right angle to tangent at *B*, *i.e.*, it is in the radical direction. Let it make angle α with horizontal. Then,

$$\cos \alpha = \frac{O_2P}{O_1O_2} = \frac{360 - O_1A - O_1D}{O_1B + O_2B} = \frac{360 - 100 - 100}{100 + 100}$$
$$= 0.8$$
$$\alpha = 36.87°$$

Equation of equilibrium for first sphere is,

$$-R_A\mathbf{i} - 100\mathbf{j} + R_B \cos 36.87°\mathbf{i} + R_B \sin 36.87°\mathbf{j} = 0$$

*i.e.*, $$(-R_A + R_B \cos 36.87°)\mathbf{i} + (R_B \sin 36.87° - 100)\mathbf{j} = 0$$

*i.e.*, $$-R_A + R_B \cos 36.87° = 0 \quad \ldots(i)$$

and $$R_B \sin 36.87° - 100 = 0 \quad \ldots(ii)$$

From Eqn. (*ii*), $$R_B = \frac{100}{\sin 36.87°} = 166.67 \text{ N} \quad \textbf{Ans.}$$

From Eqn. (*i*), $$R_A = R_B \cos 36.87° = 166.67 \cos 36.87°$$
$$= 133.33 \text{ N} \quad \textbf{Ans.}$$

Consider the equilibrium of sphere 2.

$$-R_B \cos \alpha\mathbf{i} - R_B \sin \alpha\mathbf{j} + R_D\mathbf{i} - 100\mathbf{j} + R_C\mathbf{j} = 0$$

*i.e.*, $$(-R_B \cos 36.87° + R_D)\mathbf{i} + (-R_B \sin \alpha - 100R_C)\mathbf{j} = 0$$

*i.e.*, $$R_D = R_B \cos 36.87° \quad \ldots(iii)$$

and $$-R_B \sin 36.87° + R_C = 100 \quad ...(iv)$$

From (*iii*), $R_D = 166.67 \cos 36.87° = 133.33$ kN **Ans.**

From (*iv*), $R_C = 100 + 166.67 \sin 36.87°$

$= 200$ N **Ans.**

## 3.4 EQUILIBRIUM OF CONCURRENT FORCE SYSTEM IN SPACE

As explained in Art 3.1, when a body is in equilibrium, the resultant of the force system must be zero. Mathematically,

$$\mathbf{R} = 0$$

*i.e.*,
$$\mathbf{F}_1 + \mathbf{F}_2 + \mathbf{F}_3 + .... = 0$$

$$F_{1x}\mathbf{i} + F_{1y}\mathbf{j} + F_{1z}\mathbf{k} + F_{2x}\mathbf{i} + F_{2y}\mathbf{j} + F_{2z}\mathbf{k} + .... = 0$$

$$\Sigma F_x\mathbf{i} + \Sigma F_y\mathbf{j} + \Sigma F_z\mathbf{k} = 0$$

*i.e.*,
$$\Sigma F_x = 0 \quad ...[3.7(a)]$$

$$\Sigma F_y = 0 \quad ...[3.7(b)]$$

$$\Sigma F_z = 0 \quad ...[3.7(c)]$$

It is to be noted that while writing the equations of equilibrium, applied forces, self weight and all reactions acting on the body should be considered. However in some problems, the self weight may be negligible compared to the other forces. The use of above equations is illustrated by solving few examples below.

**Example 3.12.** A tripod is resting with its legs on a horizontal plane at points *A*, *B* and *C* as shown in Fig. 3.22. Its apex point *D* is 4 m above the floor level and carries a vertically downward load of 12 kN. Determine the forces developed in the legs.

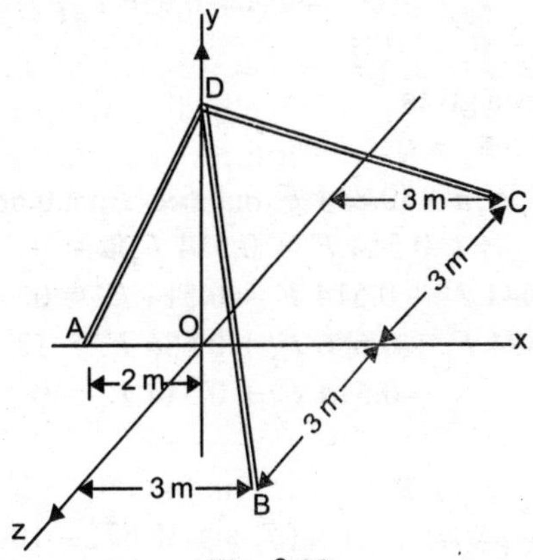


**Fig. 3.22**

**Solution.** Let the forces developed in the members *AD*, *BD* and *CD* be $\mathbf{F}_1$, $\mathbf{F}_2$ and $\mathbf{F}_3$ respectively. The point *D* is in equilibrium under the action of $\mathbf{F}_1$, $\mathbf{F}_2$, $\mathbf{F}_2$ and 12 kN load.

Taking *O*, the projection of open point *D* as origin and the coordinate system as shown in Fig. 3.22, we have

$$A(-2, 0, 0),\ B(3, 0, 3),\ C(3, 0, -3) \text{ and } D(0, 4, 0)$$

$\therefore$
$$\mathbf{r_{AD}} = [0 - (-2)]\mathbf{i} + (4 - 0)\mathbf{j} + (0 - 0)\mathbf{k} = 2\mathbf{i} + 4\mathbf{j}$$

$$\mathbf{r_{BD}} = -3\mathbf{i} + 4\mathbf{j} - 3\mathbf{k}$$

$$\mathbf{r_{CD}} = -3\mathbf{i} + 4\mathbf{j} + 3\mathbf{k}$$

$\therefore$
$$r_{AD} = \sqrt{2^2 + 4^2} = 4.472 \text{ m}$$

$$r_{BD} = \sqrt{(-3)^2 + (4)^2 + (-3)^2} = 5.831 \text{ m}$$

$$r_{CD} = \sqrt{(-3)^2 + (4)^2 + (3)^2} = 5.831 \text{ m}$$

∴ Unit vector of *AD* is,

$$\mathbf{n}_1 = \frac{2}{4.472}\mathbf{i} + \frac{4}{4.472}\mathbf{j} + 0\mathbf{k}$$
$$= 0.447\mathbf{i} + 0.894\mathbf{j}$$

Unit vector of *BD* is,

$$\mathbf{n}_2 = \frac{-3}{5.831}\mathbf{i} + \frac{4}{5.831}\mathbf{j} - \frac{3}{5.831}\mathbf{k}$$
$$= -0.514\mathbf{i} + 0.686\mathbf{j} - 0.514\mathbf{k}$$

and that of *CD* is,

$$\mathbf{n}_3 = \frac{-3}{5.831}\mathbf{i} + \frac{4}{5.831}\mathbf{j} + \frac{3}{5.831}\mathbf{k}$$
$$= -0.514\mathbf{i} + 0.686\mathbf{j} + 0.514\mathbf{k}$$

∴ Vector form of the forces in the legs are

$$\mathbf{F_1} = 0.447\ F_1\mathbf{i} + 0.894\ F_1\mathbf{j}$$
$$\mathbf{F_2} = -0.914\ F_2\mathbf{i} + 0.686\ F_2\mathbf{j} - 0.514\ F_2\mathbf{k}$$
$$\mathbf{F_3} = -\ 0.514\ F_3\mathbf{i} + 0.686\ F_3\mathbf{j} + 0.514\ F_3\mathbf{k}$$

The load is $\mathbf{F_4} = -12\mathbf{j}$

Hence the equilibrium condition gives

$$\mathbf{F_1} + \mathbf{F_2} + \mathbf{F_3} + \mathbf{F_4} = 0$$

$$(0.441\ F_1 - 0.514\ F_2 - 0.514\ F_3)\mathbf{i} + (0.894\ F_1 + 0.686\ F_2 + 0.686\ F_3 - 12)\mathbf{j} + (-0.514\ F_2 + 0.514\ F_3)\mathbf{k} = 0$$

*i.e.*,

$$0.441\ F_1 - 0.514\ F_2 - 0.514\ F_3 = 0 \qquad ...(i)$$
$$0.894\ F_1 + 0.686\ F_2 + 0.686\ F_3 = 12 \qquad ...(ii)$$
$$-0.514\ F_2 + 0.514\ F_3 = 0 \qquad ...(iii)$$

From Eqn. (*iii*),

$$F_2 = F_3 \qquad ...(iv)$$

Substituting it in Eqn. (*i*), we get

$$0.441\ F_1 - 2 \times 0.514\ F_2 = 0$$

or $F_1 = 2.331\ F_2$ ...(*v*)

Substituting for $F_1$ and $F_3$ in terms of $F_2$ in Eqn. (*ii*), we get

$$0.894 \times 2.331\ F_2 + 0.686\ F_2 + 0.686\ F_2 = 12$$

*i.e.*, $F_2 = 3.484$ kN **Ans.**

∴ $F_1 = 2.331 \times 3.484 = 8.122$ kN **Ans.**

and $F_3 = 3.484$ kN **Ans.**

**Example 3.13.** A system of 3 cables *AB*, *AC* and *AD*, as shown in Fig. 3.23, is subjected to a force of 300 kN along the *x*-direction by using turn buckle *AE*. Find the forces developed in the cables.

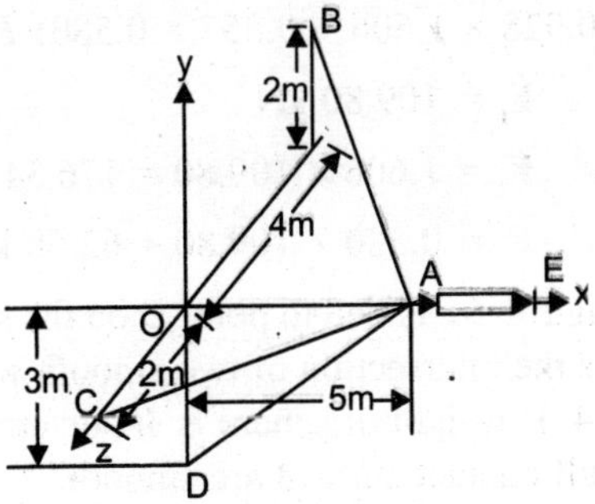


**Fig. 3.23**

**Solution.** The coordinates of points $A$, $B$, $C$ and $D$ are $A(5, 0, 0)$, $B(0, 2, -4)$, $C(0, 0, 2)$ and $D(0, -3, 0)$

$$\mathbf{r}_{AB} = -5\mathbf{i} + 2\mathbf{j} - 4\mathbf{k},\quad \mathbf{r}_{AC} = -5\mathbf{i} + 0\mathbf{j} + 2\mathbf{k},\quad \mathbf{r}_{AD} = -5\mathbf{i} - 3\mathbf{j} + 0\mathbf{k}$$

$\therefore$
$$r_{AB} = \sqrt{(-5)^2 + (2)^2 + (-4)^2} = 6.708 \text{ m}$$

$\therefore$ Unit vector in the direction of $AB$

$$\mathbf{n}_1 = -0.745\mathbf{i} + 0.298\mathbf{j} - 0.596\mathbf{k}$$

$\therefore$
$$\mathbf{F}_1 = F_1\mathbf{n}_1 = -0.745\, F_1\mathbf{i} + 0.298\, F_1\mathbf{j} - 0.596\, F_1\mathbf{k}$$

Similarly,
$$r_{AC} = \sqrt{(-5)^2 + (2)^2} = 5.385 \text{ m}$$

$$\mathbf{n}_2 = -0.921\mathbf{i} + 0\mathbf{j} + 0.371\mathbf{k}$$

$\therefore$
$$\mathbf{F}_2 = -0.928\, F_2\mathbf{i} + 0.371\, F_2\mathbf{k}$$

and
$$r_{AD} = \sqrt{(-5)^2 + (-3)^2 + 0} = 5.83 \text{ m}$$

$\therefore$
$$n_3 = -0.857\mathbf{i} - 0.514\mathbf{j}$$

and
$$\mathbf{F}_3 = (-0.857\mathbf{i} - 0.514\mathbf{j})F_3$$
$$\mathbf{F}_3 = -0.857\, F_3\mathbf{i} - 0.514\, F_3\mathbf{j}$$

Let the force applied by turn buckle be $\mathbf{F}_4$. Then

$$\mathbf{F}_4 = 300\mathbf{i}$$

From the equilibrium condition, we have

$$\mathbf{F}_1 + \mathbf{F}_2 + \mathbf{F}_3 + \mathbf{F}_4 = 0$$

*i.e.,*
$$(-0.745\, F_1 - 0.928\, F_2 - 0.857 + 300)\mathbf{i} + (0.298\, F_1 - 0.514\, F_3)\mathbf{j} + (-0.596\, F_1 + 0.371\, F_2)\mathbf{k} = 0$$

*i.e.,*
$$-0.745\, F_1 - 0.928\, F_2 - 0.857\, F_3 + 300 = 0 \quad \ldots(i)$$
$$0.298\, F_1 - 0.514\, F_3 = 0 \quad \ldots(ii)$$
$$-0.596\, F_1 + 0.371\, F_2 = 0 \quad \ldots(iii)$$

From Eqn. (*iii*),
$$F_2 = \frac{0.596}{0.371} F_1 = 1.06\, F_1$$

From Eqn. (*ii*),
$$F_3 = \frac{0.298}{0.514} F_1 = 0.580\, F_1$$

$\therefore$ From Eqn. (*i*), $(0.745 + 0.928 \times 1.606 + 0.857 \times 0.580)\ F_1 = 300$

or $$\mathbf{F_1} = 109.80 \text{ kN}$$ **Ans.**

Hence $$\mathbf{F_2} = 1.606 \times 109.80 = 176.34 \text{ kN}$$ **Ans.**

and $$\mathbf{F_3} = 0.580 \times 109.80 = 63.68 \text{ kN}$$ **Ans.**

**Example 3.14.** A wire of length 1.5 *r* is tied to point *D* on the surface of a sphere of radius *r*. Then the wire is tied to point *C* which is the intersection of two smooth walls, the walls being at right angles to each other as shown in Fig. 3.24. If weight of sphere is *W*, determine the tension in the wire and the reactions from the wall. Assume all contact surface are smooth.

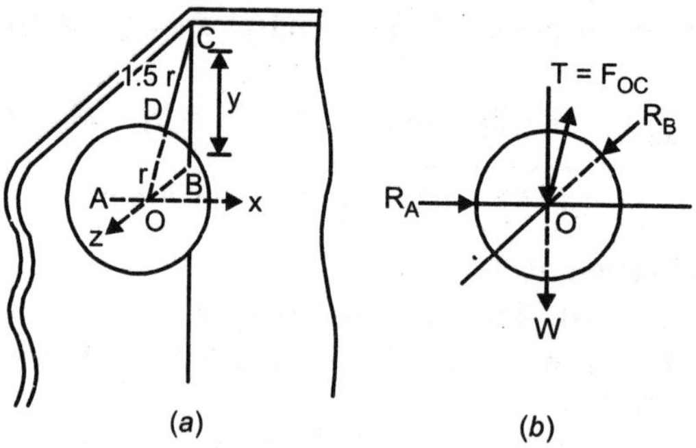


**Fig. 3.24**

**Solution.** Selecting the centre of sphere as origin and the coordinates system as shown in Fig. 3.24, the coordinates of the points of contact with walls are

$$A(-r, 0, 0),\ B(0, 0, -r) \text{ and } C(-r, y, -r)$$

Now $$OC = OD + r = 1.5r + r = 2.5r$$

$\therefore$ $$OC^2 = (r)^2 + y^2 + (-r)^2$$

$$(2.5r)^2 = r^2 + y^2 + r^2$$

$\therefore$ $$y^2 = 4.25r^2$$

$$y = 2.062\ r$$

$\therefore$ The coordinates of *C* are $(-r, 2.06r, -r)$.

Figure 3.24(*b*) shows FBD of the sphere, which is in equilibrium under the action of reaction $\mathbf{R_A} = F_{AO}$ and $\mathbf{R_B} = F_{BO}$ from walls, tension $\mathbf{T} = F_{OC}$ in wire and self weight $\mathbf{W}$.

$$\mathbf{r_{AO}} = r\mathbf{i} + 0\mathbf{j} + 0\mathbf{k}$$

$$\mathbf{r_{AO}} = r\mathbf{i}$$

$$\mathbf{n_{AO}} = \mathbf{i} + 0\mathbf{j} + 0\mathbf{k}$$

$\therefore$ $$\mathbf{F_{AO}} = F_{AO}\ (\mathbf{i} + 0\mathbf{j} + 0\mathbf{k})$$

$$\mathbf{r_{BO}} = 0\mathbf{i} + 0\mathbf{j} + r\mathbf{k}$$

$$\mathbf{r_{BO}} = r\mathbf{k}$$

$$\mathbf{n_{BO}} = 0\mathbf{i} + 0\mathbf{j} + \mathbf{k}$$

$\therefore$ $$\mathbf{F_{BO}} = F_{BO}\ (0\mathbf{i} + 0\mathbf{j} + \mathbf{k})$$

$$\mathbf{r_{OC}} = -r\mathbf{i} + 2.062\mathbf{j} - r\mathbf{k}$$

$$r_{OC} = \sqrt{(-r)^2 + (2.062r)^2 + (-r)^2} = 2.5r$$

$$\mathbf{r_{OC}} = -0.4\mathbf{i} + 0.825\mathbf{j} - 0.4\mathbf{k}$$

$$\mathbf{F_{OC}} = F_{OC}\,(-0.4\mathbf{i} + 0.825\mathbf{j} - 0.4\mathbf{k})$$
$$\mathbf{W} = W\mathbf{j}$$

∴ The equilibrium equation is,

$$\mathbf{F_{AO}} + \mathbf{F_{BO}} + \mathbf{F_{OC}} + \mathbf{W} = 0$$

*i.e.,* $(F_{AO} + 0 - 0.4\,F_{OC})\mathbf{i} + (0 + 0 + 0.825\,F_{OC} - W)\mathbf{j} + (0 + F_{BA} - 0.4F_{OC})\,\mathbf{k} = 0$

*i.e.,* $F_{AO} - 0.4\,F_{OC} = 0$ ...(*i*)

$0.825\,F_{OC} - W = 0$ ...(*ii*)

$F_{BA} - 0.4\,F_{OC} = 0$ ...(*iii*)

From Eqn. (*ii*) $F_{OC} = \dfrac{W}{0.825} = 1.212\,W$ **Ans.**

From Eqn. (*i*), $F_{BO} = 0.4\,F_{OC} = 0.485\,W$ **Ans.**

From Eqn. (*iii*), $F_{BA} = 0.4\,F_{OC} = 0.485\,W$ **Ans.**

**Example 3.15**. The system of cables supporting 100 kN load at *C* is shown in Fig. 3.25 (*a*). The coordinates of point *C* with respect to '*O*' as origin are (0.2, –0.3, 0). The points *C*, *D*, *E* and *F* are in *x-y* plane. Determine the forces in various segments of cables.

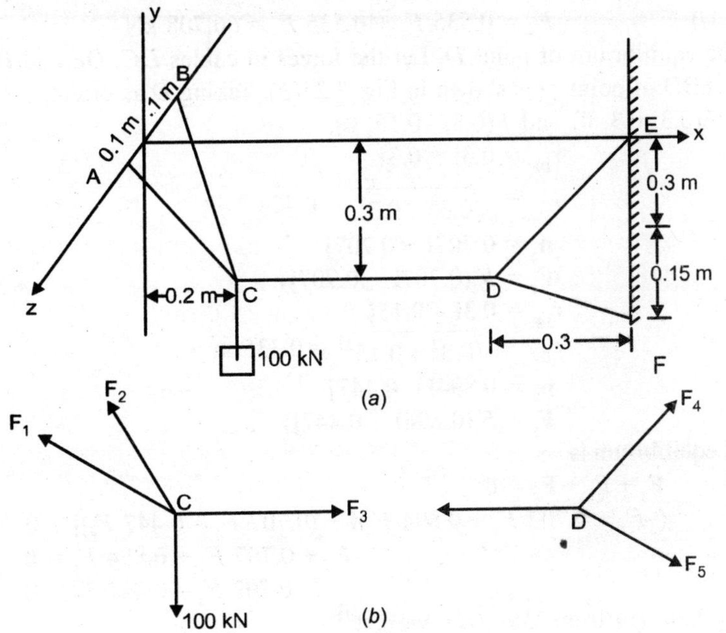


**Fig. 3.25**

**Solution.** Consider the equilibrium condition at point *C*. Let the tensile forces developed in cables *CA*, *CB* and *CD* be $\mathbf{F}_1$, $\mathbf{F}_2$, and $\mathbf{F}_3$ respectively. The FBD of point *C* is shown in Fig. 3.25(*b*).

Now, *A*(0, 0, 0.1), *B*(0, 0, –0.1) and *C*(0.2, –0.3, 0)

$$\mathbf{r_{CA}} = -0.2\mathbf{i} + 0.3\mathbf{j} + 0.1\mathbf{k}$$
$$r_{CA} = \sqrt{0.2^2 + 0.3^2 + 0.1^2} = 0.374 \text{ m}$$
$$\mathbf{n}_1 = -0.535\mathbf{i} + 0.802\mathbf{j} + 0.267\mathbf{k}$$
$$\mathbf{F}_1 = F_1(-0.535\mathbf{i} + 0.802\mathbf{j} + 0.267\mathbf{k})$$
$$\mathbf{r}_{CB} = -0.2\mathbf{i} + 0.3\mathbf{j} - 0.1\mathbf{k}$$
$$r_{CB} = \sqrt{0.2^2 + 0.3^2 + 0.1^2} = 0.374 \text{ m}$$

$$\mathbf{n}_2 = -0.535\mathbf{i} + 0.802\mathbf{j} - 0.267\mathbf{k}$$

$$\therefore \quad \mathbf{F}_2 = F_2(-0.537\mathbf{i} + 0.802\mathbf{j} - 0.267\mathbf{k})$$

$$\mathbf{F}_3 = F_3\mathbf{i} \text{ since it is in } x\text{-direction}$$

$$\mathbf{F}_4 = -100\mathbf{j}$$

∴ Equilibrium equation is

$$\mathbf{F}_1 + \mathbf{F}_2 + \mathbf{F}_3 + \mathbf{F}_4 = 0$$

$$(-0.535\,F_1 - 0.535\,F_2 + F_3)\mathbf{i} + (0.802\,F_1 + 0.802\,F_2 - 100)\mathbf{j} + (0.267\,F_1 - 0.267\,F_2)\mathbf{k} = 0$$

*i.e.,*

$$-0.535\,F_1 - 0.535\,F_2 + F_3 = 0 \qquad \ldots(i)$$

$$0.802\,F_1 + 0.802\,F_2 = 100 \qquad \ldots(ii)$$

$$0.267\,F_1 - 0.267\,F_2 = 0 \qquad \ldots(iii)$$

From Eqn. (*iii*) $\quad F_1 = F_2$

Substituting it in Eqn. (*ii*), we get

$$(0.802 + 0.802)\,F_1 = 100$$

$$F_1 = 62.344 \text{ kN} \qquad \textbf{Ans.}$$

$$F_2 = 62.344 \text{ kN} \qquad \textbf{Ans.}$$

From Eqn. (*i*) $\quad F_3 = 0.535\,F_1 + 0.535\,F_2 = 66.708 \text{ kN}$ **Ans.**

Consider the equilibrium of point *D*. Let the forces in cables *DC*, *DE* and *DF* be $\mathbf{F}_3$, $\mathbf{F}_4$ and $\mathbf{F}_5$ respectively. The FBD of point *D* is shown in Fig. 3.25(*b*). Taking *D* as origin,

*D*(0, 0, 0), *E*(0.3, 0.3, 0) and *F*(0.3, –0.15, 0),

$$\mathbf{r}_{DE} = 0.3\mathbf{i} + 0.3\mathbf{j}$$

$$r_{DE} = \sqrt{0.3^2 + 0.3^2} = 0.424$$

$$\mathbf{n}_4 = 0.707\mathbf{i} + 0.707\mathbf{j}$$

$$\mathbf{F}_4 = F_4(0.707\mathbf{i} + 0.707\mathbf{j})$$

$$\mathbf{r}_{DE} = 0.3\mathbf{i} - 0.15\mathbf{j}$$

$$r_{DE} = \sqrt{0.3^2 + 0.15^2} = 0.335 \text{ m}$$

$$\therefore \quad \mathbf{n}_4 = 0.894\mathbf{i} - 0.447\mathbf{j}$$

$$\mathbf{F}_5 = F_5(0.894\mathbf{i} - 0.447\mathbf{j})$$

Equation of equilibrium is

$$\mathbf{F}_3 + \mathbf{F}_4 + \mathbf{F}_5 = 0$$

$$(-F_3 + 0.707\,F_4 + 0.894\,F_5)\mathbf{i} + (0.707\,F_4 - 0.447\,F_5)\mathbf{j} = 0$$

*i.e.,*

$$-F_3 + 0.707\,F_4 + 0.894\,F_5 = 0 \qquad \ldots(iv)$$

$$0.707\,F_4 - 0.447\,F_5 = 0 \qquad \ldots(v)$$

Substracting Eqn. (*v*) from Eqn. (*iv*), we get

$$-F_3 + (0.894 + 0.447)\,F_5 = 0$$

$$F_5 = \frac{F_3}{1.347} = \frac{66.708}{1.347} = 49.523 \text{ kN} \qquad \textbf{Ans.}$$

From Eqn. (*v*), $\quad F_4 = \dfrac{0.447}{0.707}F_5 = 31.311 \text{ kN}$ **Ans.**

## 3.5 EQUILIBRIUM OF CO-PLANAR NON-CONCURRENT SYSTEMS

If a rigid body subjected to coplanar non-concurrent system is in equilibrium, it means that the resultant force and the moment system are zero. Mathematically,

$$\mathbf{R} = 0 \text{ and } \mathbf{M} = 0$$

Consider the equation

$$\mathbf{R} = 0$$

$$\mathbf{F_1} + \mathbf{F_2} + \ldots + \mathbf{F_N} = 0$$

*i.e.*, $$F_{1x}\mathbf{i} + F_{1y}\mathbf{j} + F_{2x}\mathbf{i} + F_{2y}\mathbf{j} + \ldots + F_{xx}\mathbf{i} + F_{xy}\mathbf{j} = 0$$

*i.e.*, $$\Sigma F_x \mathbf{i} + \Sigma F_y \mathbf{j} = 0 \quad \ldots(i)$$

Similarly $\mathbf{M} = 0$ means

$$\Sigma M_x \mathbf{i} + \Sigma M_y \mathbf{j} = 0 \quad \ldots(ii)$$

In scalar form, Eqns. (*i*) and (*ii*) mean,

$$\Sigma F_x = 0, \ \Sigma F_y = 0$$

and $$\Sigma M_z = 0$$

Noting $M_z$ as $M$ only, which is normally looked as a moment about a point in *x-y* plane, the equations of equilibrium reduce to,

$$\Sigma F_x = 0, \Sigma F_y = 0 \text{ and } \Sigma M = 0 \quad \ldots(3.8)$$

While dealing with the equilibrium of bodies under concurrent force system, the statement was made that *if a system is in equilibrium under the action of only three forces, those three forces must be concurrent*, Now that statement can be proved.

Let $\mathbf{F_1}$, $\mathbf{F_2}$ and $\mathbf{F_3}$ be the three forces acting on a body as shown in Fig. 3.26. Let $\mathbf{F_1}$ and $\mathbf{F_2}$ intesect at *A*. Then from moment equilibrium condition, we get

$$\Sigma M_A = 0$$

*i.e.*, $$F_3 d = 0$$

where '*d*' is the distance of the line of action of $\mathbf{F_3}$ from *A*. Since $\mathbf{F_3}$ is not zero, '*d*' must be zero *i.e.*, $\mathbf{F_3}$ also pass through *A*. Hence the proposition is proved.

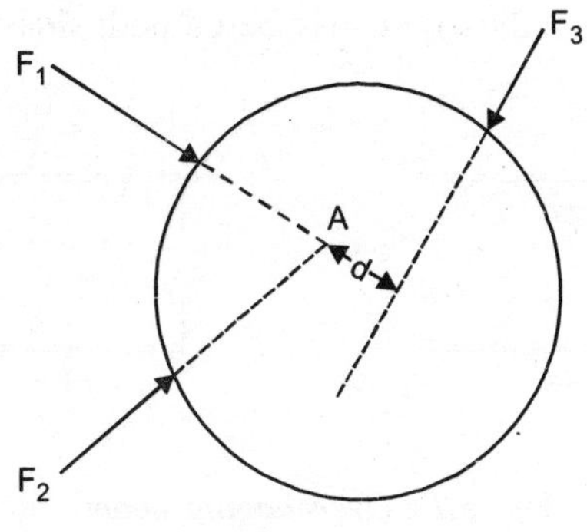


**Fig. 3.26**

## 3.6 BEAMS

Beams are common structural elements encountered in engineering which are subjected to non-concurrent system of forces. A beam may be defined as a structural element which has one dimension considerably large compared to the other two dimensions and having supports at few points. A beam is normally subjected to transverse loads and that load is transferred to supports. In this process, the beam bends. The types of beams and the reactions developed in such beams are discussed below:

### Simply Supported Beams

When both ends simply rest on supports, the beam is called simply supported beam. In this case, at both ends, reactions are at right angles to the support (Fig. 3.27). Such beams can support loads in the direction normal to its axis only.

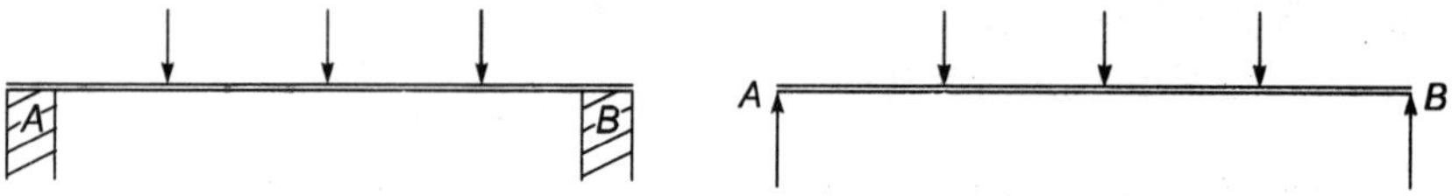


**Fig. 3.27** *Simply supported beam*

### Beam with One End Hinged and the Other on Rollers

If one end of the beam is hinged (pinned) and the other end is on rollers, the beams can resist the load in any direction (Fig. 3.28)

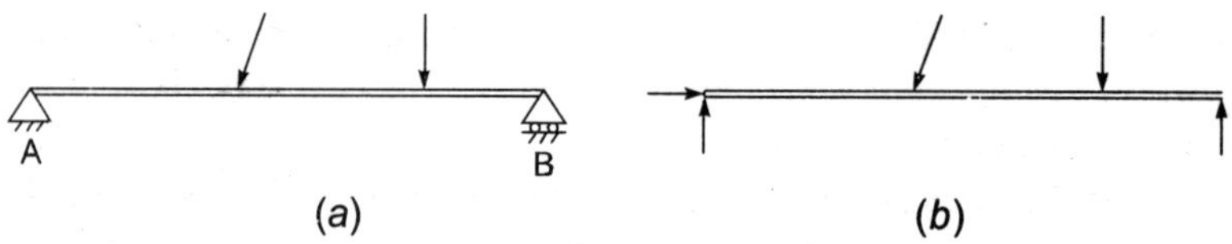


**Fig. 3.28** *Beam with one end hinged other on roller*

### Over-hanging Beams

If beams are having projection beyond the support, they are termed as overhanging beams. The overhang may be on only one side [Fig. 3.29(*a*)] or may be on both sides [Ref. Fig. 3.29(*b*)].

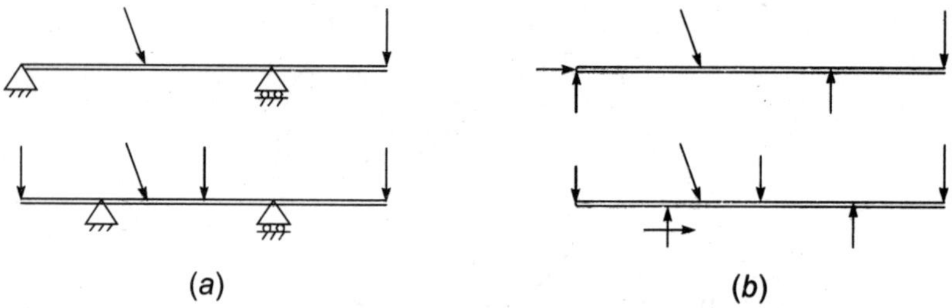


**Fig. 3.29** *Overhanging beams*

### Cantilever Beams

If the beam has fixed end at one side and a free end at the other side, it is called cantilever beam (Fig. 3.30). At fixed end, there are three reaction components and no reactions at free end.

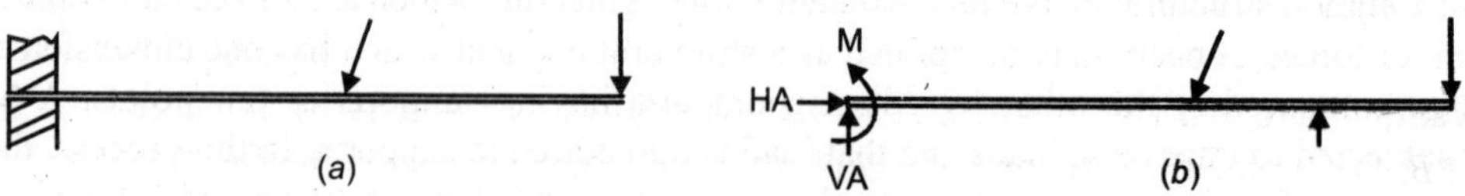


**Fig. 3.30** *Cantilever beam*

### Propped Cantilever

It is a beam with one end fixed and the other end simply supported. In this case, there are three reaction components at fixed end and one at simply supported end (Fig. 3.31).

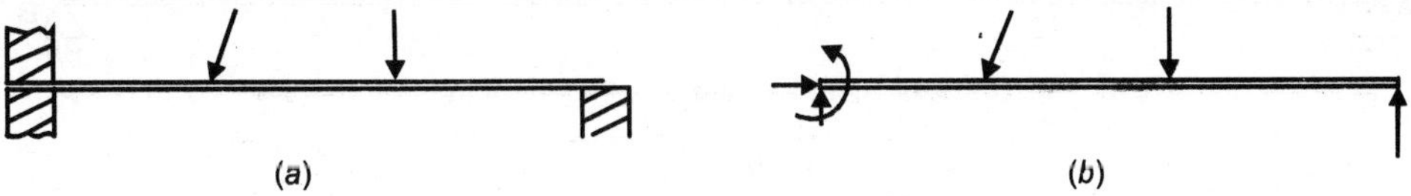


**Fig. 3.31** *Propped cantilever*

### Both Ends Hinged

In these beams, both ends are having hinged support. Hence total number of reaction components are four, two at each end (Fig. 3.32).

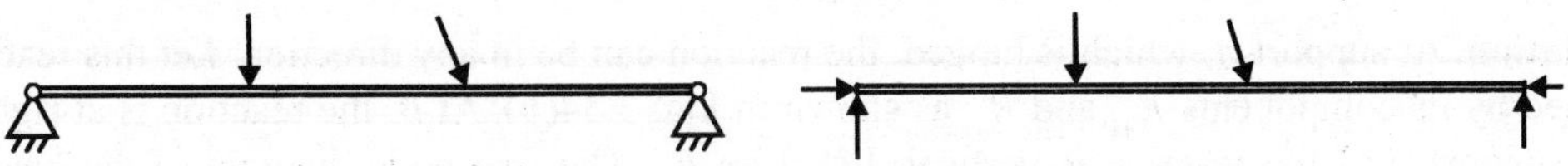

**Fig. 3.32** *Both ends hinged beam*

### Continuous Beam

A beam is said to be continuous, if it is supported at more than two points (Fig. 3.33).

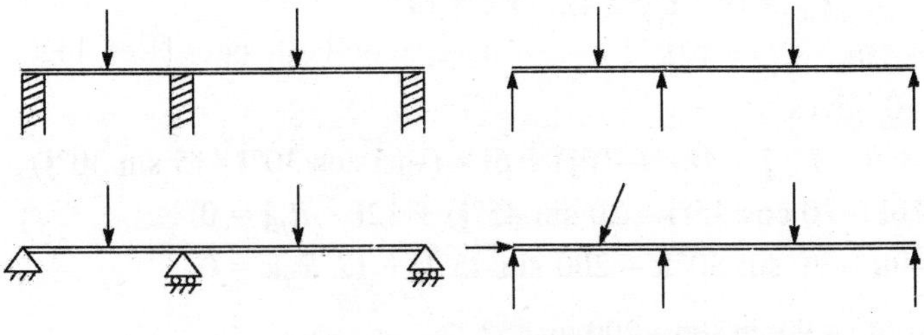

**Fig. 3.33** *Continuous beams*

In the cases of simply supported beams, beams with one end hinged and the other on rollers, cantilever and overhanging beams, it is possible to determine the reactions for given loadings by using the equations of equilibrium 3.8 only. In other cases, the number of equilibrium equations are less than the number of unkown reactions, and hence it is not possible to analyse them by using equations of equilibrium only. The beams which can be analysed using equations of equilibrium only are known as *Statically Determinate* beams and those which cannot be analysed are known as *Statically Indeterminate* beams. The anlaysis of statically indeterminate beams is beyond the scope of this book.

**Example 3.16.** The beam *AB* of span 12 m as shown in Fig. 3.34(*a*) is hinged at *A* and is on rollers at *B*. Determine the reactions of *A* and *B* for the loading shown in the figure.

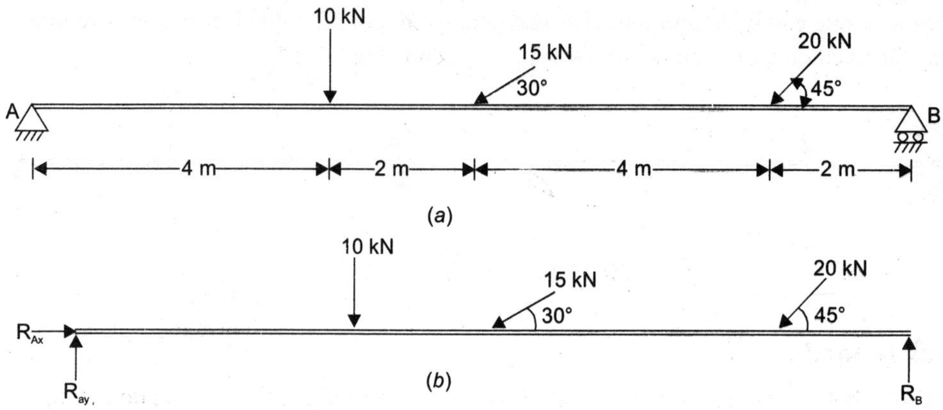


**Fig. 3.34**

**Solution.** At support *A,* which is hinged, the reaction can be in any direction. Let this reaction be represented by its components $R_{Ax}$ and $R_{Ay}$ as shown in Fig. 3.34(*b*). At *B,* the reaction is at right angle to roller support, *i.e.*, the reaction is vertical. Let it be $R_B$. The free body diagram of the beam is as shown in Fig. 3.34(*b*). Since this is a non-concurrent force system, the moment equilibrium condition should also be considered. The vector form of various forces are

$$\mathbf{F}_1 = R_{Ax}\mathbf{i}, \quad \mathbf{F}_2 = R_{Ay}\mathbf{j}, \mathbf{F}_3 = -10\mathbf{j}, \mathbf{F}_4 = -15 \cos 30°\mathbf{i} - 15 \sin 30°\mathbf{j},$$

$$\mathbf{F}_5 = -20 \cos 45°\mathbf{i} - 20 \sin 45°\mathbf{j}; F_6 = R_B\mathbf{j}$$

The position vectors of these forces from *A* are

$$\mathbf{r}_{A1} = 0, \quad \mathbf{r}_{A2} = 0, \quad \mathbf{r}_{a3} = 4\mathbf{i}, \quad \mathbf{r}_{a4} = 6\mathbf{i}$$

$$\mathbf{r}_{A5} = 10\mathbf{i}, \quad \mathbf{r}_{A6} = 12\mathbf{i}$$

$\therefore$ $\Sigma M_4 = 0$, gives

$$0 \times R_{Ax}\mathbf{i} + 0 \times F_{Ay}\mathbf{j} + 4\mathbf{i} \times (-10\mathbf{j}) + 6\mathbf{i} \times (-15 \cos 30°\mathbf{i} - 15 \sin 30°\mathbf{j})$$

$$+ 10\mathbf{i}\,(-20 \cos 45°\mathbf{i} - 20 \sin 45°\mathbf{j}) + 12\mathbf{i} \times R_B\mathbf{j} = 0$$

*i.e.*,
$$-40\mathbf{k} - 90 \sin 30°\mathbf{k} - 200 \sin 45°\mathbf{k} + 12\, R_B\mathbf{k} = 0$$

$\therefore$
$$R_B = \frac{40 + 90 \sin 30° + 200 \sin 45°}{12} = 18.868 \text{ kN}$$ **Ans.**

For equilibrium condition $\mathbf{R} = 0$, gives

$$R_{Ax}\mathbf{i} + R_{Ay}\mathbf{j} - 10\mathbf{j} - 15 \cos 30°\mathbf{i} - 15 \sin 30°\mathbf{j} - 20 \cos 45°\mathbf{i} - 20 \sin 45°\mathbf{j} + R_B\mathbf{j} = 0$$

*i.e.*,
$$(R_{Ax} - 15 \cos 30° - 20 \cos 45°)\mathbf{i} + (R_{Ay} - 10 - 15 \sin 30° - 20 \sin 45° + R_B)\mathbf{j} = 0$$

*i.e.*,
$$R_{Ax} - 15 \cos 30° - 20 \cos 45° = 0 \quad \text{...(i)}$$

and
$$R_{Ay} - 10 - 15 \sin 30° - 20 \sin 45° + R_B = 0 \quad \text{...(ii)}$$

From Eqn. (*i*), $R_{Ax} = 15 \cos 30° + 20 \cos 45° = 27.133$ kN **Ans.**

From Eqn. (*ii*),

$$R_{Ay} = 10 + 15 \sin 30° + 20 \sin 45° - R_B$$

$$= 10 + 15 \sin 30° + 20 \sin 45° - 18.868$$

$$= 12.774 \text{ kN}$$ **Ans.**

$$\therefore \qquad \mathbf{R_A} = \mathbf{R_{Ax}} + \mathbf{R_{Ay}} = 27.133\mathbf{i} + 12.774\mathbf{j}$$

$$R_A = \sqrt{27.133^2 + 12.774^2} = 29.989 \text{ kN}$$

$$\alpha_x = \cos^{-1}\frac{27.133}{29.984} = 25.21° \qquad \textbf{Ans.}$$

$$\alpha_y = \frac{12.774}{29.984} = 64.79°$$

**Example 3.17.** Determine the reactions developed at supports $A$ and $B$ in the beam as shown in Fig. 3.35($a$).

**Solution.** The reaction at $B$ will be at right angles to the support, *i.e.*, at 30° to the vertical as shown in Fig. 3.35($b$). Let the component of reaction at hinged end $A$ be $R_{Ax}$ and $R_{Ay}$. The FBD of beam is as shown in the figure.

Now
$$\mathbf{F}_1 = R_{Ax}\mathbf{i},\ \mathbf{F}_2 = R_{Ay}\mathbf{j},\ \mathbf{F}_3 = -60\mathbf{j}$$
$$\mathbf{F}_4 = -80 \cos 75°\mathbf{i} - 80 \sin 75°\mathbf{j},\ \mathbf{F}_5 = 40 \cos 60°\mathbf{i} - 40 \sin 60°\mathbf{j}$$
$$\mathbf{F}_6 = R_B \sin 30°\mathbf{i} + R_B \cos 30°\mathbf{j}$$
$$\mathbf{r}_{A1} = 0;\ \mathbf{r}_{A2} = 0,\ \mathbf{r}_{A3} = \mathbf{i},\ \mathbf{r}_{A4} = 3\mathbf{i}$$
$$\mathbf{r}_{A5} = 5.5\mathbf{i},\ \mathbf{r}_{A6} = 7\mathbf{i}$$

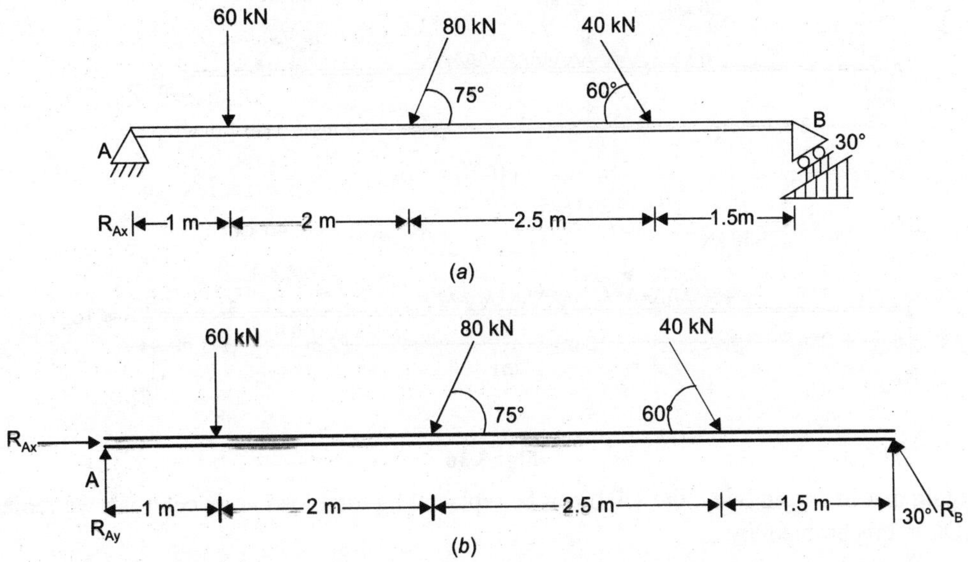


**Fig. 3.35**

$\Sigma M_A = 0$, gives

$$0 + 0 + \mathbf{i} \times (-60\mathbf{j}) + 3\mathbf{i} \times (-80 \cos 75°\mathbf{i} - 80 \sin 75°\mathbf{j})$$
$$+ 5.5\mathbf{i} \times (40 \cos 60°\mathbf{i} - 40 \sin 60°\mathbf{j}) + 7\mathbf{i} \times (-R_B \sin 30°\mathbf{i} + R_B \cos 30°\mathbf{j}) = 0$$
$$-60\mathbf{k} - 240 \sin 75°\mathbf{k} - 220 \sin 60°\mathbf{k} + 7 R_B \cos 30°\mathbf{k} = 0$$

*i.e.*,
$$R_B = \frac{60 + 240 \sin 75° + 220 + \sin 60°}{7 \cos 30°} = 79.567 \text{ kN} \qquad \textbf{Ans.}$$

$\mathbf{R} = 0$, gives

$$R_{Ax}\mathbf{i} + R_{Ay}\mathbf{j} - 60\mathbf{j} - 80 \cos 75°\mathbf{i} - 80 \sin 75°\mathbf{j} + 40 \cos 60°\mathbf{i} - 40 \sin 60°\mathbf{j} - R_B \sin 30°\mathbf{i} + R_B \cos 30°\mathbf{j} = 0$$

$$(R_{Ax} - 80 \cos 75° + 40 \cos 60° - R_B \sin 30°)\mathbf{i} + (R_{Ay} - 60 - 80 \sin 75° - 40 \sin 60° + R_B \cos 30°)\mathbf{j} = 0$$

*i.e.,*

$$R_{Ax} - 80 \cos 75° + 40 \cos 60° - R_B \sin 30° = 0 \quad ...(i)$$

and

$$R_{Ay} - 60 - 80 \sin 75° - 40 \sin 60° + R_B \cos 30° = 0 \quad ...(ii)$$

Substituting the value of $R_B$ in Eqn. (*i*), we get

$$R_{Ax} = 80 \cos 75° - 40 \cos 60° + 79.567 \sin 30°$$
$$= 40.489 \text{ kN}$$

**Ans.**

From Eqn. (*ii*),

$$R_{Ay} = 60 + 80 \sin 75° + 40 \sin 60° - 79.567 \cos 30°$$
$$= 103.008 \text{ kN}$$

**Ans.**

**Example 3.18.** Find the reactions at supports *A* and *B* of the beam as shown in Fig. 3.36(*a*).

**Solution.** The reaction at *A* is vertical. Let $R_{Bx}$ and $R_{By}$ be the horizontal and vertical components of the reactions at *B*.

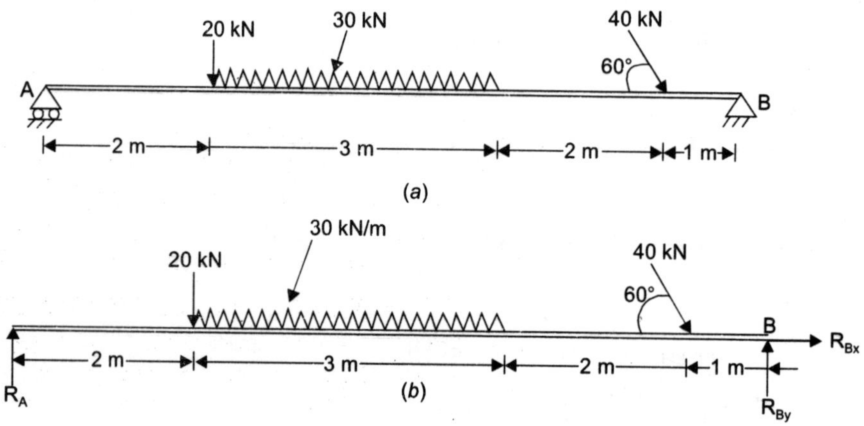


**Fig. 3.36**

Noting that in a rigid body, the UDL can be replaced by total load at its centroid, we replace the given UDL in this problem by

$$30 \times 3 = 90 \text{ kN load acting at } 2 + \frac{3}{2} = 3.5 \text{ m from } A.$$

The various forces in this system are shown in Fig. 3.36 (*b*).

$$\mathbf{F}_1 = R_A\mathbf{j},\ \mathbf{F}_2 = -20\mathbf{j},\ \mathbf{F}_3 = -30 \times 3\mathbf{j} = -90\mathbf{j},\ \mathbf{F}_4 = 40 \cos 60°\mathbf{i} - 40 \sin 60°\mathbf{j}$$

$$\mathbf{F}_5 = R_{Bx}\mathbf{i} \text{ and } \mathbf{F}_6 = R_{By}\mathbf{j},$$

$$\mathbf{r}_{B1} = -8\mathbf{i};\ \mathbf{r}_{B2} = -6\mathbf{i};\ \mathbf{r}_{B3} = (8 - 3.5)\mathbf{i} = -4.5\mathbf{i};\ \mathbf{r}_{B4} = -\mathbf{i};\ \mathbf{r}_{B5} = 0;\ \mathbf{r}_{B6} = 0$$

$\therefore \quad \Sigma M_B = 0$, gives

$$-8\mathbf{i} \times R_A\mathbf{j} - 6\mathbf{i} \times (-20\mathbf{j}) - 4.5\mathbf{i} \times (-90\mathbf{j}) - \mathbf{i} \times (40 \cos 60°\mathbf{i} - 40 \sin 60°\mathbf{j}) + 0 + 0 = 0$$

i.e.,
$$-8R_A\mathbf{k} + 120\mathbf{k} + 405\mathbf{k} + 40 \sin 60°\mathbf{k} = 0$$

$$\therefore \qquad R_A = \frac{120 + 405 + 40\sin 60°}{8} = 69.655 \text{ kN} \qquad \textbf{Ans.}$$

Resultant force of the system = 0, gives

$$\mathbf{R} = 0$$

$$R_A\mathbf{j} - 20\mathbf{j} - 90\mathbf{j} + 40 \cos 60°\mathbf{i} - 40 \sin 60°\mathbf{j} + R_{Bx}\mathbf{i} + R_{By}\mathbf{j} = 0$$

$$(40 \cos 60° + R_{Bx})\mathbf{i} + (R_A - 20 - 90 - 40 \sin 60° + R_{By})\mathbf{j} = 0$$

i.e.,
$$40 \cos 60° + R_{Bx} = 0 \qquad \ldots(i)$$

and
$$R_A - 20 - 90 - 40 \sin 60° + R_{By} = 0 \qquad \ldots(ii)$$

From Eqn. (*i*), we get

$$R_{Bx} = -40 \cos 60° = -20 \text{ kN.} \qquad \textbf{Ans.}$$

From Eqn. (*ii*),
$$R_{By} = -R_A + 20 + 90 + 40 \sin 60°$$
$$= -69.955 + 20 + 90 + 40 \sin 60°$$
$$= 74.686 \text{ kN} \qquad \textbf{Ans.}$$

**Example 3.19.** Determine the reactions developed in the cantilever beam as shown in Fig. 3.37 (*a*).

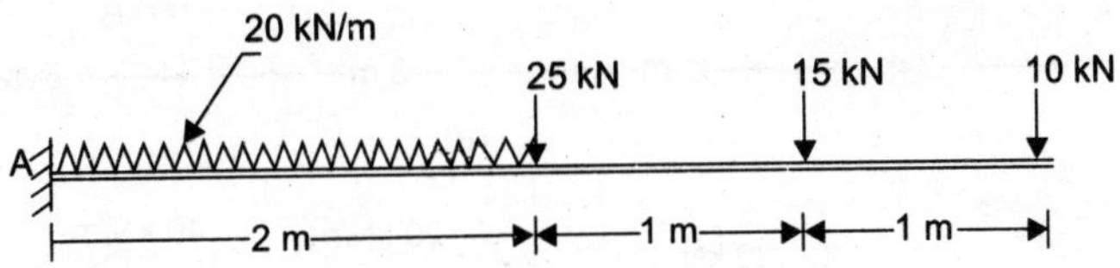


(*a*)

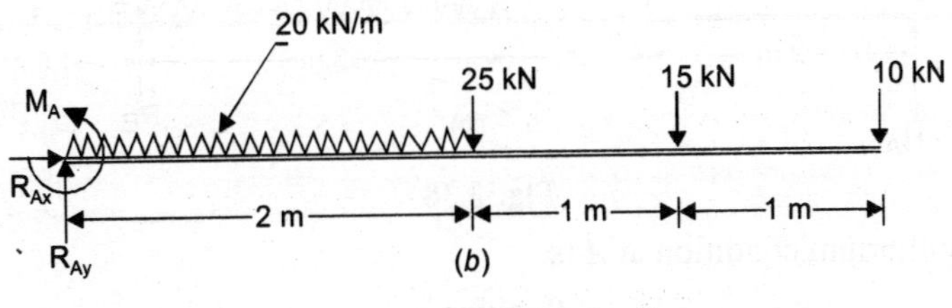


(*b*)

**Fig. 3.37**

**Solution.** Let the reactions at fixed support *A* be $R_{Ax}$, $R_{Ay}$ and $M_A$ as shown in Fig. 3.37 (*b*), which is the free body diagram of the beam. The various forces in the system are

$$\mathbf{F}_1 = R_{Ax}\mathbf{i};\ \mathbf{F}_2 = R_{Ay}\mathbf{j};\ \mathbf{F}_3 = -20 \times 2\mathbf{j} = -40\mathbf{j} \text{ acting at 1 m from } A;$$

$$\mathbf{F}_4 = -25\mathbf{j};\ \mathbf{F}_5 = -15\mathbf{j};\ \mathbf{F}_6 = -10\mathbf{j}$$

$$\mathbf{r}_{A1} = 0;\ \mathbf{r}_{A2} = 0;\ \mathbf{r}_{A3} = 1\mathbf{i};\ \mathbf{r}_{A4} = 2\mathbf{i};\ \mathbf{r}_{A5} = 3\mathbf{i} \text{ and } \mathbf{r}_{A6} = 4\mathbf{i}$$

$$\therefore \qquad \Sigma M_A = 0, \text{ gives}$$

$$M_A\mathbf{k} + 0 + 0 + 1\mathbf{i} \times (-40\mathbf{j}) + 2\mathbf{i} \times (-25\mathbf{j}) + 3\mathbf{i} \times (-15\mathbf{j}) + 4\mathbf{i} \times (-10\mathbf{j}) = 0$$

$$M_A\mathbf{k} - 40\mathbf{k} - 50\mathbf{k} - 45\mathbf{k} - 40\mathbf{k} = 0$$

or
$$M_A = 175 \text{ kN-m} \qquad \textbf{Ans.}$$

Resultant force **R** = 0, gives

$$R_{Ax}\mathbf{i} + R_{Ay}\mathbf{j} - 40\mathbf{j} - 25\mathbf{j} - 15\mathbf{j} - 10\mathbf{j} = 0$$

$$R_{Ax}\mathbf{i} + (R_{Ay} - 40 - 25 - 15 - 10)\mathbf{j} = 0$$

*i.e.*, $R_{Ax} = 0$ **Ans.**

and $R_{Ay} - 40 - 25 - 15 - 10 = 0$

*i.e.*, $R_{Ay} = 90$ kN. **Ans.**

**Example 3.20.** Determine the reactions at supports *A* and *B* of the overhanging beam as shown in Fig. 3.38(*a*).

**Solution.** The free body diagram of the beam is shown in Fig. 3.38(*b*). The 20 kN UDL may be replaced by 20 × 3 = 60 kN load acting at $3 + \frac{3}{2} = 4.5$ m from *A* and uniformly varying load acting in the overhanging portion by its equivalent of $\frac{1}{2} \times 40 \times 1.5 = 30$ kN acting at the centre of gravity of the load triangle, *i.e.*, at 0.5 m from *B*. The FBD of the beam is as shown in the figure.

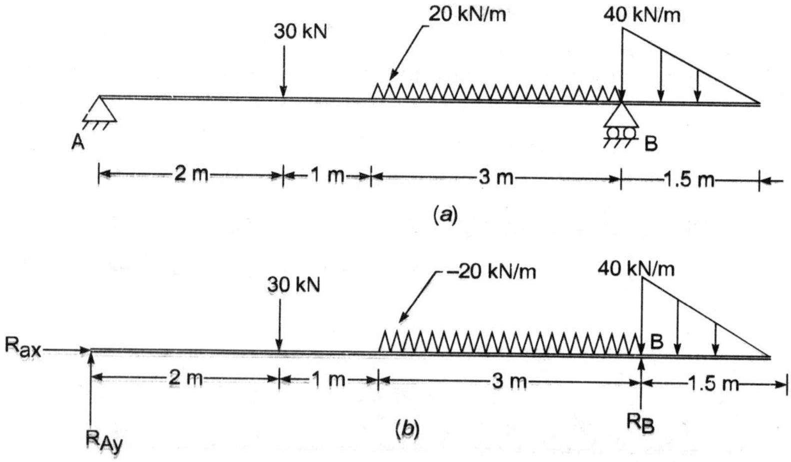


**Fig. 3.38**

The moment equilibrium condition at *A* is

$$\Sigma M_A = 0, \text{ gives}$$

$$R_B \times 6 - 30 \times 2 - 20 \times 3 \times 4.5 - \frac{1}{2} \times 1.5 \times 40 \times (6 + 0.5) = 0$$

∴ $R_B = 87.5$ kN **Ans.**

$$\Sigma F_x = 0, \text{ gives}$$

$$R_{Ax} = 0$$

$$\Sigma F_y = 0, \text{ gives}$$

$$R_{Ay} + R_B - 30 - 20 \times 3 - \frac{1}{2} \times 1.5 \times 40 = 0$$

*i.e.*, $R_{Ay} + 87.5 - 30 - 60 - 30 = 0$

∴ $R_{Ay} = 32.5$ kN **Ans.**

**Example 3.21.** Determine the reactions in the overhanging beam as shown in Fig. 3.39(*a*).

**Solution.** The FBD of the beam is as shown in Fig. 3.39 (*b*).

$\Sigma M_A = 0$, gives

$$-60 - 40 \sin 60° \times 5 - 30 \times 2 \times (6 + 1) + R_B \times 6 = 0$$

$$\therefore \quad R_B = \frac{60 + 200 \sin 60° + 420}{6} = 108.868 \text{ kN}$$

**Note:** Inclined force of 40 kN is split into its vertical and horizontal components at its intersection point with beam. Vertical component 40 sin 60° gives clockwise moment 40 sin 60° × 5, whereas horizontal component 40 cos 60° does not give any moment about *A*, since it passes through *A*, *i.e.*, its distance from *A* is zero.

$\Sigma F_x = 0$, gives

$$R_{Ax} - 40 \cos 60° = 0$$

$$R_{Ax} = 40 \cos 60° = 20 \text{ kN} \qquad \textbf{Ans.}$$

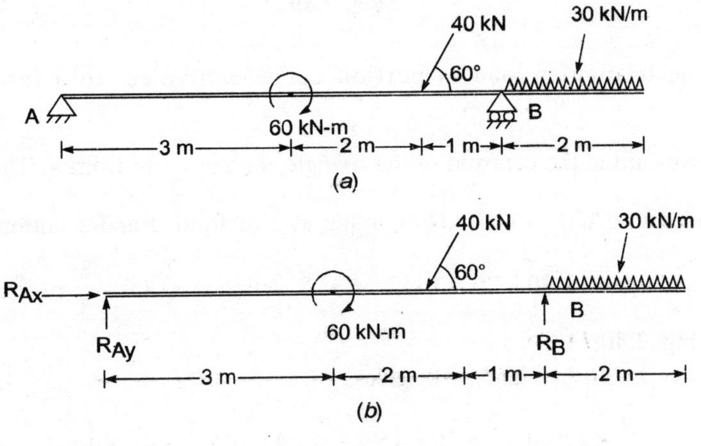


**Fig. 3.39**

$\Sigma F_y = 0$, gives

$$R_{Ay} + R_B - 40 \sin 60° - 30 \times 2 = 0$$

$$\therefore \quad R_{Ay} + 108.8 - 40 \sin 60° - 60 = 0$$

or

$$R_{Ay} = -14.265 \text{ kN} \qquad \textbf{Ans.}$$

(*i.e.*, $R_{Ay}$ = 14.265 kN, downward).

**Example 3.22.** Determine the reactions at supports $A$ and $B$ of the loaded beam as shown in Fig. 3.40($a$).

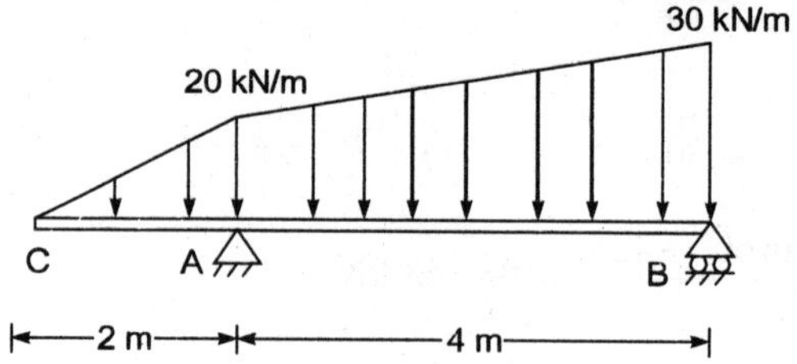


($a$)

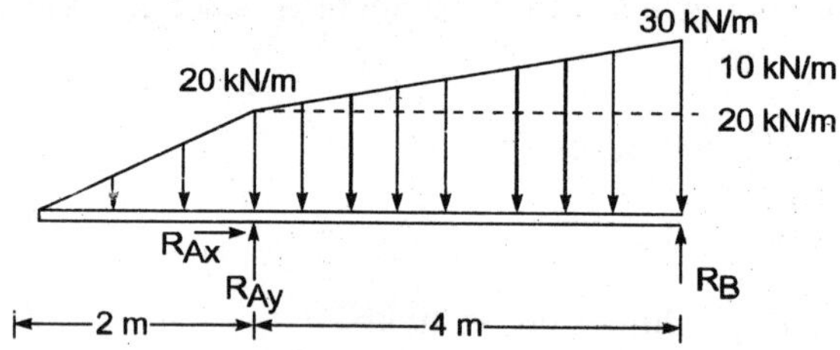


($b$)

**Fig. 3.40**

**Solution.** The triangular load in portion $CA$ is equivalent to a force of $\frac{1}{2} \times 2 \times 20$ = 20 kN acting downward at the centroid of the triangle, *i.e.*, at $\frac{2}{3}$ m from $A$. The trapezoidal load in portion $AB$ is divided into UDL of 20 kN/m acting at 2 m from $A$ and a uniformly varying load of intensity zero at $A$ and intensity 30 – 20 = 10 kN/m at $B$, acting at $\frac{2}{3} \times 4 = \frac{8}{3}$ m from $B$. The FBD of the beam is shown in Fig. 3.40($b$).

$$\Sigma M_A = 0, \text{ gives}$$

$$\frac{1}{2} \times 2 \times 20 \times \frac{2}{3} - 20 \times 4 \times 2 - \frac{1}{2} \times 4 \times 10 \times \frac{8}{3} + R_B \times 4 = 0$$

$\therefore$ $$R_B = 50 \text{ kN}$$ **Ans.**

$$\Sigma F_x = 0, \text{ gives}$$

$$R_{Ax} = 0$$ **Ans.**

$$\Sigma F_y = 0, \text{ gives}$$

$$R_{Ay} + R_B - \frac{1}{2} \times 2 \times 20 - 4 \times 20 - \frac{1}{2} \times 4 \times 10 = 0$$

$$R_{Ay} + 50 - 20 - 80 - 20 = 0$$

$\therefore$ $$R_{Ay} = 70 \text{ kN}$$ **Ans.**

**Example 3.23.** Determine the support reactions on the beam as shown in Fig. 3.41.
**Solution.** The free body diagram of the beam is as shown in Fig. 3.41($b$).

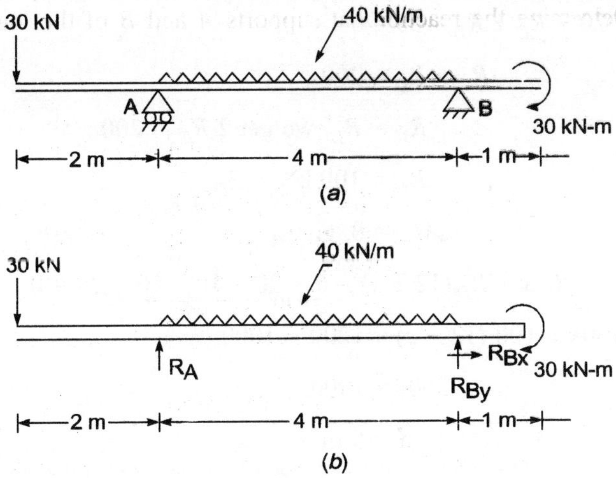


**Fig. 3.41**

$$\Sigma M_B = 0, \text{ gives}$$

$$30 \times 6 - R_A \times 4 + 40 \times 4 \times 2 - 30 = 0$$

$$\therefore \quad R_A = \frac{30 \times 6 + 40 \times 8 - 30}{4} = 117.5 \text{ kN} \qquad \textbf{Ans.}$$

$$\Sigma F_x = 0, \text{ gives}$$

$$R_{Ax} = 0 \qquad \textbf{Ans.}$$

$$\Sigma F_y = 0, \text{ gives}$$

$$-30 + R_A - 40 \times 4 + R_{By} = 0$$

$$\therefore \quad R_{By} = 30 - R_A + 40 \times 4 = 30 - 117.5 + 160$$

$$= 72.5 \text{ kN} \qquad \textbf{Ans.}$$

**Example 3.24.** Beam *AB* of 20 m long, supported on two intermediate supports *C* and *D,* 12 m apart, carries the loads as shown in Fig. 3.42. How far away should the first support *C* be located from end *A,* so that the reactions at both the supports are equal?

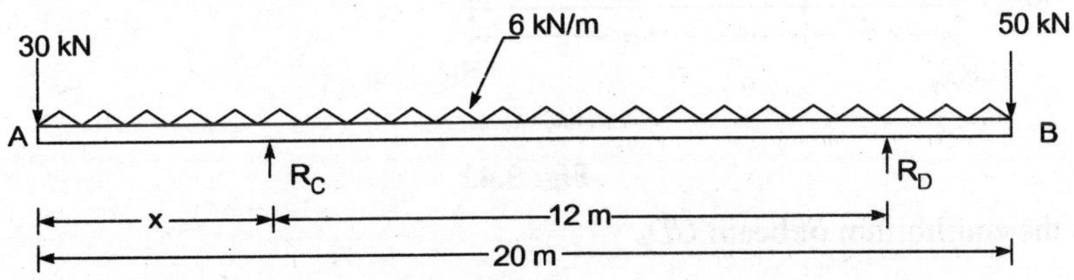


**Fig. 3.42**

**Solution.** Let the support *C* be at a distance *x* metres from the end *A*.

Now it is given that $R_C = R_D$

$$\Sigma F_y = 0, \text{ gives}$$

$$-30 - 6 \times 20 + R_C + R_D - 50 = 0$$

$\therefore$ $$R_C + R_D = 200$$

Since $R_C = R_D$, we get $2\,R_C = 200$

or $$R_C = 100 \text{ kN} = R_D$$

$$\Sigma M_A = 0, \text{ gives}$$

$$R_C x + R_D (12 + x) - 6 \times 20 \times 10 - 50 \times 20 = 0$$

$$100x + 100\,(12 + x) = 1200 + 1000$$

$$200x = 1000$$

or $$x = 5 \text{ m}$$ **Ans.**

**Example 3.25.** Determine the reactions developed at *A, B* and *D* in the compound beam as shown in Fig. 3.43(*a*).

**Solution.** The free body diagrams of beam *AB* and *CD* are as shown in Fig. 3.43(*b*). The given trapezoidal load on *CD* is divided into UDL of intensity 20 kN/m and a triangular load of intensity zero at *C* and 10 kN/m at *D*.

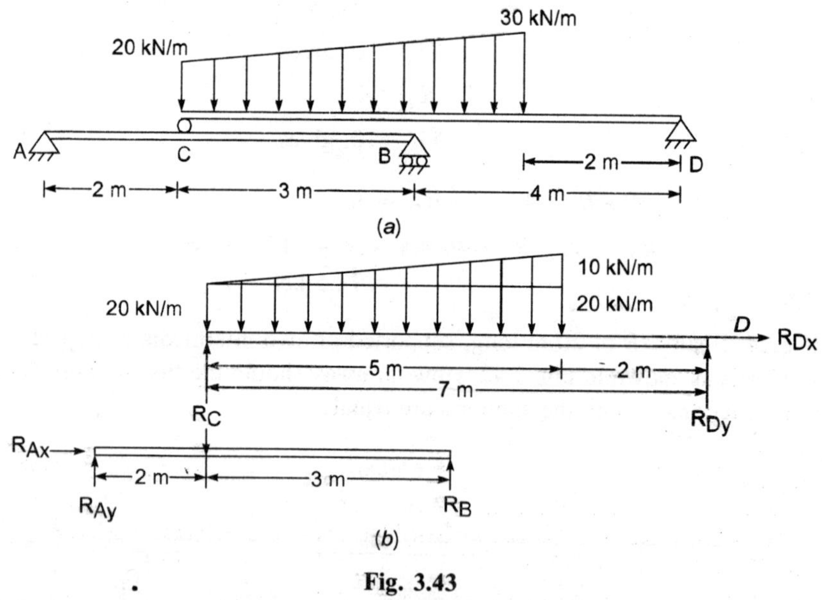


**Fig. 3.43**

Consider the equilibrium of beam *CD*.

$$\Sigma H = 0, \text{ gives}$$

$$R_{Dx} = 0$$ **Ans.**

$$\Sigma M_D = 0, \text{ gives}$$

$$R_C\, 7 - 20 \times 5 \times (7 - 2.5) - \frac{1}{2} \times 5 \times 10\,(2 + \frac{5}{3}) = 0$$

$$\therefore \quad R_C = \frac{20 \times 5 \times 4.5 + \frac{1}{2} \times 5 \times 10 \times 5.333}{7}$$

$$= 77.381 \text{ kN} \qquad \textbf{Ans.}$$

$\Sigma V = 0$, gives

$$R_C + R_D - 5 \times 20 - \frac{1}{2} \times 5 \times 10 = 0$$

$$\therefore \quad 77.381 + R_D - 100 - 25 = 0$$

or $$R_D = 47.619 \text{ kN} \qquad \textbf{Ans.}$$

Now consider the equilibrium of beam *AB*.

$\Sigma M_A = 0$, gives

$$R_B \times 5 - R_C \times 2 = 0$$

$$\therefore \quad R_B = \frac{2}{5} R_C = \frac{2}{5} \times 77.381\ 1$$

i.e., $$R_B = 30.951 \text{ kN} \qquad \textbf{Ans.}$$

$\Sigma V = 0$, gives

$$R_{Ay} + R_B - R_C = 0$$

i.e., $$R_{Ay} = R_C - R_B = 77.381 - 30.952$$

$$= 46.423 \text{ kN} \qquad \textbf{Ans.}$$

$\Sigma H = 0$, gives

$$R_{Ax} = 0 \qquad \textbf{Ans.}$$

**Example 3.26.** The compound bar *ABCD* is loaded as shown in Fig. 3.44(*a*). Determine the reactions developed at *A, C* and *D*.

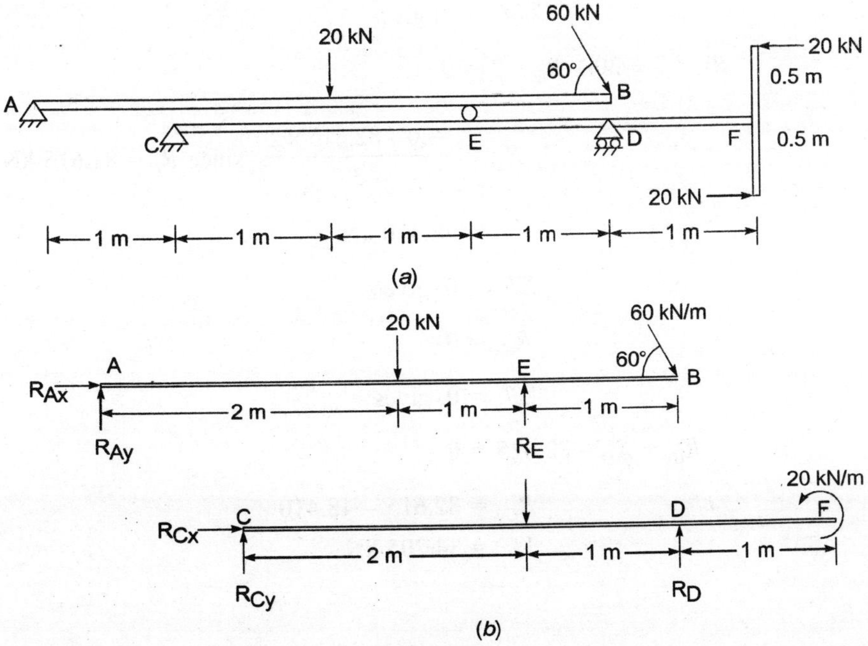


**Fig. 3.44**

**Solution.** The free body diagrams of beams *AB* and *CD* are as shown in Fig. 3.44(*b*). The two equal and opposite forces of 20 kN may be replaced by a couple moment of 20 × 1 = 20 kN-m acting at *F*.

Consider the overhanging beam *AEB*.

$$\Sigma M_A = 0, \text{ gives}$$

$$R_E \times 3 - 20 \times 2 - 60 \sin 60° \times 4 = 0$$

(**Note:** the component $R_E \cos 60°$ do not give any moment about *A*).

$$\therefore \quad R_E = \frac{20 \times 2 + 60 \sin 60° \times 4}{3} = 82.615 \text{ kN}$$

$$\Sigma F_x = 0, \text{ gives}$$

$$R_{Ax} + 60 \cos 60° = 0$$

or $$R_{Ax} = -30 \text{ kN}$$ **Ans.**

$$\Sigma F_y = 0, \text{ gives}$$

$$R_{Ay} + R_E - 20 - 60 \sin 60° = 0$$

$$R_{Ay} + 82.615 - 20 - 60 \sin 60° = 0$$

or $$R_{Ay} = -10.654 \text{ kN}$$ **Ans.**

Now consider the equilibrium of beam *CDF*.

$$\Sigma M_C = 0, \text{ gives}$$

$$R_D \times 3 + 20 - R_E \times 2 = 0$$

$$R_D = \frac{-20 + 82.615 \times 2}{3}; \text{ since } R_E = 82.615 \text{ kN}$$

$$= 48.410 \text{ kN}$$ **Ans.**

$$\Sigma F_x = 0, \text{ gives}$$

$$R_{Cx} = 0$$ **Ans.**

$$\Sigma F_y = 0, \text{ gives}$$

$$R_{Cy} + R_D - 82.625 = 0$$

$$\therefore \quad R_{Cy} = 82.615 - 48.410$$

$$= 34.205 \text{ kN}$$ **Ans.**

**Example 3.27.** Determine the reactions developed at the support in rigid cantilever plane frame as shown in Fig. 3.45.

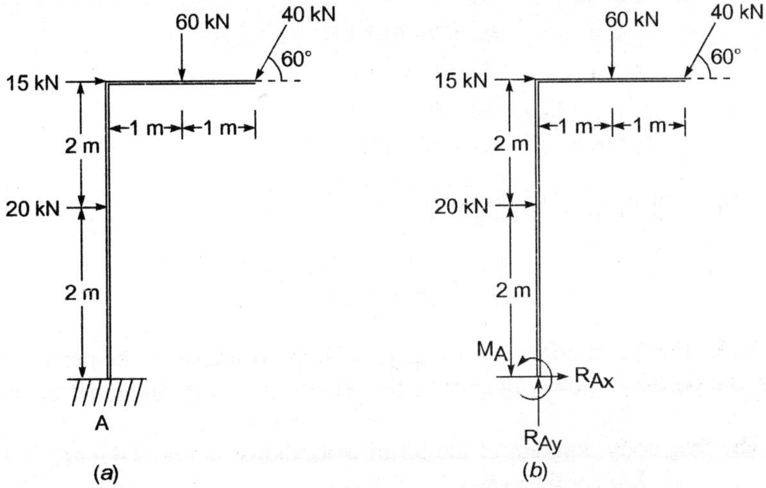


**Fig. 3.45**

**Solution.** The free body diagram of the rigid frame is as shown in Fig. 3.45(*b*).

$$\Sigma M_A = 0, \text{ gives}$$

$$M_A - 40 \sin 60° \times 2 + 40 \cos 60° \times 4 - 60 \times 1 - 15 \times 4 - 20 \times 2 = 0$$

$\therefore$ $$M_A = 149.282 \text{ kN–m}$$ **Ans.**

$$\Sigma F_x = 0, \text{ gives}$$

$$R_{Ax} - 40 \cos 60° + 15 + 20 = 0$$

or $$R_{Ax} = -15 \text{ kN}$$ **Ans.**

$$= 15 \text{ kN } (\leftarrow)$$

$$\Sigma F_y = 0, \text{ gives}$$

$$R_{Ay} - 40 \sin 60° - 60 = 0$$

$$R_{Ay} = 94.641 \text{ kN}$$ **Ans.**

**Example 3.28.** Determine the support reactions of portal frame as shown in Fig. 3.46.

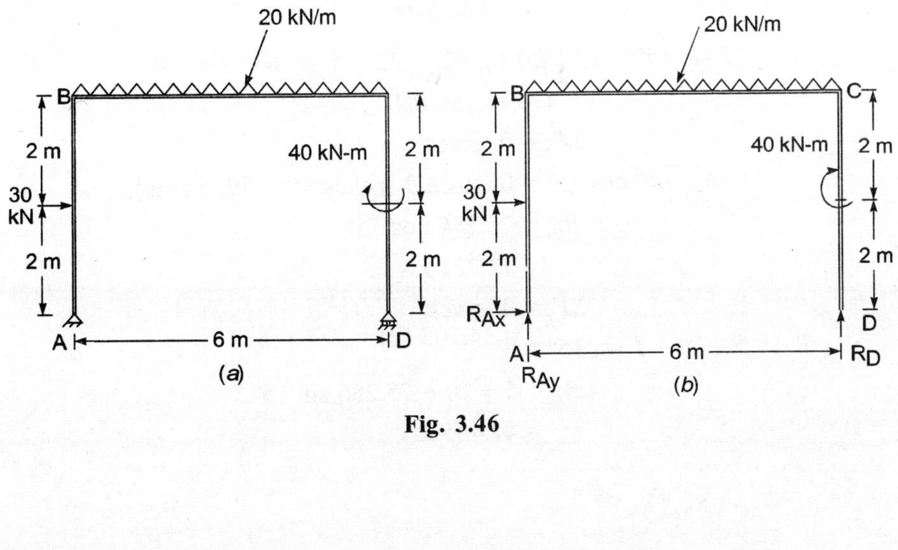


**Fig. 3.46**

**Solution.** Free body diagram of the frame is as shown in Fig. 3.46(*b*).

$$\Sigma M_A = 0, \text{ gives}$$

$$R_D \times 6 - 30 \times 2 - 20 \times 6 \times 3 - 40 = 0$$

$\therefore$ $$R_D = 76.667 \text{ kN}$$ **Ans.**

$$\Sigma F_x = 0, \text{ gives}$$

$$R_{Ax} + 30 = 0$$

or $$R_{Ax} = -30 \text{ kN}$$ **Ans.**

$$\Sigma F_y = 0, \text{ gives}$$

$$R_{Ay} + R_D - 20 \times 6 = 0$$

$\therefore$ $$R_{Ay} = 20 \times 6 - R_D = 120 - 76.667$$

*i.e.,* $$R_{Ay} = 43.333 \text{ kN}$$ **Ans.**

**Example 3.29.** The 12 m boom *AB* weighs 5 kN, the distance of the centre of gravity *G* being 6 m from *A*. For the position shown, determine the tension **T** in the cable and the reaction at *B* [Ref. Fig. 3.47].

**Solution.** The free body diagram of the beam is as shown is Fig. 3.47(*b*).

$$\Sigma M_A = 0, \text{ gives}$$

[Moments are worked after resolving forces normal to and parallel to boom.]

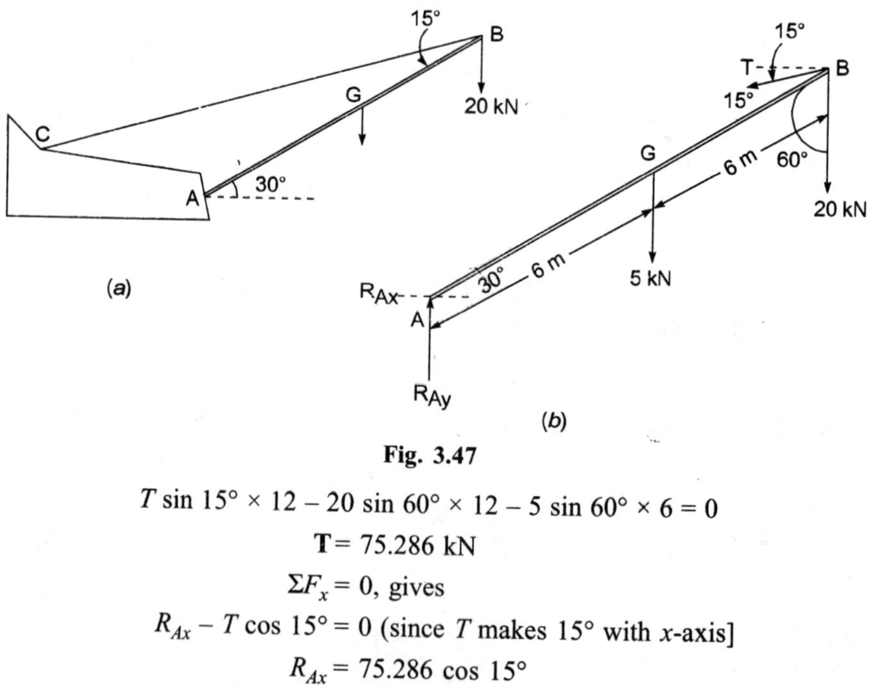


**Fig. 3.47**

$$T \sin 15° \times 12 - 20 \sin 60° \times 12 - 5 \sin 60° \times 6 = 0$$

$\therefore$ $$\mathbf{T} = 75.286 \text{ kN}$$ **Ans.**

$$\Sigma F_x = 0, \text{ gives}$$

$$R_{Ax} - T \cos 15° = 0 \text{ (since } T \text{ makes } 15° \text{ with } x\text{-axis]}$$

$$R_{Ax} = 75.286 \cos 15°$$

$$= 72.721 \text{ kN}$$ **Ans.**

$$\Sigma F_y = 0, \text{ gives}$$

$$R_{Ay} - 5 - 20 - T \sin 15° = 0$$

$$R_{Ay} = 5 + 20 + 75.286 \sin 15°$$

$$= 44.485 \text{ kN}$$ **Ans.**

**Example 3.30.** A hollow right circular cylinder of radius 800 mm is open at both ends and rests on a smooth horizontal plane as shown in Fig. 3.48(*a*). Inside the cylinder, there are two spheres weighing 1 kN and 3 kN and with radii 400 mm and 600 mm respectively. The lower sphere rests on the horizontal plane. Neglecting friction, find the minimum weight **W** of the cylinder for which it will not tip over.

**Solution.** Join the centres $O_1$ and $O_2$ and drop $O_1D$ perpendicular to horizontal through $O_2$.

Now $O_2D = 1600 - r_1 - r_2$ where $r_1$ and $r_2$ are radii of the spheres.

$$= 1600 - 400 - 600 = 600 \text{ mm}$$

If $\alpha$ is the inclination to $O_2O_1$ to horizontal, we have

$$\cos\alpha = \frac{O_2D}{O_1O_2} = \frac{600}{400+600} = 0.6$$

$$\alpha = 53.130$$

Free body diagrams of cylinder and spheres are shown in Figs. 3.48(*b*) and 3.48(*c*).

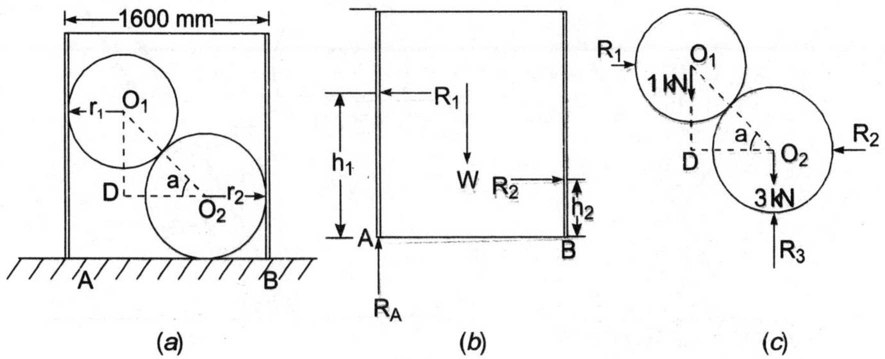


**Fig. 3.48**

Consider the equilibrium of the two spheres.

$$\Sigma M_{O2} = 0, \text{ gives}$$

$$R_1 \times O_1O_2 \sin\alpha - 1 \times O_2D = 0$$

$$R_1\ 1000 \sin 53.130° - 1 \times 600 = 0$$

$$R_1 = 0.75 \text{ kN}$$

$$\Sigma F_x = 0, \text{ gives}$$

$$R_2 = R_1 = 0.75 \text{ kN}$$

$$\Sigma F_y = 0, \text{ gives}$$

$$R_3 - 1 - 3 = 0$$

*i.e.*, $$R_3 = 4 \text{ kN}$$

Now consider the equilibrium of the cylinder. When it is about to tip over on point *A*, there is no reaction from ground at *B*. The reaction will be only at *A*.

$$\Sigma M_A = 0, \text{ gives}$$

$$R_1h_1 - R_2h_2 - W \times 800 = 0$$

$$0.75\, h_1 - 0.75\, h_2 = 800\, W, \text{ since } R_1 = R_2 = 0.75 \text{ kN}$$

$$0.75\,(h_1 - h_2) = 800\, W$$

$$0.75\, O_1D = 800\, W$$

But $$O_1D = 1000 \sin \alpha = 1000 \sin 53.130°$$

∴ $$0.75 \times 1000 \sin 53.130° = 800\, W$$

or $$W = 0.75 \text{ kN}$$ **Ans.**

**Example 3.31.** A 500 N cylinder, 1 m in diameter is placed between the two cross pieces which make an angle of 60° with each other. The two cross pieces are pinned at $C$ as shown in Fig. 3.49 (*a*). Determine the tension in the horizontal rope $DE$ assuming floor is smooth.

**Solution.** Consider the equilibrium of entire system.

Due to symmetry,

$$R_A = R_B$$

$$\Sigma F_y = 0, \text{ gives}$$

$$R_A + R_B = 500$$

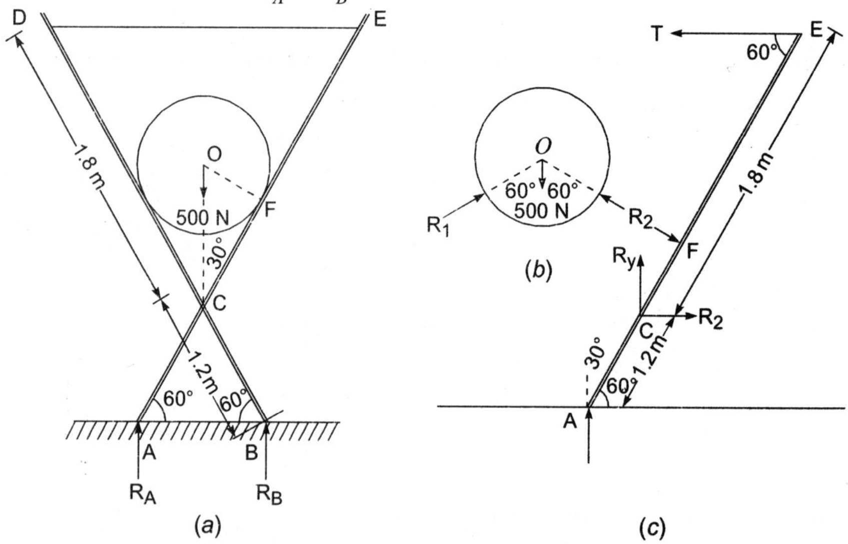


**Fig. 3.49**

or $$2R_A = 500$$

*i.e.,* $$R_A = 250 \text{ N} = R_B.$$

Now consider the equilibrium of the cylinder. [Ref. Fig. 3.49(*b*)]. The members $AE$ and $BD$ make 60° with horizontal. Hence the reactions $R_2$ and $R_1$ make 60° with the vertical

$$\Sigma F_x = 0, \text{ gives}$$

$$R_1 \sin 60° - R_2 \sin 60° = 0$$

*i.e.,* $$R_1 = R_2$$

$$\Sigma F_y = 0, \text{ gives}$$

$$R_1 \cos 60° + R_2 \cos 60° - 500 = 0$$

$$0.5\, R_1 + 0.5\, R_2 = 500$$

$$R_1 = 500 \text{ N, since } R_1 = R_2.$$

Now consider the equilibrium of the member $AE$. The free body diagram is shown in Fig. 3.49 (*c*).

$$\Sigma M_C = 0, \text{ gives}$$

$$T \sin 60° \times 1.8 - R_2\, CF - R_A \sin 30° \times 1.2 = 0$$

Now $CF = OF \cot 30° = 0.5 \cot 40° = 0.866$ m

$$\therefore \quad T = \frac{500 \times 0.866 + 250 \sin 30° \times 1.2}{1.8 \times \sin 60°}$$

*i.e.*, $T = 374.00$ N **Ans.**

**Note:** If the reactions on pin at *C* are required, the other two equilibrium equations namely $\Sigma F_x = 0$ and $\Sigma F_y = 0$ are to be considered.

**Example 3.32.** The frame as shown in Fig. 3.50(*a*), is supported by a hinge at *A* and a roller at *E*. Compute the horizontal and vertical components of the hinge forces at *B* and *C* as experienced by member *AC*.

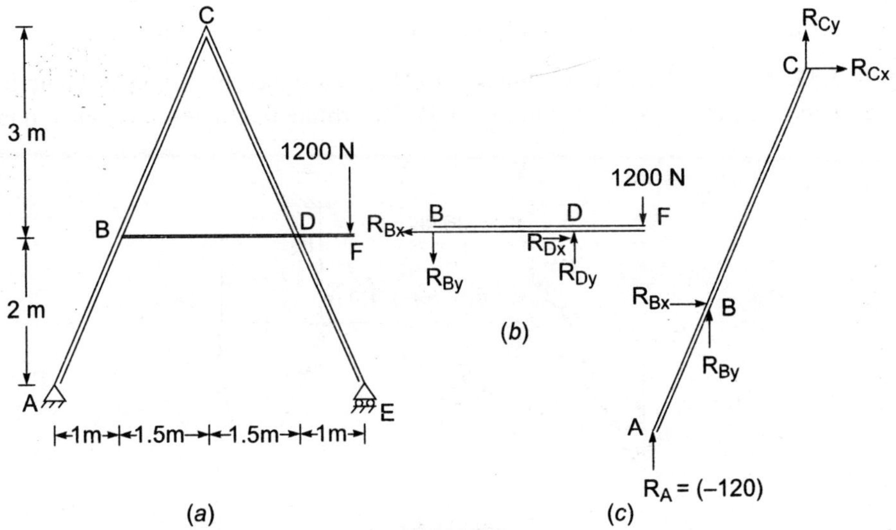


**Fig. 3.50**

**Solution.** Consider the equilibrium of the entire frame.

$$\Sigma M_A = 0, \text{ gives}$$

$$R_E \times 5 - 1200(1 + 3 + 1.5) = 0$$

$$R_E = 1320 \text{ N}$$

$$\Sigma F_x = 0, \text{ gives } R_{Ax} = 0$$

$$\Sigma F_y = 0, \text{ gives } R_{Ay} + R_E - 1200 = 0$$

$$\therefore \quad R_{Ay} = 1200 - 1320, \text{ since } R_E = 1320 \text{ N}$$

$$= -120 \text{ N}$$

Consider the equilibrium of member *BDF* [Fig. 3.51(*b*)]

$$\Sigma M_D = 0, \text{ gives}$$

$$R_{By} \times 3 - 1200 \times 1.5 = 0$$

$$R_{By} = 600 \text{ (downward)}$$

**Note:** The directions of reaction at *B* on *ABC* is opposite to that of *ABC* on member *BDF* (Actions and reactions are opposite to each other).

Consider the equilibrium of the member $AC$. [Ref. Fig. 3.50(c)].

$$\Sigma F_y = 0, \text{ gives}$$

$$R_{Cy} + (-120) + 600 = 0$$

$\therefore$ $$R_{Cy} = -480 \text{ N} = 480 \text{ N, downward.}$$ **Ans.**

$$\Sigma F_x = 0, \text{ gives}$$

$$F_{Cx} = F_{Bx} \quad ...(i)$$

$$\Sigma M_C = 0, \text{ gives}$$

$$R_{Bx} \times 3 - R_{By} \times 1.5 - (-120) \times 2.5 = 0$$

$$R_{Bx} = \frac{600 \times 1.5 - 120 \times 2.5}{3} = 200 \text{ N}$$

From Eqn. (*i*), $$R_{Cx} = R_{Bx} = 200 \text{ N.}$$ **Ans.**

**Example 3.33.** A triangular plate weighing 36 kN is suspended by lifting machine by vertical cables at the corners $A$, $B$, and $C$ as shown in Fig. 3.51. Determine the forces developed in the cables.

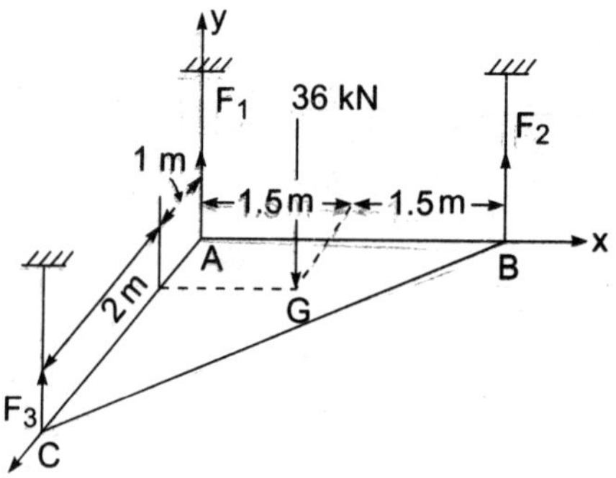


**Fig. 3.51**

**Solution.** [**Note:** This is problem of system of forces in space. Vector approach is ideally suited]. Selecting $A$ as origin, the coordinates of various points are

$$A(0, 0, 0), B(4.5, 0, 0), C(0, 0, 3.0) \text{ and } G(1.5, 0, 1.0).$$

Let $\mathbf{F}_1$, $\mathbf{F}_2$ and $\mathbf{F}_3$ be the tensile forces in cables at $A$, $B$ and $C$. Then

$$\mathbf{F}_1 = F_1\mathbf{j};\ \mathbf{F}_2 = F_2\mathbf{j} \text{ and } \mathbf{F}_3 = F_3\mathbf{j}$$

The load is $$\mathbf{F}_4 = -36\mathbf{j}$$

$$\mathbf{r}_{A1} = 0,\ \mathbf{r}_{A2} = 4.5\mathbf{i},\ \mathbf{r}_{A3} = 3\mathbf{k} \text{ and } \mathbf{r}_{A4} = 1.5\mathbf{i} + \mathbf{k}$$

The moment equilibrium equation about $A$ gives,

$$0 + 4.5\mathbf{i} \times F_2\mathbf{j} + 3\mathbf{k} \times F_3\,\mathbf{j} + (1.5\mathbf{i} + \mathbf{k}) \times (-36\,\mathbf{j}) = 0$$

*i.e.,* $$4.5\,F_2\mathbf{k} + 3F_3\,(-\mathbf{i}) - 54\mathbf{k} - 36\,(-\mathbf{i}) = 0$$

$$(-3F_3 + 36)\mathbf{i} + (4.5F_2 - 54)\mathbf{k} = 0$$

*i.e.,* $-3F_3 + 36 = 0 \quad \therefore \quad F_3 = 12kN$ **Ans.**

and $4.5F_2 - 54 = 0 \qquad F_2 = 12\ kN$ **Ans.**

From force equilibrium condition, we have

$$F_1\mathbf{j} + F_2\mathbf{j} + F_3\mathbf{j} - 36\mathbf{j} = 0$$

*i.e.*, $(F_1 + F_2 + F_3 - 36)\mathbf{j} = 0$

*i.e.*, $F_3 = 36 - F_1 - F_2$

$= 36 - 12 - 12 = 12$ kN **Ans.**

## 3.7 EQUILIBRIUM OF GENERAL SPACIAL FORCE SYSTEM

In this case, resultant force and moment should be zero, *i.e.*, $\mathbf{R} = 0$, $\mathbf{M} = 0$.

*i.e.*, $\Sigma F_x\,\mathbf{i} + \Sigma F_y\,\mathbf{j} + \Sigma F_z\mathbf{k} = 0$ ...(*i*)

and $\Sigma M_x\mathbf{i} + \Sigma M_y\mathbf{j} + \Sigma M_z\mathbf{k} = 0$ ...(*ii*)

In scalar form, Eqns. (*i*) and (*ii*) mean

$\Sigma F_x = 0, \Sigma F_y = 0$ and $\Sigma F_z = 0$ ...[3.9(*a*)]

$\Sigma M_x = 0, \Sigma M_y = 0$ and $\Sigma M_z = 0$ ...[3.9(*b*)]

Use of above equations for solving equilibrium conditions of rigid bodies subject to general spatial force system is illustrated with examples below.

**Example 3.34.** A rectangular slab of size 3 m × 6 m is to be lifted with the help of three vertical chains hooked to it as shown in Fig. 3.52. If the slab is having uniform thickness and weighs 54 kN, determine the forces developed in the chains.

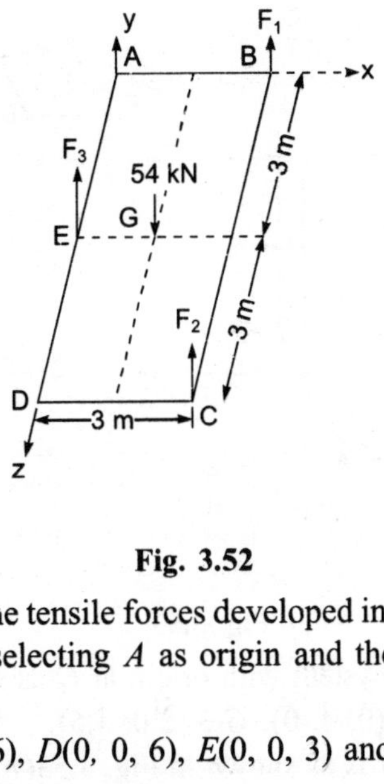


**Fig. 3.52**

**Solution.** Let $\mathbf{F}_1$, $\mathbf{F}_2$ and $\mathbf{F}_3$ be the tensile forces developed in the chains at *B*, *C* and *F*. Let *G* be the centre of gravity of the slab. Then selecting *A* as origin and the coordinate system as shown in the Fig. 3.52, we have

*A*(0, 0, 0), *B*(3, 0, 0), *C*(3, 0, 6), *D*(0, 0, 6), *E*(0, 0, 3) and *G*(1.5, 0, 3).

Vector forms of the forces to be considered are

$$\mathbf{F}_1 = F_1\,\mathbf{j};\ \mathbf{F}_2 = F_2\mathbf{j}\ ;\ \mathbf{F}_3 = F_3\mathbf{j} \text{ and } \mathbf{F}_4 = -54\mathbf{j}$$

$$\mathbf{r}_{B1} = \mathbf{r}_{BB} = 0,\ \ \mathbf{r}_{B2} = \mathbf{r}_{BC} = 6\mathbf{k},\ \ \mathbf{r}_{B3} = \mathbf{r}_{BE} = -3\mathbf{i} + 3\mathbf{k},$$

$$\mathbf{r}_{B4} = \mathbf{r}_{BG} = -1.5\mathbf{i} + 3\mathbf{k}$$

∴ $\Sigma M_B = 0$, gives

$$0 + 6\mathbf{k} \times (F_2\mathbf{j}) + (-3\mathbf{i} + 3\mathbf{k}) \times (F_3\mathbf{j}) + (-1.5\mathbf{i} + 3\mathbf{k}) \times (-54\mathbf{j}) = 0$$

*i.e.*, $6F_2\,(-\mathbf{i}) - 3F_3\mathbf{k} + 3F_3\,(-\mathbf{i}) + 81\mathbf{k} - 162\,(-\mathbf{i}) = 0$

*i.e.,* $$(-6F_2 - 3F_3 + 162)\mathbf{i} + (-3F_3 + 81)\mathbf{k} = 0$$

*i.e.,* $$-6F_2 - 3F_3 + 162 = 0 \quad \ldots(i)$$

and $$-3F_3 + 81 = 0 \quad \ldots(ii)$$

From Eqn. (*ii*), $$F_3 = \frac{81}{3} = 27 \text{ kN}$$ **Ans.**

From Eqn. (*i*),

$$6F_2 + 3F_3 = 162$$

$$F_2 = \frac{162 - 3F_3}{6} = \frac{162 - 3 \times 27}{6} = 13.5 \text{ kN}$$ **Ans.**

From force equilibrium condition, we have

$$\mathbf{F}_1 + \mathbf{F}_2 + \mathbf{F}_3 + \mathbf{F}_4 = 0$$

*i.e.,* $$(F_1 + 13.5 + 27 - 54)\mathbf{i} = 0$$

$\therefore$ $$F_1 = 13.5 \text{ kN}$$ **Ans.**

**Example 3.35.** A bar of uniform cross section, weighing 2000 N is having a ball and socket support at *A* and its end *B* is resting at the corner of smooth walls as shown in Fig. 3.53(*a*). Determine the reactions developed at the ends *A* and *B*.

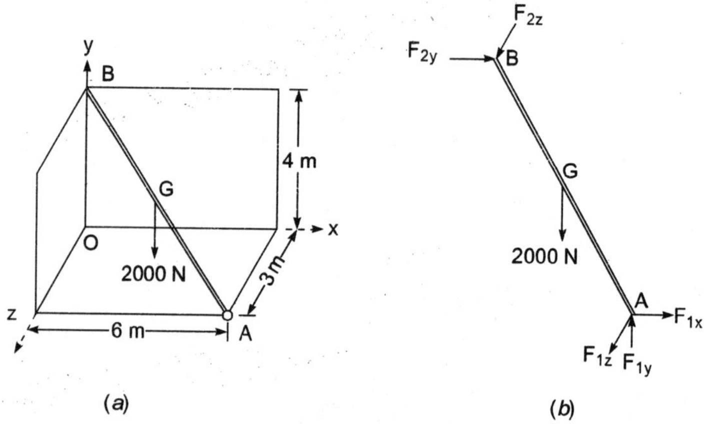


**Fig. 3.53**

**Solution.** Taking the coordinate system with origin at *O* as shown in Fig. 3.53(*a*).

$$A(6, 0, 3), \; B(0, 4, 0), \; G(3, 2.0, 1.5)$$

The free body diagram of the bar is as shown in Fig. 3.53(*b*).

The vector forms of the forces are

$$\mathbf{F}_1 = F_{1x}\mathbf{i} + \mathbf{F}_{1y}\mathbf{j} + F_{1z}\mathbf{k}$$

$$\mathbf{F}_2 = F_{2x}\mathbf{i} + \mathbf{F}_{2z}\mathbf{k}$$

and $$\mathbf{F}_3 = -2000\mathbf{j}$$

$$\mathbf{r}_{AB} = -6\mathbf{i} + 4\mathbf{j} - 3\mathbf{k} \quad \text{and} \quad \mathbf{r}_{AG} = -3\mathbf{i} + 2.0\mathbf{j} - 1.5\mathbf{k}$$

$\Sigma M_A = 0$, gives

$$\mathbf{r}_{AB} \times \mathbf{F}_2 + \mathbf{r}_{AG} \times \mathbf{F}_3 = 0$$

*i.e.,* $$(-6\mathbf{i} + 4\mathbf{j} - 3\mathbf{k}) \times (F_{2x}\mathbf{i} + F_{2z}\mathbf{k}) + (-3\mathbf{i} + 2.0\mathbf{j} - 1.5\mathbf{k}) \times (-2000\mathbf{j}) = 0$$

$$\begin{vmatrix} \mathbf{i} & \mathbf{j} & \mathbf{k} \\ -6 & 4 & -3 \\ F_{2x} & 0 & F_{2z} \end{vmatrix} + 6000\,\mathbf{k} + 3000\,(-\mathbf{i}) = 0$$

*i.e.,* $\mathbf{i}(4F_{2z}) - \mathbf{j}(-6F_{2z} + 3F_{2x}) + \mathbf{k}(0 - 4F_{2x}) + 6000\mathbf{k} - 3000\mathbf{i} = 0$

*i.e.,* $(4F_{2z} - 3000)\mathbf{i} + \mathbf{j}(6F_{2z} - 3F_{2x}) + \mathbf{k}(-4F_{2x} + 6000) = 0$

*i.e.,* $4F_{2z} - 3000 = 0$ ...(*i*)

$6F_{2z} - 3F_{2x} = 0$ ...(*ii*)

and $-4F_{2x} + 6000 = 0$ ...(*iii*)

From Eqn. (*i*), $F_{2z} = 750\ N$ **Ans.**

From Eqn. (*ii*), $F_{2x} = 2F_{2z} = 1500$ N **Ans.**

From Eqn. (*iii*), $F_{2x} = 1500$, same as from Eqn. (*ii*).

Now consider the force equilibrium equation:

$$\mathbf{F}_1 + \mathbf{F}_2 + \mathbf{F}_3 = 0$$

$$(F_{1x}\mathbf{i} + F_{1y}\mathbf{j} + F_{1z}\mathbf{k}) + F_{2x}\mathbf{i} + F_{2z}\mathbf{k} - 2000\mathbf{j} = 0$$

*i.e.,* $(F_{1x} + F_{2x})\mathbf{i} + (F_{1y} + F_{2z} - 2000)\mathbf{j} + (F_{1z} + F_{2z})\mathbf{k} = 0$

*i.e.,* $F_{1x} + F_{2x} = 0$ ...(*iv*)

$F_{1y} + F_{2z} - 2000 = 0$ ...(*v*)

$F_{1z} + F_{2z} = 0$ ...(*vi*)

From Eqn. (*iv*), $F_{1x} = -F_{2x} = -1500$ N **Ans.**

From Eqn. (*v*), $F_{1y} = 2000 - F_{2z} = 2000 - 750 = 1250$ N **Ans.**

From Eqn. (*vi*), $F_{1z} = -F_{2z} = -750$ N **Ans.**

**Example 3.36.** A block of size 4 m × 3 m × 2 m is having a uniform mass and is weighing 20 kN. It is supported by a ball and socket at *d'* and by a roller support on the horizontal plane at *A* and by two cables *Ba* and *Ca'* which are in the directions of the *ac'* and *a'c* as shown in Fig. 3.54(*a*). Determine the support reactions.

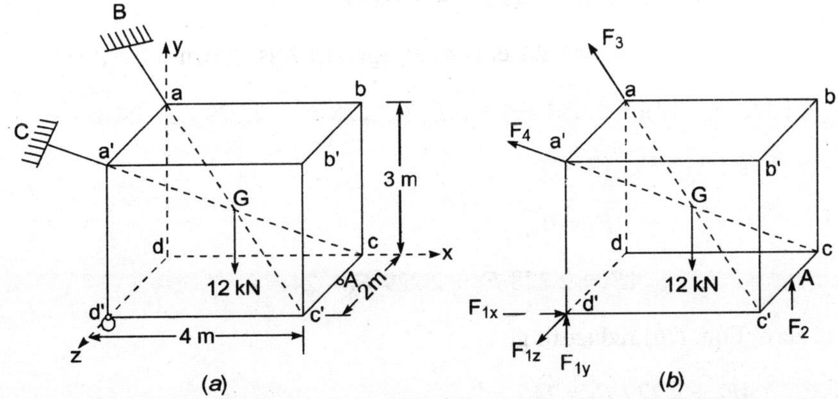


Fig. 3.54

**Solution.** Selecting $d$ as the origin, the coordinates of various points are

$a(0, 3, 0)$, $b(4, 3, 0)$, $c(4, 0, 0)$, $d(0, 0, 0)$

$a'(0, 3, 2)$, $b'(4, 3, 2)$, $c'(4, 0, 2)$, $d'(0, 0, 2)$

$G(2, 1.5, 1.0)$, $A(4, 0, 1)$.

Vector forms of reactions:

$$\mathbf{F}_1 = F_{1x}\mathbf{i} + F_{1y}\mathbf{j} + F_{1z}\mathbf{k}$$

$$\mathbf{F}_2 = F_2\mathbf{j}$$

$$\mathbf{r}_{c'a} = -4\mathbf{i} + 3\mathbf{j} - 2\mathbf{k}$$

$$r_{c'a} = \sqrt{4^2 + 3^2 + 2^2} = 5.385 \text{ m}$$

$$\mathbf{n}_{c'a} = -0.743\mathbf{i} + 0.557\mathbf{j} - 0.371\mathbf{k}$$

$$\mathbf{F}_3 = F_3(-0.743\mathbf{i} + 0.557\mathbf{j} - 0.371\mathbf{k})$$

$$\mathbf{r}_{ca'} = -4\mathbf{i} + 3\mathbf{j} + 2\mathbf{k}$$

$$r_{ca'} = \sqrt{4^2 + 3^2 + 2^2} = 5.385 \text{ m}$$

$$\therefore \quad \mathbf{n}_{ca'} = -0.743\mathbf{i} + 0.557\mathbf{j} - 0.371\mathbf{k}$$

$$\therefore \quad \mathbf{F}_4 = F_4(-0.743\mathbf{i} + 0.557\mathbf{j} - 0.371\mathbf{k})$$

$$\mathbf{F}_5 = \text{self wt.} = -12\mathbf{j}$$

Moment equilibrium condition about $d'$ is to be considered.

$$\mathbf{r}_{d'1} = \mathbf{r}_{da'} = 0, \; \mathbf{r}_{d'2} = \mathbf{r}_{d'A} = 4\mathbf{i} - \mathbf{k}$$

$$\mathbf{r}_{d'3} = \mathbf{r}_{d'c'} = 4\mathbf{i}, \; \mathbf{r}_{d'4} = \mathbf{r}_{d'a'} = 3\mathbf{j}$$

$$\mathbf{r}_{d'5} = 2\mathbf{i} + 1.5\mathbf{j} - \mathbf{k}$$

$\therefore$ The moment equilibrium condition about $d'$ gives

$$0 + (4\mathbf{i} - \mathbf{k}) \times F_2\mathbf{j} + 4\mathbf{i} \times (-0.743\mathbf{i} + 0.557\mathbf{j} - 0.371\mathbf{k})\, F_3$$
$$+ 3\mathbf{j} \times (-0.743\mathbf{i} + 0.557\mathbf{j} + 0.371\mathbf{k})F_4$$
$$+ (2\mathbf{i} + 1.5\mathbf{j} - \mathbf{k}) \times (-12\mathbf{j}) = 0$$

*i.e.,*

$$4\mathbf{k}\, F_2 - F_2\,(-\mathbf{i}) + 2.228\, F_3\mathbf{k} - 1.484\, F_3\,(-\mathbf{j})$$
$$-2.229\,(-\mathbf{k})\, F_4 + 1.113\, F_4\mathbf{i} - 24\mathbf{k} + 12\,(-\mathbf{i}) = 0$$

$$(F_2 + 1.113\, F_4 - 12)\mathbf{i} + 1.484\, F_3\mathbf{j} + (4F_2 + 2.228\, F_3 + 2.229\, F_4 - 24)\mathbf{k} = 0$$

*i.e.,*

$$F_2 + 1.113\, F_4 - 12 = 0 \qquad \ldots(i)$$

$$F_3 = 0 \qquad \ldots(ii)$$

and

$$4F_2 + 2.228\, F_3 + 2.229\, F_4 - 24 = 0 \qquad \ldots(iii)$$

Since $F_3$ is zero, Eqn. (*iii*) reduces to

$$4F_2 + 2.229\, F_4 = 24 \qquad \ldots(iv)$$

Subtracting 4 times Eqn. (*i*) from Eqn. (*iv*), we get

$$(2.229 - 4 \times 1.113)\, F_4 = 24 - 4 \times 12 = -24$$

$$\therefore \quad F_4 = 10.796 \text{ kN} \qquad \textbf{Ans.}$$

∴ From Eqn. (*i*), we get

$$F_2 = 12 - 1.113\, F_4 = 12 - 1.113F \times 10.796$$

$$= 0.016 \text{ kN} \qquad \textbf{Ans.}$$

For equilibrium condition, gives

$$\mathbf{F}_1 + \mathbf{F}_2 + \mathbf{F}_3 + \mathbf{F}_4 + \mathbf{F}_5 = 0$$

$$F_{1x}\mathbf{i} + F_{1y}\mathbf{j} + F_{1z}\mathbf{k} + F_2\mathbf{j} + (-0.743\mathbf{i} + 0.557\mathbf{j} - 0.371\mathbf{k})F_3$$
$$+ (-0.743\mathbf{i} + 0.557\mathbf{j} + 0.371\mathbf{k})\, F_4 - 12\mathbf{j} = 0$$

Substituting the values of $\mathbf{F}_2$, $\mathbf{F}_3$ and $\mathbf{F}_4$, we get

$$(F_{1x} - 0 - 0.743 \times 10.796)\mathbf{i} + (F_{1y} + 0.016 + 0 + 0.557 \times 10.796)\mathbf{j}$$
$$+ (F_{1z} - 0.371 \times 0 + 0.371 \times 10.796)\mathbf{k} = 0$$

*i.e.*, $$F_{1x} = 0.743 \times 10.796 = 4.00 \text{ kN}$$

$$F_{1y} = -6.029 \text{ kN} \qquad \textbf{Ans.}$$

and $$F_{1z} = -0.371 \times 10.796 = 4.00 \text{ kN}$$

**Example 3.37.** A derric girder is in the position as shown in Fig. 3.55(*a*). Its boom weigh 6 kN. Determine the forces developed in the cables and in the boom, if the load is 18 kN.

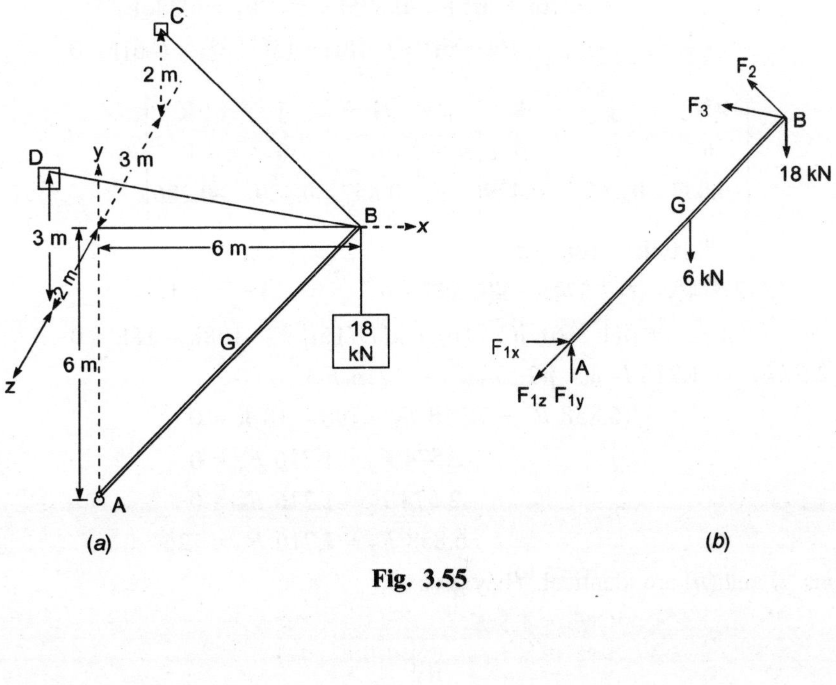


**Fig. 3.55**

**Solution.** Now,

$$A(0, -6, 0);\ B(6, 0, 0);\ C(0, 2, -3);\ D(0, 3, 2)$$

and $G(3, -3, 0)$.

The free body diagram of the boom is as shown in Fig. 3.55(*b*). Let $\mathbf{F}_1$ be the force developed at $A$ and $\mathbf{F}_2$, $\mathbf{F}_3$ be the tensile forces developed in the cables $BC$ and $BD$ respectively.

$$\mathbf{F}_1 = F_{1x}\mathbf{i} + F_{1y}\mathbf{j} + F_{1z}\mathbf{k}$$

$$\mathbf{r}_{BC} = -6\mathbf{i} + 2\mathbf{j} - 3\mathbf{k}$$

$$\therefore \quad r_{BC} = \sqrt{6^2 + 2^2 + 3^3} = 7.0 \text{ m}$$

$$\therefore \quad \mathbf{n}_{BC} = -0.857\mathbf{i} + 0.286\mathbf{j} - 0.429\mathbf{k}$$

$$\mathbf{F}_2 = (-0.857\mathbf{i} + 0.286\mathbf{j} - 0.429\mathbf{k})\ F_2$$

$$\mathbf{r}_{\mathbf{BD}} = -6\mathbf{i} + 3\mathbf{j} + 2\mathbf{k}$$

$$\therefore \quad r_{BD} = \sqrt{6^2 + 3^3 + 2^2} = 7 \text{ m}$$

$$\therefore \quad \mathbf{n}_{BD} = -0.857\mathbf{i} + 0.429\mathbf{j} + 0.286\mathbf{k}$$

$$\therefore \quad \mathbf{F}_3 = (-0.875\mathbf{i} + 0.429\mathbf{j} + 0.286\mathbf{k})\ F_3$$

$$\mathbf{F}_4 = -18\mathbf{j},\ \mathbf{F}_5 = -6\mathbf{j}$$

The moment equilibrium condition about $A$ is to be considered.

$$\mathbf{r}_{A1} = \mathbf{r}_{\mathbf{AA}} = 0,\quad \mathbf{r}_{A2} = \mathbf{r}_{\mathbf{AB}} = 6\mathbf{i} + 6\mathbf{j},\quad \mathbf{r}_{A3} = \mathbf{r}_{\mathbf{AB}} = 6\mathbf{i} + 6\mathbf{j},\quad \mathbf{r}_{A4} = \mathbf{r}_{\mathbf{AB}} = 6\mathbf{i} + 6\mathbf{j},$$

$$\mathbf{r}_{A5} = \mathbf{r}_{\mathbf{AG}} = 3\mathbf{i} + 3\mathbf{j},$$

$$\therefore \quad \Sigma M_A = 0, \text{ gives}$$

$$0 + (6\mathbf{i} + 6\mathbf{j}) \times (-0.857\mathbf{i} + 0.286\mathbf{j} - 0.429\mathbf{k})\ F_2$$
$$+ (6\mathbf{i} + 6\mathbf{j}) \times (-0.875\mathbf{i} + 0.429\mathbf{j} + 0.286\mathbf{k})\ F_3$$
$$+ (6\mathbf{i} + 6\mathbf{j}) \times (-18\mathbf{j}) + (3\mathbf{i} + 3\mathbf{j}) \times (-6\mathbf{j}) = 0$$

*i.e.,*

$$\begin{vmatrix} \mathbf{i} & \mathbf{j} & \mathbf{k} \\ 6 & 6 & 0 \\ -0.857 & 0.286 & -0.429 \end{vmatrix} F_2 + \begin{vmatrix} \mathbf{i} & \mathbf{j} & \mathbf{k} \\ 6 & 6 & 0 \\ -0.857 & 0.429 & -0.286 \end{vmatrix} F_3$$

$$-108\mathbf{k} - 18\mathbf{k} = 0$$

*i.e.,*

$$[\mathbf{i}(-2.574) - \mathbf{j}(-2.574) + \mathbf{k}(6.858)]\ F_2$$
$$+ [\mathbf{i}(1.716) - \mathbf{j}(1.716) + \mathbf{k}(7.716)]\ F_3 - 108\mathbf{k} - 18\mathbf{k} = 0$$

$$(-2.574\ F_2 + 1.716\ F_3)\mathbf{i} + \mathbf{j}(2.574\ F_2 - 1.716\ F_3)$$
$$+ (6.858\ F_2 + 7.716\ F_3 - 108 - 18)\mathbf{k} = 0$$

*i.e.,*

$$-2.574\ F_2 + 1.716\ F_3 = 0 \quad \text{...(i)}$$

$$2.574\ F_2 - 1.716\ F_3 = 0 \quad \text{...(ii)}$$

$$6.858\ F_2 + 7.716\ F_3 = 126 \quad \text{...(iii)}$$

Equations (*i*) and (*ii*) are identical. They give,

$$F_3 = 1.5\ F_2 \quad \text{...(iv)}$$

Substituting it in Eqn. (*iii*), we get

$$6.858\ F_2 + 1.5 \times 7.716\ F_2 = 126$$

$$\therefore \quad F_2 = 6.836 \text{ kN} \qquad \textbf{Ans.}$$

$$\therefore \quad F_3 = 1.5\ F_2 = 10.254 \text{ kN} \qquad \textbf{Ans.}$$

From force equilibrium condition, we have

$$\mathbf{F}_1 + \mathbf{F}_2 + \mathbf{F}_3 + \mathbf{F}_4 + \mathbf{F}_5 = 0$$

$$(F_{1x}\mathbf{i} + F_{1y}\mathbf{j} + F_{1z}\mathbf{k}) + (-0.857\mathbf{i} + 0.286\mathbf{j} - 0.429\mathbf{k})\ F_2 + (-0.857\mathbf{i} + 0.429\mathbf{j} + 0.286\mathbf{k})\ F_3 - 18\mathbf{j} - 6\mathbf{j} = 0$$

*i.e.*,
$$(F_{1x} - 0.857\ F_2 - 0.857\ F_3)\mathbf{i} + (F_{1y} + 0.286\ F_2 + 0.429\ F_3 - 18 - 6)\mathbf{j} + (F_{1z} - 0.429\ F_2 + 0.286\ F_3)\mathbf{k} = 0$$

*i.e.*,
$$F_{1x} - 0.857\ F_2 - 0.857\ F_3 = 0 \qquad ...(v)$$
$$F_{1y} + 0.286\ F_2 + 0.429\ F_3 = 26 \qquad ...(vi)$$

and
$$F_{1z} - 0.429\ F_2 + 0.268\ F_3 = 0 \qquad ...(vii)$$

From Eqn. (*v*) $\quad F_{1x} = 0.857\ (F_2 + F_3)$

Substituting the values of $F_2$ and $F_3$, we get

$$F_{1x} = 0.857\ (6.836 + 10.254) = 14.646 \text{ kN} \qquad \textbf{Ans.}$$

From Eqn. (*vi*),

$$F_{1y} = 24 - 0.286F_2 - 0.429F_3$$
$$= 24 - 0.286 \times 6.836 - 0.429 \times 10.254$$
$$= 17.646 \text{ kN} \qquad \textbf{Ans.}$$

From Eqn. (*vii*),

$$F_{1z} = 0.429F_2 - 0.286F_3$$

*i.e.*, $\quad F_{1z} = 0 \qquad$ **Ans.**

## QUESTIONS

**3.1** A spherical ball weighing 100 N is held in the position by applying a force of 200 N as shown in Fig. 3.56. Determine the tension in the string and angle α made by the string with the vertical. **[Ans. T** = 400 N, θ = 63.435°]

**3.2** An electric light fixture weighing 150 N is supported by two wires as shown in Fig. 3.57. Determine the tensile forces developed in the wires.

**[Ans.** $\mathbf{F_{OA}}$ = 77.646 N, $\mathbf{F_{OB}}$ = 109.808 N]

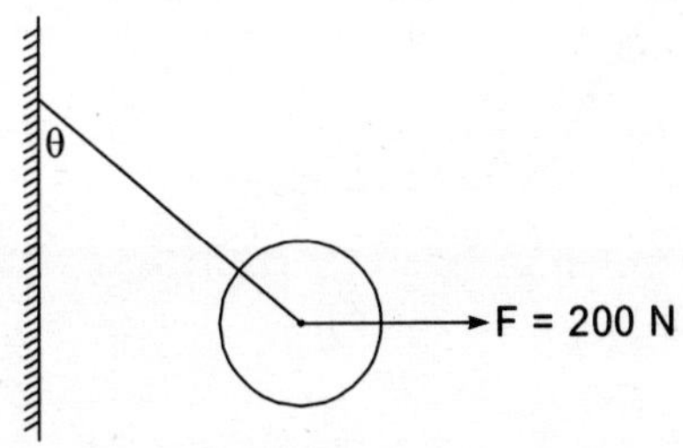


**Fig. 3.56**

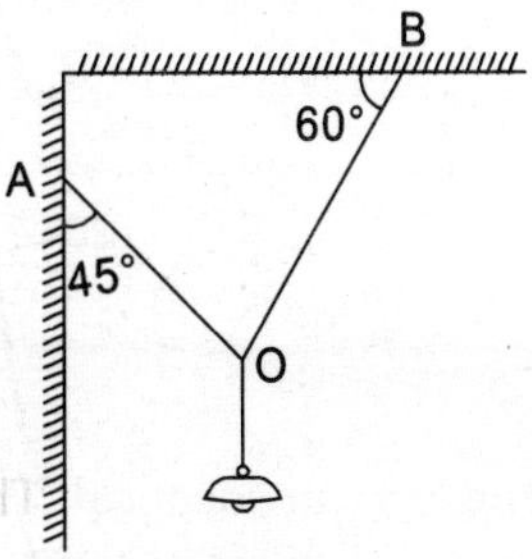


**Fig. 3.57**

**3.3** The frictionless pulley $A$, as shown in Fig. 3.58, is supported by two bars $AB$ and $AC$ which are hinged at $B$ and $C$ to a vertical wall. The flexible cable $DA$ hinged at $D$, goes over the pulley and supports a load of 25 kN at $A$. The angles between the various members are shown in the figure. Determine the forces in the bars $AB$ and $AC$. Neglect the size of pulley and treat it as frictionless. **[Ans.** $F_{AB} = 0$, $F_{AC} = 43.301$ kN]

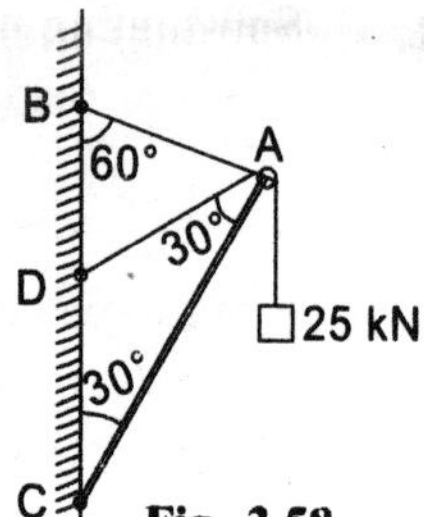

**Fig. 3.58**

**3.4** A three bar link mechanism is in equilibrium under the action of the forces shown in Fig. 3.59. Determine the magnitude of force **F** and the forces developed in each bar.

**[Ans.** $F = 304.72$ N; $F_{BA} = 163.30$ N, $F_{BC} = 223.07$ N and $T_{CD} = 273.20$ N]

**3.5** Two cylinders of diameter 100 mm and 50 mm, weighing 200 N and 50 N respectively are placed in a trough as shown in Fig. 3.60. Assuming all contact surfaces smooth, find the reactions developed at contact surfaces 1, 2, 3, and 4.

**[Ans.** $\mathbf{R_1} = 37.5$ N, $\mathbf{R_2} = 62.5$ N, $\mathbf{R_3} = 287.5$ N and $\mathbf{R_4} = 353.5$ N]

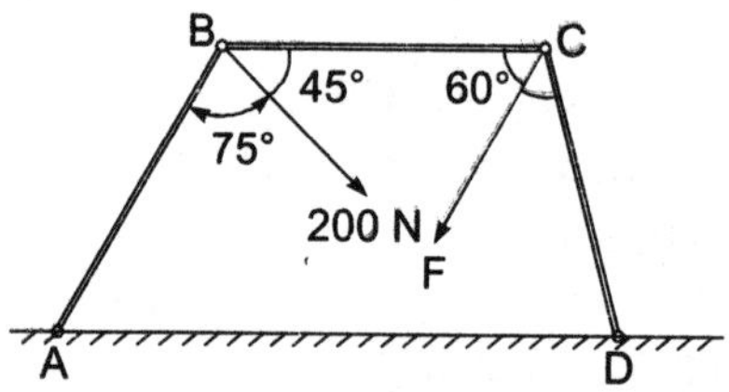

**Fig. 3.59**

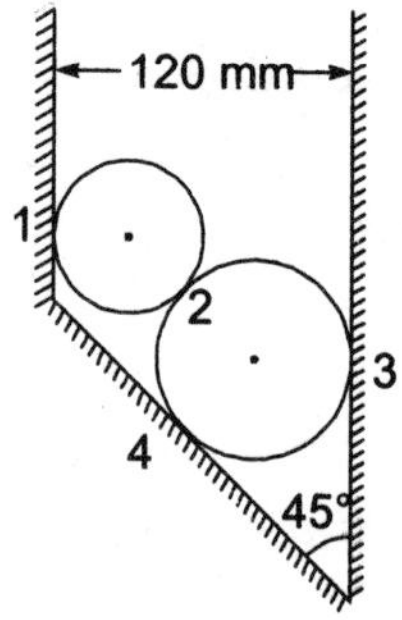

**Fig. 3.60**

**3.6** The tripod shown in Fig. 3.61 is subjected to a vertical force of 40 kN and a horizontal force of 10 kN in $x$-direction at its apex point $D$. Determine the forces developed in its legs. Given $D$ is 3 m above the horizontal plane and $A$, $B$, $C$ are on the circumference of a horizontal circle of radius 3 m. **[Ans.** $F_{AD} = 9.429$ kN, $F_{BD} = 23.574$ kN, $F_{CD} = 23.574$ kN]

[**Note:** $A$ (3, 0, 0), $B$ (–1.5, 0, –2.598), $C$ (–1.5, 0, 2.598), $D$ (0.3, 0)]

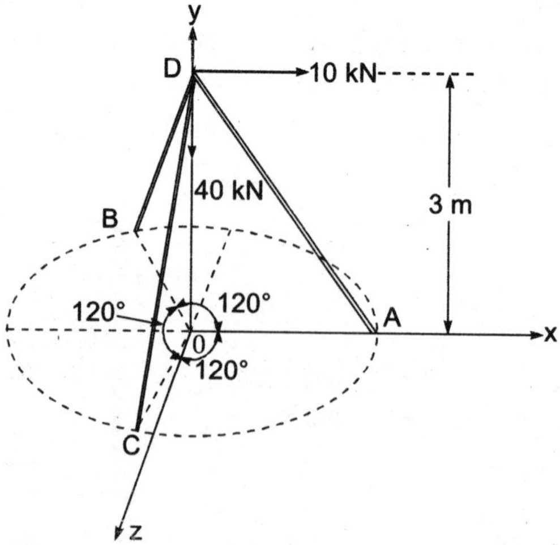

**Fig. 3.61**

**3.7** Beam *AB* 8 m long is hinged at *A* and is supported on roller at *B*. The roller support is inclined at 45° to the horizontal. Determine the reactions developed at supports for the loading shown in Fig. 3.62.

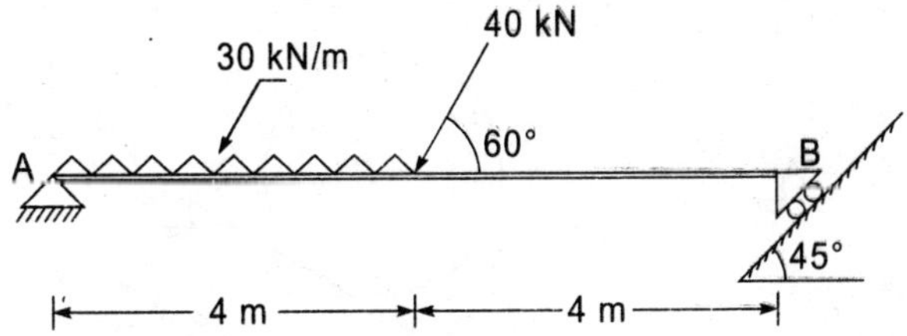


**Fig. 3.62**

[**Ans.** $R_{Ax}$ = 67.32 kN, $R_{Ay}$ = 107.32, $R_B$ = 66.921 kN, normal to support]

**3.8** Determine the reactions at *A* and *B* in the overhanging beam as shown in Fig. 3.63.

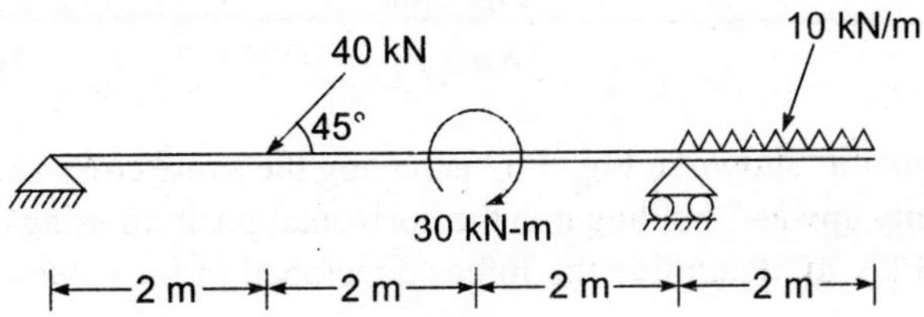


**Fig. 3.63**

[**Ans.** $R_{Ax}$ = 28.28 kN; $R_{Ay}$ = 10.523 kN; $R_B$ = 37.761 kN]

**3.9** Determine the reactions developed at the supports in the beam shown in Fig 3.64.

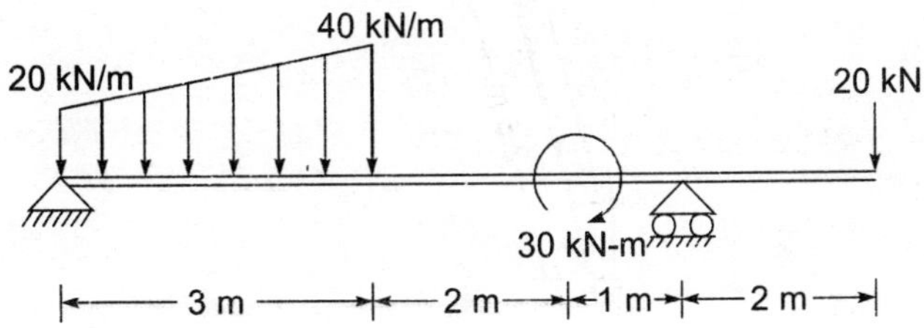


**Fig. 3.64**

[**Ans.** $R_{Ax}$ = 0; $R_{Ay}$ = 53.333 kN; $R_B$ = 56.667 kN]

**3.10** Determine the reactions at supports *A*, *C* and *B* in the compound beam as shown in Fig. 3.65.

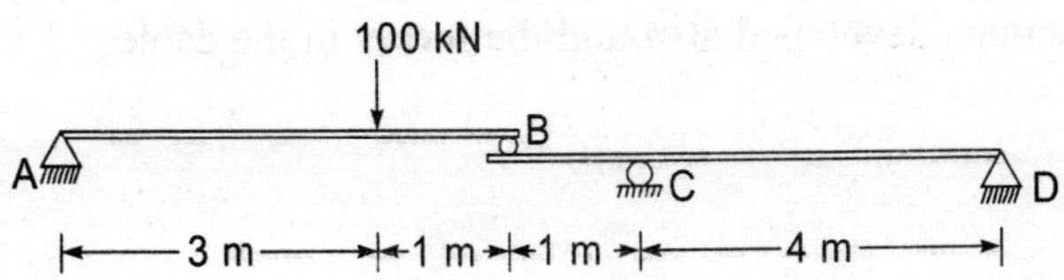


**Fig. 3.65**

[**Ans.** $R_A$ = 25 kN, $R_C$ = 93.75 kN, $R_D$ = –18.75 kN]

**3.11** A 100 N cylinder is supported by a horizontal rod $AB$ and a smooth uniform rod $CD$ which weighs 500 N as shown in Fig. 3.66. Assuming the pins at $A$, $B$, $C$ and $D$ to be frictionless and weight of rod $AB$ is negligible, find the reaction at $C$ and $D$.

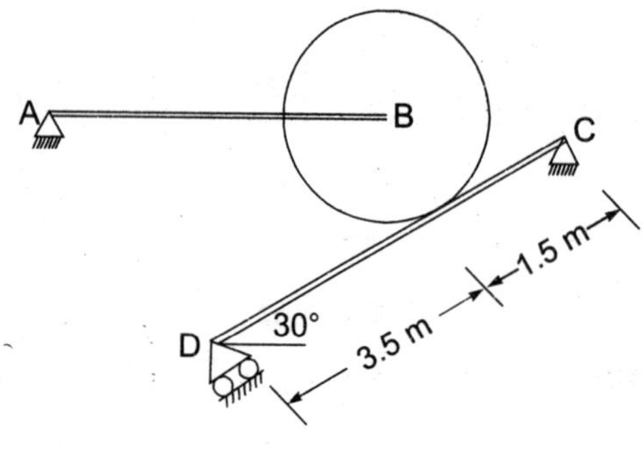


**Fig. 3.66**

[**Ans.** $R_{Cx} = -295.9$ N, $R_{Cy} = 1012.51$ $N$, $R_D = 562.9$ $N$, normal to plane]

**3.12** The $T$ shaped member as shown in Fig. 3.67 is having the same cross-section for web and flange. It was lifted vertically upward, holding it in the horizontal position using three cables. If the weight of the member is 3 kN/m, determine the forces developed in the cables.

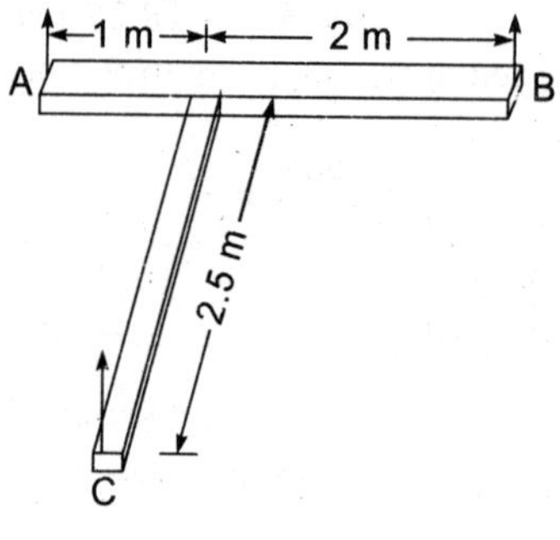


**Fig. 3.67**

[**Ans.** $F_A = 7$ kN; $F_B = 5.75$ kN; $F_C = 3.75$ kN]

**3.13** Boom $AB$ of a crane is having uniform cross-section and weighs 5 kN. It is supported with ball and socket at $A$. It is held in position shown in Fig. 3.68 with the help of cables $BC$ and $DE$. Determine the reactions developed at $A$ and the forces in the cables.

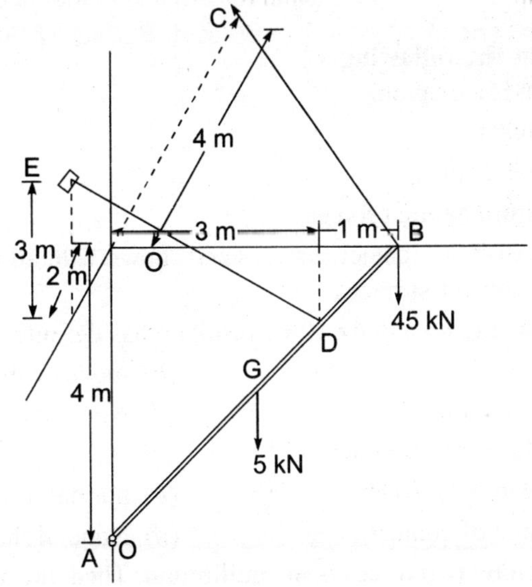


**Fig. 3.68**

[**Ans.** $F_{Ax} = 31.582$ kN, $F_{Ay} = 21.837$ kN, $F_{Az} = -3.515$ kN, $F_{BC} = 14.916$ kN, $F_{OE} = 37.902$ kN]

**3.14** The rigid body, as shown in Fig. 3.69, is having the ball and socket joint at $d'$ and a roller support $b$ which gives reactions only in the horizontal directions. A load of 30 kN is applied at $c'$ and the body is held in the position shown by using a cable connection $CE$. Neglecting the self weight, determine the reactions developed.

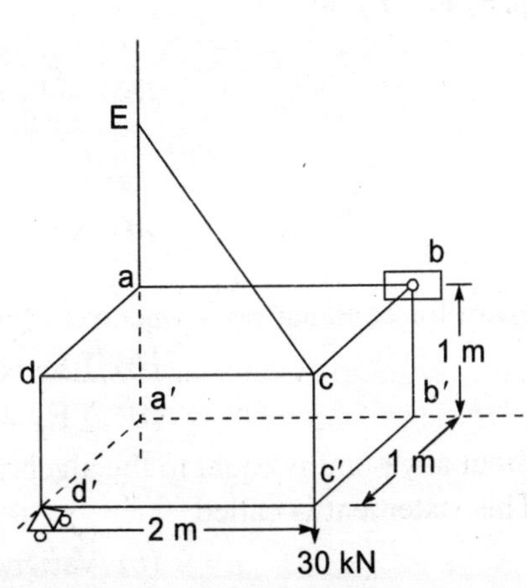


**Fig. 3.69**

[**Ans.** $F_{d'x} = 24$ kN; $F_{d'y} = 12$ kN; $F_{d'z} = 0$; $F_{bx} = 0$; $F_{b'z} = 12$ kN; $F_{CE} = 32.31$ kN]

## OBJECTIVE TYPE QUESTIONS

**Select the correct option in the following:**

**3.1** A free body diagram is a diagram
(*a*) drawn by free hand
(*b*) of a freely suspended body
(*c*) of a body free from applied forces
(*d*) of a body freed from its contact surfaces and shown with all forces acting on it, including reactions from contact surfaces

**3.2** A system of forces acting on a body, if the resultant has definite value, the body
(*a*) must be at rest
(*b*) moves with uniform motion
(*c*) moves with accelerations
(*d*) None of the above

**3.3** From a smooth surface, reaction acts
(*a*) parallel to the plane of contact
(*b*) normal to the plane of contact
(*c*) at 45° to the plane of contact
(*d*) None of the above

**3.4** A particle acted upon by two forces is in equilibrium. Then, the angle between the two forces is equal to
(*a*) 0°
(*b*) 45°
(*c*) 90°
(*d*) 180°

**3.5** If a body is in equilibrium, it is concluded that
(*a*) no force is acting on it
(*b*) the resultant force on it is zero
(*c*) the moment of forces about any point is zero
(*d*) both (*b*) and (*c*)

**3.6** If the angle between $F_1$ and $F_2$ is $\alpha$, $F_2$ and $F_3$ is $\beta$ and $F_3$ and $F_1$ is $\gamma$, then Lami's theorem for the concurrent forces $F_1$, $F_2$ and $F_3$ is

(*a*) $\dfrac{F_1}{\sin\alpha} = \dfrac{F_2}{\sin\beta} = \dfrac{F_3}{\sin\gamma}$

(*b*) $\dfrac{F_1}{\sin\beta} = \dfrac{F_2}{\sin\lambda} = \dfrac{F_3}{\sin\alpha}$

(*c*) $\dfrac{F_1}{\sin\gamma} = \dfrac{F_2}{\sin\alpha} = \dfrac{F_3}{\sin\beta}$

(*d*) None of the above

**3.7** The condition of equilibrium for coplanar non-concurrent forces is
(*a*) $\Sigma F_x = 0, \Sigma F_y = 0$
(*b*) $\Sigma F_x = 0, \Sigma M = 0$
(*c*) $\Sigma F_y = 0, \Sigma M = 0$
(*d*) $\Sigma F_x = 0, \Sigma F_y = 0, \Sigma M = 0$

**3.8** The moment of a force about any point is equal to the algebraic sum of the components of the forces about that point. This statement is called
(*a*) Lami's theorem
(*b*) Varignon's theorem
(*c*) Law of transmissibility
(*d*) Parallelogram law

**3.9** If a system of three forces is in equilibrium, they must be
(*a*) Collinear
(*b*) Concurrent
(*c*) Non-concurrent
(*d*) None of the above

# CHAPTER

# 4

# ANALYSIS OF PINJOINTED FRAMES

## 4.1 INTRODUCTION

A truss is a structure made up of slender members pin-connected at ends and is capable of taking loads at joints. They are used as roof trusses to support sloping roofs and as bridge trusses to support deck. In many machines, steel trusses are used. Transmission towers are also the examples of trusses. In the case of wooden trusses, the ends are connected by making suitable joints or by nailing and bolting, whereas in steel trusses, ends are connected by bolting or welding. The trusses are also known as '*pinjointed frames*'.

A truss in which all the members lie in a single plane is called as a *plane truss*. In such trusses, loads act in the plane of the truss only. Roof trusses and bridge trusses can be considered as plane trusses. If all the members of a truss do not lie in a single plane, then it is called a space truss. Tripod and transmission towers are the examples of space trusses. In this chapter, the analysis of only plane trusses is considered.

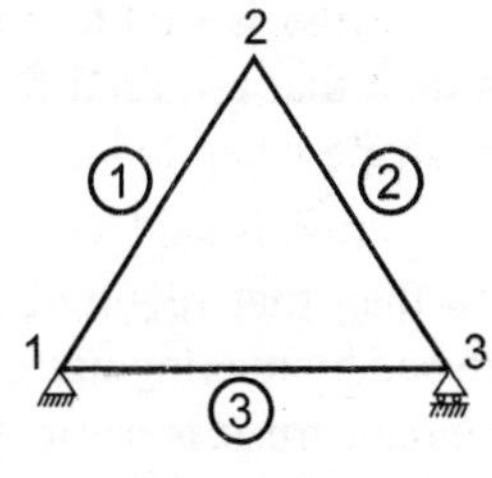


Fig. 4.1

## 4.2 PERFECT, DEFICIENT AND REDUNDANT TRUSSES

A pinjointed truss which has got just sufficient number of members to resist the loads without undergoing appreciable deformation in shape is called a *perfect truss*. Triangular truss is the simplest perfect truss and it has three joints and three members (Fig. 4.1). Perfect trusses with four and five joints are shown in Figs. 4.2 and 4.3 respectively.

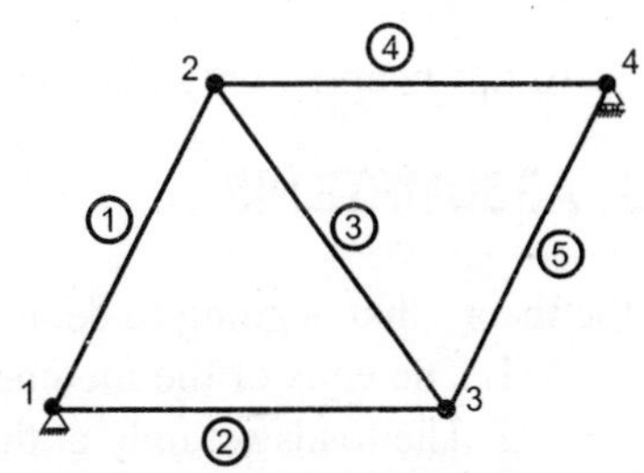


Fig. 4.2

It may be observed that to increase one joint in a perfect truss, two more members are required. Hence, the following expression may be written as the relationship between number of joints $j$, and the number of members $m$, in a perfect truss,

$$m = 2j - 3 \quad \text{...(4.1)}$$

However, the above equation gives only a necessary, but not a sufficient condition of a perfect truss. For example, the two trusses shown in Figs. 4.4 (*a*) and (*b*) have the same number of members and joints. The truss shown in Fig. 4.4 (*a*) is perfect, whereas the one shown in Fig. 4.4 (*b*) is not capable of retaining its shape if loaded at the joint marked 6. Therefore, the only necessary and sufficient condition of a perfect truss is that it should retain its shape when load is applied at any joint in any direction.

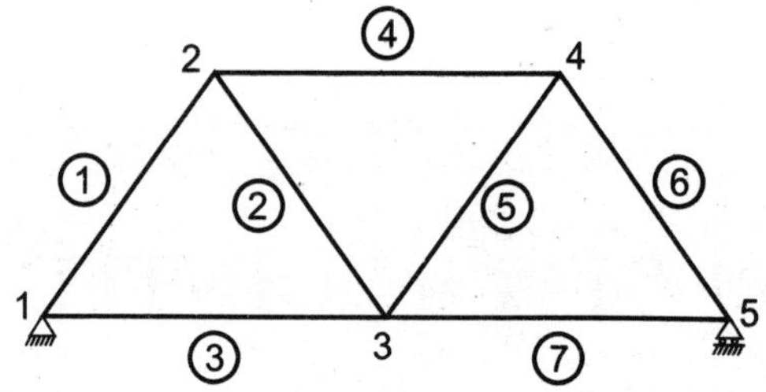


**Fig. 4.3**

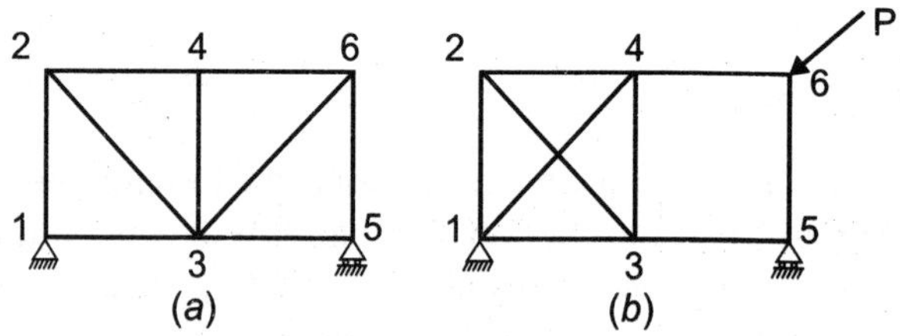


**Fig. 4.4**

A truss is said to be *deficient* if the number of members in it is less than that required for a perfect truss. Such trusses cannot retain their shape when loaded. A deficient truss is shown in Fig. 4.5.

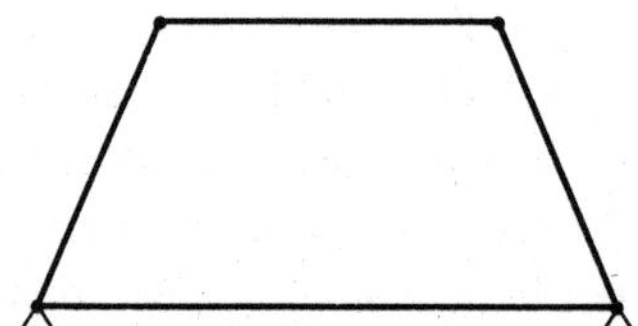

**Fig. 4.5**

A truss is said to be *redundant* if the number of members in it is more than that required in a perfect truss. Such trusses cannot be analysed by making use of the equations of equilibrium alone. Thus, a redundant truss is statically indeterminate. Each extra member adds one degree of indeterminancy. For the analysis of such members, the consistency of deformations is to be considered. The truss shown in Fig. 4.6 is a typical redundant truss. In this truss, one diagonal member in each panel is extra. Hence, it is a two-degree redundant truss.

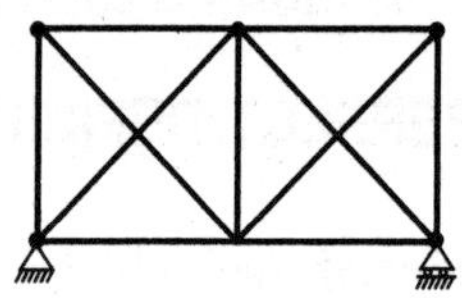

**Fig. 4.6**

In this chapter, only the analysis of perfect frames is considered.

## 4.3 ASSUMPTIONS

In the theory that is going to be developed in this chapter, the following assumptions are made:

1. The ends of the members are pin-connected (hinged);
2. The loads act only at the joints;
3. Self-weights of the members are negligible;
4. Cross-sections of the members are uniform;
5. If the cross-section of a member in different panels along a line varies, the centre of gravity of the section is assumed to be located along the same longitudinal line.

In reality, the members are connected by bolting, riveting or by welding. No special care is taken to ensure perfect pin-connections. However, experiments have shown that assuming pin-connected ends is quite satisfactory, since the members used are slender.

In most of the frames, the loads act at the joints. Even if a load is not acting at a joint, it can be replaced by its reaction at the joint and a local bending effect on the member. The frame may be analysed for the joint loads and the local bending effect on the member superposed in the design of that member.

In most of the trusses, the self-weight is really small compared to the loads they carry. Hence, self-weight of the members may be neglected.

It is the duty of construction engineer to see that the centroid of all cross-sections lie along a single axis so that the member is held in equilibrium by the two forces acting at its ends.

Because of the assumption of pin-connected ends, it is more appropriate to call the theory that is going to be developed in this chapter as *analysis of pin-connected plane trusses.* Analysis of rigid frames is not covered in this book.

## 4.4 NATURE OF FORCES IN MEMBERS

The members of a truss are subjected to either tensile or compressive forces. A typical truss *ABCDE* loaded at joint *E* is shown in Fig. 4.7 (*a*). The member *BC* is subjected to compressive force **C** as shown in Fig. 4.7 (*b*). Effect of this force on the joint *B* (or *C*) is equal and opposite to the force **C** as shown in the figure.

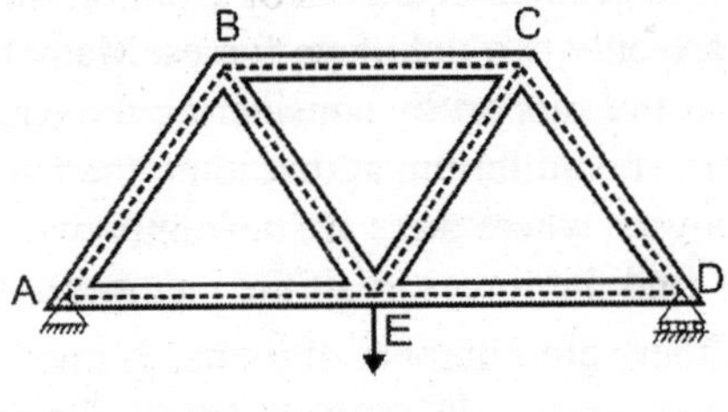


**Fig. 4.7(*a*)**

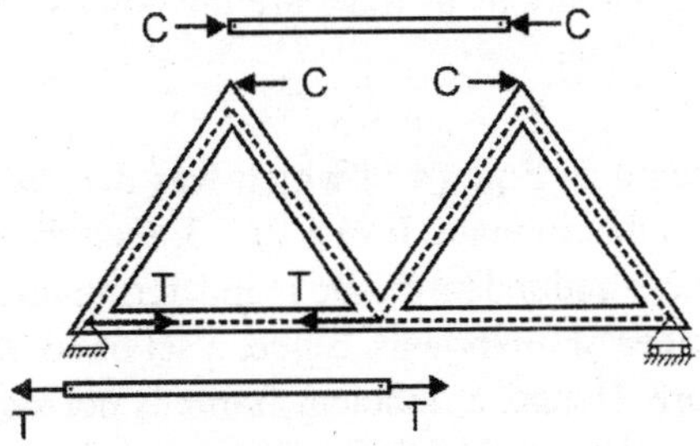


**Fig. 4.7(*b*)**

The member *AE* is subjected to tensile force **T**. Its effect on the joints *A* and *E* are as shown in Fig. 4.7 (*b*). In the analysis of frame, we mark the forces on the joints, instead of the forces in the members as shown in Fig. 4.7 (*c*). It may be noted that compressive force in a member is represented in a figure by two arrows going away from each other and a tensile force by two arrows coming towards each other. This is quite logical considering the fact that the markings on the members represent the internal reactive forces developed which are opposite in direction to the applied forces.

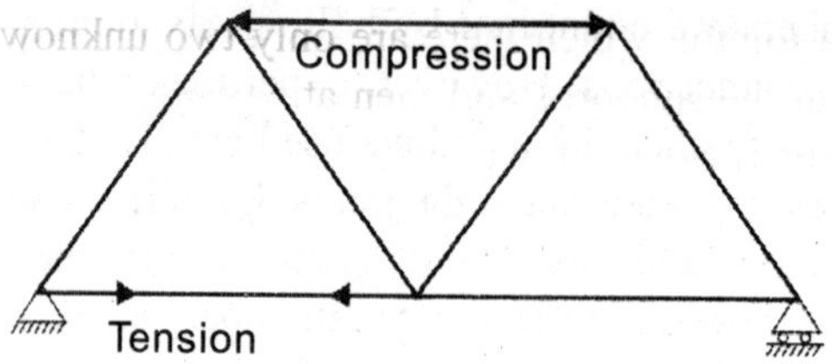


Fig. 4.7(*c*)

## 4.5 METHODS OF ANALYSIS

The following three methods are available for the analysis of pin-connected frames:

1. Method of joints
2. Method of section
3. Graphical method.

The first two are analytical methods and they are dealt in this chapter. For graphical method, any book on graphic statics may be referred.

## 4.6 METHOD OF JOINTS

At each joint, the forces in the members meeting and the loads acting, if any, constitute a system of concurrent forces. Hence, two independent equations of equilibrium can be formed at each joint. First, a joint is selected where there are only two unknown forces. Many times such a joint can be identified only after finding the reaction at the support by considering the equilibrium of the entire frame. Then making use of the two equations of equilibrium at that joint, the two unknown forces are found. Then, the next joint is selected for analysis, where there are now only two unknown forces. Thus, the analysis proceeds from joints to joint to find the forces in all the members.

It may be noted that if there are $j$ number of joints, $2j$ number of the equations can be formed. There will be three reactions in a general determinate truss. The force in each member is unknown. Hence, if there are $m$ number of members, the total number of unknowns will be $m + 3$. A problem can be analysed if there are as many equations as there are unknowns. Hence, a frame analysis problem is determinate if:

$$2j = m + 3 \qquad \text{...(4.2)}$$

This equation is the same as Eqn. (4.1) which was derived on the consideration of a perfect frame. Hence, a perfect frame is determinate. If $m > 2j - 3$, then the number of unknowns is more than the number of equations. Hence, a redundant frame is indeterminate. If $m < 2j - 3$, then the number of equations is more than the number of unknowns. Since, a set of solutions can satisfy such equations, it shows instability of the structure. Hence, a deficient frame is not stable.

The method of joints is illustrated with Examples 1 to 7.

**Example 4.1.** Find the forces in all the members of the truss shown in Fig. 4.8(*a*). Tabulate the results.

**Solution. Step 1:** Determine the inclinations of all inclined members. In this case,

$$\tan\theta = \frac{3}{3} = 1$$

$$\therefore \quad \theta = 45°$$

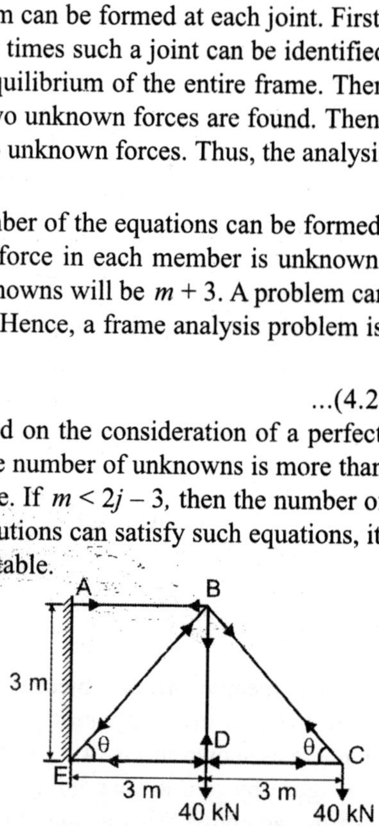


Fig. 4.8(*a*)

**Step 2:** Look for a joint at which there are only two unknowns. If such a joint is not available, determine the reactions at the supports, and then at the supports, unknowns may reduce to only two.

Now at joint *C*, there are only two unknowns, *i.e.*, forces in members *CB* and *CD*, say $F_{CB}$ and $F_{CD}$.

**Note:** Usually in cantilever type frames, we find such joints without the need to find reactions.

**Step 3:** Now there are two equations of equilibrium for the forces meeting at the joint and two unknown forces. Hence, the unknown forces can be determined.

At joint *C* [Ref. Fig. 4.8 (*b*)], $\Sigma V = 0$, condition shows that the force $\mathbf{F}_{CB}$ should act away from the joint *C* so that its vertical component balances the vertical downward load

$$F_{CB} \sin 45° = 40$$

$$\therefore \quad F_{CB} = 40\sqrt{2} \text{ kN}$$

Now $\Sigma H = 0$ indicates that $F_{CD}$ should act towards *C*.

$$F_{CD} - F_{CB} \cos 45° = 0$$

$$F_{CD} = F_{CB} \cos 45° = 40\sqrt{2} \times \frac{1}{\sqrt{2}} = 40 \text{ kN}$$

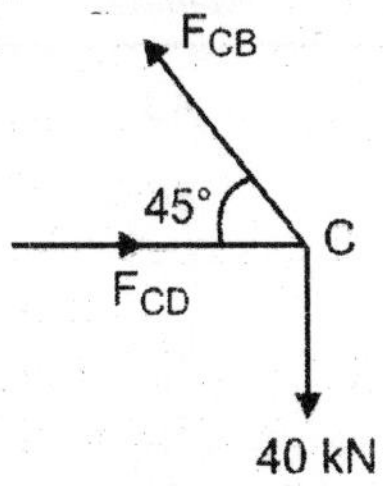

**Fig. 4.8(*b*)**

**Note:** If the assumed direction of unknown force is opposite, the value will be negative. Then reverse the direction and proceed.

**Step 4:** On the diagram of the truss, mark arrows on the members near the joint analysed to indicate the forces on the joint. At the other end, mark the arrows in the reverse direction.

In the present case, near the joint *C*, the arrows are marked on the members *CB* and *CD* to indicate forces $\mathbf{F}_{CB}$ and $\mathbf{F}_{CD}$ directions as found in the analysis of joint *C*. Then reversed directions are marked in the members *CB* and *CD* near joints *B* and *D*, respectively.

**Step 5:** Look for the next joint where there are only two unknown forces and analyse that joint.

In this case, there are only two unknown forces at the joint *D* as shown in Fig. 4.8 (*c*).

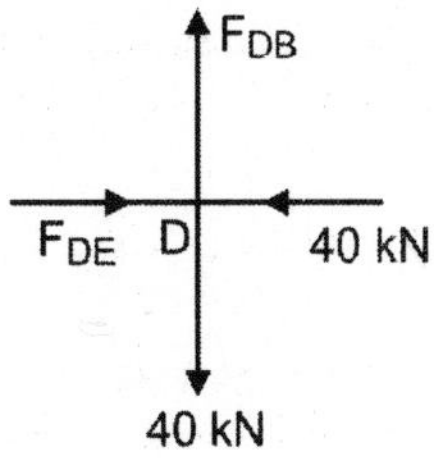

**Fig. 4.8(*c*)**

$$\Sigma V = 0, \text{ gives}$$

$$F_{DB} = 40 \text{ kN}$$

$$\Sigma H = 0, \text{ gives}$$

$$F_{DE} = 40 \text{ kN}$$

**Step 6:** Repeat Steps 4 and 5, till forces in all the members are found.

In the present case, after marking the forces in the members *DB* and *DE*, we find that analysis of joint *B* can be taken up. Referring to Fig. 4.8 (*d*).

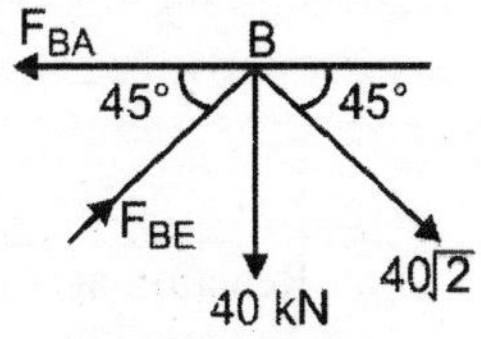

**Fig. 4.8(*d*)**

$$\Sigma V = 0, \text{ gives}$$

$$F_{BE} \sin 45° - 40 - 40\sqrt{2} \times \sin 45° = 0$$

$$\therefore \quad F_{BE} = 80\sqrt{2} \text{ kN}$$

$$\Sigma H = 0, \text{ gives}$$

$$F_{BA} - F_{BE} \cos 45° - 40\sqrt{2} \times \cos 45° = 0$$

$$F_{BA} = 80\sqrt{2} \times \frac{1}{\sqrt{2}} + 40\sqrt{2} \times \frac{1}{\sqrt{2}}$$

$$\therefore \quad F_{BA} = 120 \text{ kN}$$

The directions of these forces are marked on the diagram. Now the analysis is complete, since, the forces in all the members are determined.

**Step 7:** Determine the nature of forces in each member and tabulate the results. Note that if the arrow marks on a member are towards each other, then the member is in tension and if the arrow marks are away from each other, the member is in compression [Ref. Fig. 4.8 (*e*)]. In this case,

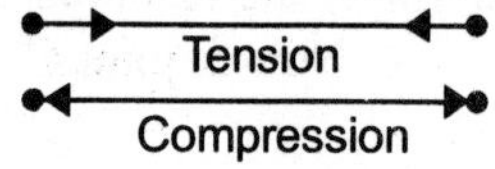


**Fig. 4.8(*e*)**

| Member | Magnitude of force in kN | Nature |
|---|---|---|
| AB | 120 | Tension |
| BC | $40\sqrt{2}$ | Tension |
| CD | 40 | Compression |
| DE | 40 | Compression |
| BE | $80\sqrt{2}$ | Compression |
| BD | 40 | Tension |

**Example 4.2.** Determine the forces in all the members of the truss as shown in Fig. 4.9(*a*) and indicate the magnitude and nature of forces on the diagram of the truss. All inclined members are at 60° to horizontal and length of each member is 2 m.

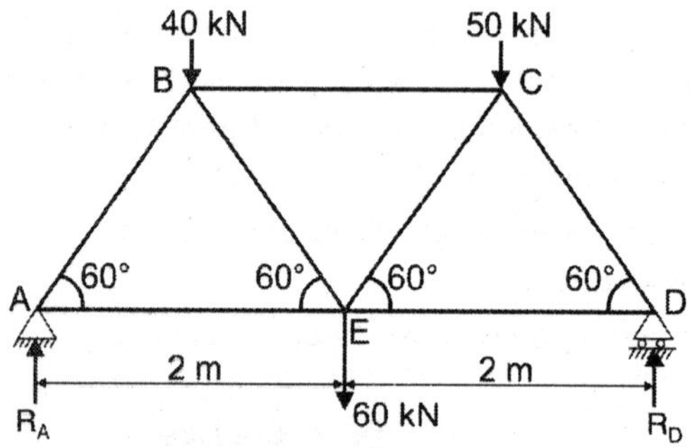


**Fig. 4.9(*a*)**

**Solution:** Now, we cannot find a joint with only two unknown forces without finding reactions.

Consider the equilibrium of the entire frame.

$$\Sigma M_A = 0, \text{ gives}$$

$$R_D \times 4 - 40 \times 1 - 60 \times 2 - 50 \times 3 = 0$$

$$\therefore \quad R_D = 77.5 \text{ kN}$$

$$\Sigma H = 0, \text{ gives}$$

$$\therefore \quad H_A = 0$$

$\therefore$ Reaction at $A$ is vertical only

$$\Sigma V = 0, \text{ gives}$$

$$R_A + 77.5 = 40 + 60 + 50$$

$$\therefore \quad R_A = 72.5 \text{ kN}$$

Joint $A$: $\quad \Sigma V = 0$, gives

$$F_{AB} \sin 60° = R_A = 72.5$$

$$F_{AB} = 83.716 \text{ kN (Comp.)}$$

$$\Sigma H = 0, \text{ gives}$$

$$F_{AE} - 83.716 \cos 60° = 0$$

$$F_{AE} = 41.858 \text{ kN (Tension)}$$

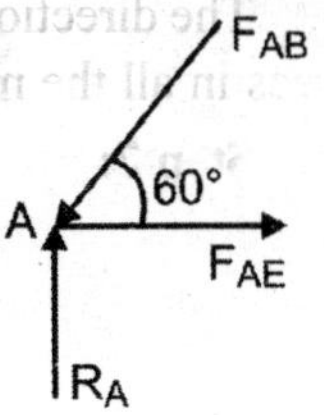


Fig. 4.9(*b*)

Joint $D$: $\Sigma V = 0$, gives

$$F_{DC} \sin 60° = R_D = 77.5$$

$$\therefore \quad F_{DC} = 89.489 \text{ kN (Comp.)}$$

$$\Sigma H = 0, \text{ gives}$$

$$F_{DE} - 87.489 \cos 60° = 0$$

$$\therefore \quad F_{DE} = 44.745 \text{ kN (Tension)}$$

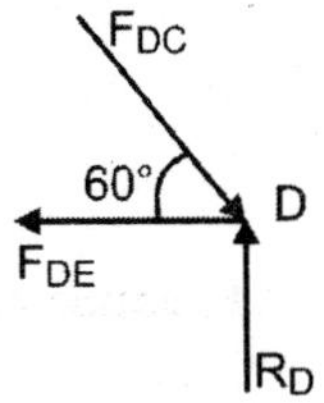


Fig. 4.9(*c*)

Joint $B$: $\Sigma V = 0$, gives

$$F_{BE} \sin 60° - F_{AB} \sin 60° + 40 = 0$$

$$\therefore \quad F_{BE} = \frac{72.5 - 40}{\sin 60°} = 37.528 \text{ (Tension)}$$

$$\Sigma H = 0, \text{ gives}$$

$$F_{BC} - F_{AB} \cos 60° - F_{BE} \cos 60° = 0$$

$$F_{BC} = (83.716 + 37.527) \times 0.5$$

$$F_{BC} = 60.624 \text{ kN (Comp.)}$$

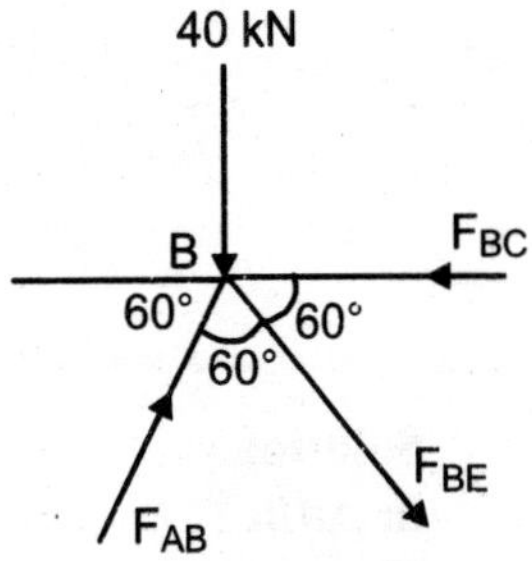


Fig. 4.9(*d*)

Joint $C$: $\Sigma V = 0$, gives

$$F_{CE} \sin 60° + 50 - F_{DC} \sin 60° = 0$$

$$F_{CE} = \frac{77.5 - 50}{\sin 60°} = 31.754 \text{ kN (Tension)}$$

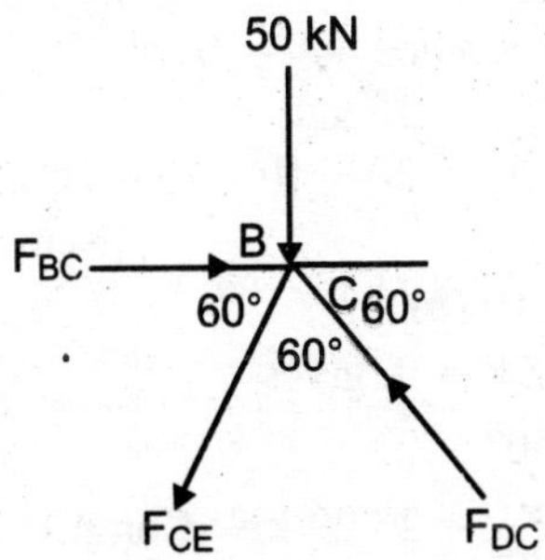


Fig. 4.9(*e*)

Now the forces in all the members are known. If joint $E$ is analysed, it will give the check for the analysis. The results are shown on the diagram of the truss in Fig. 4.9 (*f*).

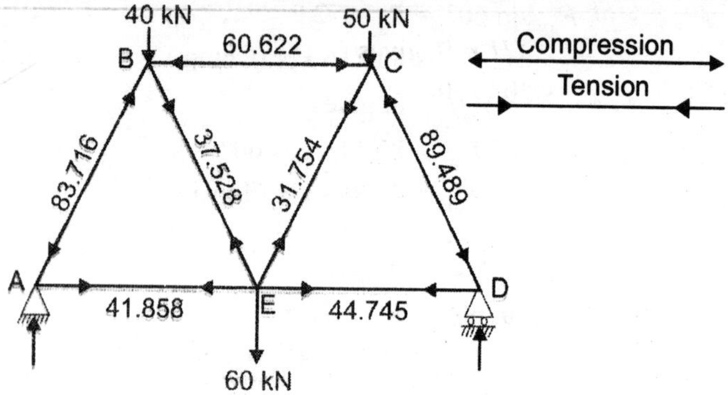


**Fig. 4.9(*f*)**

**Example 4.3.** Determine the forces in all the members of the truss shown in Fig. 4.10 (*a*).

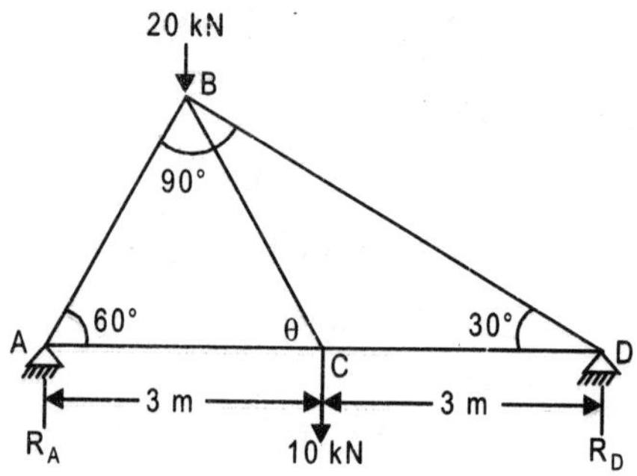


**Fig. 4.10(*a*)**

**Solution.** $AB = 6 \sin 30° = 3$ m; $BD = 6 \cos 30° = 23.09$ m

In $\Delta ABC$, $AC = AB = 3$ m

$\therefore$ $\angle ABC = \angle BCA$ = say $\theta$

$\therefore$ $2\theta = 180 - 60 = 120$

or $\theta = 60°$

Taking moment about $A$, we get

$$R_D \times 6 - 10 \times 3 - 20 \times 3 \cos 60° = 0$$

$$R_D = 10 \text{ kN}$$

$\therefore$ $R_A = 20 + 10 - 10 = 20$ kN

[**Note:** $H_A = 0$]

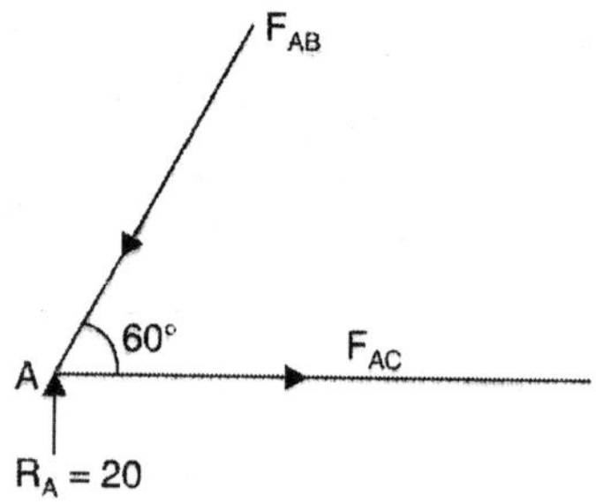


**Fig. 4.10(*b*)**

Joint $A$: Referring to Fig. 4.10 (*b*)

$\Sigma V = 0$, gives

$$20 - F_{AB} \sin 60° = 0$$

$\therefore$ $F_{AB} = 23.09$ kN (Comp.)

$\Sigma H = 0$, gives

$F_{AC} - F_{AB} \cos 60° = 0$ $\therefore$ $F_{AC} = 23.09 \cos 60° = 11.55$ kN (Tensile)

Joint $D$:

$\Sigma V = 0$, gives

$F_{DB} \sin 30° = 10$ or $F_{DB} = 20$ kN (Comp.)

$$\Sigma H = 0, \text{ gives}$$

$$-F_{DC} + F_{DB} \cos 30° = 0$$

*i.e.,* $$F_{DC} = 20 \cos 30° = 17.32 \text{ kN (Tensile)}$$

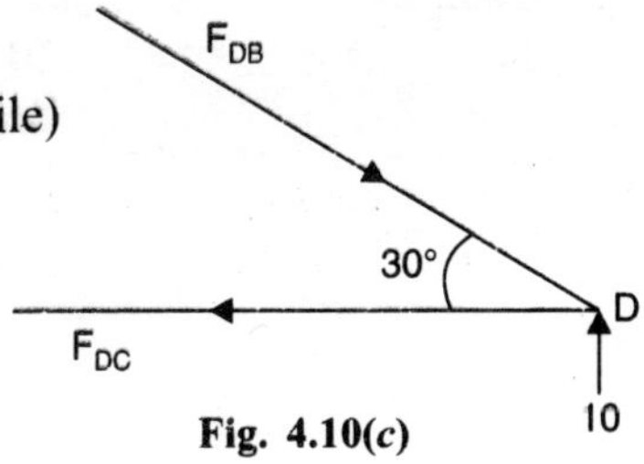


**Fig. 4.10(c)**

Joint $C$: Referring to Fig. 4.10($d$),

$$\Sigma V = 0, \text{ gives}$$

$$F_{CB} \sin 60° = 10$$

$\therefore$ $$F_{CB} = 11.55 \text{ kN}$$

Check $$\Sigma H = 0, \text{ gives}$$

$$F_{CB} \cos 60° = F_{CD} - F_{CA} = 17.32 - 11.55$$

$\therefore$ $$F_{CB} = 11.55, \quad \text{Checked}$$

These forces are marked in Fig. 4.10 ($e$).

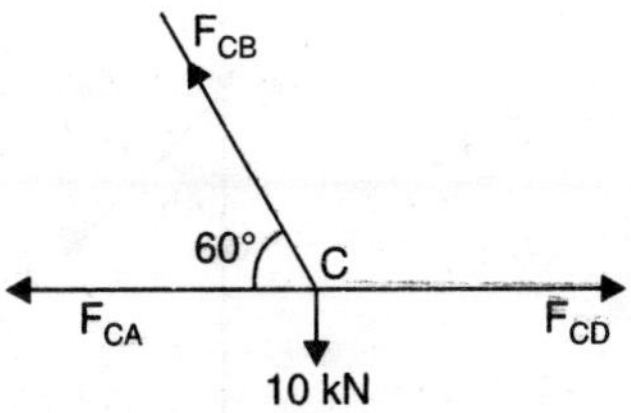


**Fig. 4.10($d$)**

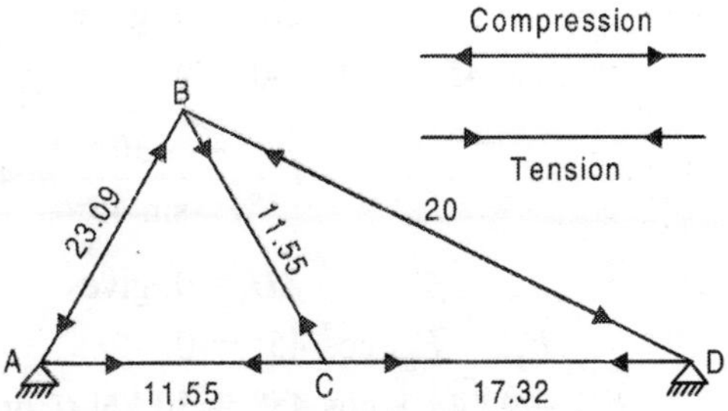


**Fig. 4.10($e$)**

**Example 4.4.** Determine the forces in the members of truss shown in Fig. 4.11($a$).

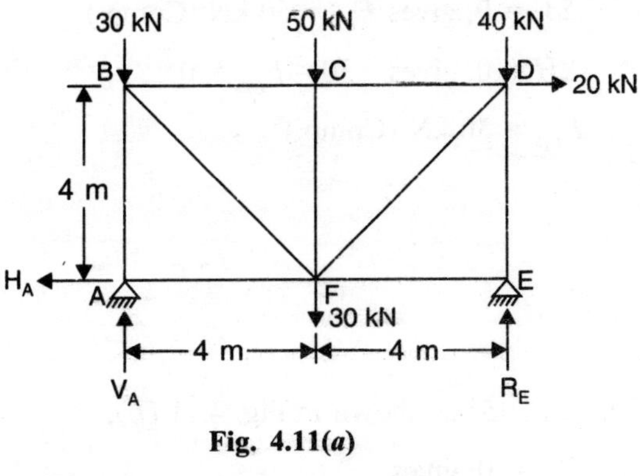


**Fig. 4.11($a$)**

**Solution.** Sum of forces in horizontal direction = 0, gives

$$H_A = 20 \text{ kN}$$

$$\Sigma M_A = 0, \text{ gives}$$

$$-R_E \times 8 + 50 \times 4 + 40 \times 8 + 20 \times 4 + 30 \times 4 = 0$$

$\therefore$ $$R_E = 90 \text{ kN}$$

$$\Sigma V_A = 0, \text{ gives}$$

$$V_A + R_E = 30 + 50 + 40 + 30$$

$\therefore$ $$V_A = 150 - R_E = 150 - 90 = 60 \text{ kN}$$

Joint $A$: $\Sigma V = 0$, gives $F_{AB} = 60$ kN (Comp.)

$\Sigma H = 0$, gives $F_{AF} = 20$ kN (Tensile)

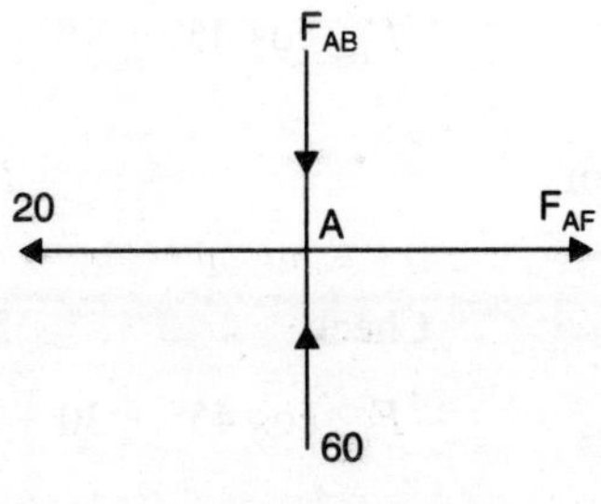


**Fig. 4.11($b$)**

Joint $E$:

$$\Sigma V = 0, \text{ gives, } \quad F_{ED} = 90 \text{ kN (Comp.)}$$
$$\Sigma H = 0, \text{ gives, } \quad F_{EF} = 0$$

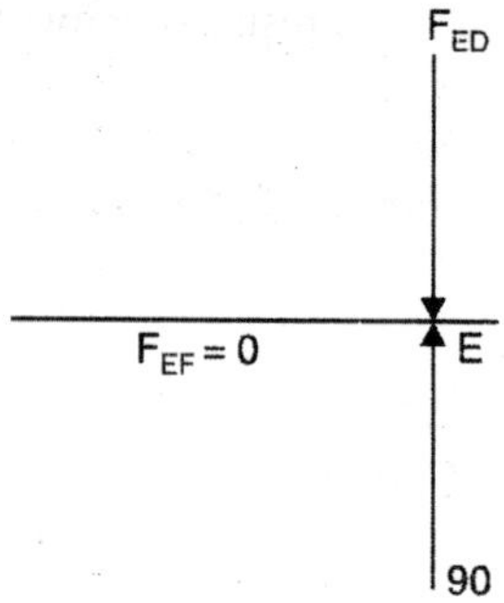


**Fig. 4.11(*c*)**

Joint $B$: Note that inclined member is at 45°,

$$\Sigma V = 0; \text{ gives,}$$
$$-F_{BF} \sin 45° - 30 + 60 = 0$$

or
$$F_{BF} = \frac{30}{\sin 45°} = 42.43 \text{ kN (Tensile)}$$

$$\Sigma H = 0, \text{ gives}$$
$$-F_{BC} + F_{BF} \cos 45° = 0$$

or
$$F_{BC} = 42.43 \times \cos 45° = 30 \text{ kN (Comp.)}$$

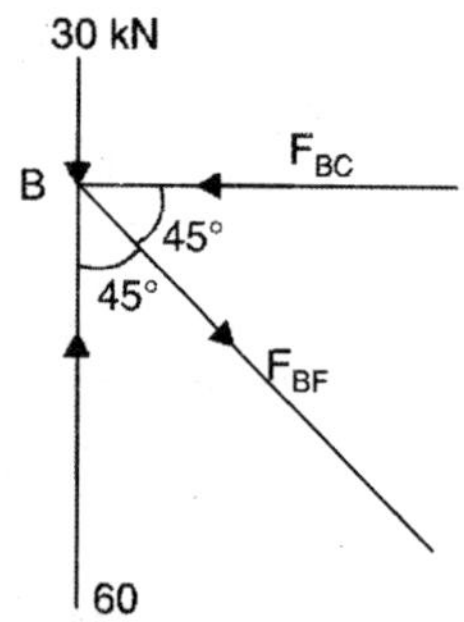


**Fig. 4.11(*d*)**

Joint $C$:
$$\Sigma V = 0, \text{ gives } F_{CF} = 50 \text{ kN (Comp.)}$$
$$\Sigma H = 0, \text{ gives } \quad 30 - F_{CD} = 0$$

or
$$F_{CD} = 30 \text{ kN (Comp.)}$$

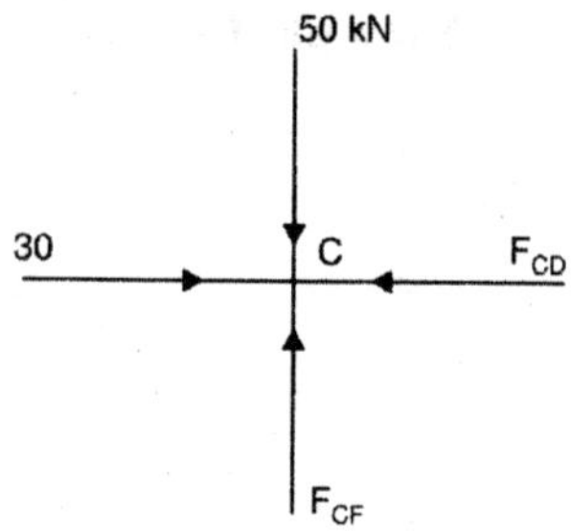


**Fig. 4.11(*e*)**

Joint $D$: Note that $DF$ is at 45° as shown in Fig. 4.11 (*f*),

$$\Sigma V = 0, \text{ gives}$$
$$-F_{DF} \cos 45° + 90 - 40 = 0$$

or
$$F_{DF} = \frac{50}{\cos 45°} = 70.71 \text{ kN (Tensile)}$$

Check
$$\Sigma H = 0, \text{ gives}$$
$$-F_{DF} \cos 45° + 30 + 20 = 0 \text{ or } F_{DF} = 70.71 \text{ kN (Checked.)}$$

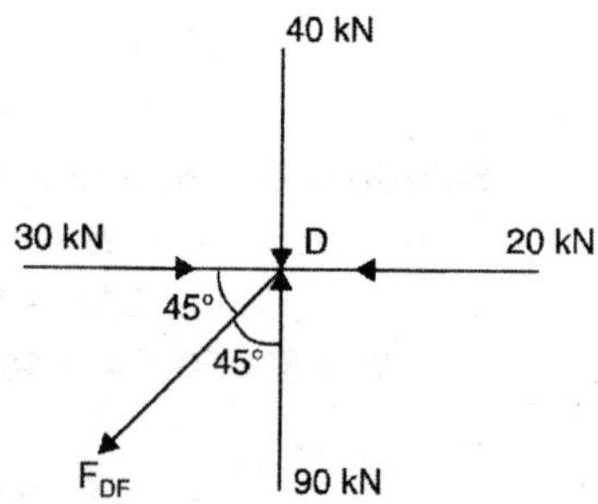


**Fig. 4.11(*f*)**

Final result is shown in Fig. 4.11($g$).

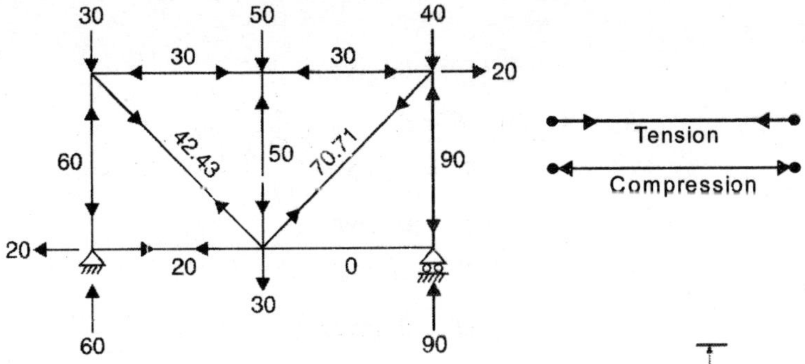


**Fig. 4.11($g$)**

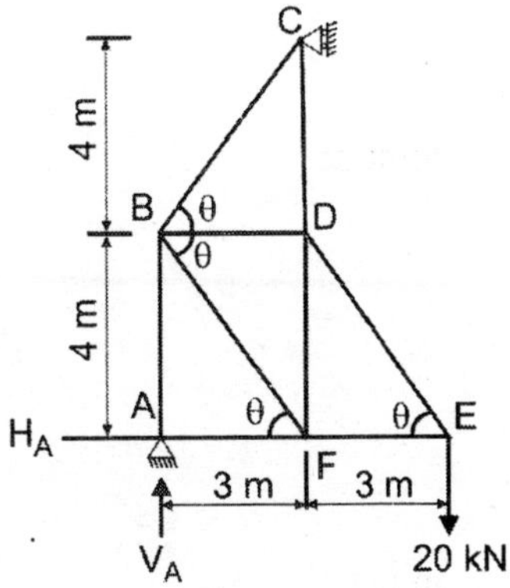


**Fig. 4.12($a$)**

**Example 4.5.** Analyse the truss shown in Fig. 4.12($a$).

**Solution.** All inclined members have the same inclination to horizontal. Now, length of an inclined member is

$$= \sqrt{3^2 + 4^2} = 5 \text{ m}$$

$$\therefore \qquad \sin\theta = \frac{4}{5}$$

$$\cos\theta = \frac{3}{5}$$

As soon as a joint is analysed, the forces on the joint are marked on members [Fig. 4.12 ($b$)]

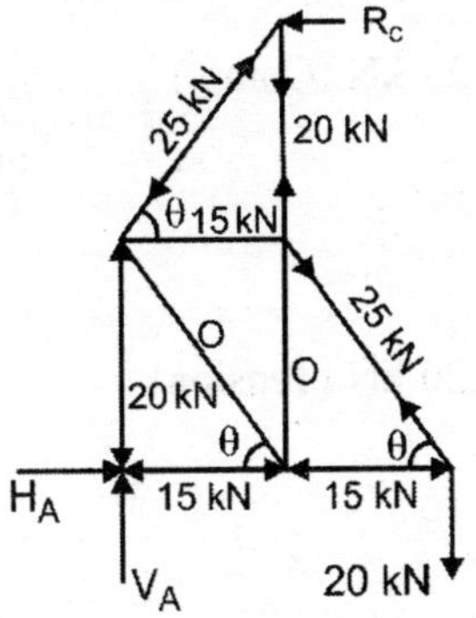


**Fig. 4.12($b$)**

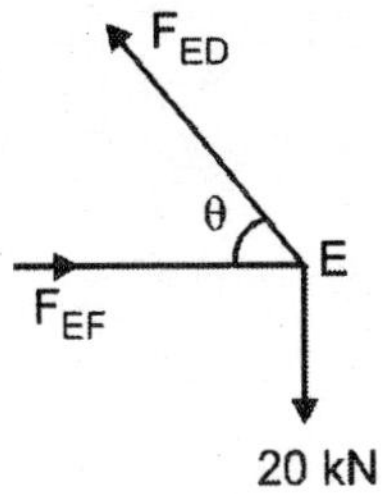


**Fig. 4.12($c$)**

Joint $E$: $\Sigma V = 0$, gives $F_{ED} \times \frac{4}{5} - 20 = 0$

$$\therefore \qquad F_{ED} = 25 \text{ kN (Tension)}$$

$$\Sigma H = 0, \text{ gives}$$

$$F_{EF} - F_{ED} \cos\theta = 0$$

$$\therefore \qquad F_{EF} = 25 \times \frac{3}{5} = 15 \text{ kN (Comp.)}$$

At this stage as no other joint is having only two unknowns, no further progress is possible. Let us find the reactions at the supports considering the whole structure. Let the reaction be as shown in Fig. 4.12 (*b*).

$$\Sigma M_A = 0, \text{ gives}$$

$$R_C \times 8 - 20 \times 6 = 0$$

$$R_C = 15 \text{ kN}$$

$$\Sigma V = 0, \text{ gives}$$

$$V_A = 20 \text{ kN}$$

$$\Sigma H = 0, \text{ gives}$$

$$H_A = R_C = 15 \text{ kN}$$

Joint *A*:

$$\Sigma V = 0, \text{ gives}$$

$$F_{AB} - V_A = 0$$

$$F_{AB} = 20 \text{ kN (Comp.)}$$

$$\Sigma H = 0, \text{ gives}$$

$$F_{AF} - H_A = 0$$

$$F_{AF} = 15 \text{ kN (Comp.)}$$

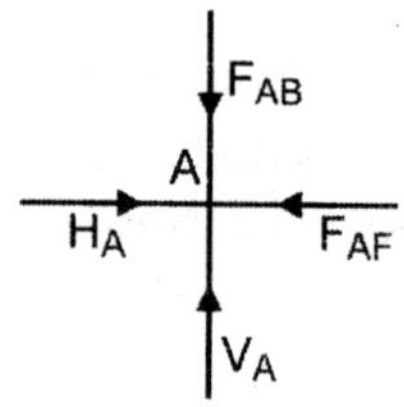


**Fig. 4.12(*d*)**

Joint *C*:

$$\Sigma H = 0, \text{ gives}$$

$$F_{CB} \times \frac{3}{5} - R_C = 0$$

$$F_{CB} = 15 \times \frac{5}{3} = 25 \text{ kN (Comp.)}$$

$$\Sigma V = 0, \text{ gives}$$

$$F_{CD} = F_{CB} \sin \theta$$

$$= 25 \times \frac{4}{5} = 20 \text{ kN (Tension)}$$

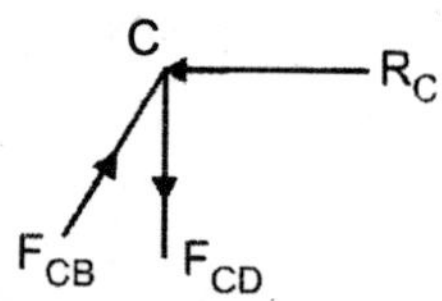


**Fig. 4.12(*e*)**

Joint *B*:

$$\Sigma V = 0, \text{ gives}$$

$$F_{BF} \times \frac{4}{5} - F_{BC} \times \frac{4}{5} + F_{AB} = 0$$

$$F_{BF} \times \frac{4}{5} = 25 \times \frac{4}{5} - 20 = 0$$

$\therefore$

$$F_{BF} = 0$$

$$\Sigma H = 0, \text{ gives}$$

$$F_{BD} - 25 \times \frac{3}{5} = 0$$

$$F_{BD} = 15 \text{ kN (Tension)}$$

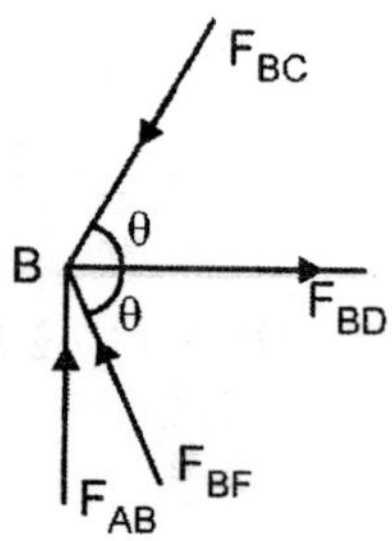


**Fig. 4.12(*f*)**

Joint *F*:

$$\Sigma V = 0$$

$$F_{FD} = 0 \text{ (since } F_{BF} = 0)$$

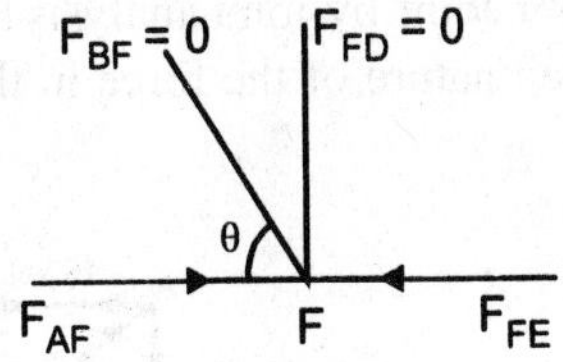


**Fig. 4.12(*g*)**

**Note:** When three members are meeting at an unloaded joint and out of them two are collinear, then the force in third member will be zero. Such situations are illustrated in Figs. 4.12 (*h*) and (*i*).

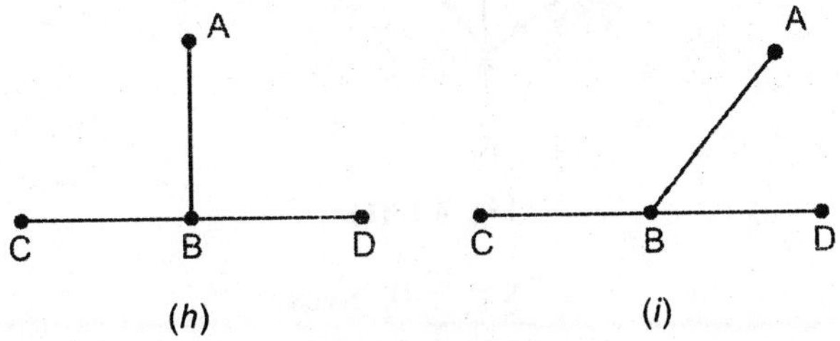


**Figs. 4.12(*h*) and (*i*)**

**Example 4.6.** Find the forces in all the members of the truss shown in Fig. 4.13(*a*).

**Solution.**

$$\tan \theta_1 = \frac{4}{6} = 33.69°$$

$$\theta_2 = \tan^{-1}\left(\frac{8}{3} \times \frac{1}{2}\right) = 53.13°$$

$$\theta_3 = \tan^{-1}\left(\frac{4}{3}\right) = 53.13°$$

$$\theta_2 = \theta_3 = \theta$$

$$\sin \theta = \frac{4}{5} \text{ and } \cos \theta = \frac{3}{5}$$

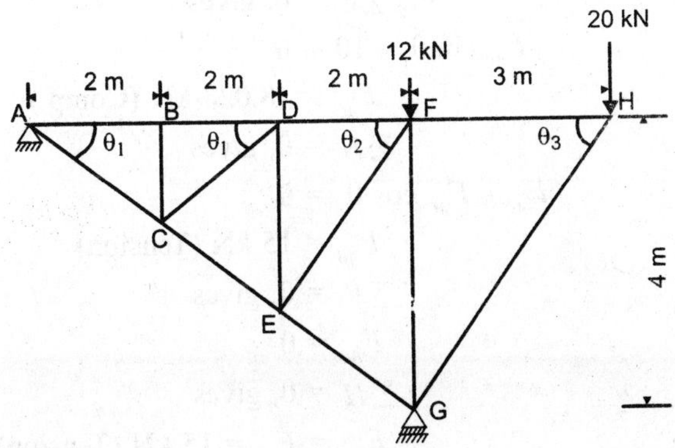


**Fig. 4.13(*a*)**

Joint-by-joint analysis is carried out as given below and the joint forces are marked in Fig. 4.13(*b*). Then nature of the force in the members is determined.

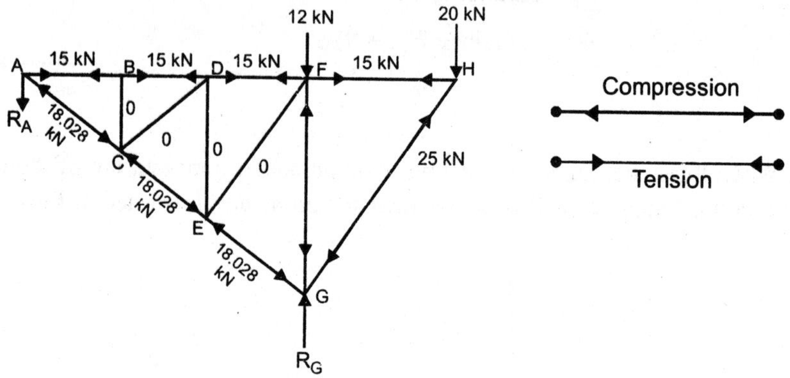


**Fig. 4.13(*b*)**

Joint $H$:

$$\Sigma V = 0, \text{ gives}$$

$$F_{HG} \sin \theta_3 = 20$$

$$\therefore \quad F_{HG} = 20 \times \frac{5}{4} = 25 \text{ kN (Comp.)}$$

$$\Sigma H = 0, \text{ gives}$$

$$F_{HF} - F_{HG} \cos \theta_2 = 0$$

$$F_{HF} = 25 \times \frac{3}{5} = 15 \text{ kN (Tension)}$$

Now

$$\Sigma M_G = 0, \text{ gives}$$

$$R_A \times 6 - 20 \times 3 = 0$$

$$R_A = 10 \text{ kN (Downward)}$$

$$\Sigma V = 0, \text{ gives}$$

$$R_G - 10 - 12 - 20 = 0$$

$$R_G = 42 \text{ kN}$$

Joint $A$:

$$\Sigma V = 0, \text{ gives}$$

$$F_{AC} \sin \theta_1 - 10 = 0$$

$$F_{AC} = 18.028 \text{ kN (Comp.)}$$

$$\Sigma H = 0, \text{ gives}$$

$$F_{AB} - F_{AC} \cos \theta_1 = 0$$

$$F_{AB} = 15 \text{ kN (Tension)}$$

Joint $B$:

$$\Sigma V = 0, \text{ gives}$$

$$F_{BC} = 0$$

$$\Sigma H = 0, \text{ gives}$$

$$F_{BD} = F_{BA} = 15 \text{ kN (Tension)}$$

Joint $C$: Sum of forces normal to $AC = 0$, gives

$$F_{CD} = 0, \text{ since } F_{BC} = 0$$

Sum of forces parallel to $CE = 0$

$$F_{CE} = F_{CA} = 18.028 \text{ (Comp.)}$$

Joint *D*: $\Sigma V = 0$, gives

$$F_{DE} = 0$$

$$\Sigma H = 0, \text{ gives}$$

$$F_{DF} = F_{DB} = 15 \text{ kN (Tension)}$$

Joint *E*: Sum of forces normal to $CG = 0$, gives

$$F_{EF} = 0 \text{ and}$$

Sum of forces in the direction of $CG = 0$, gives

$$F_{EG} = F_{CE} = 18.028 \text{ kN (Comp.)}$$

Joint *F*: $\Sigma V = 0$, gives

$$F_{FG} - 12 = 0$$

$$F_{FG} = 12 \text{ kN (Tension)}$$

**Example 4.7.** Analyse the truss shown in Fig. 4.14(*a*). All the members are of 3 m length.

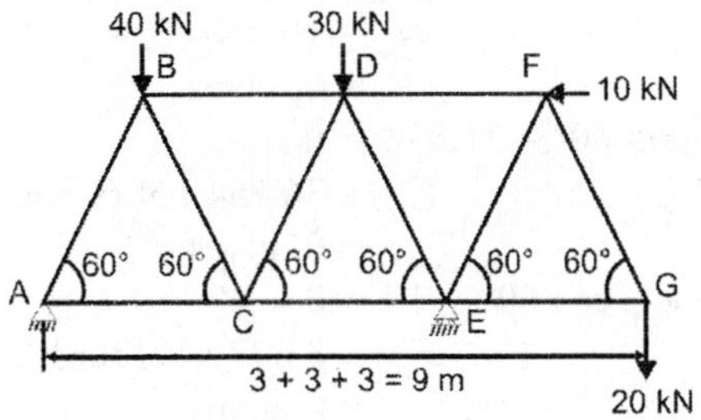


**Fig. 4.14(*a*)**

**Solution.** Since, all members are 3 m long, all triangles are equilateral and hence all inclined members are at 60° to horizontal. Joint-by-joint analysis is carried out and the forces are represented in Fig. 4.14 (*b*). Then nature of the force is determined.

Joint *G*: $\Sigma V = 0$, gives

$$F_{GF} \sin 60° = 20$$

$$F_{GF} = 23.094 \text{ kN (Tension)}$$

$$\Sigma H = 0, \text{ gives}$$

$$F_{GE} - F_{GE} \cos 60° = 0$$

$$F_{GE} = 11.547 \text{ kN (Comp.)}$$

Joint *F*: $\Sigma V = 0$, gives

$$F_{FG} \sin 60° - F_{GF} \sin 60° = 0$$

$$F_{FG} = F_{GF} = 23.094 \text{ kN (Comp.)}$$

$$\Sigma H = 0, \text{ gives}$$

$$F_{FD} + 10 - F_{GF} \cos 60° - F_{FE} \cos 60° = 0$$

$$F_{FD} = 13.094 \text{ kN (Tension)}$$

Now, without finding reaction, we cannot proceed. Hence, consider equilibrium of the entire truss.

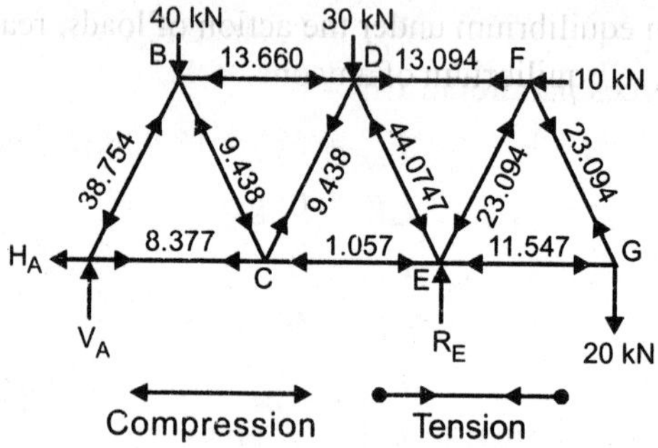


**Fig. 4.14(*b*)**

$$\Sigma M_A = 0$$

$$R_E \times 6 + 10 \times 3 \times \sin 60° - 40 \times 1.5 - 30 \times 4.5 - 20 \times 9 = 0$$

$\therefore$
$$R_E = 58.17 \text{ kN}$$

$$\Sigma V = 0, \text{ gives}$$

$$V_A = 40 + 30 + 20 - R_E = 31.830 \text{ kN}$$

$$\Sigma H = 0, \text{ gives}$$

$$H_A = 10 \text{ kN}$$

Joint *A*:
$$\Sigma V = 0, \text{ gives}$$

$$F_{AB} \sin 60° - 31.830 = 0$$

$$F_{AB} = 36.754 \text{ kN (Comp.)}$$

$$\Sigma H = 0, \text{ gives}$$

$$F_{AC} - F_{AB} \cos 60° + 10 = 0$$

$$F_{AC} = 8.377 \text{ kN (Tension)}$$

Joint *B*:
$$\Sigma V = 0, \text{ gives}$$

$$F_{BC} \sin 60° + F_{AB} \sin 60° - 40 = 0$$

$$F_{BC} = 9.434 \text{ kN (Comp.)}$$

$$\Sigma H = 0, \text{ gives}$$

$$F_{BD} + F_{BC} \cos 60° - F_{BA} \cos 60° = 0$$

$$F_{BD} = 13.660 \text{ kN (Comp.)}$$

Joint *C*:
$$\Sigma V = 0, \text{ gives}$$

$$F_{CD} \sin 60° - F_{BC} \sin 60° = 0$$

$$F_{CD} = F_{BC} = 9.434 \text{ kN (Tension)}$$

$$\Sigma H = 0, \text{ gives}$$

$$F_{CE} + F_{AC} - F_{CD} \cos 60° - F_{BC} \cos 60 = 0$$

$$F_{CE} = 2 \times 9.434 \times \frac{1}{2} - 8.377 = 1.057 \text{ kN (Comp.)}$$

Joint *D*:
$$\Sigma V = 0, \text{ gives}$$

$$F_{DE} \sin 60° - F_{CD} \sin 60° - 30 = 0$$

$$F_{DE} = 44.075 \text{ kN (Comp.)}$$

## 4.7 METHOD OF SECTION

In the method of section, after determining the reactions, a section line is drawn passing through not more than three members in which forces are not known such that the frame is cut into two separate parts. Each part should be in equilibrium under the action of loads, reactions and forces in the members that are cut by the section line. Equilibrium of any one of these two parts is considered and the unknown forces in the members cut by the section line are determined. The system of forces acting on either part of truss constitutes a non-concurrent force system. Since, there are only three independent equations of equilibrium, there should be only three unknown forces. Hence, in this method, it is an essential condition that, **the section line should pass through not more than three members in which forces are not known and it should separate the frame into two parts.**

Thus, the method of section is the application of non-concurrent force system analysis; whereas the method of joints, described in previous article, was the application of analysis of concurrent force system.

Under the following two conditions, the method of section is preferred over the method of joints:

1. In a large truss in which forces in only few members are required;
2. In the situation where the method of joints fails to start/proceed with analysis.

The method of section is illustrated with the Examples 8 to 11. Examples 8 and 9 are the cases in which method of section is advantageous, since, forces in only few members are required. Examples 10 and 11 are the cases in which method of joints fails to start/proceed to get the solution. In practice, the frames may be analysed partly by method of section and partly by method of joints as illustrated in Example 11.

**Example 4.8.** Determine the forces in the members *FH*, *HG* and *GI* in the truss as shown in Fig. 4.15(*a*). Each load is 10 kN and all triangles are equilateral with sides 4 m.

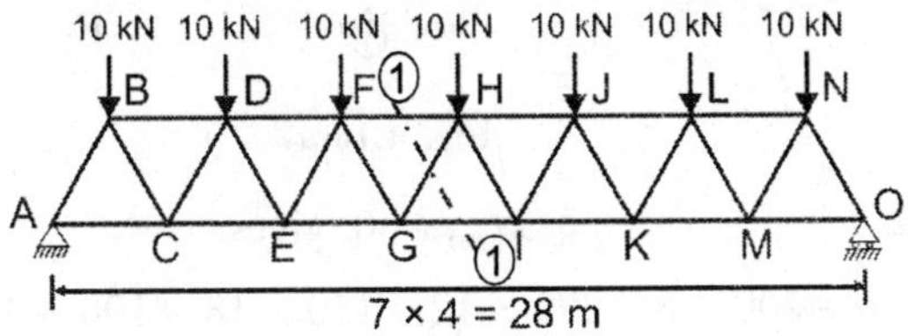


**Fig. 4.15(*a*)**

**Solution.** Due to symmetry,

$$R_A = R_0 = \frac{1}{2} \times 10 \times 7 = 35 \text{ kN}$$

Take section (1)–(1), which cuts the members *FH*, *GH* and *GI* and separates the truss into two parts. Consider the equilibrium of left hand side part as shown in Fig. 4.15 (*b*) (Prefer part in which number of forces are less).

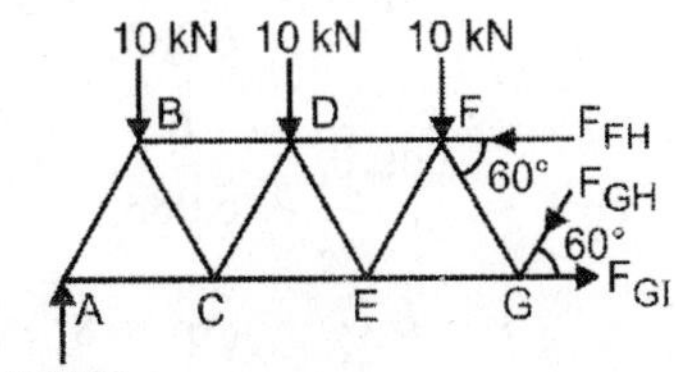


**Fig. 4.15(*b*)**

$$\Sigma M_G = 0, \text{ gives}$$

$$F_{FH} \times 4 \sin 60° - 35 \times 12 + 10 \times 10 + 10 \times 6 + 10 \times 2 = 0$$

$$F_{FH} = 69.282 \text{ kN (Comp.)}$$

$$\Sigma V = 0, \text{ gives}$$

$$F_{GH} \sin 60° + 10 + 10 + 10 - 35 = 0$$

$$F_{GH} = 5.776 \text{ kN (Comp.)}$$

$$\Sigma H = 0, \text{ gives}$$

$$F_{GI} - F_{FH} - F_{GH} \cos 60° = 0$$

$$F_{GI} = 69.282 + 5.776 \cos 60°$$

$$= 72.169 \text{ kN (Tension)}$$

**Example 4.9.** Find the magnitude and nature of the forces in the members $U_3U_4$, $L_3L_4$ and $U_4L_3$ of the loaded truss shown in Fig. 4.16($a$).

**Solution.** To determine reactions, consider equilibrium equations.

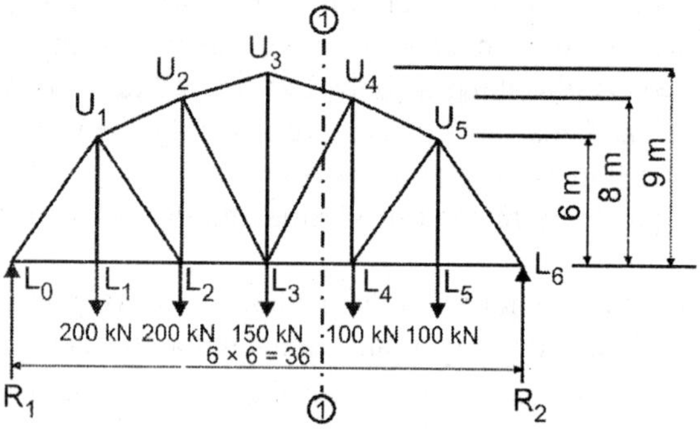


**Fig. 4.16($a$)**

Now,
$$\Sigma M_{LO} = 0, \text{ gives}$$

$$R_2 \times 36 - 200 \times 6 - 200 \times 12 - 150 \times 18 - 100 \times 24 - 100 \times 30 = 0$$

$$R_2 = 325 \text{ kN}$$

$$\Sigma V = 0, \text{ gives}$$

$$R_1 = 200 + 200 + 150 + 100 + 100 - 325 = 425 \text{ kN}$$

Take the section (1)–(1) and consider the right hand side part.

Now,
$$U_3U_4 = \sqrt{1^2 + 6^2} = 6.083$$

$$\sin \theta_1 = \frac{1}{6.083} = 0.164$$

$$\cos \theta_1 = \frac{6}{6.083} = 0.986$$

$$L_3U_4 = \sqrt{6^2 + 8^2} = 10$$

$$\sin \theta_2 = 0.6 \text{ and } \cos \theta_2 = 0.8$$

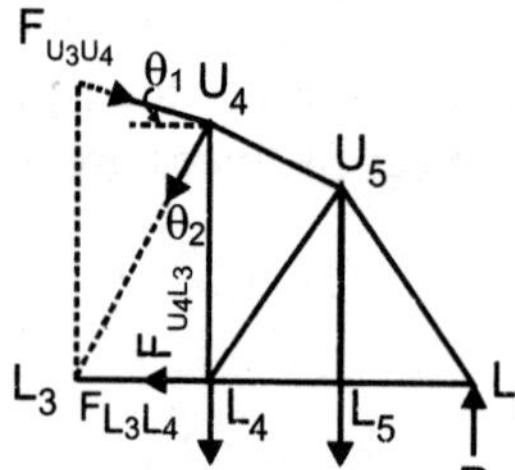


**Fig. 4.16($b$)**

$$\Sigma M_{U_4} = 0, \text{ gives}$$

$$F_{L_3L_4} \times 8 - 325 \times 12 + 100 \times 6 = 0$$

$$F_{L_3L_4} = 412.5 \text{ kN (Tension)}$$

$$\Sigma M_{L_3} = 0, \text{ gives}$$

$$F_{U_4U_3} \times \cos\theta_1 \times 9 + 100 \times 6 + 100 \times 12 - 325 \times 18 = 0$$

$$F_{U_4U_3} = 456.207 \text{ kN (Comp.)}$$

$$\Sigma H = 0, \text{ gives}$$

$$F_{U_4U_3} \sin\theta_2 - F_{U_4U_3} \cos\theta_1 + F_{L_4L_3} = 0$$

$$F_{U_4U_3} = \frac{456.207 \times 0.986 - 412.5}{0.6}$$

$$= 62.5 \text{ kN (Tension)}$$

**Example 4.10.** Find the forces in the members (1), (2) and (3) of French truss as shown in Fig. 4.17(*a*).

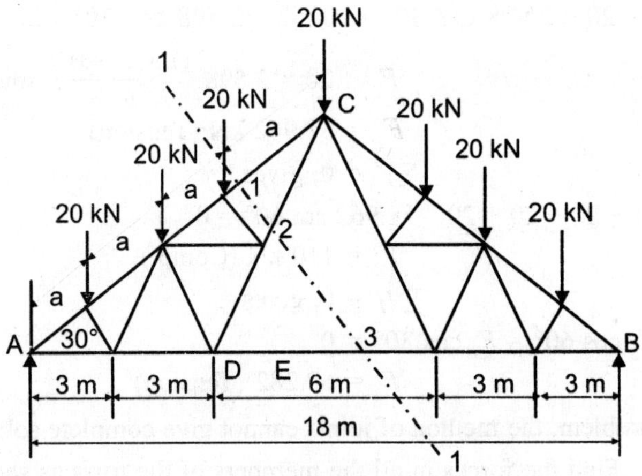


**Fig. 4.17(*a*)**

**Solution.** Due to symmetry

$$R_A = R_B = \frac{1}{2} \times 20 \times 7 = 70 \text{ kN}$$

Now

$$AC = 4 \times a = \frac{9}{\cos 30°}$$

$\therefore$ $$a = 2.598 \text{ m.}$$

Take section (1)–(1) and consider the equilibrium of left hand side part of the French truss as shown in Fig. 4.17 (*b*).

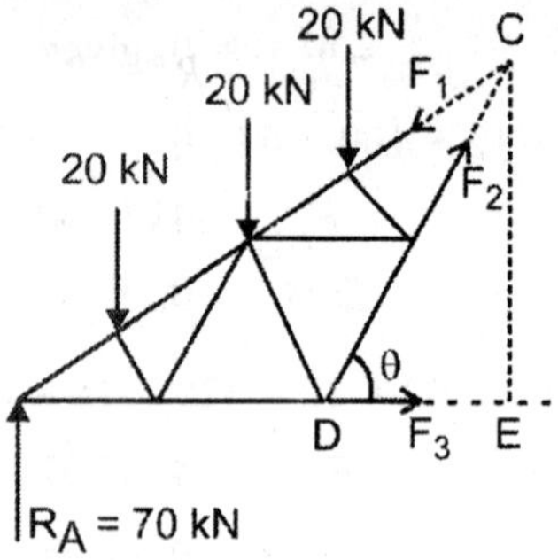


**Fig. 4.17(*b*)**

Drop perpendicular *CE* on *AB*.

Now $CE = 9 \tan 30°$ and $DE = 3$ m

$$\therefore \quad \tan\theta = \frac{9 \tan 30°}{3} = 3 \times \frac{1}{\sqrt{3}} = \sqrt{3}$$

$$\therefore \quad \theta = 60°$$

$$\Sigma M_A = 0, \text{ gives}$$

$$F_2 \sin 60° \times 6 - 20 \times 2.598 \cos 30° - 20 \times 2 \times 2.598 \cos 30° - 20 \times 3 \times 2.598 \cos 30° = 0$$

$$F_2 = 20 \times 2.598 \frac{(1+2+3)}{6}, \text{ since } \sin 60° = \cos 30°$$

$$F_2 = 51.962 \text{ kN (Tension)}$$

$$\Sigma V = 0, \text{ gives}$$

$$F_1 \sin 30° - 70 + 20 + 20 + 20 - 51.962 \sin 60° = 0$$

$$F_1 = 110 \text{ kN (Comp.)}$$

$$\Sigma H = 0, \text{ gives}$$

$$F_3 + F_2 \cos 60° - F_1 \cos 30° = 0$$

$$F_3 = 69.282 \text{ (Tension)}$$

**Note:** In this problem, the method of joints cannot give complete solution.

**Example 4.11.** Find the forces in all the members of the truss as shown in Fig. 4.18(*a*).

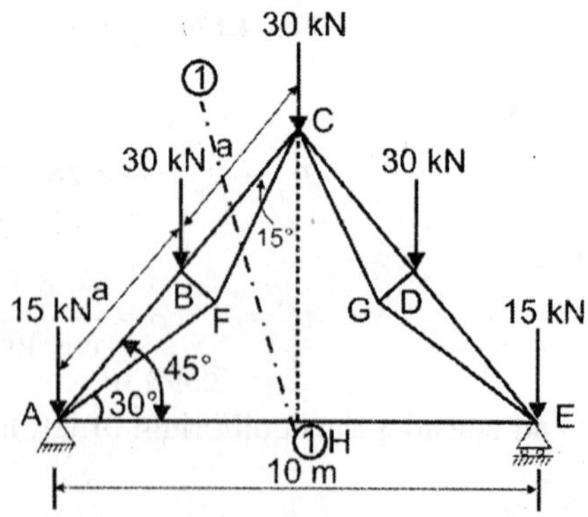


**Fig. 4.18(*a*)**

**Solution.** Due to symmetry, the reactions are equal.

$$R_A = R_E = \frac{1}{2} \times \text{Total load}$$

$$= \frac{1}{2}(15 + 30 + 30 + 30 + 15) = 60 \text{ kN}$$

Drop perpendicular $CH$ on $AF$.

In $\Delta ACH$, $\angle ACH = 45°$

$\therefore$ $FC$ is inclined at 30° to vertical, *i.e.*, 60° to horizontal and $CH = 5$ m.

It is not possible to find a joint where there are only two unknowns. Hence, consider section (1)–(1).

For left hand side part of the frame:

$$\Sigma M_C = 0, \text{ gives}$$

$$F_{AE} \times 5 - 60 \times 5 + 15 \times 5 + 30 \times 2.5 = 0$$

$$\therefore \quad F_{AE} = 30 \text{ kN (Tension)}$$

Assuming the directions for $F_{FC}$ and $F_{BC}$ as shown in Fig. 4.18 (*b*)

$$\Sigma V = 0, \text{ gives}$$

$$F_{FC} \sin 60° - F_{BC} \sin 45° + 60 - 15 - 30 = 0 \quad \text{...(4.3)}$$

$$\Sigma H = 0, \text{ gives}$$

$$F_{FC} \cos 60° - F_{BC} \cos 45° + F_{AE} = 0$$

$$F_{FC} \times \frac{1}{2} - F_{BC} \frac{1}{\sqrt{2}} + 30 = 0$$

$$\therefore \quad F_{FC} = F_{BC}\sqrt{2} - 60 \quad \text{...(4.4)}$$

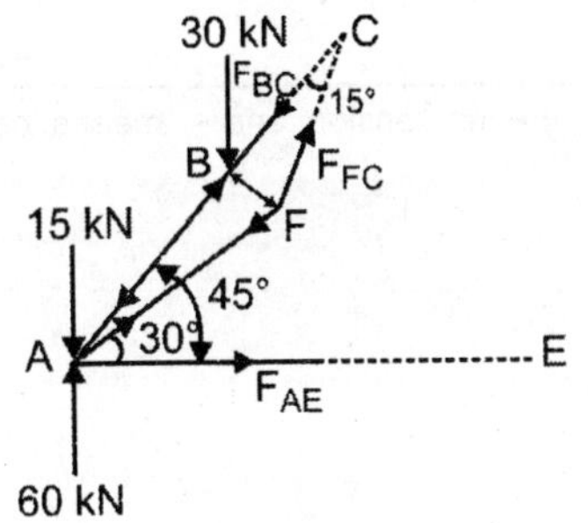


**Fig. 4.18(*b*)**

Substituting this value of $F_{FC}$ in Eqn. (4.3), we get

$$(F_{BC}\sqrt{2} - 60)\sin 60° - F_{BC}\frac{1}{\sqrt{2}} + 15 = 0$$

$$F_{BC}\left(\sqrt{2}\sin 60° - \frac{1}{\sqrt{2}}\right) = 60 \sin 60° - 15$$

$$F_{BC} = 71.404 \text{ kN (Comp.)}$$

$$\therefore \quad F_{FC} = 71.404 \times \sqrt{2} - 60$$

$$= 40.981 \text{ kN (Tension)}$$

Assumed directions of $F_{BC}$ and $F_{FC}$ are correct.

Therefore, $F_{BC}$ is in compression and $F_{FC}$ is in tension.

Now we can proceed with method of joints to find the forces in other members. Since, it is a symmetric truss, analysis of half the truss is sufficient. Other values may be written by making use of symmetry.

Joint $B$:

$$\Sigma \text{ Forces normal to } AC = 0, \text{ gives}$$

$$F_{BF} - 30 \cos 45° = 0$$

$$F_{BF} = 21.213 \text{ kN (Comp.)}$$

$$\Sigma \text{ Forces parallel to } AC = 0, \text{ gives}$$

$$F_{AB} - F_{BC} - 30 \sin 45° = 0$$

$$F_{AB} = 71.404 + 21.213$$

$$= 92.617 \text{ kN (Comp.)}$$

Joint $A$:

$$\Sigma V = 0, \text{ gives}$$

$$F_{AF} \sin 30° - F_{AB} \sin 45° - 15 + 60 = 0$$

$$F_{AF} = 40.981 \text{ kN (Tension)}$$

The results are tabulated below:

| Members | Force in kN |
|---|---|
| AB and ED | – 92.617 |
| BC and DC | – 71.404 |
| BF and DG | – 21.213 |
| AF and EG | + 40.981 |
| FC and GC | + 40.981 |
| AE | + 30.00 |

+ means tension and – means compression

## IMPORTANT FORMULA

1. In a perfect truss

$$m = 2j - 3$$

## THEORY QUESTIONS

**4.1** Bring out the differences among perfect, defficient and redundant trusses.

**4.2** List the assumptions made in the analysis of pin-jointed trusses.

**4.3** When do you prefer method of section over method of joint?

## EXERCISES

**4.1 to 4.6** Determine the forces in all the members of the frames as shown in Figs. 4.19 to 4.24. Indicate the nature of the forces also. (Tension as +ve and compression as –ve).

**4.1** [**Ans.** $F_{AB} = +67.5$ kN; $F_{BC} = +15$ kN; $F_{CD} = -25$ kN; $F_{DE} = -30$ kN; $F_{EF} = -105$ kN; $F_{AE} = +62.5$ kN; $F_{BE} = -62.5$ kN; $F_{BD} = +25$ kN]

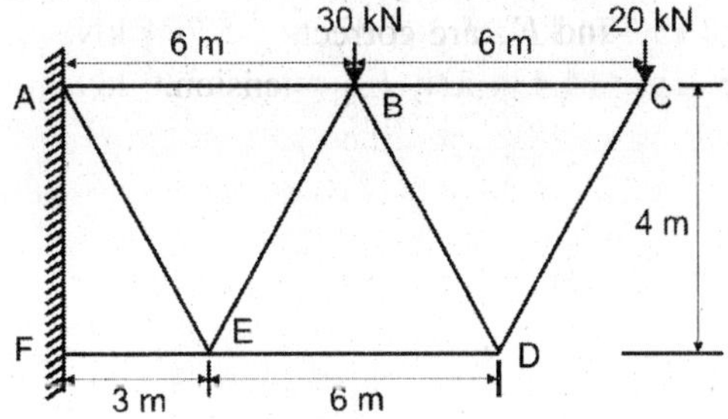


Fig. 4.19

4.2 [**Ans.** $F_{AB}$ = +82.074 kN; $F_{BC}$ = +73.866 kN; $F_{CD}$ = 49.244 kN; $F_{DE}$ = –45 kN; $F_{EF}$ = –45 kN; $F_{FG}$ = –67.5 kN; $F_{BG}$ = –10.0 kN; $F_{FC}$ = +24.622 kN; $F_{CE}$ = 0; $F_{BF}$ = 10 kN]

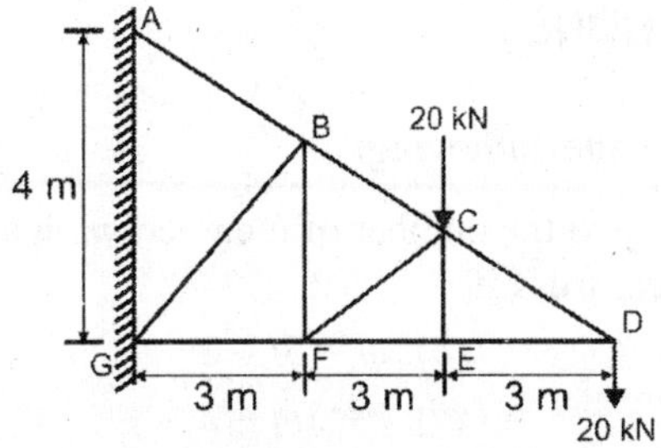


Fig. 4.20

4.3 [**Ans.** $F_{AC}$ = $F_{CE}$ = $F_{EG}$ = +193.185 kN; $F_{BD}$ = $F_{DE}$ = $F_{FG}$ = –193.185 kN; All others are zero members]

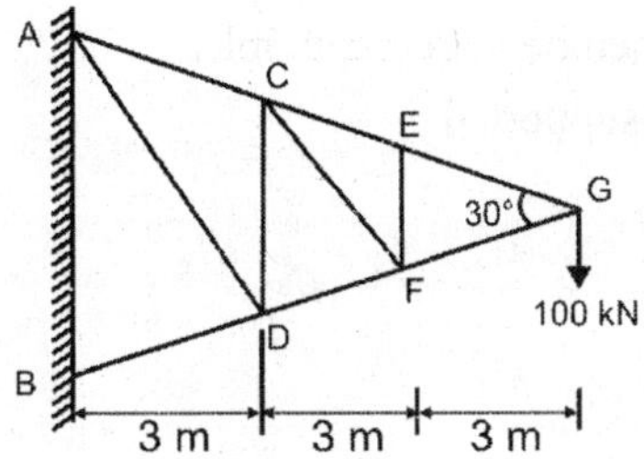


Fig. 4.21

4.4 [**Ans.** $F_{EC}$ = +447.214 kN; $F_{CA}$ = +400 kN; $F_{AB}$ = –447.214 kN; $F_{BD}$ = –400 kN; $F_{CD}$ = 0; $F_{CB}$ = –200 kN]

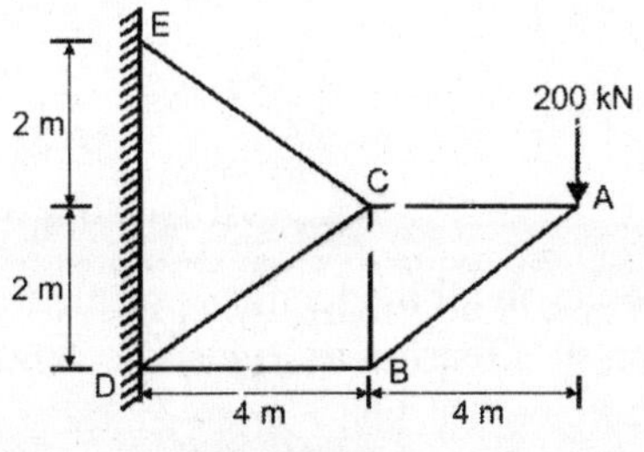


Fig. 4.22

**4.5** [**Ans.** $F_{DB} = F_{BA} = +5.774$ kN; $F_{BC} = F_{DE} = -5.774$ kN; $F_{AC} = -2.887$ kN; $F_{CE} = -14.434$ kN; $F_{DC} = +17.320$ kN; $F_{DF} = +20.0$ kN]

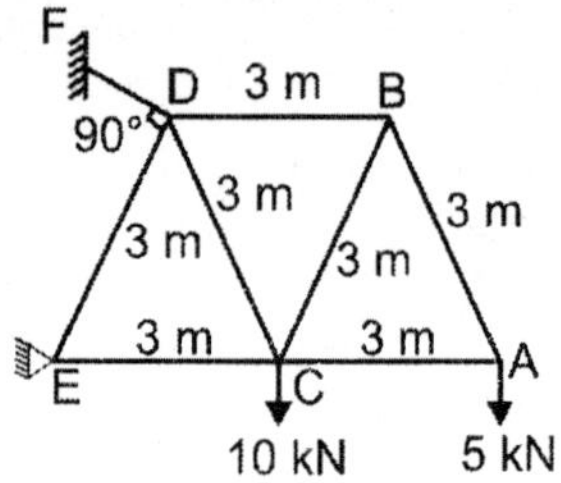


**Fig. 4.23**

**4.6** [**Ans.** $F_{AB} = -30$ kN; $F_{AC} = -160$ kN; $F_{BC} = +50$ kN; $F_{BD} = -200$ kN; $F_{CD} = -50$ kN; $F_{CE} = -120$ kN; $F_{DF} = -266.67$ kN; $F_{DE} = +83.33$ kN]

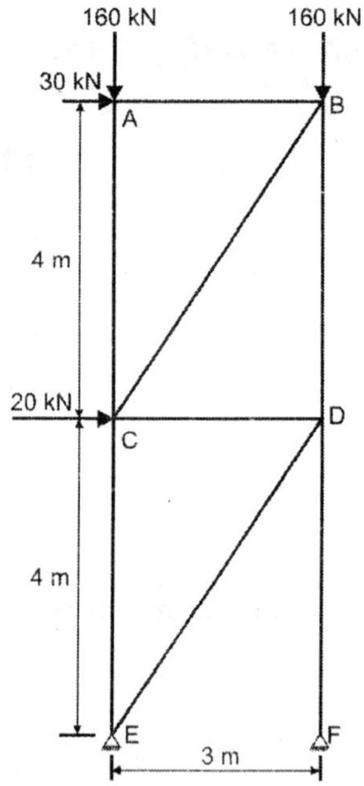


**Fig. 4.24**

## OBJECTIVE TYPE QUESTIONS

**Select the correct options in the following:**

**4.1** If number of joints are $j$ and the number of members $m$ in a plane truss, it is a perfect truss if

(*a*) $m = 2j - 3$ (*b*) $m = 3j - 2$

(*c*) $j = 2m - 3$ (*d*) $j = 3m - 2$

**4.2** Which one of the following is not correct as far as pinjointed trusses are considered:

(*a*) The ends of the members are pinconnected.

(*b*) Loads act only at the joints.

(*c*) Self weight of the members are negligible.

(*d*) Trusses are simply supported.

CHAPTER

# 5

# FRICTION

## 5.1 INTRODUCTION

When a body moves or tends to move over another body, a force opposing the motion develops at the contact surfaces. This is due to the resistance offered to motion by minutely projecting particles at the contact surfaces. The force that opposes the movement or the tendency of movement is called *frictional force* or simply *friction.*

Frictional force has the remarkable property of adjusting its magnitude to the force tending to produce the motion of the body so that motion is prevented. However, there is a limit beyond which the magnitude of frictional force cannot increase. Hence, if the applied force causing motion is more than this maximum frictional force, there will be movement of one body over the other. This maximum value of frictional force, which comes into play, when motion is *impending,* is known as *Limiting Friction.* The frictional force developed till the body starts moving is called *Static Friction.* It may have any value between zero and limiting friction. Frictional resistance developed at contact surface, when a body moves over the other is known as *Dynamic Friction.*

It is experimentally found that the magnitude of limiting friction bears a constant ratio to the normal reaction between the two surfaces and this ratio is called *Coefficient of Friction.* Thus, in Fig. 5.1,

$$\text{Coefficient of friction } \mu = \frac{F}{N} \qquad \text{... (5.1)}$$

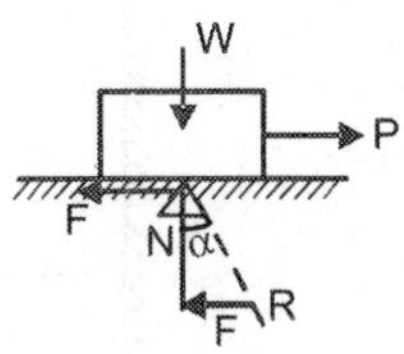

Fig. 5.1

where $F$ = Limiting friction

and $N$ = Normal reaction

The angle made by the resultant of $F$ and $N$ with respect to normal reaction $N$ is called angle of friction. Thus,

$$\tan\alpha = \frac{F}{N} = \mu \qquad \text{... (5.2)}$$

## 5.2 LAWS OF COULOMB FRICTION

The principles discussed above are mainly due to the experimental studies by Coulomb (1781) and by Morin (1831). These principles are applicable when the contact surfaces are not wet and hence are known as *Laws of Dry Friction* or as *Laws of Coulomb Friction.* These laws are listed below:

1. The force of friction always acts in a direction opposite to that in which the body tends to move.
2. Till the limiting value is reached, the magnitude of friction is exactly equal to the force which tends to move the body.
3. The magnitude of limiting friction bears a constant ratio to the normal reaction between the two contact surfaces.
4. The force of friction depends upon the roughness/smoothness of the surfaces.
5. The force of friction is independent of the area of contact between the two surfaces.
6. After the body starts moving, the dynamic friction comes into play, the magnitude of which is less than that of limiting friction and it bears a constant ratio with normal force. This ratio is called coefficient of dynamic friction.

## 5.3 EQUILIBRIUM OF BODIES INVOLVING DRY FRICTION

Various bodies we come across in engineering mechanics problems may be grouped under the following three categories:

(*a*) Equilibrium of Blocks,
(*b*) Equilibrium of Wedges, and
(*c*) Equilibrium of Ladders.

**(*a*) Equilibrium of Blocks**

The system of forces acting on these bodies may be treated as concurrent. Hence, using the system of equilibrium equations for concurrent forces and the laws of friction, these problems can be solved. This procedure is illustrated with Examples 5.1 to 5.8 below:

**Example 5.1.** Block *A* weighing 1000 N rests over block *B* which weighs 2000 N as shown in Fig. 5.2(*a*). Block *A* is tied to wall with a horizontal string. If the coefficient of friction between *A* and *B* is 1/4 and between *B* and the floor is 1/3, what should be the value of **P** to move the block *B* if (*a*) **P** is horizontal? (*b*) and **P** acts 30° upwards to horizontal?

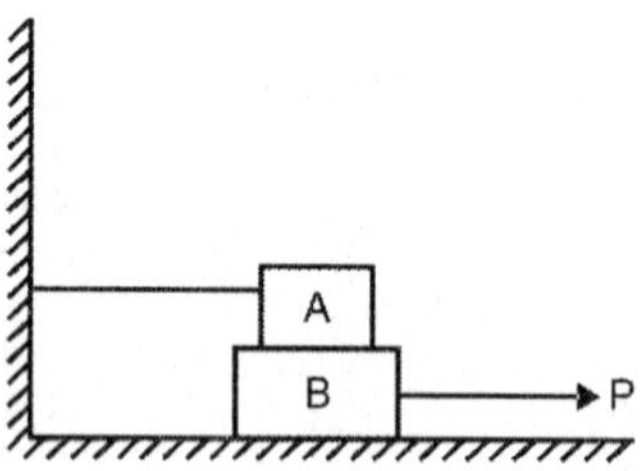


(*a*)

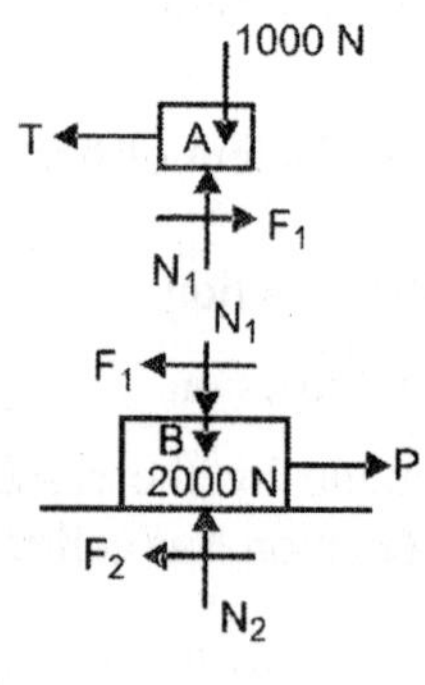


(*b*)

**Fig. 5.2**

**Solution.** (*a*) *When P is horizontal*

The free body diagrams of the two blocks are shown in Fig. 5.2 (*b*). (Note that the frictional forces $\mathbf{F}_1$ and $\mathbf{F}_2$ are to be marked in the opposite directions of impending relative motion.) Considering block *A*,

$$\Sigma V = 0, \text{ gives}$$
$$N_1 = 1000 \text{ N}$$

Since, $F_1$ is limiting friction, $\dfrac{F_1}{N_1} = \dfrac{1}{4}$

$$\therefore \quad F_1 = 250 \text{ N}$$
$$\Sigma H = 0, \text{ gives}$$
$$T = F_1 = 250 \text{ N}$$

Considering block *B*, $\Sigma V = 0$, gives

$$N_2 - 2000 - N_1 = 0$$
$$N_2 = 3000 \text{ N, since } N_1 = 1000 \text{ N}$$

Since, $F_2$ is the limiting friction $F_2 = \mu_2 N_2$

$$= \frac{1}{3} \times 3000 = 1000 \text{ N}$$
$$\Sigma H = 0, \text{ gives}$$
$$P - F_1 - F_2 = 0$$
$$\mathbf{P} = F_1 + F_2 = 250 + 1000 = \mathbf{1250\ N.} \qquad \textbf{Ans.}$$

(*b*) *When P is inclined*

Free body diagram for this case is shown in Fig. 5.2 (*c*).

As in the previous case, here also $N_1 = 1000$ N and $F_1 = 250$ N. Consider the equilibrium of block *B*.

$$\Sigma V = 0, \text{ gives}$$
$$N_2 - 2000 - N_1 + P \sin 30° = 0$$
$$N_2 + 0.5\,P = 3000 \text{ N, since } N_1 = 1000 \text{ N}$$

From law of friction, $F_2 = \dfrac{1}{3} N_2 = \dfrac{1}{3}(3000 - 0.5\,P)$

$$= 1000 - \frac{0.5}{3} P$$
$$\Sigma H = 0, \text{ gives}$$
$$P \cos 30° - F_1 - F_2 = 0$$
$$P \cos 30° - 250 - \left(1000 - \frac{0.5}{3} P\right) = 0$$
$$\therefore \quad \mathbf{P} = \mathbf{1210.43\ N.} \qquad \textbf{Ans.}$$

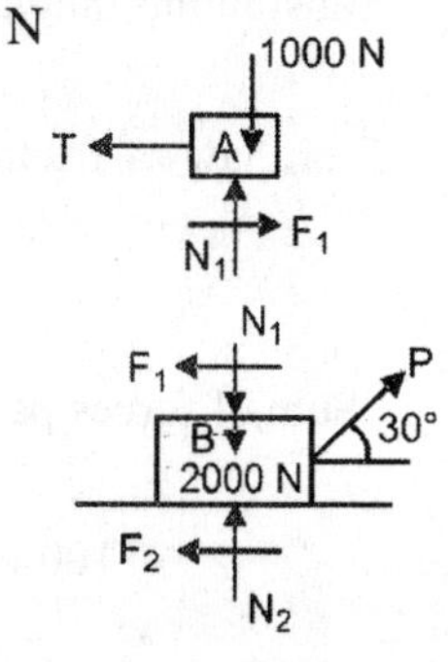


**Fig. 5.2 (*c*)**

**Example 5.2.** What should be the value of θ in Fig. 5.3 (*a*) which will make the motion of 900 N block down the plane to impend? The coefficient of friction for all contact surfaces is $\dfrac{1}{3}$.

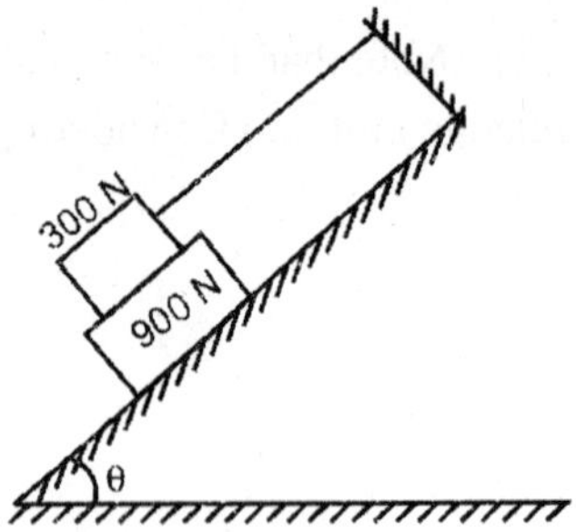


(a)

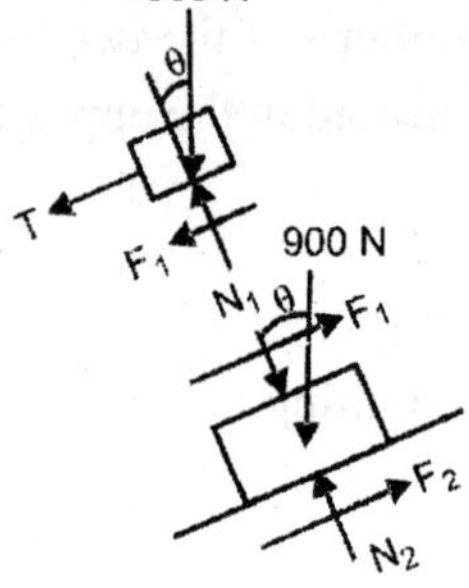


(b)

**Fig. 5.3**

**Solution:** . 900 N block is on the verge of moving downward. Hence, frictional forces $\mathbf{F_1}$ and $\mathbf{F_2}$ [Fig. 5.3(*b*)] act up the plane on 900 N block. Free body diagram of the blocks is as shown in Fig. 5.3 (*b*).

For 300 N block:

Sum of forces normal to the plane = 0, gives

$$N_1 - 300 \cos \theta = 0$$

or
$$N_1 = 300 \cos \theta \quad ...(5.3)$$

From law of friction,
$$F_1 = \frac{1}{3} N_1 = 100 \cos \theta \quad ...(5.4)$$

For 900 N block:

Sum of forces normal to the plane = 0, gives

$$N_2 - N_1 - 900 \cos \theta = 0$$

or
$$N_2 = N_1 + 900 \cos \theta$$

Substituting the value of $N_1$ from (5.3), we get

$$N_2 = 1200 \cos \theta \quad ...(5.5)$$

From law of friction,

$$F_2 = \frac{1}{3} N_2 = 400 \cos \theta \quad ...(5.6)$$

Sum of forces parallel to the plane = 0, gives

$$F_1 + F_2 - 900 \sin \theta = 0$$

*i.e.,*
$$100 \cos \theta + 400 \cos \theta = 900 \sin \theta$$

$$\tan \theta = \frac{5}{9}$$

$\therefore$
$$\boldsymbol{\theta = 29.05°}. \quad \textbf{Ans.}$$

**Example 5.3.** A weight 500 N just starts moving down a rough inclined plane when it is supported by a force of 200 N acting parallel to the plane in upward direction and it is at the point of moving up the plane when pulled by a force of 300 N parallel to the plane. Find the inclination of the plane and the coefficient of friction between the inclined plane and the weight.

**Solution.** Free body diagram of the block when it just starts moving down the plane is shown in Fig. 5.4 (*a*) and when just starts moving up is shown in Fig. 5.4 (*b*). Now, frictional forces oppose the direction of the movement of the block and since it is limiting case, $\frac{F}{N} = \mu$.

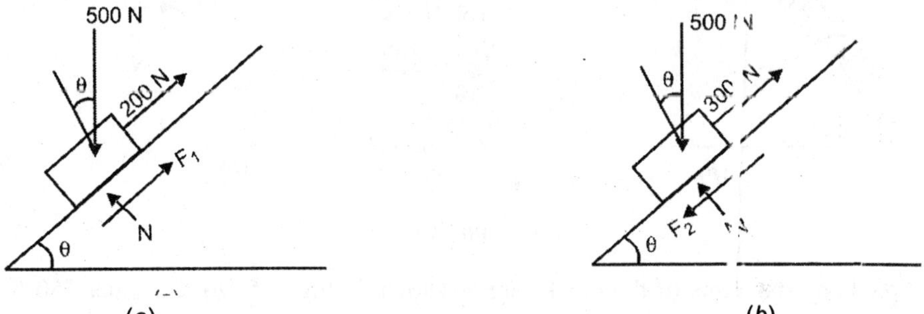


**Fig. 5.4**

When the block just starts moving down the plane [Fig. 5.4 (*a*)]:

Sum of forces perpendicular to the plane = 0, gives

$$N = 500 \cos \theta \quad ...(5.7)$$

From law of friction, $F_1 = \mu N$

*i.e.,* $$F_1 = 500 \mu \cos \theta \quad ...(5.8)$$

Sum of forces parallel to the plane = 0, gives

$$500 \sin \theta - F_1 - 200 = 0$$

*i.e.,* $$500 \sin \theta - 500 \mu \cos \theta = 200 \quad ...(5.9)$$

When the block just starts moving up the plane [Fig. 5.4 (*b*)]:

Sum of forces perpendicular to the plane = 0, gives

$$N = 500 \cos \theta \quad ...(5.10)$$

From the law of friction, $$F_2 = 500 \mu \cos \theta \quad ...(5.11)$$

Sum of forces parallel to the plane = 0, gives

$$500 \sin \theta + F_2 - 300 = 0$$

*i.e.,* $$500 \sin \theta + 500 \mu \cos \theta = 300 \quad ...(5.12)$$

Adding Eqns. (5.9) and (5.12), we get

$$1000 \sin \theta = 500$$

$$\sin \theta = \frac{1}{2}$$

or $$\boldsymbol{\theta = 30°}$$ **Ans.**

Substituting it in Eqn. (5.12), we get

$$500 \mu \cos 30° = 300 - 500 \sin 30° = 50$$

$\therefore$ $$\boldsymbol{\mu = 0.11547.}$$ **Ans.**

**Example 5.4.** What is the value of **P** in the system, as shown in Fig. 5.5 (*a*) to cause the motion to impend? Assume the pulley is smooth and coefficient of friction between the other contact surfaces is 0.20.

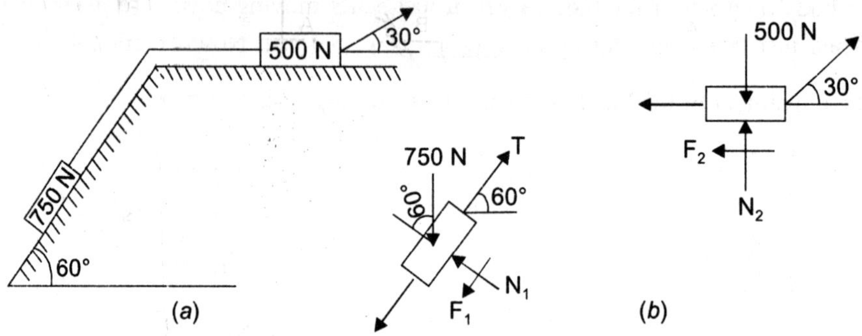


**Fig. 5.5**

**Solution.** Free body diagrams of the blocks are as shown in Fig. 5.5 (*b*). Consider 750 N block:

Sum of forces normal to the plane = 0, gives

$$N_1 - 750 \cos 60° = 0$$

$$N_1 = 375 \text{ N}$$

Since, the motion is impending, from law of friction,

$$F_1 = \mu N_1 = 0.2 \times 375 = 75 \text{ N}$$

Sum of forces parallel to the plane = 0, gives

$$T - F_1 - 750 \sin 60° = 0$$

$$T = 75 + 750 \sin 60° = 724.52 \text{ N}$$

Consider 500 N body:

$$\Sigma V = 0, \text{ gives}$$

$$N_2 - 500 + P \sin 30° = 0$$

$$N_2 + 0.5P = 500$$

From law of friction,

$$F_2 = 0.2\, N_2$$

$$= 0.2\,(500 - 0.5P) = 100 - 0.1P$$

$$\Sigma H = 0, \text{ gives}$$

$$P \cos 30° - T - F_2 = 0$$

$$P \cos 30° - 724.52 - 100 + 0.1P = 0$$

$$\mathbf{P = 853.52\ N.}$$ **Ans.**

**Example 5.5.** Two blocks connected by a horizontal link *AB* are supported on two rough planes as shown in Fig. 5.6 (*a*). The coefficient of friction for the block on the horizontal plane is 0.4. The limiting angle of friction for block *B* on the inclined plane is 20°. What is the smallest weight *W* of the block *A* for which equilibrium of the system can exist if weight of block *B* is 5 kN?

**Solution.** Free body diagrams for blocks *A* and *B* are as shown in Fig. 5.6 (*b*).

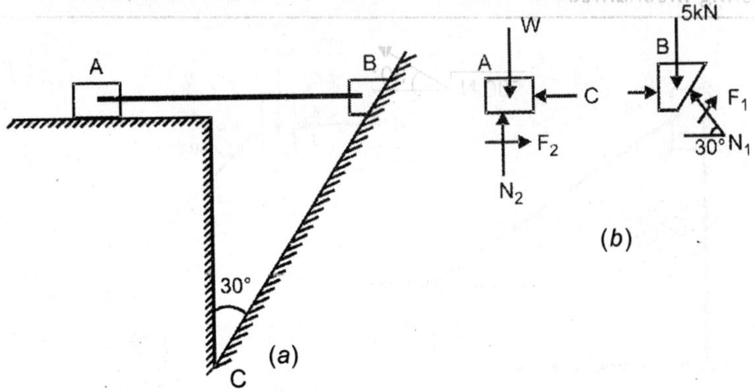


**Fig. 5.6**

Consider block $B$:

From law of friction,

$$F_1 = N_1 \tan 20° \qquad \text{[Since, } \mu = \tan 20°\text{]}$$

$$\Sigma V = 0, \text{ gives}$$

$$N_1 \sin 30° + F_1 \sin 60° - 5 = 0$$

$$0.5\, N_1 + N_1 \tan 20° \sin 60° = 5$$

$$\therefore \quad N_1 = 6.133 \text{ kN}$$

$$\therefore \quad F_1 = 6.133 \tan 20° = 2.232 \text{ kN}$$

$$\Sigma H = 0, \text{ gives}$$

$$C + F_1 \cos 60° - N_1 \cos 30° = 0$$

$$C = 6.133 \cos 30° - 2.232 \cos 60° = 4.196 \text{ kN}$$

Now consider the equilibrium of block $A$:

$$\Sigma H = 0, \text{ gives}$$

$$F_2 = C = 4.196 \text{ kN}$$

From law of friction, $F_2 = \mu N_2$

*i.e.*, $$N_2 = \frac{4.196}{0.4} = 10.49 \text{ kN}$$

$$\Sigma V = 0, \text{ gives}$$

$$\mathbf{W} = \mathbf{N_2} = \mathbf{10.49\ kN} \qquad \textbf{Ans.}$$

**Example 5.6.** Two blocks $A$ and $B$ weighing 2000 $N$ each are to be held from slipping by the thrust of two weightless link rods each of which is connected by pin joints at one end to the blocks and interconnected by a pinjoint at other end $O$, and subjected to a horizontal force $P$ needed to keep the blocks from slipping as shown in Fig. 5.7. The coefficient of friction is 0.25 for all contact surfaces.

**Solution.** Let $C_1$ be the force in link $AO$ and $C_2$ in link $OB$. The free body diagrams of blocks $A$, $B$ and the hinge $O$ are shown in Fig. 5.7 ($b$). Block $A$ is on the verge of slipping.

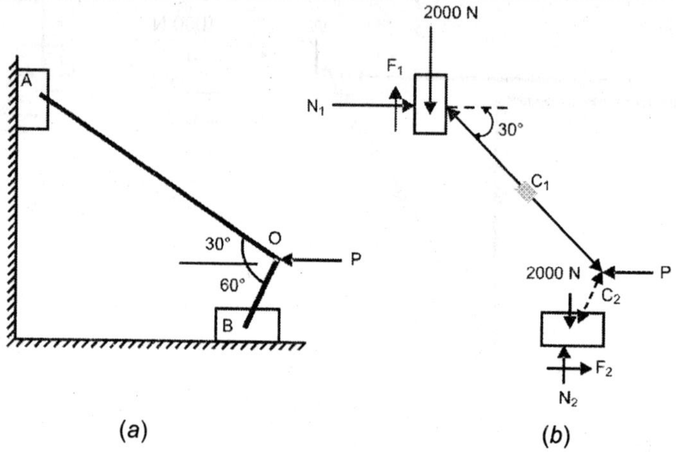


**Fig. 5.7**

Hence, from law of friction, $F_1 = \mu N_1 = 0.25\ N_1$

Consider block $A$: $\Sigma H = 0$, gives

$$N_1 = C_1 \cos 30°$$

$$\Sigma V = 0, \text{ gives}$$

$$F_1 + C_1 \sin 30° - 2000 = 0$$

$$0.25\ N_1 + C_1 \times 0.5 = 2000$$

$$(0.25 \cos 30° + 0.5)\ C_1 = 2000$$

$$C_1 = 2791.32 \text{ N}$$

Applying Lami's theorem to the equilibrium of the joint,

$$\frac{P}{\sin 90°} = \frac{C_2}{\sin 150°} = \frac{C_1}{\sin 120°}$$

∴ $\mathbf{P = 3223.14\ N.}$ **Ans.**

$\mathbf{C_2 = 1611.57\ N.}$ **Ans.**

The above solution holds good provided block $B$ is not slipping. To verify this fact, consider the block $B$.

$$\Sigma H = 0, \text{ gives}$$

$$F_2 = C_2 \cos 60° = 805.79 \text{ N}$$

$$\Sigma V = 0, \text{ gives}$$

$$N_2 = 2000 + C_2 \sin 60° = 3395.60 \text{ N}$$

$$\text{Limiting friction} = \mu\ N_2 = 0.25\ N_2 = 848.92 \text{ N}$$

Since, the actual frictional force $\mathbf{F_2}$ developed is less than the limiting frictional force, block $B$ is stationary and hence **P = 3223.14 N** is correct answer. [Note that if $\mathbf{F_2}$ calculated is more than the limiting friction $\mu N_2$, there is no possibility of maintaining equilibrium condition in the position shown.]

**Example 5.7.** Two planes $AC$ and $BC$ inclined at 60° and 30° to the horizontal meet at $C$. A load of 1000 N rests on the inclined plane $BC$ and is tied by a rope passing over a pulley to a block weighing $W$ Newtons and resting on the plane $AC$ as shown in Fig. 5.8 ($a$). If the coefficient of friction between the load and the plane $BC$ is 0.28 and that between the block and the plane $AC$ is 0.20, find the least and the greatest value of **W** for the equilibrium of the system.

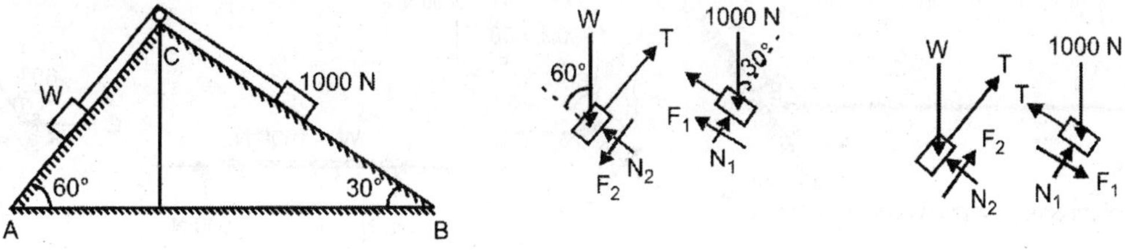


(*a*) (*b*) (*c*)

**Fig. 5.8**

**Solution.** For the least value of **W** for equilibrium, the motion of 1000 N block is impending downward. For such a case, the free body diagram of blocks are shown in Fig. 5.8 (*b*). Considering the 1000 N block:

Sum of forces normal to the plane = 0, gives

$$N_1 = 1000 \cos 30° = 866.03 \text{ N}$$

From the law of friction,
$$F_1 = 0.28\, N_1 = 242.49 \text{ N}$$

Sum of forces parallel to the plane = 0

$$T = -F_1 + 1000 \sin 30° = 257.51 \text{ N}$$

Now consider the equilibrium of block of weight, *W*:

Sum of forces normal to the plane = 0, gives

$$N_2 = W \cos 60° = 0.5W$$

$$\therefore \quad F_2 = 0.2\, N_2 = 0.1W$$

Sum of forces parallel to the plane = 0, gives

$$F_2 + W \sin 60° = T$$

$$0.1\, W + W \sin 60° = 257.51$$

$$\therefore \quad \mathbf{W = 266.57 \text{ N.}}$$ **Ans.**

For the greatest value of **W**, the 1000 N block is on the verge of moving up the plane. For such a case, the free body diagrams of the blocks are as shown in Fig. 5.8 (*c*).

Consider block of 1000 N,

$$N_1 = 866.03 \text{ N}$$

$$F_1 = 242.49 \text{ N}$$

$$T = 1000 \sin 30° + F_1 = 742.49 \text{ N}$$

Consider block of weight, *W*

$$N_2 = W \cos 60° = 0.5\, W$$

$$F_2 = 0.2\, N_2 = 0.1\, W$$

and
$$W \sin 60° - F_2 = T$$

$$W (\sin 60° - 0.1) = 742.49$$

$$\mathbf{W = 969.28 \text{ N.}}$$ **Ans.**

**Example 5.8.** Two blocks *A* and *B* each weighing 1500 N are connected by a uniform horizontal bar which weighs 1000 N. If the angle of limiting friction under each block is 15°, find the force **P** directed parallel to the 60° inclined plane that will cause motion impending to the right [Ref. Fig. 5.9 (*a*)].

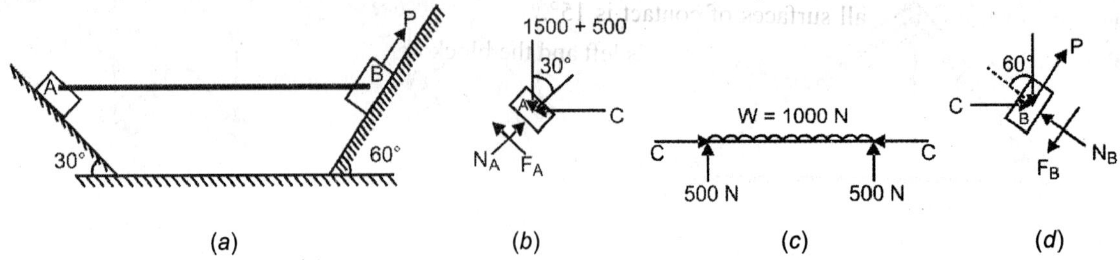


**Fig. 5.9**

**Solution.** Free body diagrams of block *A*, beam *AB* and block *B* are as shown in Figs. 5.9 (*b*), (*c*) and (*d*), respectively.

From vertical equilibrium of beam *AB*, it may be found that 500 N force is transferred at *A* and *B* which may be directly added to self-weights of the blocks.

Now consider block *A*:

$$\Sigma V = 0, \text{ gives}$$

$$N_A \cos 30° + F_A \sin 30° - 2000 = 0$$

But from the law of friction, $F_A = N_A \tan 15°$

$$\therefore \quad N_A (\cos 30° + \sin 30° \tan 15°) = 2000$$

$$N_A = 2000 \text{ N}$$

$$\therefore \quad F_A = 2000 \tan 15° = 535.90 \text{ N}$$

$$\Sigma H = 0, \text{ gives}$$

$$C - N_A \sin 30° + F_A \cos 30° = 0$$

$$C = 2000 \times \sin 30° - 535.9 \cos 30° = 535.90 \text{ N}$$

Consider block *B*:

Sum of forces normal to the inclined plane = 0, gives

$$N_B - 2000 \cos 60° - C \cos 30° = 0$$

$$N_B = 1464.10 \text{ N}$$

$$\therefore \quad F_B = N_B \tan 15° = 392.30 \text{ N}$$

Sum of forces parallel to the inclined plane = 0, gives

$$P - F_B - 2000 \sin 60° + C \sin 30° = 0$$

$$\mathbf{P = 1856.40 \text{ N}.}$$ **Ans.**

**(*b*) Equilibrium of Wedges**

Wedges are small pieces of materials with two of their opposite surfaces not parallel. They are used to lift slightly, heavy blocks, machinery, precast beams, etc., required for final alignment or to make place for inserting lifting devices. The weight of the wedge is very small compared to the weight lifted, hence, in all problems, the weight of wedges will be neglected. In the analysis of many problems, instead of treating normal reaction and frictional force independently, it is advantageous to treat their resultant.

If **F** is limiting friction, then resultant makes limiting frictional angle α with the normal. Its direction should be marked correctly. Note that the tangential component of the reaction **R** is the frictional force and it will always oppose impending motion.

**Example 5.9.** Determine force **P** required to start the movement of the wedge as shown in Fig. 5.10 (*a*). The angle of friction for all surfaces of contact is 15°.

**Solution.** As wedge is driven, it moves towards left and the block upwards. When motion is impending, limiting friction develops. Hence, resultant force makes limiting angle 15° with normal. The free body diagrams of block and wedge are shown in Fig. 5.10 (*b*). The forces on block and wedge are redrawn in Figs. 5.10 (*c*) and (*d*) so that Lami's theorem can be applied conveniently.

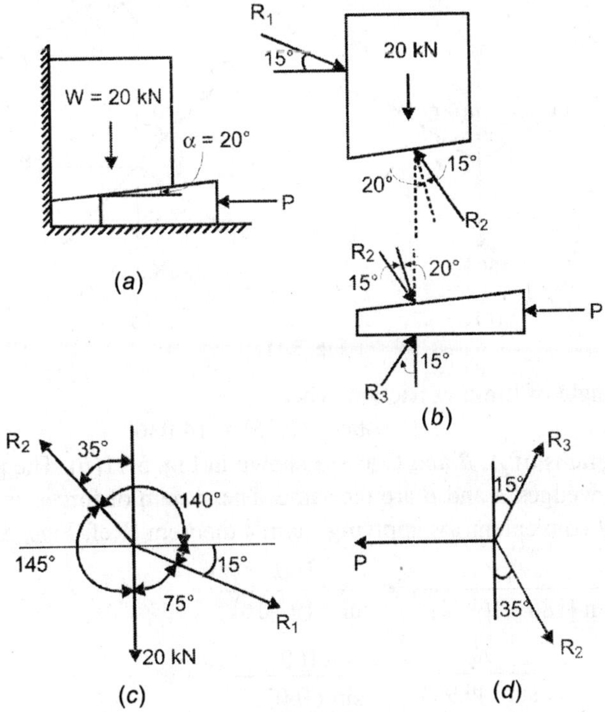


**Fig. 5.10**

Applying Lami's theorem to the system of forces on block, we have

$$\frac{R_1}{\sin 145°} = \frac{R_2}{\sin 75°} = \frac{20}{\sin 140°}$$

$$R_1 = 17.847 \text{ kN}$$

$$R_2 = 30.054 \text{ kN}$$

Applying Lami's theorem to system of forces on the wedge, we get

$$\frac{P}{\sin 130°} = \frac{R_2}{\sin 105°}$$

$$\mathbf{P = 23.835 \text{ kN.}}$$ **Ans.**

**Example 5.10.** A weight of 160 kN is to be raised by means of the wedges *A* and *B* as shown in Fig. 5.11(*a*). Find the value of force **P** for impending motion of block *C* upwards, if coefficient of friction is 0.25 for all surfaces. Weights of the block *C* and the wedges may be neglected.

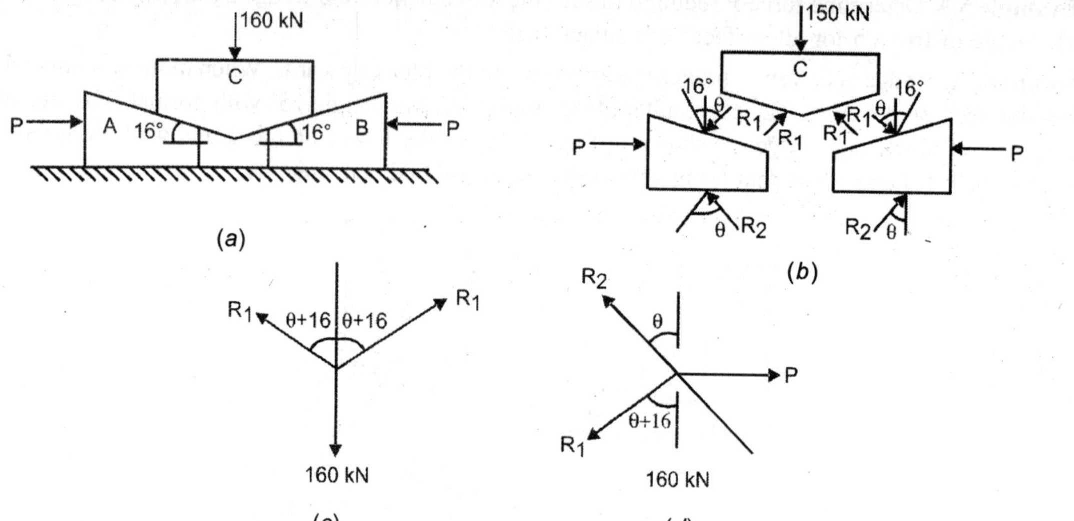


**Fig. 5.11**

**Solution.** Let θ be the angle of limiting friction. Then,

$$\theta = \tan^{-1}(0.25) = 14.036°$$

The free body diagrams of *A*, *B* and *C* are as shown in Fig. 5.11(*b*). The problem being symmetric, forces $\mathbf{R_1}$ and $\mathbf{R_2}$ on wedges *A* and *B* are the same. The system of forces on block *C* and on wedge *A* are shown in the form convenient for applying Lami's theorem [Ref. Figs. 5.11(*c*) and (*d*)].

$$\frac{R_1}{\sin(180-16-\theta)°} = \frac{160}{\sin 2(\theta+16)°}$$

*i.e.,*

$$\frac{R_1}{\sin 149.96°} = \frac{160}{\sin 60.072°}$$

$$R_1 = 92.41 \text{ kN}$$

Consider the equilibrium of wedge *A*:

$$\frac{P}{\sin(180-\theta-\theta-16)°} = \frac{R_1}{\sin(90+\theta)°}$$

$$\mathbf{P = 66.256\ kN.}$$ **Ans.**

**Example 5.11.** Block *C*, as shown in Fig. 5.12 (*a*), has a weight of 1000 *N* and identical blocks *A* and *B* weigh 750 N each. The blocks are placed on the floor as shown. If coefficient of friction is μ = 0.2 for all contacting surfaces, can the arrangement shown in the diagram remain in equilibrium ?

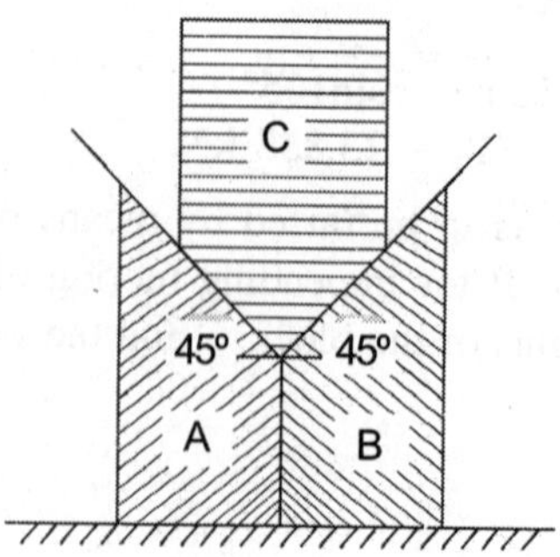


**Fig. 5.12 (*a*)**

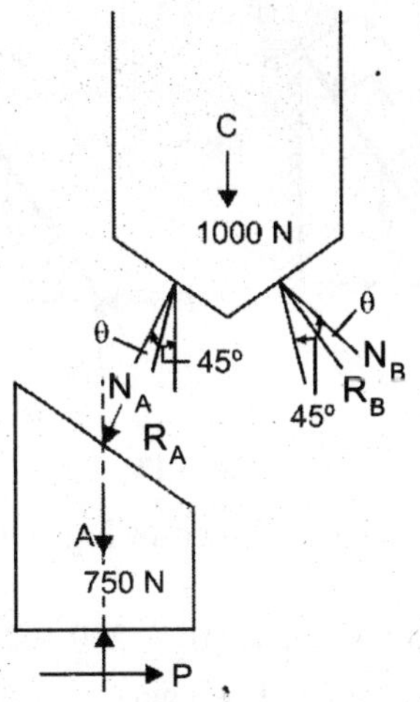


Fig. 5.12 (*b*)

**Solution.** The FBDs of block *C* and block *A*, when the blocks are on the verge of sliding are as shown in Fig. 5.12 (*b*). As block *C* tries to slide down if equilibrium does not exist, reaction **R** makes angle of 45 – θ with vertical. Reaction from block *B* on *C* is symmetric with reaction from block *A*. Considering equilibrium of block *C*, we get

$$\Sigma V = 0, \text{ gives}$$

$$2R \cos (45 - \theta)^\circ = 1000$$

where $\theta = \tan^{-1} 0.2 = 11.31°$

$\therefore$ $2R \cos (45 - 11.31)^\circ = 1000$

*i.e.*, $R = 600.92$ N.

Consider the equilibrium of block *A*,

$$\Sigma V = 0, \text{ gives}$$

$$600.92 \cos (45 - 11.31)^\circ + 750 = N_A$$

$\therefore$ $N_A = 1250$ N.

Let *P* be the horizontal force required to prevent sliding. Then

$$\Sigma H = 0, \text{ gives}$$

$$P = R \sin (45 - 11.31)^\circ = 333.3 \text{ N}.$$

Maximum frictional resistance

$$F = \mu N_A = 0.2 \times 1250 = 250 \text{ N}.$$

Since, $P > \mathbf{F}_{\text{lim.}}$, the blocks slide.

### (*c*) Equilibrium of Ladders

The system of forces acting on ladder constitutes non-concurrent system. Hence, in these frictional problems, one has to use Laws of Friction and equations of equilibrium of non-concurrent forces. The procedure is illustrated with the three examples given below:

**Example 5.12.** A ladder of length 4 m weighing 200 *N* is placed against a vertical wall as shown in Fig. 5.13(*a*). The coefficient of friction between the wall and the ladder is 0.2 and that between the floor and the ladder is 0.3. The ladder in addition to its own weight has to support a man weighing 600 *N* at a distance of 3 m from *A*. Calculate the minimum horizontal force to be applied at *A* to prevent slipping.

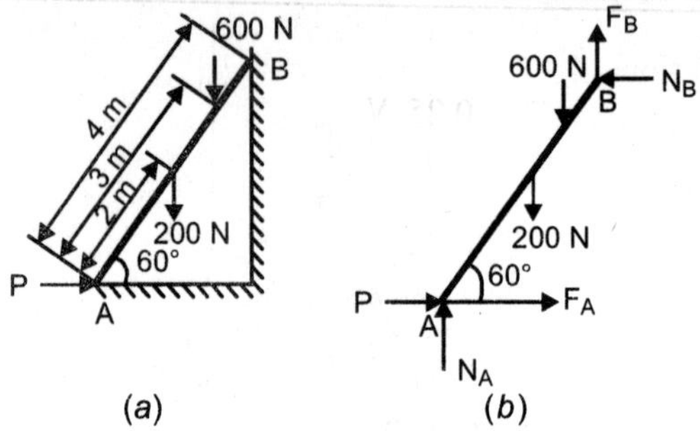


**Fig. 5.13**

**Solution.** The free body diagram of ladder is shown in Fig. 5.13(*b*).

$$\Sigma M_A = 0, \text{ gives}$$

$$N_B\, 4 \sin 60° + F_B\, 4 \cos 60° - 600 \times 3 \cos 60° - 200 \times 2 \cos 60° = 0$$

Dividing throughout by 4 and rearranging, we get

$$N_B\, 0.866 + 0.5\, F_B = 275$$

From the law of friction, $F_B = 0.2\, N_B$

$$\therefore \quad N_B\, (0.866 + 0.5 \times 0.2) = 275$$

$$N_B = 284.68 \text{ N}$$

$$\therefore \quad F_B = 56.934 \text{ N}$$

$$\Sigma V = 0, \text{ gives}$$

$$N_A - 200 - 600 + 56.934 = 0$$

$$N_A = 743.066 \text{ N}$$

$$\therefore \quad F_A = 0.3\, N_A$$

$$\therefore \quad F_A = 222.92 \text{ N}$$

$$\Sigma H = 0, \text{ gives}$$

$$P + F_A - N_B = 0$$

$$P = N_B - F_A = 284.68 - 222.92$$

**P = 61.76 N.** **Ans.**

**Example 5.13.** The ladder shown in Fig. 5.14 (*a*) is 6 m long and is supported by a horizontal floor and vertical wall. The coefficient of friction between the floor and the ladder is 0.25 and between wall and ladder is 0.4. The weight of ladder is 200 N and may be considered as concentrated at *G*. The ladder also supports a vertical load of 900 N at *C* which is at a distance of 1 m from *B*. Determine the least value of α at which the ladder may be placed without slipping. Determine the reaction at that stage.

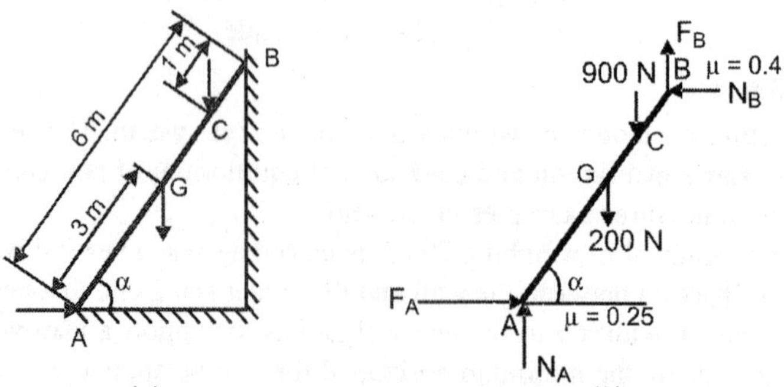


**Fig. 5.14**

**Solution.** From the law of friction,

$$F_A = 0.25\ N_A \qquad ...(5.13)$$

$$F_B = 0.4\ N_B \qquad ...(5.14)$$

and $\qquad \Sigma V = 0$, gives

$$N_A - 200 - 900 + F_B = 0$$

*i.e.,*

$$N_A + 0.4\ N_B = 1100 \qquad ...(5.15)$$

$$\Sigma H - 0, \text{ gives}$$

$$F_A - N_B = 0$$

$$0.25\ N_A = N_B \qquad ...(5.16)$$

From Eqns. (5.15), and (5.16), we get

$$N_A (1 + 0.4 \times 0.25) = 1100$$

$$\mathbf{N_A = 1000\ N.} \qquad \textbf{Ans.}$$

$$\mathbf{F_A = 250\ N.} \qquad \textbf{Ans.}$$

$$\mathbf{N_B = 250\ N.} \qquad \textbf{Ans.}$$

$$\mathbf{F_B = 0.4 \times 250 = 100\ N.} \qquad \textbf{Ans.}$$

$$\Sigma M_A = 0$$

$$N_B \times 6 \sin \alpha + F_B \times 6 \cos \alpha - 200 \times 3 \cos \alpha - 900 \times 5 \cos \alpha = 0$$

$$250 \times 6 \sin \alpha = (-100 \times 6 + 600 + 4500) \cos \alpha$$

$$\tan \alpha = \frac{4500}{1500} = 3$$

$$\boldsymbol{\alpha = 71.565°.} \qquad \textbf{Ans.}$$

**Example 5.14.** A ladder of length *L* rests against a wall, the angle of inclination being 45°. If the coefficient of friction between the ladder and the ground and that between ground and the wall is 0.5 each, what will be the maximum distance on ladder to which a man whose weight is 1.5 times the weight of ladder may ascend before the ladder begins to slip?

**Solution.** Figure 5.15(*a*) shows the ladder when it is about to slip when the man weighing 1.5 *W* is at a distance, '*aL*' from the end *A*. Its free body diagram is shown in Fig. 5.15(*b*).

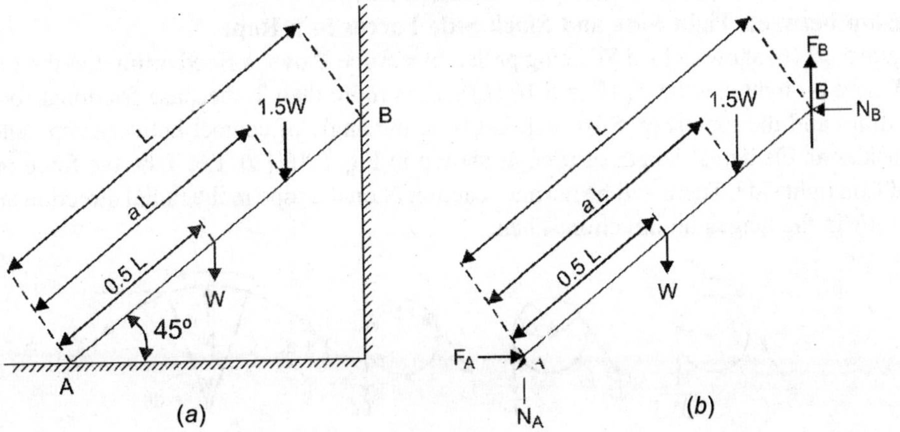


**Fig. 5.15**

Since, ladder is on the verge of slipping, from law of friction,

$$F_A = \mu N_A = 0.5\, N_A \qquad \text{...(5.17)}$$

and
$$F_B = \mu N_B = 0.5\, N_B \qquad \text{...(5.18)}$$

$$\Sigma H = 0, \text{ gives}$$

$$F_A - N_B = 0 \quad \text{or} \quad N_B = F_A = 0.5\, N_A \qquad \text{...(5.19)}$$

$\therefore$
$$F_B = 0.5\, N_B = 0.25\, N_A \qquad \text{...(5.20)}$$

$$\Sigma V = 0, \text{ gives}$$

$$N_A + F_B = W + 1.5W$$

i.e.,
$$N_A + 0.25\, N_A = 2.5W$$

or
$$N_A = \frac{2.5}{1.25} W = 2W \qquad \text{...(5.21)}$$

$$\Sigma M_A = 0, \text{ gives}$$

$$-F_B\, L \cos 45° - N_B\, L \sin 45° + 1.5\, W\, aL \cos 45° + W\, 0.5\, L \cos 45° = 0$$

Since, sin 45° = cos 45°, we get

$$F_B + N_B = 1.5\, aW + 0.5W$$

$$0.25\, N_A + 0.5\, N_A = 1.5\, aW + 0.5W$$

i.e.,
$$0.75 \times 2W = 1.5\, aW + 0.5W$$

i.e.,
$$1.5 = 1.5\, a + 0.5$$

or
$$a = \frac{1.0}{1.5} = 0.667$$

**Thus in this case, the man can ascend up to '0.667L' of ladder.** **Ans.**

## 5.4 BELT FRICTION

The transmission of power by means of belt or rope drives is possible because of friction which exists between the wheels and the belt. Similarly, band brakes stop the rotating discs because of friction between the belt and the disc. All along the contact surface, the frictional resistance develops. Hence, the tension in the rope is more on the side it is pulled and is less on the other side. Accordingly, the two sides of the rope may be called as tight side and slack side.

**Relationship between Tight Side and Slack Side Forces in a Rope**

Figure 5.16 (*a*) shows a load **W** being pulled by a force **P** over a fixed drum. Let the force on slack side be $\mathbf{T_1}$ and on tight side be $\mathbf{T_2}$ [Fig. 5.16 (*b*)]. $\mathbf{T_2}$ is more than $\mathbf{T_1}$ because frictional force develops between drum and the rope [Fig. 5.16 (*c*)]. Let $\theta$ be the angle of contact between rope and the drum. Now, consider an elemental length of rope as shown in Fig. 5.16 (*d*). Let **T** be the force on slack side and **T** + **dT** on tight side. There will be normal reaction **N** on the rope in the radial direction and frictional force **F** = $\mu N$ in the tangential direction. Then,

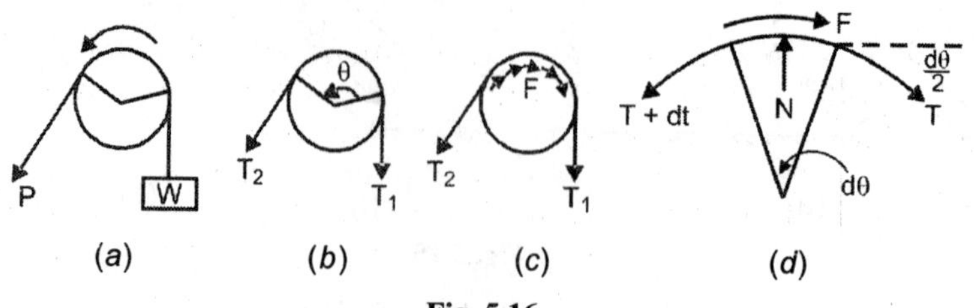


**Fig. 5.16**

Sum of forces in radial direction = 0, gives

$$N - T\sin\frac{d\theta}{2} - (T + dT)\sin\frac{d\theta}{2} = 0$$

Since, $d\theta$ is small, $\sin\frac{d\theta}{2} = \frac{d\theta}{2}$

$$\therefore \quad N - T\frac{d\theta}{2} - (T + dT)\frac{d\theta}{2} = 0 \qquad ...(5.22)$$

*i.e.,*

$$N = \left(T + \frac{dT}{2}\right)d\theta$$

From the law of friction,

$$F = \mu N = \mu\left(T + \frac{dT}{2}\right)d\theta \qquad ...(5.23)$$

where $\mu$ is coefficient of friction.

Now, Sum of forces in tangential direction = 0, gives

$$(T + dT)\cos\frac{d\theta}{2} = F + T\cos\frac{d\theta}{2}$$

Since, $\frac{d\theta}{2}$ is small, $\cos\frac{d\theta}{2} = 1$

$$\therefore \quad T + dT = F + T$$

or

$$dT = F \qquad ...(5.24)$$

From Eqns. (5.23) and (5.24), $dT = \mu\left(T + \frac{dT}{2}\right)d\theta$

Neglecting small quantity of higher order, we get

$$dT = \mu T d\theta$$

$$\frac{dT}{T} = \mu d\theta$$

Integrating both sides over 0 to $\theta$,

$$\int_{T_1}^{T_2} \frac{dT}{T} = \int_0^{\theta} \mu d\theta$$

$$\therefore \quad [\log T]_{T_1}^{T_2} = \mu[\theta]_0^{\theta}$$

$$\log\frac{T_2}{T_1} = \mu\theta$$

$$\frac{T_2}{T_1} = e^{\mu\theta}$$

*i.e.,*

$$T_2 = T_1 e^{\mu\theta} \qquad ...(5.25)$$

**Note:** $\theta$ should be in radians.

**Example 5.15.** A rope making $1\frac{1}{4}$ turns around a stationary horizontal drum is used to support a weight **W** (Fig. 5.17). If the coefficient of friction is 0.3, what range of weight can be supported by exerting a 600 N force at the other end of the rope?

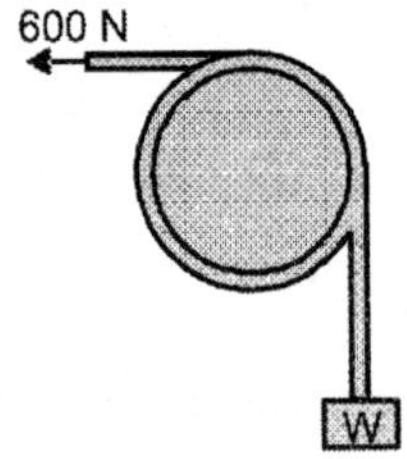

**Fig. 5.17**

**Solution.** Angle of contact $= 1.25 \times 2\pi = 2.5\pi$

1. Let the impending motion of weight be downwards.

Then,

$$\mathbf{T}_1 = 600\text{ N};\ \mathbf{T}_2 = W$$

$$\frac{W}{600} = e^{\mu 2.5\pi} = e^{0.3 \times 2.5\pi} = e^{0.75\pi}$$

$$\mathbf{W} = 6330.43\text{ N}$$

2. Let the impending motion of weight be upwards. Then

$$\mathbf{T}_1 = W;\qquad \mathbf{T}_2 = 600\text{ N}$$

$$\mathbf{T}_2 = \mathbf{T}_1 e^{\mu\theta}$$

$$600 = W e^{0.75\pi}$$

$$\mathbf{W} = 56.87\text{ N}$$

Thus, a 600 N force can support a range of loads between 56.87 N to 6330.43 N weight on the other side of drum. **Ans.**

**Example 5.16.** In Fig. 5.18 (*a*), the coefficient of friction is 0.20 between the rope and the fixed drum, and between other surface of contact $\mu = 0.3$. Determine the minimum weight **W** to prevent downward motion of the 1000 N body.

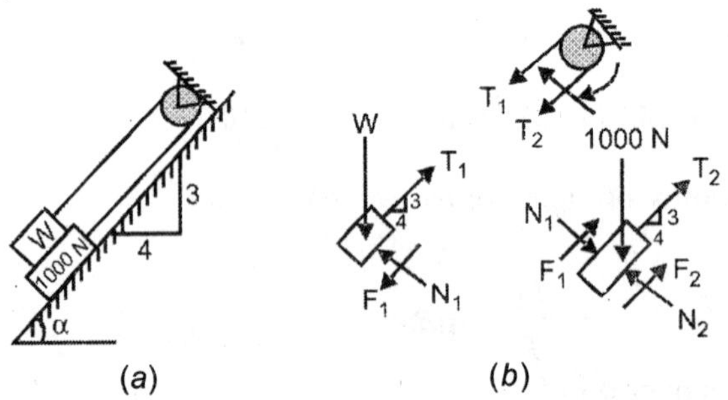

**Fig. 5.18**

**Solution.** Since, 1000 N weight is on the verge of sliding downwards, the rope connecting it is the tight side and the rope connecting $W$ is the slack side. Free body diagrams for $W$ and 1000 N body are shown in Fig. 5.18 (*b*).

Now,

$$\cos\alpha = 0.8$$

$$\sin\alpha = 0.6$$

Consider the equilibrium of weight $W$,

Sum of forces perpendicular to the plane = 0, gives

$$N_1 = W\cos\alpha$$

$$N_1 = 0.8\,W \qquad \text{...(5.26)}$$

$$\therefore \quad F_1 = \mu N_1 = 0.3 \times 0.8\,W$$

$$F_1 = 0.24\,W \qquad \text{...(5.27)}$$

Sum of forces parallel to the plane = 0, gives

$$T_1 = F_1 + W\sin\alpha = 0.24\,W + 0.6\,W$$

$$= 0.84\,W \qquad \text{...(5.28)}$$

Angle of contact of rope with the pulley = 180° = π radians

Applying friction equation, we get

$$T_2 = T_1 e^{\mu\theta} = T_1\, e^{0.3\pi}$$

$$T_2 = 2.566\ T_1$$

Substituting the value of $T_1$ from Eqn. (5.28)

$$T_2 = 2.156\ W \qquad \text{...(5.29)}$$

Now, consider 1000 N body,

Sum of forces perpendicular to the plane = 0, gives

$$N_2 - N_1 - 1000 \cos \alpha = 0$$

Substituting the value of $N_1$ from Eqn. (5.26),

$$N_2 = 0.8\ W + 1000 \times 0.8 = 0.8W + 800$$

$$\therefore \quad F_2 = 0.3\ N_2 = 0.24\ W + 240 \qquad \text{...(5.30)}$$

Sum of forces parallel to the plane = 0, gives

$$F_1 + F_2 - 1000 \sin \alpha + T_2 = 0$$

Substituting the values from Eqns. (5.27), (5.29) and (5.30),

$$0.24\ W + 0.24\ W + 240 - 1000 \times 0.6 + 2.156\ W = 0$$

$$\mathbf{W = 136.57\ N.}$$ **Ans.**

**Example 5.17.** A torque of 300 N-m acts on the brake drum shown in Fig. 5.19 (*a*). If the brake band is in contact with the brake drum through 250° and the coefficient of friction is 0.3 determine the force **P** applied at the end of the brake lever for the position shown in the figure.

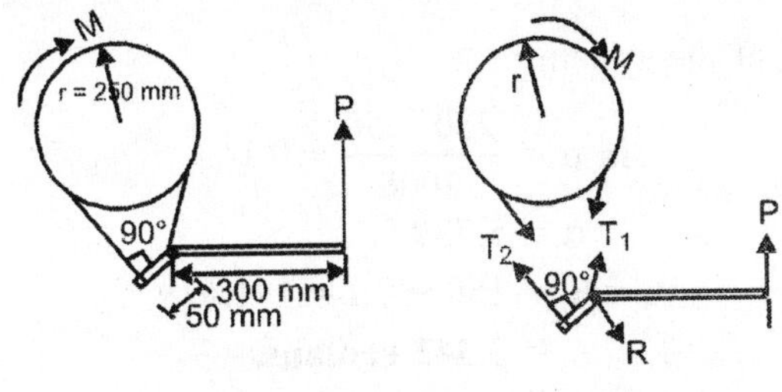


**Fig. 5.19**

**Solution.** Figure 5.19 (*b*) shows free body diagrams of brake drum and the lever arm.

Now $$T_2 = T_1 e^{\mu\theta}$$

$$\theta = \frac{250\pi}{180} \text{ radians and } \mu = 0.3$$

$$\therefore \quad \mu\,\theta = 1.309$$

$$\therefore \quad T_2 = T_1\, e^{1.309} = 3.7025\ T_1$$

Now, $$(T_2 - T_1)\, r = M$$

$$(3.7025 - 1)\ T_1 \times 250 = 300 \times 10^3$$

$$\therefore \quad T_1 = 444.04 \text{ N}$$

$$\therefore \quad T_2 = 1644.058 \text{ N}$$

Consider the lever arm. Taking moment about the hinge, we get

$$T_2 \times 50 = P \times 300$$

$$\mathbf{P = 274.0\ N.}$$ **Ans.**

**Example 5.18.** Two parallel shafts 3 m apart are to be connected by a belt running over the pulleys of diamter 500 mm and 100 mm respectively. Determine the length of belt required and the angle of contacts between belt and each pulley if it is crossed. What power can be transferred if the larger pulley rotates at 220 revolutions per minute? Given: Maximum permissible tension in the belt = 1 kN and coefficient of friction between the belt and the pulley is 0.25.

**Solution.** The crossed belt drive system is shown in Fig. 5.20.

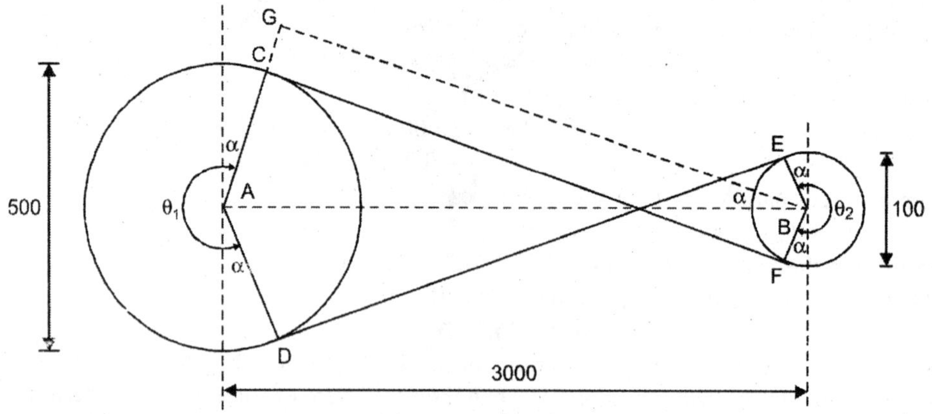


**Fig. 5.20**

From the geometry of the system,

$$\sin \alpha = \frac{250 + 50}{3000} = 0.1$$

$$\therefore \quad \alpha = 5.739°$$

$$\therefore \quad \boldsymbol{\theta_2 = \theta_1} = 180 + 2\alpha = 191.478°$$

$$= \mathbf{3.342\ radians.} \qquad \textbf{Ans.}$$

Length of belt = Arc length $DC$ + Arc length $FE$ + $2BG$

$$= 250\,\theta_1 + 50\,\theta_2 + 2 \times 3000 \cos \alpha$$

$$= 250 \times 3.342 + 50 \times 3.342 + 2 \times 3000 \cos 5.734°$$

$$= \mathbf{6972.5\ mm.} \qquad \textbf{Ans.}$$

Max. Tension in the belt = 1 kN = 1000 N.

From rope friction formula,

$$\frac{T_1}{T_2} = e^{\mu\theta} = e^{0.25 \times 3.342} = e^{0.6684} = 2.306$$

In this case, $T_1$ = Maximum tension = 1000 N

$$\therefore \quad T_2 = \frac{1000}{2.306} = 433.66 \text{ N}$$

Velocity of the belt = $r\,\theta$

$$= 250 \times 220 \times 2\pi \text{ mm/min.}$$

$$= \frac{250 \times 220 \times 2\pi}{60} \text{ mm/s.}$$

$= 5799.6$ mm/s.

$\therefore$ Power transmitted $= (T_2 - T_1) \times$ Velocity

$= (1000 - 433.66) \times 5759.6$

$= 3261884$ N-mm/s.

$=$ **3.261884 kW.** **Ans.**

**Example 5.19.** In the above example, if the belt is connected parallel find angle of contact, length of belt and maximum power that can be transmitted.

**Solution.** The parallel drive system is shown in Fig. 5.21.

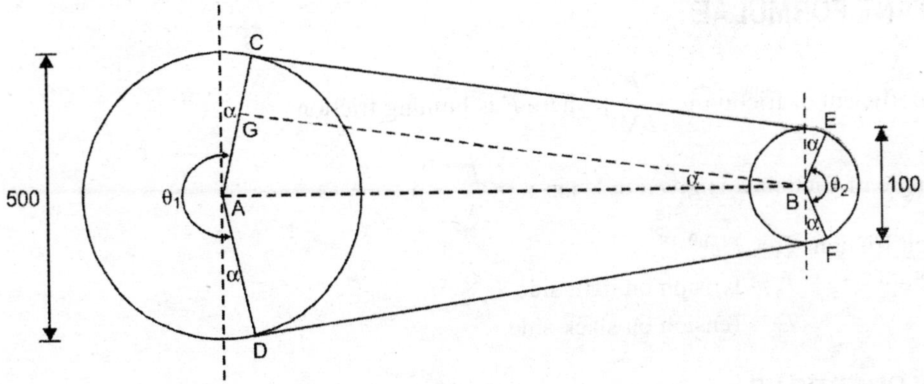


**Fig. 5.21**

From the geometry of the system,

$$\sin \alpha = \frac{250 - 50}{3000}$$

$\therefore$ $\alpha = 3.822°$

$\therefore$ $\theta_1 = \pi + 2\alpha = 180 + 2 \times 3.822$

$= 187.644°$

$=$ **3.275 radians.** **Ans.**

$\theta_2 = \pi - 2\alpha = 180 - 2 \times 3.822$

$= 172.356$

$=$ **3.008 radians.** **Ans.**

Length of belt required

$=$ Arc length $CD$ + Arc length $EF + 2BG$

$= 250\,\theta_1 + 50\,\theta_2 + \sqrt{(3000)^2 - (250 - 50)^2}$

$= 250 \times 3.275 + 50 \times 3.008 + 2 \times 2993.325$

$=$ **6955.8 mm.** **Ans.**

As frictional force is less on smaller pulley, slippage takes place on smaller pulley. Hence

$$\frac{T_1}{T_2} = e^{\mu\theta_2} = e^{0.25 \times 3.008} = 2.1212$$

$$\therefore \quad T_2 = \frac{T_1}{2.1212} = \frac{1000}{2.1212} = 471.4 \text{ Newton}$$

$$\text{Velocity of belt} = r \times \frac{2\pi \times 220}{60} = 250 \times \frac{2\pi \times 220}{60}$$

$$= 5759.6 \text{ mm/s}$$

∴ Maximum power that can be transferred without slippage

$$= (T_1 - T_2) \times \text{Velocity}$$

$$= (1000 - 471.4) \times 5759.6 = 3044517 \text{ N mm/s}$$

= **3.044517 kW.** **Ans.**

## IMPORTANT FORMULAE

1. Coefficient of friction $\mu = \frac{F}{N}$, where $F$ is limiting friction.
2. Angle of friction $\alpha$ is given by $\tan \alpha = \frac{F}{N} = \mu$.
3. Belt friction: $T_2 = T_1 e^{\mu\theta}$

   where $T_2$ = Tension on tight side

   $T_1$ = Tension on slack side

## THEORY QUESTIONS

**5.1** Define and explain:

(*i*) Coefficient of friction

(*ii*) Angle of friction.

**5.2** State the laws of dry friction.

**5.3** Derive the relationship between tensions on tight side and slack side in a rope.

## EXERCISES

**5.1** A pull of 180 N applied upward at 30° to a rough horizontal plane was required to just move a body resting on the plane, while a push of 220 N applied along the same line of action was required to just move the same body downwards. Determine the weight of the body and the coefficient of friction. [**Ans. W** = 990 N; μ = 0.1722]

**5.2** The block *A*, as shown in Fig. 5.22, weighs 2000 N. The cord attached to *A* passes over a frictionless pulley and supports a weight equal to 800 N. The value of coefficient of friction between *A* and the horizontal plane is 0.35. Solve for horizontal force **P**: (*a*) if motion is impending towards the left, and (*b*) if the motion is impending towards the right.

[**Ans.** (*a*) 1252.82 N; (*b*) 132.82 N]

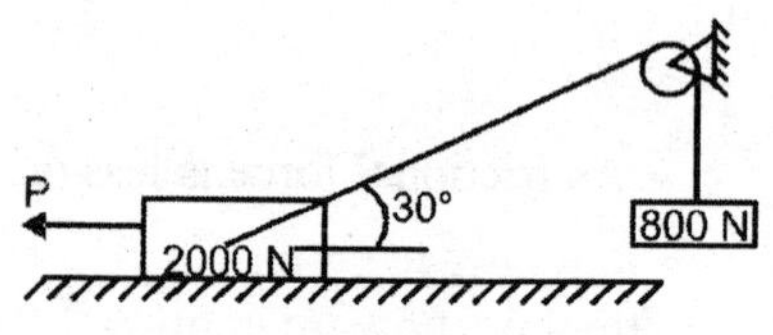


**Fig. 5.22**

**5.3** A 3000 N block is placed on an inclined plane as shown in Fig. 5.23. Find the maximum value of **W** for equilibrium if tipping does not occur. Assume coefficient of friction as 0.2. [**Ans.** 2636.15N]

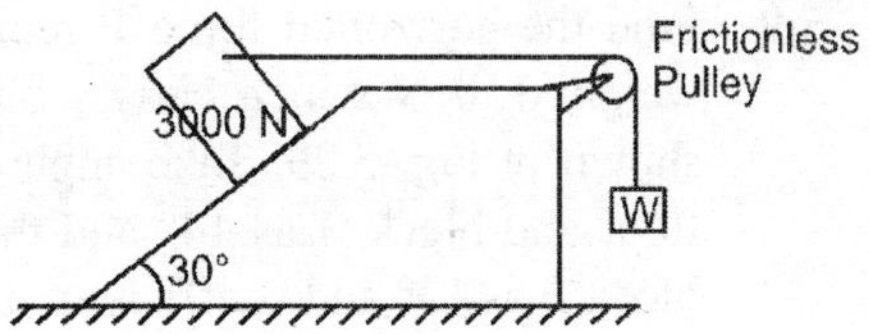


**Fig. 5.23**

**5.4** Find whether block $A$ is moving up or down the plane in Fig. 5.24 for the data given below: Weight of block $A$ = 300 N. Weight of block $B$ = 600 N. Coefficient of limiting friction between plane $AB$ and block $A$ is 0.2. Coefficient of limiting friction between plane $BC$ and block $B$ is 0.25. Assume pulley as smooth.

[**Ans.** The block $A$ is stationary, since, $F$ developed < $F_{lim}$]

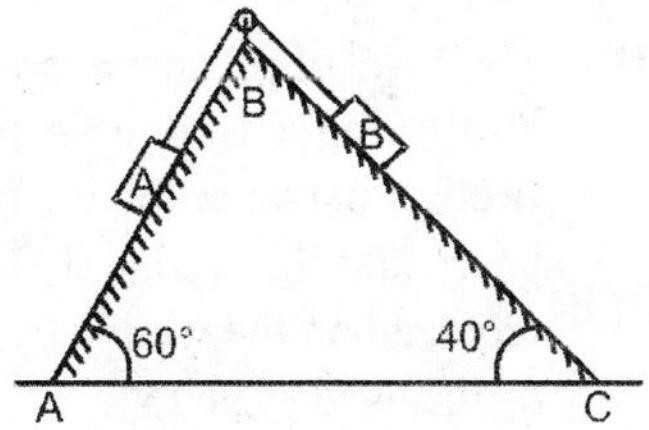


**Fig. 5.24**

**5.5** Two identical blocks $A$ and $B$ are connected by a rod and they rest against vertical and horizontal planes respectively as shown in Fig. 5.25. If sliding impends when θ = 45°, determine the coefficient of friction, assuming it to be same for both floor and wall.

[**Ans.** 0.414]

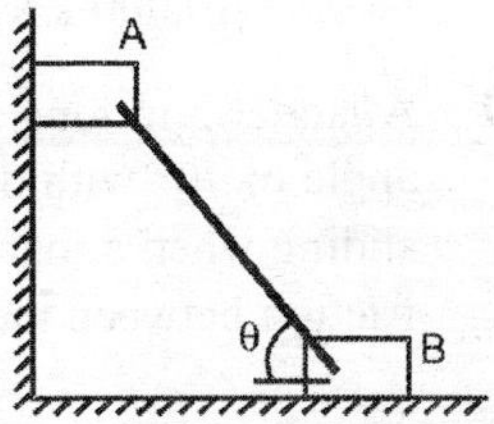


**Fig. 5.25**

**5.6** Determine force **P** required to start the wedge as shown in Fig. 5.26. The angle of friction for all surfaces of contact is 15°. [**Ans.** 26.6784 kN]

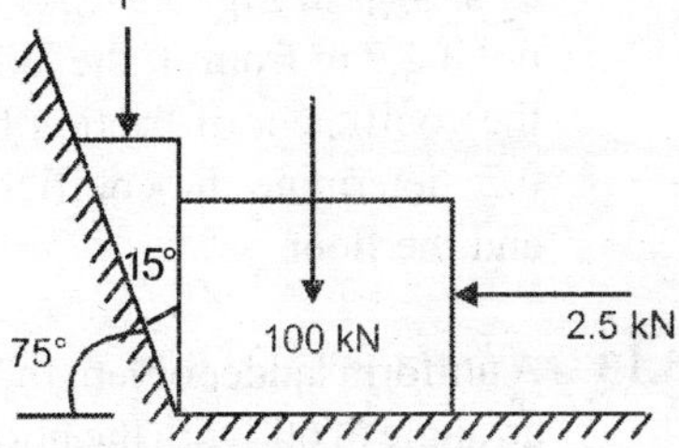


**Fig. 5.26**

**5.7** Two blocks $A$ and $B$ weighing 15 kN and 3 kN, respectively, are held in position against an inclined plane by applying a horizontal force **P** as shown in Fig. 5.27. Find the least value of $P$ which will induce motion of the block $A$ upwards. Angle of friction for all contact surfaces is 12°. [**Ans.** 14.025 kN]

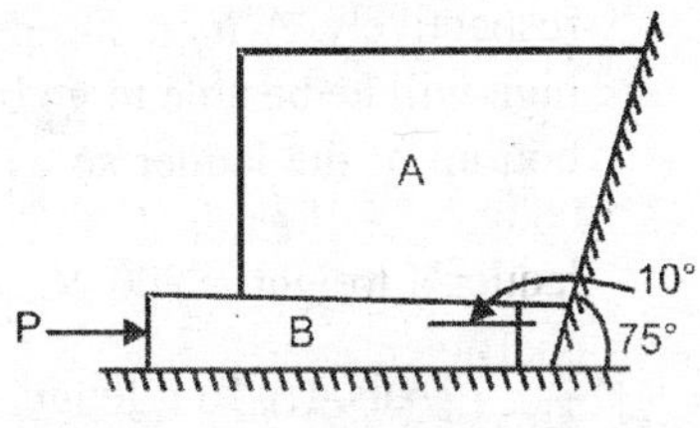


**Fig. 5.27**

**5.8** In Fig. 5.28, $C$ is a stone block weighing 6 kN. It is being raised slightly by means of two wooden wedges $A$ and $B$ with a force **P** on wedge $B$. The angle between the contacting surfaces of the wedge is 5°. If coefficient of friction is 0.3 for all contacting surfaces, compute the value of $P$ required to impend upward motion of the block $C$. Neglect weight of the wedges.

[**Ans.** 2.344 kN]

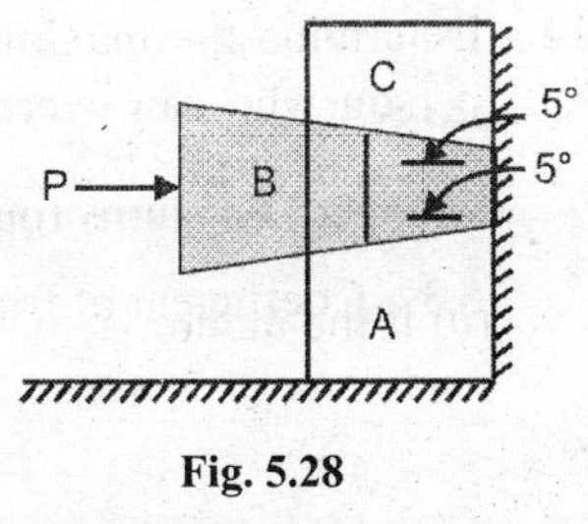


**Fig. 5.28**

**5.9** Find the horizontal force **P** required to push the block $A$ of weight 150 N which carries block $B$ of weight 1280 N as shown in Fig. 5.29. Take angle of limiting friction between floor and block $A$ as 14° and that between vertical wall and block $B$ as 13° and coefficient of limiting friction between the blocks as 0.3. [**Ans. P** = 1294.2 N]

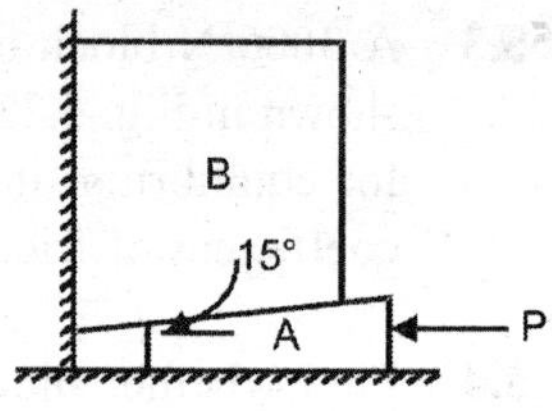


**Fig. 5.29**

**5.10** The level of precast beam weighing 20,000 N is to be adjusted by driving a wedge as shown in Fig. 5.30. If coefficient of friction between the wedge and pier is 0.35 and that between beam and the wedge is 0.25, determine the minimum force **P** required on the wedge to make adjustment of the beam. Angle of the wedge is 15°.

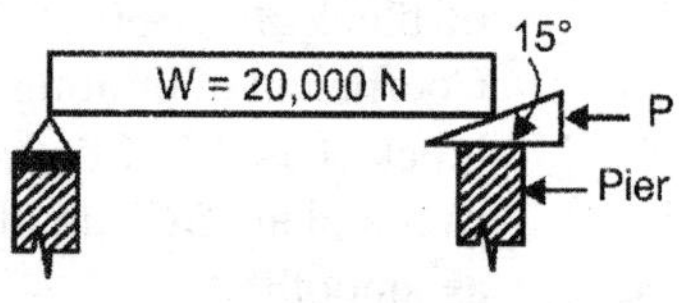


**Fig. 5.30**

[**Hint:** Vertical component of reaction on wedge at contact with beam = 1/2 vertical load on beam = 10,000 kN] [**Ans.** 9057.4 N]

**5.11** A ladder 5 m long rests on a horizontal ground and leans against a smooth vertical wall at an angle of 70° with the horizontal. The weight of ladder is 300 N. The ladder is on the verge of sliding when a man weighing 750 N stands on a rung 1.5 m high. Calculate the coefficient of friction between the ladder and the floor. [**Ans.** $\mu$ = 0.1837]

**5.12** A 4 m ladder weighing 200 N is placed against a vertical wall as shown in Fig. 5.31. As a man weighing 800 N, reaches a point 2.7 m from $A$, the ladder is about to slip. Assuming that the coefficient of friction between the ladder and the wall is 0.2, determine the coefficient of friction between the ladder and the floor. [**Ans.** 0.3548]

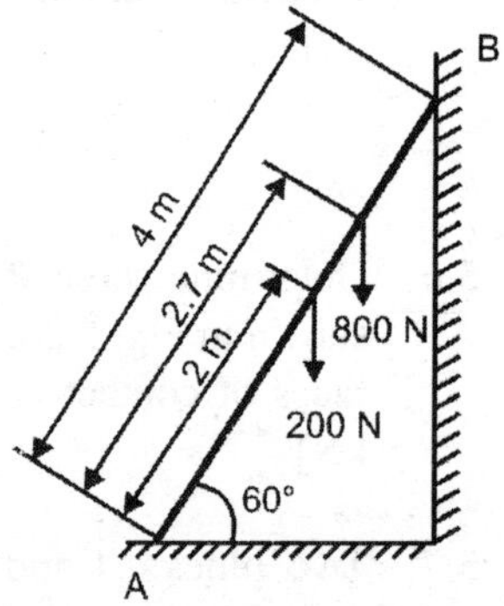


**Fig. 5.31**

**5.13** A uniform ladder of length '$L$' rests against a vertical wall making an angle of 60° with the horizontal. Coefficient of friction between wall and ladder and ground and ladder are 0.3 and 0.25 respectively. A man weighing 650 N ascends the ladder. How high will he be able to go before the ladder slips? Find the magnitude of weight to be put at the bottom of the ladder so as to make it just sufficient to permit the man to go to the top. Take ladder's weight = 900 N. [**Hint.** Find $P$ as found in Example 11. Then $W = \frac{P}{\mu}$]

[**Ans.** (*i*) To a length of 0.4345 $L$, *i.e.*, to a height of 0.435 $L$ sin 60° = 0.367 $L$ metres.
(*ii*) $W$ = 777.68 N]

**5.14** Determine the maximum weight that can be lowered by a person who can exert a 300 N pull on rope if the rope is wrapped $2\frac{1}{2}$ turns round a horizontal spur as shown in Fig. 5.32. Coefficient of friction between spur and the rope is 0.3. [**Ans.** 33395.33 N]

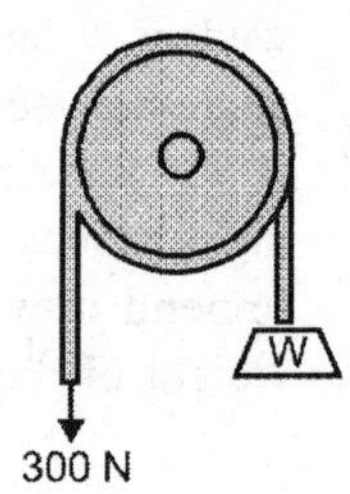


**Fig. 5.32**

**5.15** Determine the minimum value of **W** required to cause motion of blocks *A* and *B* towards right (Ref. Fig. 5.33). Each block weighs 3000 N and coefficient of friction between blocks and the planes is 0.2. Coefficient of friction between the drum and rope is 0.1. Angle of wrap over the drum is 90°. **[Ans.** 3065.18 N]

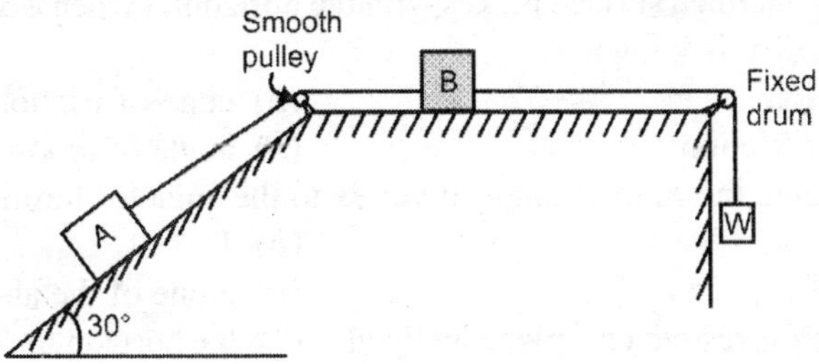


**Fig. 5.33**

**5.16** The dimension of a brake drum is as shown in Fig. 5.34. Determine the torque **M** exerted on the drum if the load **P** = 50 N. Assume coefficient of kinetic friction between rope and drum to be 0.15. **[Ans.** 747.685 N-m]

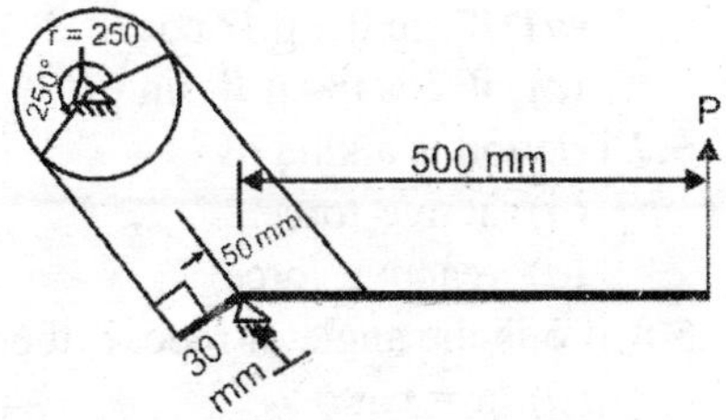


**Fig. 5.34**

**5.17** A belt drive is required to transmit 12 kW power. The velocity of belt is 9.6 m/s. If coefficient of friction is 0.25 and the angle of contact is 150°, determine the maximum tension in the belt.

**[Ans. $T_1$** = 2.602 kN]

**5.18** Two parallel shafts are having pulleys of diameters 300 mm and 500 mm. The distance between the shafts is 2.5 m. They are connected by crossed belt drive system. The speed of larger pulley is 500 rpm and coefficient of friction μ = 0.3, determine the maximum power that can be transmitted, if maximum permissible tension in the belt is 2.4 kN. **[Ans.** *P* = 20.3 kN]

## OBJECTIVE TYPE QUESTIONS

**Select the correct option in the following:**

**5.1** Coefficient of friction is

(*a*) $\dfrac{\text{Frictional force}}{\text{Normal reaction}}$ (*b*) $\dfrac{\text{Limiting friction}}{\text{Normal reaction}}$

(*c*) $\dfrac{\text{Frictional force}}{\text{Total reaction}}$ (*d*) $\dfrac{\text{Limiting friction}}{\text{Total reaction}}$

**5.2** The coefficient of friction depends upon

(*a*) Area of contact (*b*) Roughness of the surface

(*c*) The shape of contact area (*d*) All of the above

**5.3** If **F** is the limiting friction, **N** is the normal reaction and **R** is the resultant of **F** and **N**, the angle of limiting friction is

(*a*) angle between **F** and **N**
(*b*) angle between **F** and **R**
(*c*) angle between **N** and **R**
(*d*) None of the above

**5.4** The angle which an inclined surface makes with the horizontal when a body placed on it is on the verge at moving down, is known as

(*a*) angle of friction
(*b*) cone of friction
(*c*) coefficient of friction
(*d*) angle of repose

**5.5** In the cone of friction, the ratio of angle at vertex to the angle of limiting friction is

(*a*) 2
(*b*) 1
(*c*) 0.5
(*d*) None of the above

**5.6** A block weighing **W** is resting on a plane inclined at $\theta$ to the horizontal. The coefficient of friction is $\mu$. The force **P** required to pull it up the plane is

(*a*) $W \sin \theta - \mu W \cos \theta$
(*b*) $W \sin \theta + \mu W \cos \theta$
(*c*) $W \cos \theta - \mu W \sin \theta$
(*d*) $W \cos \theta + \mu W \cos \theta$

**5.7** Friction is a kind of

(*a*) active force
(*b*) body force
(*c*) reactive force
(*d*) None of the above

**5.8** If $\phi$ is the angle of repose, then coefficient of friction $\mu$ is given by

(*a*) $\mu = \tan \phi$
(*b*) $\mu = \cot \phi$
(*c*) $\mu = \sec \phi$
(*d*) None of the above

**5.9** Compared to static friction, dynamic friction is

(*a*) more
(*b*) same
(*c*) less
(*d*) None of these

# CHAPTER 6

# CENTROID AND MOMENT OF INERTIA

Under this topic, first we will see how to find the areas of given figures and the volumes of given solids. Then the terms centre of gravity and centroids are explained. Though the title of this topic does not indicate the centroid of line segment, that term is also explained, since the centroid of line segment will be useful in finding the surface area and volume of solids using theorems of Pappus and Guidinus. Then the term first moment of area is explained and the method of finding centroid of plane areas and volumes is illustrated. After explaining the term second moment of area, the method of finding moment of inertia of plane figures about $x$-$x$ or $y$-$y$ axis is illustrated. The term product moment of inertia is defined and the mehtod of finding principal moment of inertia is presented. At the end, the method of finding mass moment of inertia is presented.

## 6.1 DETERMINATION OF AREAS AND VOLUMES

In the school education, methods of finding areas and volumes of simple cases are taught by many methods. Here we will see the general approach which is common to all cases, *i.e.*, by the method of integration. In this method, the expression for an elemental area will be written, then suitable integrations are carried out so as to take care of entire surface/volume. This method is illustrated with standard cases below, first for finding the areas and latter for finding the volumes:

**(A) Area of Standard Figures**

(*i*) *Area of a rectangle*

Let the size of rectangle be $b \times d$, as shown in Fig. 6.1., $dA$ is an elemental area of side $dx \times dy$.

$$\text{Area of rectangle, } A = \oint dA = \int_{-b/2}^{b/2}\int_{-d/2}^{d/2} dx\,dy$$

$$= [x]_{-b/2}^{b/2}\,[y]_{-d/2}^{d/2}$$

$$= bd.$$

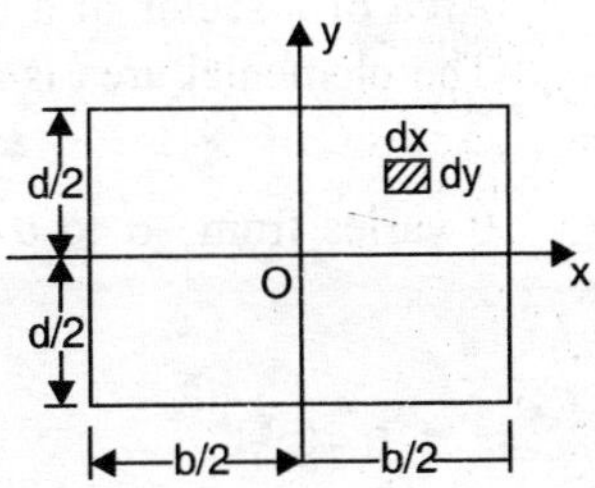


Fig. 6.1

If we take element as shown in Fig. 6.2,

$$A = \int_{-d/2}^{d/2} dA = \int_{-d/2}^{d/2} b\,dy$$

$$= b\left[y\right]_{-d/2}^{d/2}$$

$$= bd$$

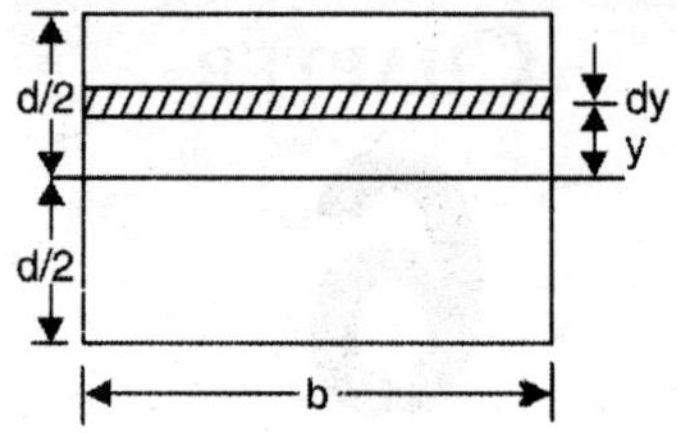


Fig. 6.2

(ii) *Area of a triangle of base width 'b' height 'h'*. Referring to Fig. 6.3, let the element be selected as shown by hatched lines.

Then $$dA = b'dy = b\frac{y}{h}dy$$

$$A = \int_0^h dA = \int_0^h b\frac{y}{h}dy$$

$$= \frac{b}{h}\left[\frac{y^2}{2}\right]_0^h = \frac{bh}{2}$$

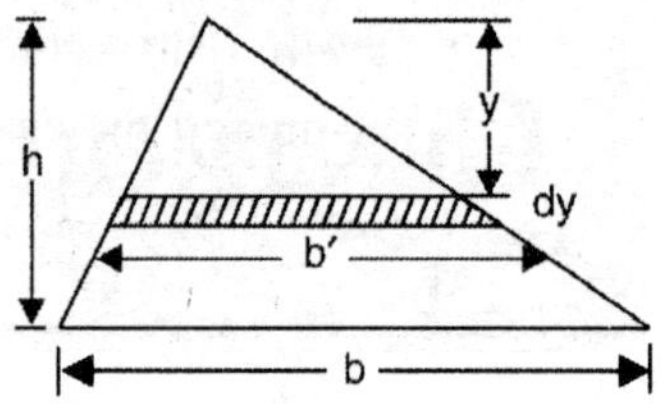


Fig. 6.3

(iii) *Area of a circle*

Consider the elemental area $dA = rd\theta\ dr$ as shown in Fig. 6.4. Now,

$$dA = rd\theta\ dr$$

$r$ varies from 0 to $R$ and $\theta$ varies from 0 to $2\pi$

$$\therefore \quad A = \int_0^{2\pi}\int_0^R rd\theta\, dr$$

$$= \int_0^{2\pi}\left[\frac{r^2}{2}\right]_0^R d\theta$$

$$= \int_0^{2\pi}\frac{R^2}{2}d\theta$$

$$= \frac{R^2}{2}\left[\theta\right]_0^{2\pi}$$

$$= \frac{R^2}{2}.2\pi = \pi R^2$$

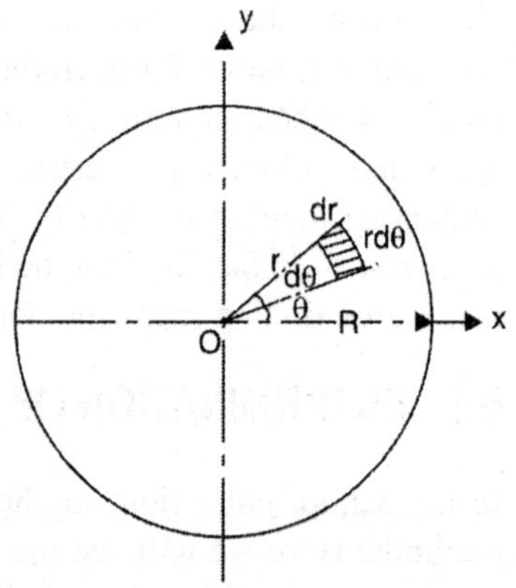


Fig. 6.4

In the above derivation, if we take variation of $\theta$ from 0 to $\pi$, we get the area of semicircle as $\frac{\pi R^2}{2}$ and if the limit is from 0 to $\pi/2$ the area of quarter of a circle is obtained as $\frac{\pi R^2}{4}$.

(iv) *Area of a sector of a circle*

Area of a sector of a circle with included angle $2\alpha$, as shown in Fig. 6.5, is to be determined. The elemental area is as shown in the figure

$$dA = rd\theta \cdot dr$$

$\theta$ varies from $-\alpha$ to $\alpha$ and $r$ varies from 0 to $R$

$$\therefore \quad A = \oint dA = \int_{-\alpha}^{\alpha}\int_0^R r\,d\theta\,dr$$

$$= \int_{-\alpha}^{\alpha}\left[\frac{r^2}{2}\right]_0^R d\theta = \int_{-\alpha}^{\alpha}\frac{R^2}{2}d\theta$$

$$= \left[\frac{R^2}{2}\theta\right]_{-\alpha}^{\alpha} = \frac{R^2}{2}(2\alpha) = R^2\alpha$$

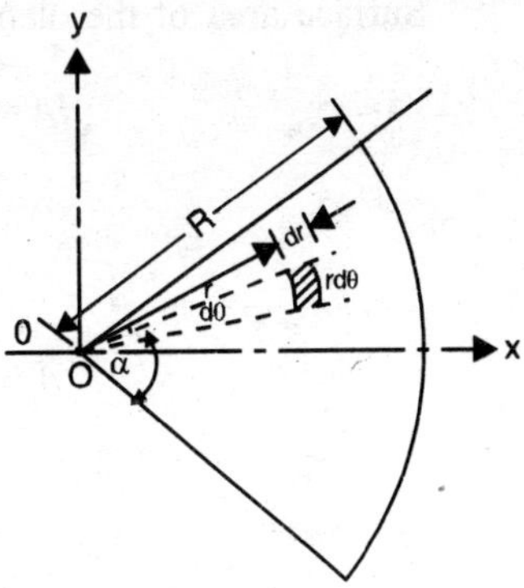


Fig. 6.5

(v) *Area of a parabolic spandrel*

Two types of parabolic curves are possible

(a) $y = kx^2$

(b) $y^2 = kx$

*Case a:* This curve is shown in Fig. 6.6.

The area of the element

$$dA = y\,dx = kx^2\,dx$$

$$\therefore \quad A = \int_0^a dA = \int_0^a kx^2\,dx$$

$$= k\left[\frac{x^3}{3}\right]_0^a = \frac{ka^3}{3}$$

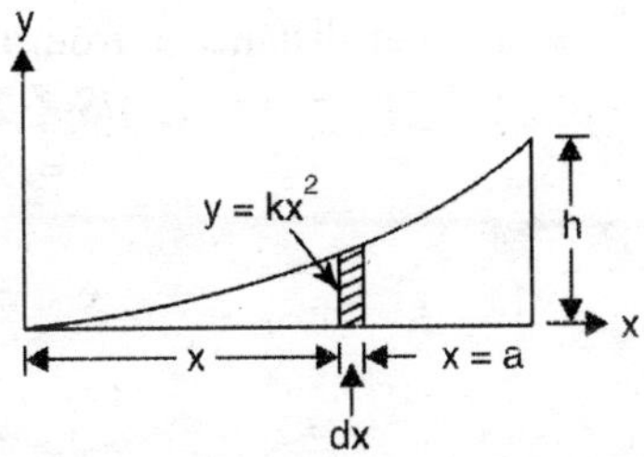


Fig. 6.6

We know, when $x = a$, $y = h$

i.e., $\quad h = ka^2$ or $k = \dfrac{h}{a^2}$

$$\therefore \quad A = \frac{ka^3}{3} = \frac{h}{a^2}\frac{a^3}{3} = \frac{1}{3}ha = \frac{1}{3}\text{rd the area of rectangle of size } a \times h$$

*Case b:* In this case, $y^2 = kx$

Referring to Fig. 6.7

$$dA = y\,dx = \sqrt{kx}\,dx$$

$$A = \int_0^a y\,dx = \int_0^a \sqrt{kx}\,dx$$

$$= \sqrt{k}\left[x^{3/2}\frac{2}{3}\right]_0^a = \sqrt{k}\,\frac{2}{3}\,a^{3/2}$$

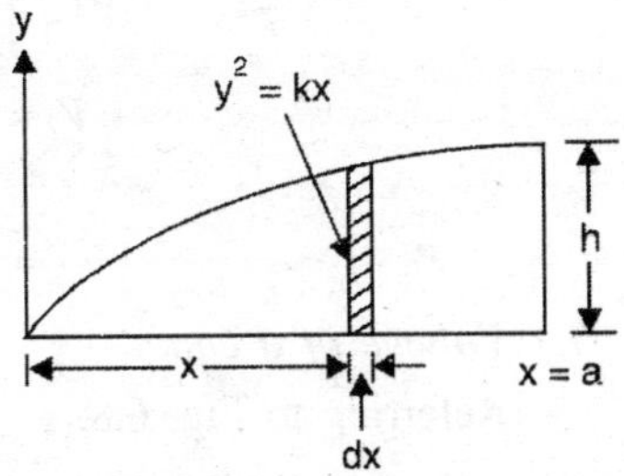


Fig. 6.7

We know, when $x = a$, $y = h$

$$\therefore \quad h^2 = ka \quad \text{or} \quad k = \frac{h^2}{a}$$

Hence, $\quad A = \dfrac{h}{\sqrt{a}}\cdot\dfrac{2}{3}\cdot a^{3/2}$

i.e., $\quad A = \dfrac{2}{3}ha = \dfrac{2}{3}$rd the area of rectangle of size $a \times h$

(vi) *Surface area of a cone*

Consider the cone as shown in Fig. 6.8. Now,

$$y = \frac{x}{h}R$$

Surface area of the element,

$$dA = 2\pi y\, dl = 2\pi \frac{x}{h} R\, dl$$

$$= 2\pi \frac{x}{h} R \frac{dx}{\sin\alpha}$$

$$\therefore \quad A = \frac{2\pi R}{h \sin\alpha}\left[\frac{x^2}{2}\right]_0^h$$

$$= \frac{\pi R h}{\sin\alpha} = \pi R l$$

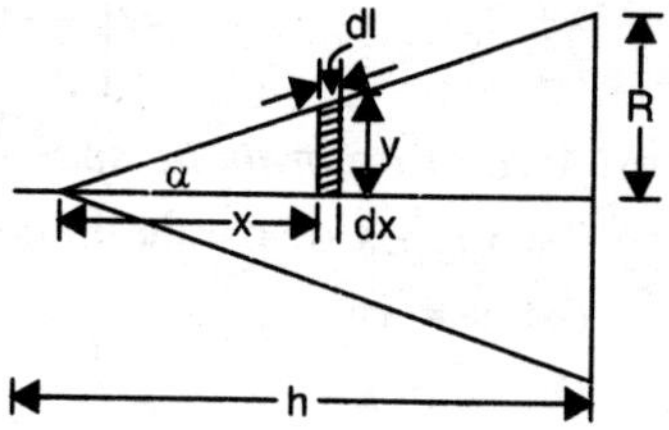


**Fig. 6.8**

(*vii*) *Surface area of a sphere*

Consider the sphere of radius $R$ as shown in Fig. 6.9. The element considered is the parallel circle at distance $y$ from the diametral axis of sphere.

$$dS = 2\pi x\, Rd\theta$$

$$= 2\pi\, R \cos\theta\, Rd\theta, \text{ since } x = R \cos$$

$$\therefore \quad S = 2\pi R^2 \int_{-\pi/2}^{\pi/2} \cos\theta\, d\theta$$

$$= 2\pi\, R^2 [\sin\theta]_{-\pi/2}^{\pi/2}$$

$$= 4\pi R^2$$

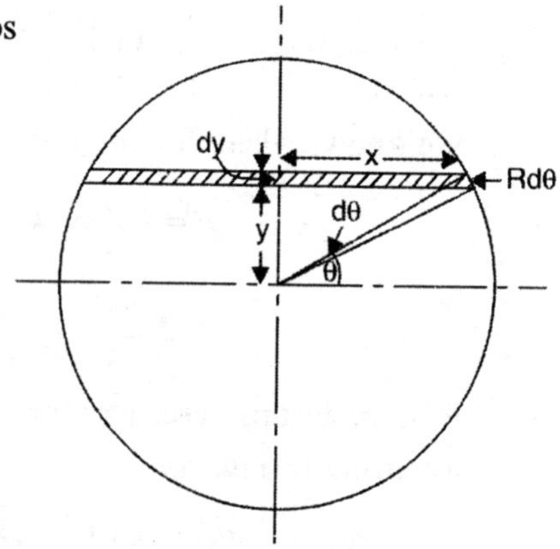


**Fig. 6.9**

**(B) Volume of Standard Solids**

(*i*) *Volume of a parallelopiped*

Let the size of the parallelopiped be $a \times b \times c$.

The volume of the element is

$$dV = dx\, dy\, dz$$

$$V = \int_0^a\int_0^b\int_0^c dx\, dy\, dz$$

$$= [x]_0^a\, [y]_0^b\, [z]_0^c = abc$$

(*ii*) *Volume of a cone*

Referring to Fig. 6.8,

$$dV = \pi y^2 \,.\, dx = \pi \frac{x^2}{h^2} R^2 dx, \quad \text{since } y = \frac{x}{h} R$$

$$V = \frac{\pi}{h^2} R^2 \int_0^h x^2 dx = \frac{\pi}{h^2} R^2 \left[\frac{x^3}{3}\right]_0^h$$

$$= \frac{\pi}{h^2} R^2 \frac{h^3}{3} = \frac{\pi R^2 h}{3}$$

(*iii*) *Volume of a sphere*

Referring to Fig. 6.9,

$$dV = \pi x^2\, dy$$

But $$x^2 + y^2 = R^2$$

*i.e.*, $$x^2 = R^2 - y^2$$

$\therefore$ $$dV = \pi\,(R^2 - y^2)dy$$

$$V = \int_{-R}^{R} \pi\left(R^2 - y^2\right)dy$$

$$= \pi\left[R^2 y - \frac{y^3}{3}\right]_{-R}^{R}$$

$$= \pi\left[R^2 \cdot R - \frac{R^3}{3} - \left\{-R^3 - \frac{(-R)^3}{3}\right\}\right]$$

$$= \pi R^3\left[1 - \frac{1}{3} + 1 - \frac{1}{3}\right] = \frac{4}{3}\pi R^3$$

The surface areas and volumes of solids of revolutions such as cone, sphere, may be easily found using theorems of Pappus and Guldinus. This will be taken up latter in this chapter, since it needs the term centroid of generating lines.

## 6.2 CENTRE OF GRAVITY AND CENTROIDS

Consider the suspended body as shown in Fig. 6.10(*a*). The self weight of various parts of this body is acting vertically downward. The only upward force is the force **T** in the string. To satisfy the equilibrium condition, the resultant weight of the body **W** must act along the line of string 1–1. Now, if the position is changed and the body is suspended again [Fig. 6.10(*b*)], it will reach equilibrium condition in a particular position. Let the line of action of the resultant weight be 2–2 intersecting 1–1 at *G*. It is obvious that if the body is suspended in any other position, the line of action of resultant weight **W** passes through *G*. This point is called the centre of gravity of the body. Thus, *centre of gravity can be defined as the point through which the resultant of force of gravity of the body acts*.

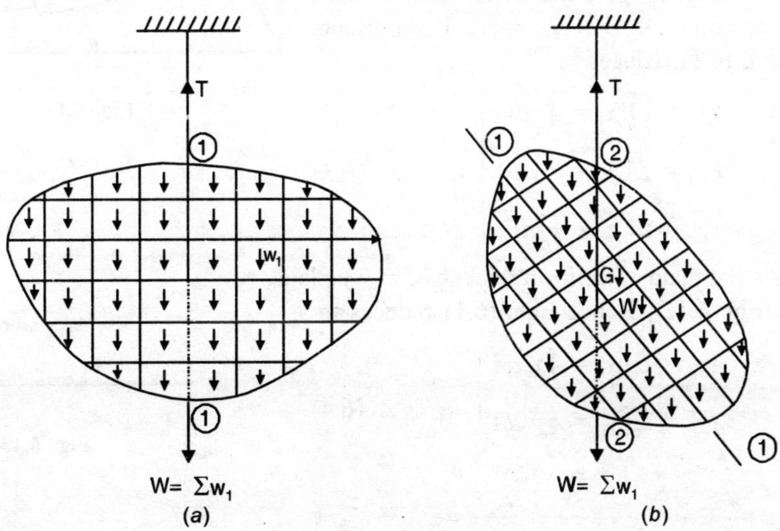


**Fig. 6.10**

The above method of locating centre of gravity is the practical method. If one desires to locate centre of gravity of a body analytically, it is to be noted that the resultant of weight of various portions of the body is to be determined. For this Varignon's theorem, which states that the moment of resultant force is equal to the sum of moments of component forces, can be used.

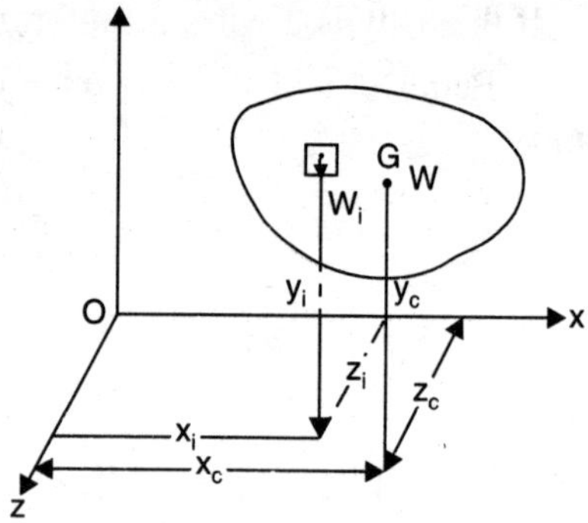

**Fig. 6.11**

Referring to Fig. 6.11, let $\mathbf{W_i}$ be the weight of an element in the given body. $\mathbf{W}$ be the total weight of the body. Let the coordinates of the element be $x_i$, $y_i$, $z_i$ and that of centroid $G$ be $x_c$, $y_c$, $z_c$. Since, $\mathbf{W}$ is the resultant of $\mathbf{W_i}$ forces,

$$\mathbf{W} = \mathbf{W_1} + \mathbf{W_2} + \mathbf{W_3} \ldots$$

$$= \Sigma \mathbf{W_i}$$

and $$Wx_c = W_1x_1 + W_2x_2 + W_3x_3 + \ldots$$

$$\left.\begin{aligned} \therefore \quad Wx_c &= \Sigma W_ix_i = \oint x dw \\ \text{Similarly,} \quad Wy_c &= \Sigma W_iy_i = \oint y dw \\ \text{and} \quad Wz_c &= \Sigma W_iz_c = \oint z dw \end{aligned}\right\} \quad \ldots (6.1)$$

If $M$ is the mass of the body and $m_i$ that of the element, then

$$M = \frac{W}{g} \quad \text{and} \quad m_i = \frac{W_i}{g}, \text{ hence we get}$$

$$\left.\begin{aligned} Mx_c &= \Sigma m_ix_i = \oint x_i dm \\ My_c &= \Sigma m_iy_i = \oint y_i dm \\ \text{and} \quad Mz_c &= \Sigma m_iz_i = \oint z_i dm \end{aligned}\right\} \quad \ldots (6.2)$$

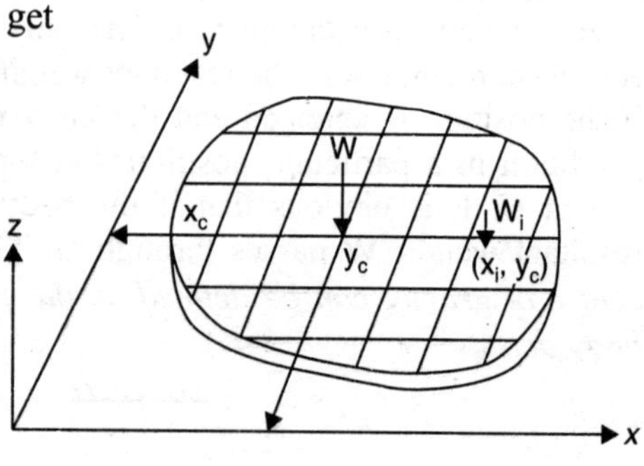

**Fig. 6.12**

If the body is made up of uniform material of unit weight γ, then we know $W_i = V_i\gamma$, where $V$ represents volume, then Eqn. (6.1) reduces to

$$\left.\begin{aligned} Vx_c &= \Sigma V_ix_i = \oint x dV \\ Vy_c &= \Sigma V_iy_i = \oint y dV \\ Vz_c &= \Sigma V_iz_i = \oint z dV \end{aligned}\right\} \quad \ldots (6.3)$$

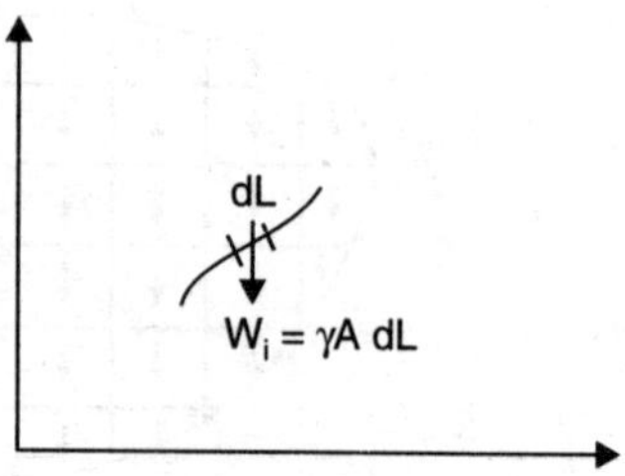

**Fig. 6.13**

If the body is a flat plate of uniform thickness, in $x$-$y$ plane, $W_i = \gamma\ A_it$ (Ref. Fig. 6.12). Hence Eqn. (6.1) reduces to

$$\left.\begin{aligned} Ax_c &= \Sigma A_ix_i = \oint x\ dA \\ Ay_c &= \Sigma A_iy_i = \oint y\ dA \end{aligned}\right\} \quad \ldots (6.4)$$

If the body is a wire of uniform cross-section in plane $x$, $y$ (Ref. Fig. 6.13), Eqn. (6.1) reduces to

$$\left.\begin{aligned} Lx_c &= \Sigma L_i x_i = \oint x\, dL \\ Ly_c &= \Sigma L_i y_i = \oint y\, dL \end{aligned}\right\} \quad \text{...(6.5)}$$

The term centre of gravity is used only when the gravitational forces (weights) are considered. This term is applicable to solids. Equation (6.2) in which only masses are used, the point obtained is termed as *centre of mass*. The central points obtained for volumes, surfaces and line segments [Obtained by Eqns. (6.3), (6.4) and (6.5)] are termed as *centroids*.

## 6.3 CENTROID OF A LINE

Centroid of a line can be determined using Eqn. (6.5). Method of finding the centroid of a line for some standard cases is illustrated below:

(*i*) *Centroid of a straight line*

Selecting the $x$-coordinate along the line (Fig. 6.14)

$$Lx_c = \int_0^L x\, dx = \left[\frac{x^2}{2}\right]_0^L = \frac{L^2}{2}$$

$$\therefore \quad x_c = \frac{L}{2}$$

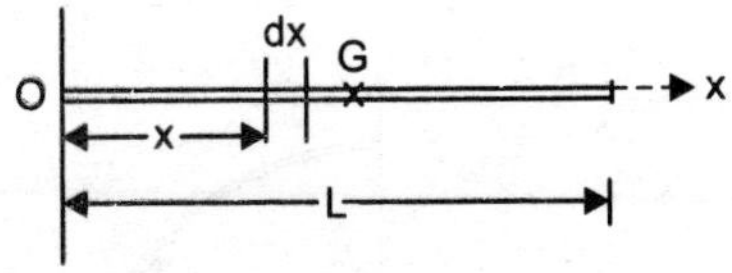


Fig. 6.14

Thus, the centroid lies at midpoint of a straight line, whatever be the orientation of line (Ref. Fig. 6.15).

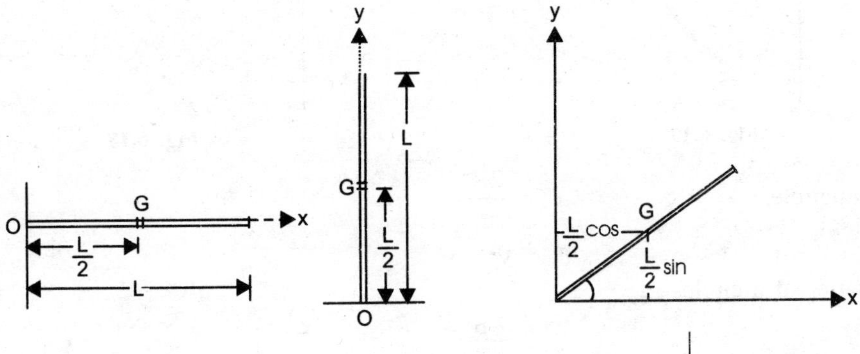


Fig. 6.15

(*ii*) *Centroid of an arc of a circle*

Referring to Fig. 6.16,

$$L = \text{Length of arc} = R\, 2\alpha$$

$$dL = R d\theta$$

Hence, from Eqn. (6.5)

$$x_c L = \int_{-\alpha}^{\alpha} x dL$$

*i.e.*,

$$x_c\, R\, 2\alpha = \int_{-\alpha}^{\alpha} R \cos\theta \cdot R d\theta$$

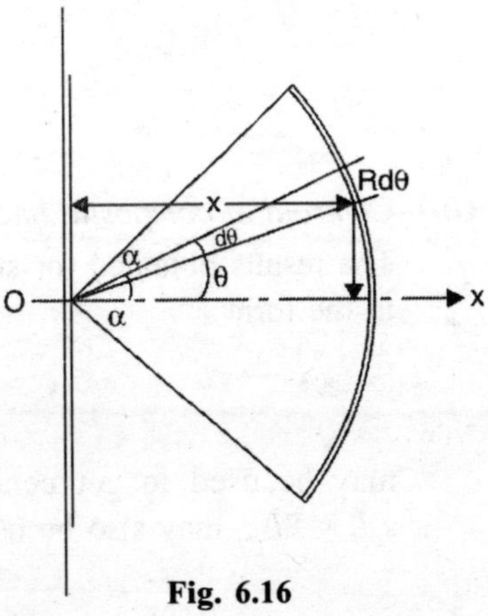


Fig. 6.16

$$= R^2 \left[\sin\theta\right]_{-\alpha}^{\alpha} \qquad \ldots(i)$$

$$\therefore \qquad x_c = \frac{R^2 \times 2\sin\alpha}{2R\alpha} = \frac{R\sin\alpha}{\alpha}$$

and
$$y_c L \int_{-\alpha}^{\alpha} y\, dL = \int_{-\alpha}^{\alpha} R \sin\theta \cdot R d\theta$$

$$= R^2 \left[-\cos\theta\right]_{-\alpha}^{\alpha} \qquad \ldots(ii)$$

$$= 0$$

$$\therefore \qquad y_c = 0$$

From Eqns. (*i*) and (*ii*), we can get the centroid of semicircle, as shown in Fig. 6.17 by putting $\alpha = \pi/2$ and for quarter of a circle, as shown in Fig. 6.18, by putting $\alpha$ varying from zero to $\pi/2$.

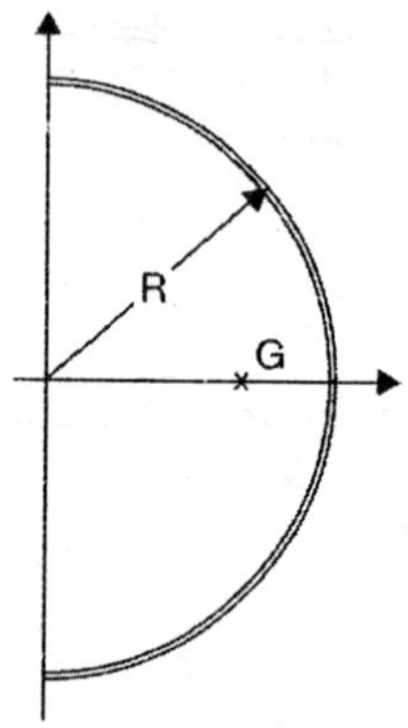


**Fig. 6.17**

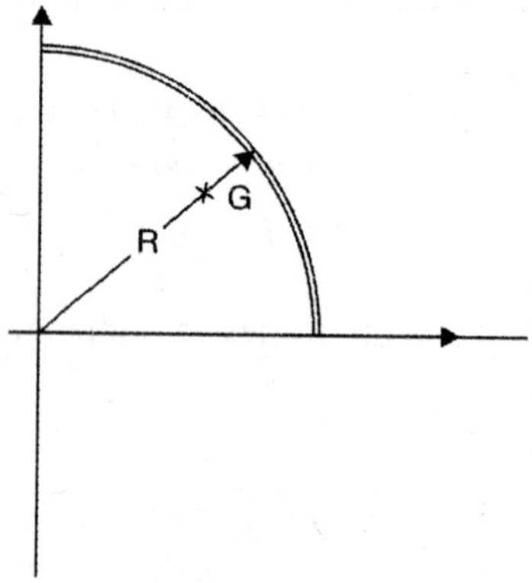


**Fig. 6.18**

For semicircle,
$$x_c = \frac{2R}{\pi}$$
$$y_c = 0$$

For quarter of a circle,
$$x_c = \frac{2R}{\pi}$$
$$y_c = \frac{2R}{\pi}$$

(*iii*) *Centroid of composite line segments*

The results obtained for standard cases may be used for various segments and then Eqn. (6.5) in the form

$$x_c L = \Sigma L_i x_i$$
$$y_c L = \Sigma L_i y_i$$

may be used to get centroid $x_c$ and $y_c$. If the line segments is in space the expression $z_c L = \Sigma L_i z_i$ may also be used. The method is illustrated with few examples below:

**Example 6.1.** Determine the centroid of the wire as shown in Fig. 6.19.

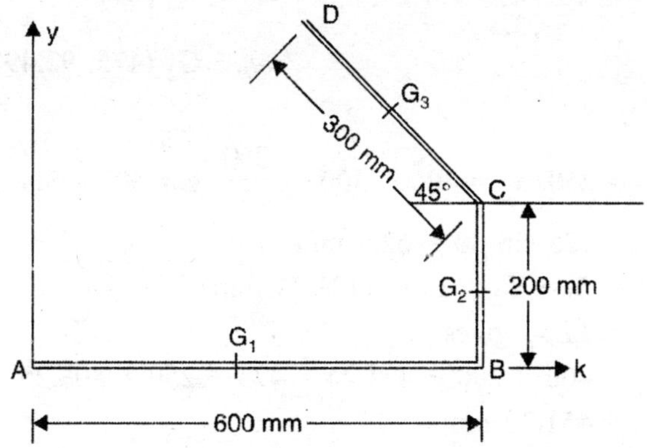


**Fig. 6.19**

**Solution.** The wire is divided into three segments *AB, BC* and *CD*. Taking *A* as origin, the coordinates of the centroids of *AB, BC* and *CD* are

$G_1(300, 0)$; $G_2(600, 100)$ and $G_3(600 - 150 \cos 45°; 200 + 150 \sin 45°)$

*i.e.*, $G_3(493.93, 306.07)$

$$L_1 = 600 \text{ mm}, L_2 = 200 \text{ mm}, L_3 = 300 \text{ mm}$$

$\therefore$ Total length $L = 600 + 200 + 300 = 1100$ mm

$\therefore$ From the equation $Lx_c = \Sigma L_i x_i$, we get

$$1100\, x_c = L_1x_1 + L_2x_2 + L_3x_3$$
$$= 600 \times 300 + 200 \times 600 + 300 \times 493.93$$

$\therefore$ $x_c = 407.44$ mm **Ans.**

Now, $Ly_c = \Sigma L_i y_i$

$$1100\, y_c = 600 \times 0 + 200 \times 100 + 300 \times 306.07$$
$$y_c = 101.66 \text{ mm}$$ **Ans.**

**Example 6.2.** Locate the centroid of the uniform wire bent as shown in Fig. 6.20.

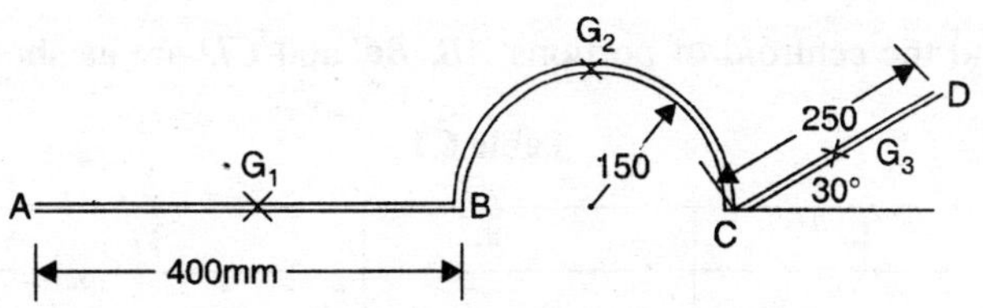


All dimensions in mm

**Fig. 6.20**

**Solution.** The composite figure is divided into 3 simple figures and taking *A* as origin coordinates of their centroids noted as shown below:

*AB* is a straight line

$L_1 = 400$ mm, $G_1$ (200, 0)

*BC* is a semicircle

$$L_2 = 150\ \pi = 471.24, \qquad G_2\left(475, \frac{2\times 150}{\pi}\right)$$

$$i.e., \quad G_2\ (475, 92.49)$$

*CD* is a straight line

$$L_3 = 250;\ x_3 = 400 + 300 + \frac{250}{2} \cos 30° = 808.25 \text{ mm}$$

$$y_3 = 125 \sin 30 = 62.5 \text{ mm}$$

$\therefore$ Total length $\quad L = L_1 + L_2 + L_3 = 1121.24$ mm

$\therefore \quad Lx_c = \Sigma L_i x_i$, gives

$$1121.24\ x_c = 400 \times 200 + 471.24 \times 475 + 250 \times 808.25$$

$$x_c = 451.20 \text{ mm}$$ **Ans.**

$Ly_c = \Sigma L_i y_i$, gives

$$1121.24\ y_c = 400 \times 0 + 471.24 \times 95.49 + 250 \times 62.5$$

$$y_c = 54.07 \text{ mm.}$$ **Ans.**

**Example 6.3.** Locate the centroid of uniform wire as shown in Fig. 6.21. **Note:** portion *AB* is in *x-z* plane, *BC* in *y-z* plane and *CD* in *x-y* plane. *AB* and *BC* are semicircular in shape.

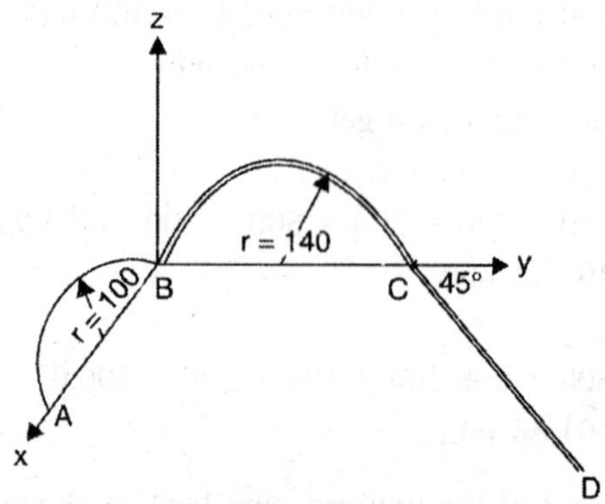


**Fig. 6.21**

**Solution.** The length and the centroid of portions *AB, BC* and *CD* are as shown in Table 6.1 below:

**Table 6.1**

| Portion | $L_i$ | $x_i$ | $y_i$ | $z_i$ |
|---|---|---|---|---|
| *AB* | 100π | 100 | 0 | $\frac{2\times 100}{\pi}$ |
| *BC* | 140π | 0 | 140 | $\frac{2\times 140}{\pi}$ |
| *CD* | 300 | 300 sin 45° | 280 + 300 cos 45° = 492.13 | 0 |

$$\therefore \quad L = 100\pi + 140\pi + 300 = 1053.98 \text{ mm}$$

From Equation $\quad Lx_c = \Sigma L_i x_i$, we get

$$1053.98\ x_c = 100\pi \times 100 + 140\pi \times 0 + 300 \times 300 \sin 45°$$

$x_c = 90.19$ mm **Ans.**

Similarly, $1053.98\, y_c = 100\pi \times 0 + 140\pi \times 140 + 300 \times 492.13$

$y_c = 198.50$ mm **Ans.**

and $1053.98\, z_c = 100\pi \times \frac{200}{\pi} + 140\pi \times \frac{2 \times 140}{\pi} + 300 \times 0$

$z_c = 56.17$ mm **Ans.**

## 6.4 FIRST MOMENT OF AREA AND CENTROID

From Eqn. (6.1), we have

$$x_c = \frac{\sum W_i x_i}{W}, \quad y_c = \frac{\sum W_i y_i}{W} \quad \text{and} \quad z_c = \frac{\sum W_i z_i}{W}$$

From the above equation, we can make the statement that distance of centre of gravity of a body from an axis is obtained by dividing moment of the gravitational forces acting on the body, about the axis, by the total weight of the body. Similarly, from Eqn. (6.4), we have

$$x_c = \frac{\sum A_i x_i}{A}, \quad y_c = \frac{\sum A_i y_i}{A}$$

By terming $\Sigma A_i x$: As the moment of area about the axis, we can say centroid of plane area from any axis is equal to moment of area about the axis divided by the total area. The moment of area $\Sigma A_i x_i$ is termed as first moment of area also just to differentiate this from the term $\Sigma A_i x_i^2$, which will be dealt later. It may be noted that, since, the moment of area about an axis divided by total area gives the distance of the centroid from that axis, the moment of area is zero about any centroidal axis.

**Difference between Centre of Gravity and Centroid**

From the above discussion, we can draw the following differences between centre of gravity and centroid:

1. The term centre of gravity applies to bodies with weight, and centroid applies to lines, plane areas and volumes.
2. Centre of gravity of a body is a point through which the resultant gravitational force (weight) acts for any orientation of the body, whereas centroid is a point in a line, plane area or volume such that the moment of area about any axis through that point is zero.

**Use of Axis of Symmetry**

*Centroid of an area lies on the axis of symmetry if it exits.* This is useful theorem to locate the centroid of an area.

This theorem can be proved as follows:

Consider the area shown in Fig. 6.22. In this figure, *y-y* is the axis of symmetry. From Eqn. (6.4), the distance of centroid from this axis is given by:

$$\frac{\sum A_i x_i}{A}$$

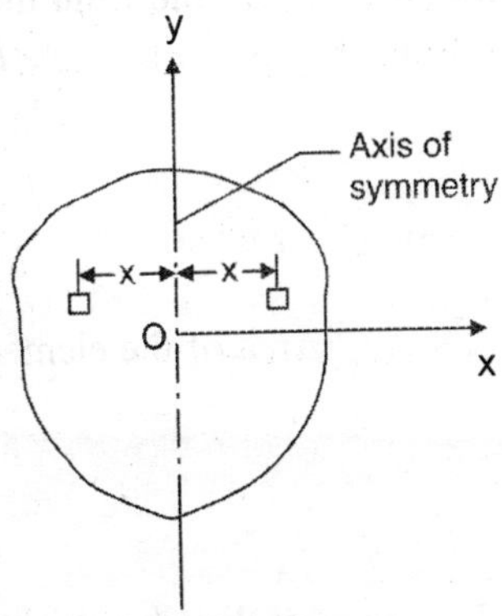


**Fig. 6.22**

Consider the two elemental areas, as shown in Fig. 6.22, which are equal in size and are equidistant from the axis, but on either side. Now the sum of moments of these areas cancel each other, since, the areas and distances are the same, but signs of distances are opposite. Similarly, we can go on considering an area on one side of symmetric axis and corresponding image area on the other side, and prove that total moments of area ($\Sigma A_i x_i$) about the symmetric axis is zero. Hence, the distance of centroid from the symmetric axis is zero, *i.e.*, centroid always lies on symmetric axis.

Making use of the symmetry, we can conclude that:

1. Centroid of a circle is its centre (Fig. 6.23).
2. Centroid of a rectangle of sides $b$ and $d$ is at distance $\frac{b}{2}$ and $\frac{d}{2}$ from the corner, as shown in Fig. 6.24.

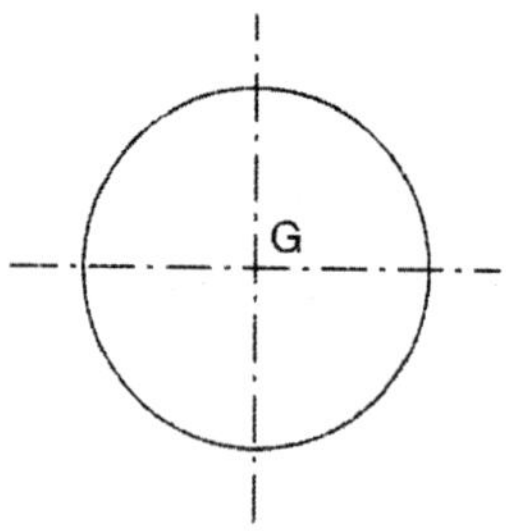


**Fig. 6.23**

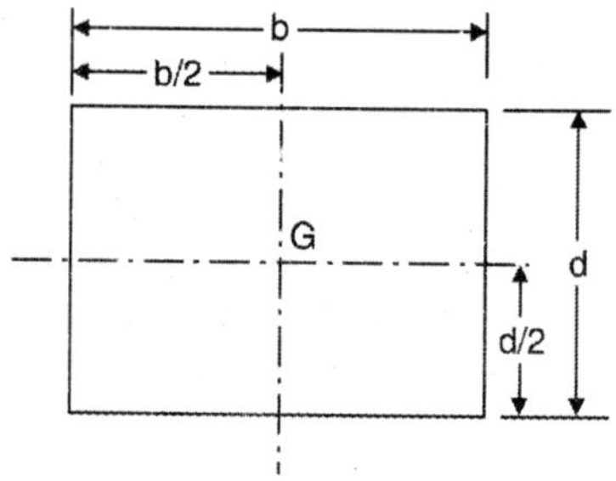


**Fig. 6.24**

### Determination of Centroid of Simple Figures From the First Principles

For simple figures such as triangle and semicircle, we can write general expression for the elemental area and its distance from an axis. Then Eqns. (6.4) are given as

$$\bar{y} = \frac{\int y dA}{A}$$

$$\bar{x} = \frac{\int x dA}{A}$$

The location of the centroid using the above equations may be considered as finding centroid from the first principles. Now, let us find centroid of some standard figures from the first principles.

### Centroid of a Triangle

Consider the triangle *ABC* of base width $b$ and height $h$ as shown in Fig. 6.25. Let us locate the distance of centroid from the base. Let $b_1$ be the width of elemental strip of thickness $dy$ at a distance $y$ from the base. Since, $\Delta AEF$ and $\Delta ABC$ are similar triangles, we can write:

$$\frac{b_1}{b} = \frac{h-y}{h}$$

$$b_1 = \left(\frac{h-y}{h}\right)b = \left(1-\frac{y}{h}\right)b$$

$\therefore$ Area of the element

$$= dA = b_1 dy$$

$$= \left(1-\frac{y}{h}\right)b\, dy$$

Area of the triangle $A = \frac{1}{2}bh$

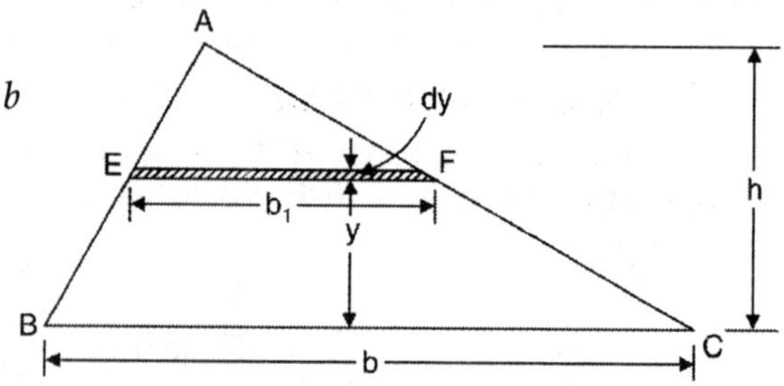


**Fig. 6.25**

$\therefore$ From Eqn. (6.4),

$$\bar{y} = \frac{\text{Moment of area}}{\text{Total area}} = \frac{\int y dA}{A}$$

Now,

$$\int y dA = \int_0^h y\left(1-\frac{y}{h}\right) b\, dy$$

$$= \int_0^h \left(y-\frac{y^2}{h}\right) b\, dy$$

$$= b\left[\frac{y^2}{2}-\frac{y^3}{3h}\right]_0^h$$

$$= \frac{bh^2}{6}$$

$$\therefore \quad \bar{y} = \frac{\int y dA}{A} = \frac{bh^2}{6} \times \frac{1}{\frac{1}{2}bh}$$

$$\therefore \quad \bar{y} = \frac{h}{3}$$

Thus, the centroid of a triangle is at a distance $\frac{h}{3}$ from the base (or $\frac{2h}{3}$ from the apex) of the triangle where $h$ is the height of the triangle.

**Centroid of a Semicircle**

Consider the semicircle of radius $R$ as shown in Fig. 6.26. Due to symmetry, centroid must lie on $y$-axis. Let its distance from diametral axis be $\bar{y}$. To find $\bar{y}$, consider an element at a distance $r$ from the centre $O$ of the semicircle, radial width being $dr$ and bound by radii at $\theta$ and $\theta + d\theta$.

Area of element $= r\, d\theta\, dr$.

Its moment about diametral axis $x$ is given by:

$$r d\theta \times dr \times r \sin\theta = r^2 \sin\theta\, dr\, d\theta$$

$\therefore$ Total moment of area about diametral axis,

$$\int_0^\pi \int_0^R r^2 \sin\theta\, dr\, d\theta = \int_0^\pi \left[\frac{r^3}{3}\right]_0^R \sin\theta\, d\theta$$

$$= \frac{R^3}{3}\left[-\cos\theta\right]_0^\pi$$

$$= \frac{R^3}{3}[1+1] = \frac{2R^3}{3}$$

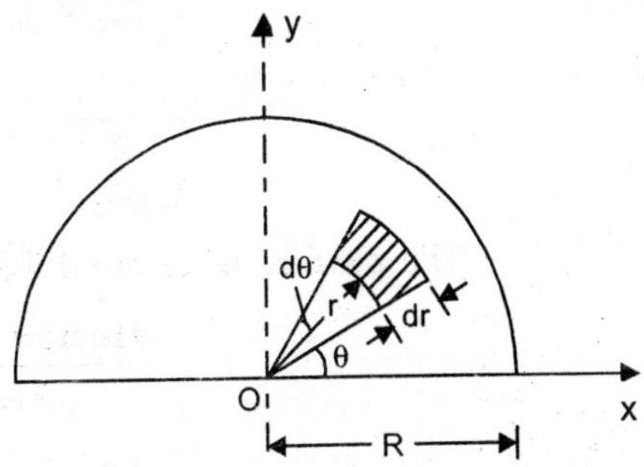


Fig. 6.26

Area of semicircle $A = \frac{1}{2}\pi R^2$

$$\therefore \quad \bar{y} = \frac{\text{Moment of area}}{\text{Total area}} = \frac{\frac{2R^3}{3}}{\frac{1}{2}\pi R^2}$$

$$= \frac{4R}{3\pi}$$

Thus, the centroid of circle is at a distance $\frac{4R}{3\pi}$ from the diametral axis.

**Centroid of Sector of a Circle**

Consider the sector of a circle of angle $2\alpha$ as shown in Fig. 6.27. Due to symmetry, centroid lies on $x$-axis. To find its distance from the centre $O$, consider the elemental area as shown in the figure.

Area of the element $= rd\theta\, dr$

Its moment about $y$-axis

$$= rd\theta \times dr \times r \cos\theta$$
$$= r^2 \cos\theta\, drd\theta$$

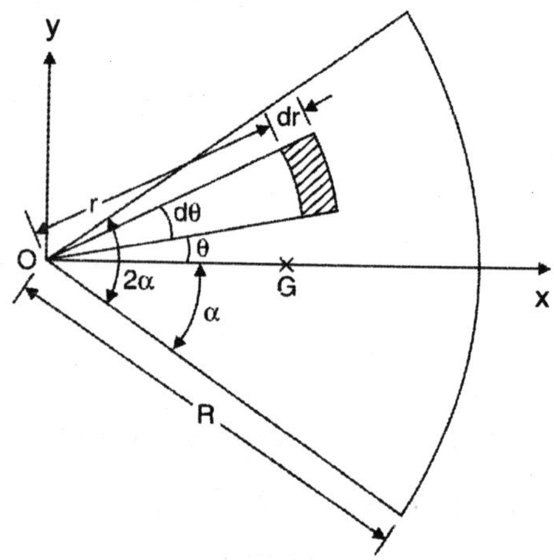


Fig. 6.27

$\therefore$ Total moment of area about $y$-axis

$$= \int_{-\alpha}^{\alpha}\int_{0}^{R} r^2 \cos\theta\, drd\theta$$

$$= \left[\frac{r^3}{3}\right]_0^R [\sin\theta]_{-\alpha}^{\alpha}$$

$$= \frac{R^3}{3}\, 2\sin\alpha$$

Total area of the sector

$$= \int_{-\alpha}^{\alpha}\int_{0}^{R} rdrd\theta$$

$$= \int_{-\alpha}^{\alpha}\left[\frac{r^2}{2}\right]_0^R d\theta$$

$$= \frac{R^2}{2}[\theta]_{-\alpha}^{\alpha}$$

$$= R^2\alpha$$

$\therefore$ The distance of centroid from centre $O$

$$= \frac{\text{Moment of area about } y\text{-axis}}{\text{Area of the figure}}$$

$$= \frac{\frac{2R^3}{3}\sin\alpha}{R^2\alpha} = \frac{2R}{3\alpha}\sin\alpha$$

**Centroid of Parabolic Spandrel**

Consider the parabolic spandrel as shown in Fig. 6.28. Height of the element at a distance $x$ from $O$ is $y = kx^2$.

Width of element $= dx$

$\therefore$ Area of the element $= kx^2 dx$

$\therefore$ Total area of spandrel $= \int_0^a kx^2 dx$

$$= \left[\frac{kx^3}{3}\right]_0^a = \frac{ka^3}{3}$$

Moment of area about $y$-axis

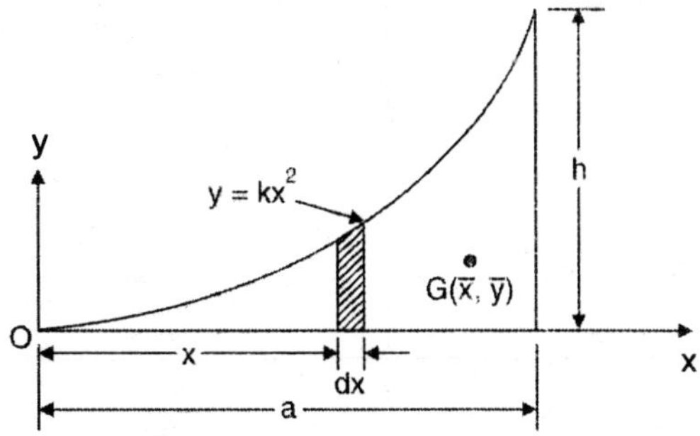


Fig. 6.28

$$= \int_0^a kx^2 dx \times x$$

$$= \int_0^a kx^3\, dx$$

$$= \left[\frac{kx^4}{4}\right]_0^a$$

$$= \frac{ka^4}{4}$$

Moment of area about $x$-axis $= \int_0^\alpha dAy/2$

$$= \int_0^a kx^2 dx \frac{kx^2}{2} = \int_0^a \frac{k^2x^4}{2} dx = \left[\frac{k^2x^5}{2\times 5}\right]_0^a$$

$$= \frac{k^2a^5}{10}$$

$\therefore$ $$\bar{x} = \frac{ka^4}{4} \div \frac{ka^3}{3} = \frac{3a}{4}$$

$$\bar{y} = \frac{k^2a^5}{10} \div \frac{ka^3}{3} = \frac{3}{10}\, ka^2$$

From Fig. 6.28, at $x = a$, $y = h$

$\therefore$ $$h = ka^2 \text{ or } k = \frac{h}{a^2}$$

$\therefore$ $$\bar{y} = \frac{3}{10} \times \frac{h}{a^2} a^2 = \frac{3h}{10}$$

Thus, centroid of spandrel is $\left(\dfrac{3a}{4}, \dfrac{3h}{10}\right)$

Centroids of some common figures are shown in Table 6.2.

**Table 6.2:** *Centroid of Some Common Figures*

| Shape | Figure | $\bar{x}$ | $\bar{y}$ | Area |
|---|---|---|---|---|
| Triangle | 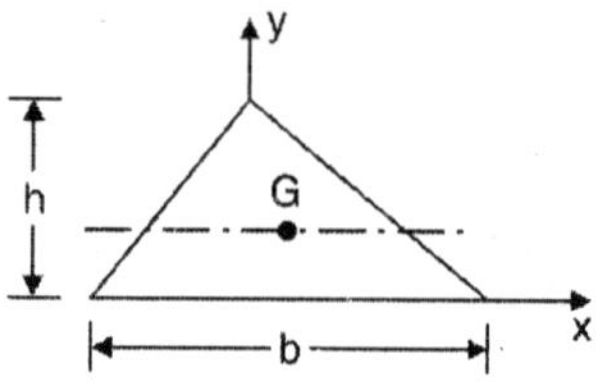 | — | $\frac{h}{3}$ | $\frac{bh}{2}$ |
| Semicircle | 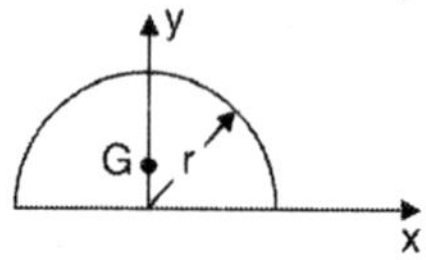 | 0 | $\frac{4R}{3\pi}$ | $\frac{\pi R^2}{2}$ |
| Quarter circle | 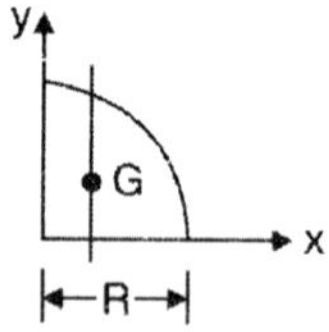 | $\frac{4R}{3\pi}$ | $\frac{4R}{3\pi}$ | $\frac{\pi R^2}{4}$ |
| Sector of a circle | 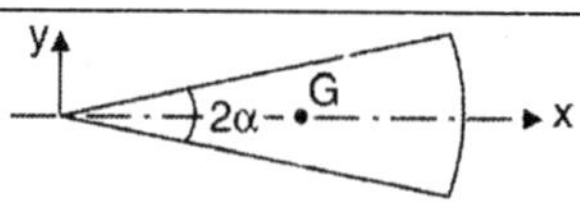 | $\frac{2R}{3\alpha}\sin\alpha$ | 0 | $\alpha R^2$ |
| Parabola | 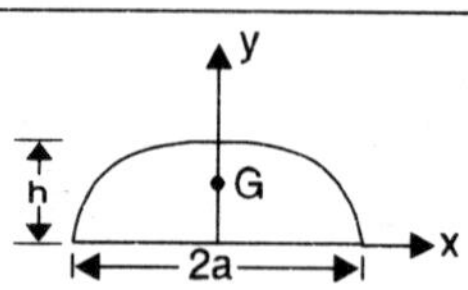 | 0 | $\frac{3h}{5}$ | $\frac{4ah}{3}$ |
| Semi parabola | 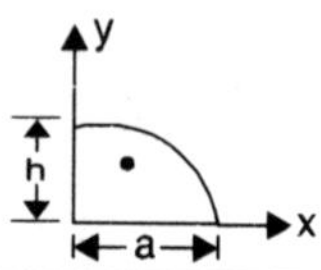 | $\frac{3a}{8}$ | $\frac{3h}{5}$ | $\frac{2ah}{3}$ |
| Parabolic spandrel | 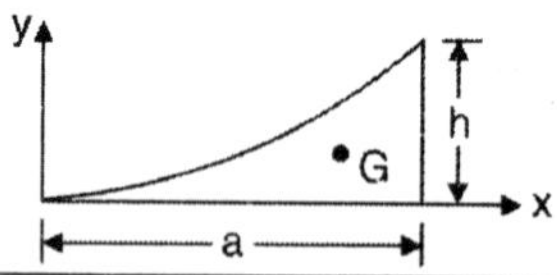 | $\frac{3a}{4}$ | $\frac{3h}{10}$ | $\frac{ah}{3}$ |

## Centroid of Composite Sections

So far, the discussion was confined to locating the centroid of simple figures such as rectangle, triangle, circle, semicircle, etc. In engineering practice, use of sections which are built up of many simple sections is very common. Such sections may be called as built-up sections or composite sections. To locate the centroid of composite sections, one need not go for the first principles (method of integration). The given composite section can be split into suitable simple figures and then the centroid of each simple figure can be found by inspection or using the standard formulae listed in Table 6.2. Assuming the area of the simple figure as concentrated at its centroid, its moment about an axis can be found by multiplying the area with distance of its centroid from the reference axis. After determining moment of each area about reference axis, the distance of centroid from the axis is obtained by dividing total moment of area by total area of the composite section.

**Example 6.4.** Locate the centroid of the *T*-section as shown in the Fig. 6.29.

**Solution.** Selecting the axis as shown in Fig. 6.29, we can say due to symmetry centroid lies on *y*-axis, *i.e.*, $\bar{x} = 0$. Now the given *T*-section may be divided into two rectangles $A_1$ and $A_2$ each of size 100 × 20 and 20 × 100. The centroid of $A_1$ and $A_2$ are $g_1(0, 10)$ and $g_2(0, 70)$ respectively.

∴ The distance of centroid from top is given by:

$$\bar{y} = \frac{100 \times 20 \times 10 + 20 \times 100 \times 70}{100 \times 20 + 20 \times 100}$$

$$= 40 \text{ mm}$$

Hence, *centroid of T-section is on the symmetric axis at a distance 40 mm from the top*.

**Ans.**

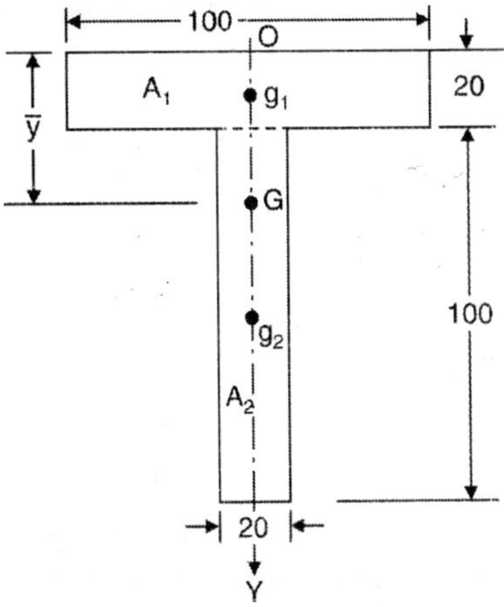


**Fig. 6.29**

**Example 6.5.** Find the centroid of the unequal angle 200 × 150 × 12 mm, as shown in Fig. 6.30.

**Solution.** The given composite figure can be divided into two rectangles:

$$A_1 = 150 \times 12 = 1800 \text{ mm}^2$$

$$A_2 = (200 - 12) \times 12 = 2256 \text{ mm}^2$$

Total area $\quad A = A_1 + A_2 = 4056 \text{ mm}^2$

Selecting the reference axes *x* and *y* as shown in Fig. 6.30. The centroid of $A_1$ is $g_1$ (75, 6) and that of $A_2$ is:

$$g_2\left[6, 12 + \frac{1}{2}(200 - 12)\right]$$

*i.e.*, $\quad g_2$ (6, 106)

$$\therefore \quad \bar{x} = \frac{\text{Moment about } y\text{-axis}}{\text{Total area}}$$

$$= \frac{A_1x_1 + A_2x_2}{A}$$

$$= \frac{1800 \times 75 + 2256 \times 6}{4056} = 36.62 \text{ mm}$$

$$\bar{y} = \frac{\text{Moment about } x\text{-axis}}{\text{Total area}}$$

$$= \frac{A_1y_1 + A_2y_2}{A}$$

$$= \frac{1800 \times 6 + 2256 \times 106}{4056} = 61.62 \text{ mm}$$

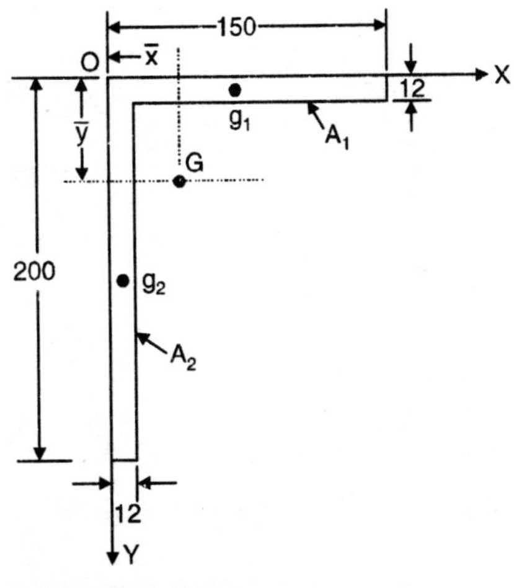


**Fig. 6.30**

Thus, the centroid is at $\bar{x}$ = 36.62 mm and $\bar{y}$ = 61.62 mm as shown in the figure. **Ans.**

**Example 6.6.** Locate the centroid of the *I*-section as shown in Fig. 6.31.

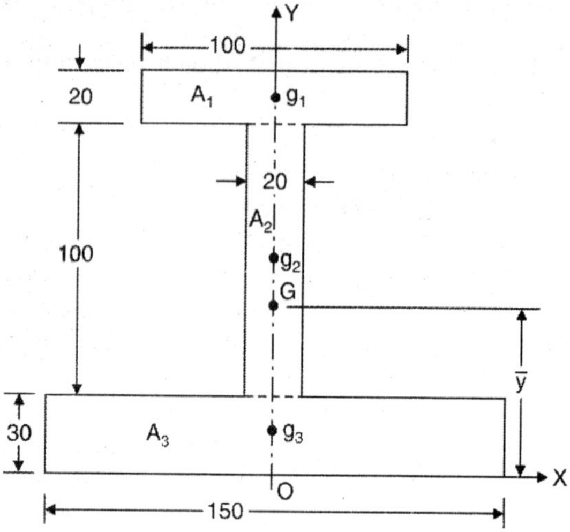


**Fig. 6.31**

**Solution.** Selecting the co-ordinate system as shown in Fig. 6.31, due to symmetry, centroid must lie on *y*-axis,

*i.e.*, $\bar{x} = 0$

Now, the composite section may be split into three rectangles

$$A_1 = 100 \times 20 = 2000 \text{ mm}^2$$

Centroid of $A_1$ from the origin is:

$$y_1 = 30 + 100 + \frac{20}{2} = 140 \text{ mm}$$

Similarly,
$$A_2 = 100 \times 20 = 2000 \text{ mm}^2$$

$$y_2 = 30 + \frac{100}{2} = 80 \text{ mm}$$

$$A_3 = 150 \times 30 = 4500 \text{ mm}^2, \text{ and}$$

$$y_3 = \frac{30}{2} = 15 \text{ mm}$$

$$\therefore \quad \bar{y} = \frac{A_1 y_1 + A_2 y_2 + A_3 y_3}{A}$$

$$= \frac{2000 \times 140 + 2000 \times 80 + 4500 \times 15}{2000 + 2000 + 4500}$$

$$= 59.71 \text{ mm}$$

Thus, *the centroid is on the symmetric axis at a distance 59.71 mm from the bottom* as shown in Fig. 6.31. **Ans.**

**Example 6.7.** Determine the centroid of the section of the concrete dam as shown in Fig. 6.32.

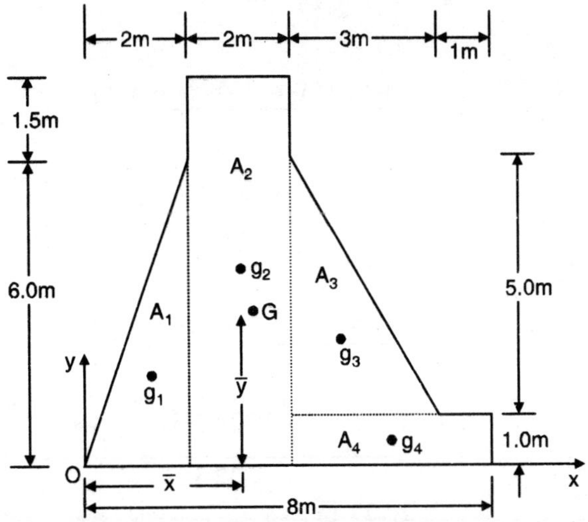


Fig. 6.32

**Solution.** Let the axis be selected as shown in Fig. 6.32. (Note that it is convenient to take axis in such a way that the centroids of all simple figures are having positive coordinates. If coordinate of any simple figure comes out to be negative, one should be careful in assigning the sign of moment of area of that figure).

The composite figure can be conveniently divided into two triangles and two rectangles, as shown in Fig. 6.32.

Now,
$$A_1 = \frac{1}{2} \times 2 \times 6 = 6 \text{ m}^2$$
$$A_2 = 2 \times 7.5 = 15 \text{ m}^2$$
$$A_3 = \frac{1}{2} \times 3 \times 5 = 7.5 \text{ m}^2$$
$$A_4 = 1 \times 4 = 4 \text{ m}^2$$
$$A = \text{Total area} = 32.5 \text{ m}^2$$

Centroids of simple figures are:
$$x_1 = \frac{2}{3} \times 2 = \frac{4}{3} \text{ m}$$
$$y_1 = \frac{1}{3} \times 6 = 2 \text{ m}$$
$$x_2 = 2 + 1 = 3 \text{ m}$$
$$y_2 = \frac{7.5}{2} = 3.75 \text{ m}$$
$$x_3 = 2 + 2 + \frac{1}{3} \times 3 = 5 \text{ m}$$
$$y_3 = 1 + \frac{1}{3} \times 5 = \frac{8}{3} \text{ m}$$

$$x_4 = 4 + \frac{4}{2} = 6 \text{ m}$$

$$y_4 = 0.5 \text{ m}$$

$$\bar{x} = \frac{A_1x_1 + A_2x_2 + A_3x_3 + A_4x_4}{A}$$

$$= \frac{6 \times \frac{4}{3} + 15 \times 3 + 7.5 \times 5 + 4 \times 6}{32.5}$$

$$= 3.523 \text{ m}$$

$$\bar{y} = \frac{A_1y_1 + A_2y_2 + A_3y_3 + A_4y_4}{A}$$

$$= \frac{6 \times 2 + 15 \times 3.75 + 7.5 \times \frac{8}{3} + 4 \times 0.5}{32.5}$$

$$= 2.777 \text{ m}$$

The centroid is at $\bar{x} = 3.523$ m **Ans.**

and $\bar{y} = 2.777$ m **Ans.**

**Example 6.8.** Determine the centroid of the area, as shown in Fig. 6.33, with respect to the axis shown.

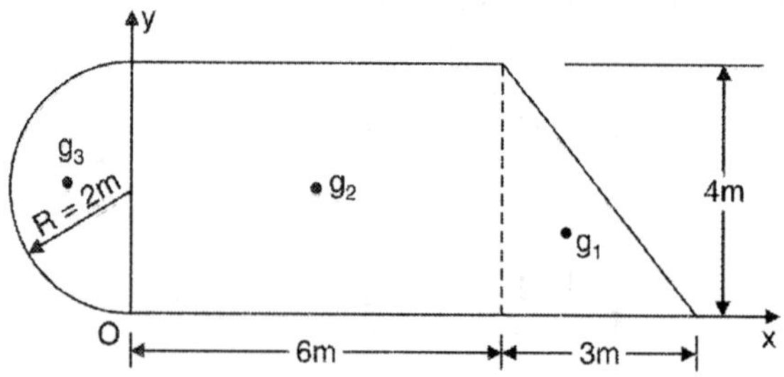


**Fig. 6.33**

**Solution.** The composite section is divided into three simple figures, a triangle, a rectangle and a semicircle.

Now, Area of triangle $A_1 = \frac{1}{2} \times 3 \times 4 = 6 \text{ m}^2$

Area of rectangle $A_2 = 6 \times 4 = 24 \text{ m}^2$

Area of semicircle $A_3 = \frac{1}{2} \times \pi \times 2^2 = 6.2832 \text{ m}^2$

$\therefore$ Total area $A = 36.2832 \text{ m}^2$

The coordinates of centroids of these three simple figures are:

$$x_1 = 6 + \frac{1}{3} \times 3 = 7 \text{ m}$$

$$y_1 = \frac{4}{3} \text{ m}$$

$$x_2 = 3 \text{ m}$$

$$y_2 = 2 \text{ m}$$

$$x_3 = \frac{-4R}{3\pi} = -\frac{4 \times 2}{3\pi} = -0.8488 \text{ m}$$

$$y_3 = 2 \text{ m}$$

(Note carefully the sign of $x_3$).

$$\bar{x} = \frac{A_1x_1 + A_2x_2 + A_3x_3}{A}$$

$$= \frac{6 \times 7 + 24 \times 3 + 6.2832 \times (-0.8488)}{36.2832}$$

*i.e.,* $\bar{x} = 2.995$ m **Ans.**

$$\bar{y} = \frac{A_1y_1 + A_2y_2 + A_3y_3}{A}$$

$$= \frac{\frac{6 \times 4}{3} + 24 \times 2 + 6.2832 \times 2}{36.2832}$$

*i.e.,* $\bar{y} = 1.890$ m **Ans.**

**Example 6.9.** In a gusset plate, there are six rivet holes of 21.5 mm diameter as shown in Fig. 6.34. Find the position of the centroid of the gusset plate.

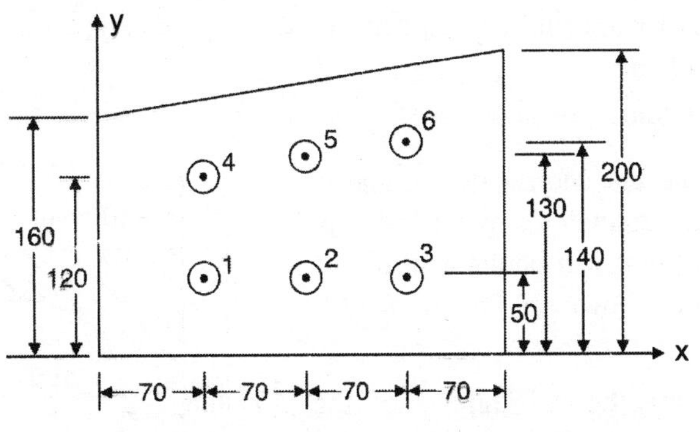


**Fig. 6.34**

**Solution.** The composite area is equal to a rectangle of size 160 × 280 mm plus a triangle of size 280 mm base width and 40 mm height and minus areas of six holes. In this case also, Eqn. (6.4) can be used for locating centroid by treating area of holes as negative. The areas of simple figures and their centroids are as shown in Table 6.3.

**Table 6.3:** *Calculation table for Example 6.9*

| Figure | Area in mm² | $x_i$ in mm | $y_i$ in mm |
|---|---|---|---|
| Rectangle | $160 \times 280 = 44{,}800$ | 140 | 80 |
| Triangle | $\frac{1}{2} \times 280 \times 40 = 5600$ | $\frac{560}{3}$ | $160 + \frac{40}{3} = 173.33$ |
| 1st hole | $\frac{-\pi \times 21.5^2}{4} = -363.05$ | 70 | 50 |
| 2nd hole | –363.05 | 140 | 50 |
| 3rd hole | –363.05 | 210 | 50 |
| 4th hole | –363.05 | 70 | 120 |
| 5th hole | –363.05 | 140 | 130 |
| 6th hole | –363.05 | 210 | 140 |

$$\therefore \quad A = \Sigma A_i = 48{,}221.70$$

$$\therefore \quad \Sigma A_i x_i = 44{,}800 \times 140 + 5600 \times \frac{560}{3} - 363.05\,(70 + 140 + 210 + 70 + 140 + 210)$$

$$= 70{,}12{,}371.3 \text{ mm}^3$$

$$\bar{x} = \frac{\Sigma A_i x_i}{A} = 145.42 \text{ mm}$$

$$\Sigma A_i y_i = 44{,}800 \times 80 + 5600 \times 173.33 - 363.05\,(50 \times 3 + 120 + 130 + 140)$$

$$= 43{,}58{,}601 \text{ mm}^3$$

$$\bar{y} = \frac{\Sigma A_i y_i}{A} = \frac{43,58,601}{48221.70}$$

$$= 90.39 \text{ mm}$$

Thus, the coordinates of centroid of composite figure is given by:

$\bar{x} = 145.42$ mm **Ans.**

$\bar{y} = 90.39$ mm **Ans.**

**Example 6.10.** Determine the coordinates $x_c$ and $y_c$ of the centre of a 100 mm diameter circular hole cut in a thin plate so that this point will be the centroid of the remaining shaded area shown in Fig. 6.35 (All dimensions are in mm).

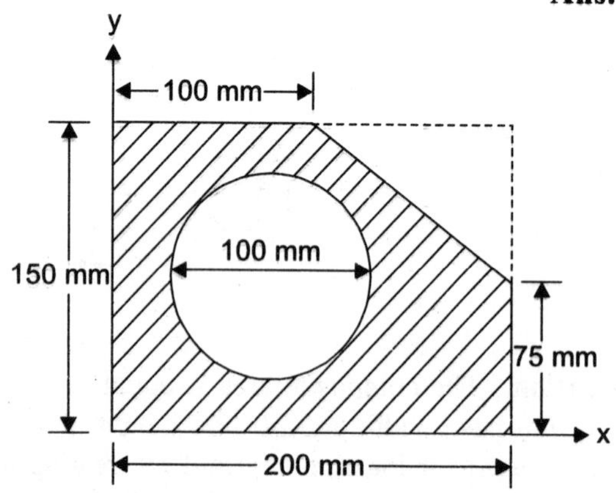


**Fig. 6.35**

**Solution.** If $x_c$ and $y_c$ are the coordinates of the centre of the circle, centroid also must have the coordinates $x_c$ and $y_c$ as per the condition laid down in the problem. The shaded area may be considered as a rectangle of size 200 mm × 150 mm *minus* a triangle of sides 100 mm × 75 mm and a circle of diameter 100 mm.

$$\therefore \quad \text{Total area} = 200 \times 150 - \frac{1}{2} \times 100 \times 75 - \left(\frac{\pi}{4}\right) 100^2 = 18{,}396 \text{ mm}^2$$

$$\bar{x} = x_c = \frac{200 \times 150 \times 100 - \frac{1}{2} \times 100 \times 75 \times \left[200 - \left(\frac{100}{3}\right)\right] - \frac{\pi}{4} \times 100^2 \times x_c}{18,396}$$

$$\therefore \quad x_c(18,396) = 200 \times 150 \times 100 - \frac{1}{2} \times 100 \times 75 \times 166.67 - \frac{\pi}{4} \times 100^2\, x_c$$

$$x_c = \frac{23,75000}{26,250} = 90.48 \text{ mm} \qquad \textbf{Ans.}$$

Similarly,

$$18,396\, y_c = 200 \times 150 \times 75 - \frac{1}{2} \times 100 \times 75 \times (150 - 25) - \frac{\pi}{4} \times 100^2\, y_c$$

$$\therefore \quad y_c = \frac{17,81250.0}{26,250} = 67.86 \text{ mm} \qquad \textbf{Ans.}$$

Centre of the circle should be located at (90.48, 67.86), so that, this point will be the centroid of the remaining shaded area as shown in Fig. 6.35.

**Note:** The centroid of the given figure will coincide with the centroid of the figure without circular hole. Hence, the centroid of the given figure may be obtained by determining the centroid of the figure without the circular hole also.

**Example 6.11.** Determine the coordinates of the centroid of the plane area, as shown in Fig. 6.36, with reference to the axis shown. Take $x$ = 40 mm.

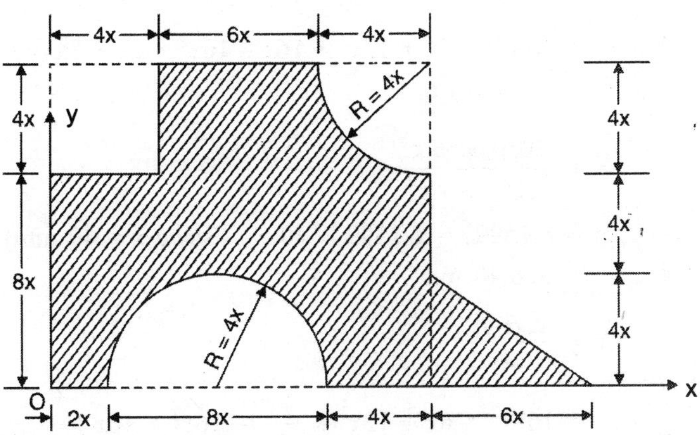


**Fig. 6.36**

**Solution.** The composite figure is divided into the following simple figures:

1. A rectangle $\quad A_1 = (14x) \times (12x) = 168x^2$

$\quad x_1 = 7x;\ y_1 = 6x$

2. A triangle $\quad A_2 = \frac{1}{2}(6x) \times (4x) = 12x^2$

$\quad x_2 = 14x + 2x = 16x$

$\quad y_2 = \frac{4x}{3}$

3. A rectangle

$$A_3 = (-4x) \times (4x) = -16x^2$$
$$x_3 = 2x;\ y_3 = 8x + 2x = 10x$$

4. A semicircle

$$A_4 = -\frac{1}{2}\pi\,(4x)^2 = -8\pi x^2$$
$$x_4 = 6x$$
$$y_4 = \frac{4R}{3\pi} = 4 \times \frac{(4x)}{3\pi} = \frac{16x}{3\pi}$$

5. A quarter of a circle

$$A_5 = -\frac{1}{4} \times \pi\,(4x)^2 = -4\pi\,x^2$$
$$x_5 = 14x - \frac{4R}{3\pi} = 14x - (4)\left(\frac{4x}{3\pi}\right) = 12.3023\,x$$
$$y_5 = 12x - 4 \times \left(\frac{4x}{3\pi}\right) = 10.3023x$$

Total area $\quad A = 168x^2 + 12x^2 - 16x^2 - 8\pi x^2 - 4\pi x^2$

$$= 126.3009\,x^2$$
$$\bar{x} = \frac{\Sigma A_i x_i}{A}$$
$$\sum A_i y_i = 168x^2 \times 7x + 12x^2 \times 16x - 16x^2 \times 2x - 8\pi x^2 \times 6x - 4\pi x^2 \times 12.3023x$$
$$= 1030.6083x$$

$\therefore$
$$\bar{x} = \frac{1030.6083\,x^3}{126.3009\,x^2}$$
$$= 8.1599x = 8.1599 \times 40 \qquad \text{(Since } x = 40 \text{ mm)}$$
$$= 326.40 \text{ mm}$$
$$\bar{y} = \frac{\Sigma A_i y_i}{A}$$
$$\sum A_i x_i = 168x^2 \times 6x + 12x^2 \times \frac{4x}{3} - 16x^2 \times 10x - 8\pi\,x^2 \times \frac{16x}{3\pi} - 4\pi\,x^2 \times 10.3023x$$
$$= 691.8708x^3$$

$\therefore$
$$\bar{y} = \frac{691.8708x^3}{126.3009x^2}$$
$$= 5.4780x$$
$$= 219.12 \text{ mm} \qquad \text{(Since } x = 40 \text{ mm)}$$

Centroid is at (326.40, 219.12). **Ans.**

## 6.5 SECOND MOMENTS OF PLANE AREA

Consider the area as shown in Fig. 6.37(*a*). $\delta A$ is an elemental area with coordinates as $x$ and $y$. The term $\Sigma y_i^2 \, dA_i$ is called *moment of inertia* of the area about $x$-axis and is denoted as $I_{xx}$. Similarly, the moment of inertia about $y$-axis is

$$I_{yy} = \Sigma x_i^2 \, dA_i$$

In general, if $r$ is the distance of elemental area $\delta A$ from the axis $AB$ [Fig. 6.37(*b*)], the sum of the terms $\Sigma r^2 \, \delta A$ to cover the entire area is called moment of inertia of the area about the axis $AB$. If $r$ and $\delta A$ can be expressed in general terms, for any element, then the sum becomes an integral. Thus,

$$I_{AB} = \Sigma r_i^2 \, \delta A_i = \int r^2 \, dA \qquad \text{...(6.6)}$$

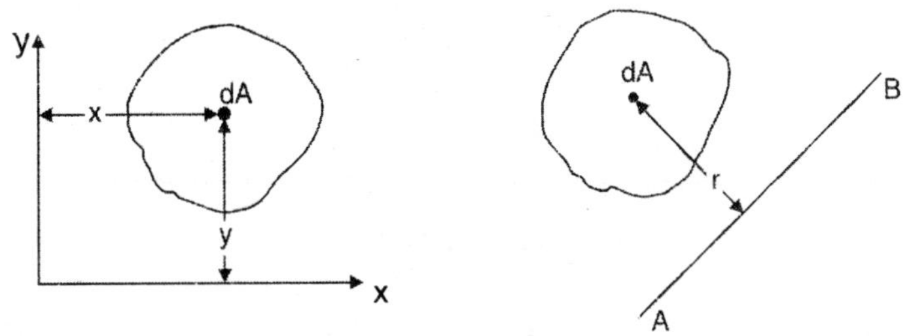


**Fig. 6.37**

The term $rdA$ may be called as moment of area, similar to moment of a force, and hence $r^2dA$ may be called as moment of area or the second moment of area. Thus, the moment of inertia of area is nothing but second moment of area. In fact, the term *'second moment of area'* appears to correctly signify the meaning of the expression $\Sigma r^2 \, dA$. The term 'moment of inertia' is rather a misnomer. However, the term moment of inertia has come to stay for long time, and hence, it will be used in this book also.

Though moment of inertia of plane area is a purely mathematical term, it is one of the important properties of areas. The strength of members subject to bending depends on the moment of inertia of its cross-sectional area. Students will find this property of area very useful when they study subjects such as strength of materials, structural design and machine design.

The moment of inertia is the fourth dimensional term, since, it is a term obtained by multiplying area by the square of the distance. Hence, in SI units, if metre (m) is the unit for linear measurements used then $m^4$ is the unit of moment of inertia. If millimetre (mm) is the unit used for linear measurements, then $mm^4$ is the unit of moment of inertia. In MKS system, $m^4$ or $cm^4$ and in FPS system, $ft^4$ or $in.^4$ are commonly used as units for moment of inertia.

### Polar Moment of Inertia

Moment of inertia about an axis perpendicular to the plane of an area is known as *polar moment of inertia*. It may be denoted as $J$ or $I_{zz}$. Thus, the moment of inertia about an axis perpendicular to the plane of the area at $O$ in Fig. 6.38 is called polar moment of inertia at point $O$, and is given by

$$I_{zz} = \Sigma r^2 dA \qquad \text{...(6.7)}$$

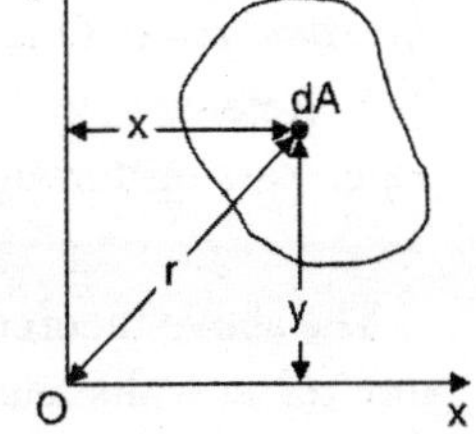


**Fig. 6.38**

## Radius of Gyration

Radius of gyration is a mathematical term defined by the relation

$$k = \sqrt{\frac{I}{A}} \quad ...(6.8)$$

where $k$ = Radius of gyration

$I$ = Moment of inertia

$A$ = The cross-sectional area

Suffixes with moment of inertia $I$ also accompany the term radius of gyration $k$. Thus, we can have

$$k_{xx} = \sqrt{\frac{I_{xx}}{A}}$$

$$k_{yy} = \sqrt{\frac{I_{yy}}{A}}$$

$$k_{AB} = \sqrt{\frac{I_{AB}}{A}}$$

and so on.

The relation between radius of gyration and moment of inertia can be put in the form:

$$I = Ak^2 \quad ...(6.9)$$

From the above relation, a geometric meaning can be assigned to the term 'radius of gyration.' We can consider $k$ as the distance at which the complete area is squeezed and kept as a strip of negligible width (Fig. 6.39) such that there is no change in the moment of inertia.

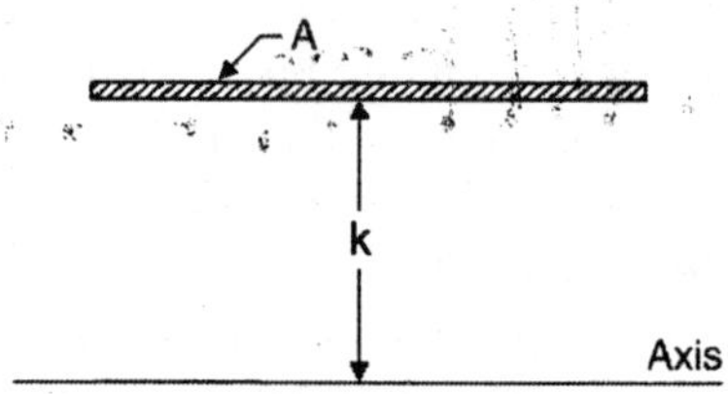


Fig. 6.39

## Theorems of Moments of Inertia

There are two theorems of moment of inertia:

1. Perpendicular axis theorem, and
2. Parallel axis theorem.

These are explained and proved below.

### 1. *Perpendicular Axis Theorem*

*The moment of inertia of an area about an axis perpendicular to its plane (polar moment of inertia) at any point O is equal to the sum of moments of inertia about any two mutually perpendicular axes through the same point O and lying in the plane of the area.*

Referring to Fig. 6.40, if $z$-$z$ is the axis normal to the plane of paper passing through point $O$, as per this theorem,

$$I_{zz} = I_{xx} + I_{yy} \quad ...(6.10)$$

The above theorem can be easily proved. Let us consider an elemental area $dA$ at a distance $r$ from $O$. Let the coordinates of $dA$ be $x$ and $y$. Then from definition:

$$I_{zz} = \Sigma r^2 dA$$

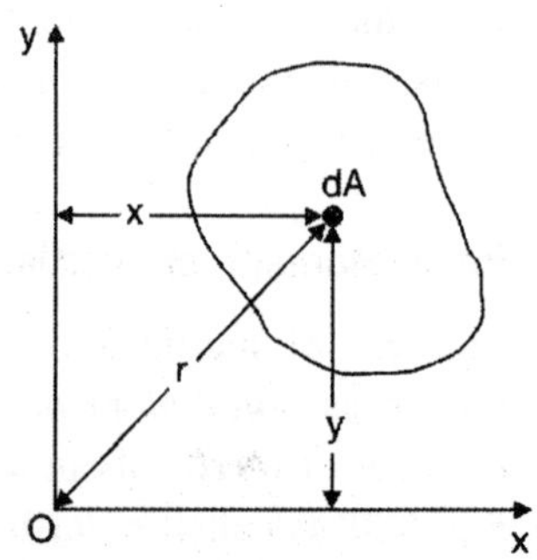


Fig. 6.40

$$= \Sigma(x^2 + y^2)dA$$
$$= \Sigma x^2 dA + \Sigma y^2 dA$$
$$I_{zz} = I_{xx} + I_{yy}$$

**2. *Parallel Axis Theorem***

*Moment of inertia about any axis in the plane of an area is equal to the sum of moment of inertia about a parallel centroidal axis and the product of area and square of the distance between the two parallel axes.* Referring to Fig. 6.41, the above theorem means:

$$I_{AB} = I_{GG} + Ay_c^2 \quad \text{...(6.11)}$$

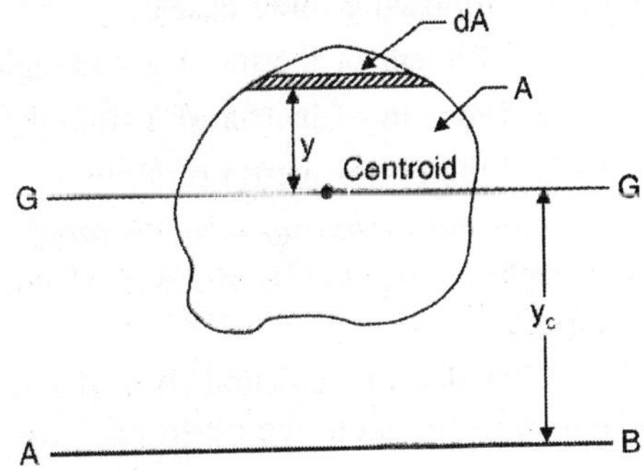


**Fig. 6.41**

where

$I_{AB}$ = Moment of inertia about axis $AB$.

$I_{GG}$ = Moment of inertia about centroidal axis $GG$ parallel to $AB$.

$A$ = The area of the plane figure given.

$y_c$ = The distance between the axis $AB$ and the parallel centroidal axis $GG$.

**Proof:** Consider an elemental parallel strip $dA$ at a distance $y$ from the centroidal axis (Fig. 6.41).

Then,
$$I_{AB} = \Sigma(y + y_c)^2 dA$$
$$= \Sigma(y^2 + 2y\, y_c + y_c^2)dA$$
$$= \Sigma y^2 \delta A + \Sigma 2y\, y_c\, dA + \Sigma\, y_c^2\, dA$$

Now,
$$\Sigma y^2 dA = \text{Moment of inertia about the axis } GG$$
$$= I_{GG}$$
$$\Sigma 2yy_c\, dA = 2y_c\, \Sigma y\, dA$$
$$= 2y_c A \frac{\Sigma\, ydA}{A}$$

In the above term, $2y_c A$ is constant and $\frac{\Sigma\, ydA}{A}$ is the distance of centroid from the reference axis $GG$. Since, $GG$ is passing through the centroid itself $\frac{ydA}{A}$ is zero and hence the term $\Sigma\, 2y\, y_c\, \delta A$ is zero.

Now, the third term,

$$\Sigma y_c^2 dA = y_c^2 \Sigma dA$$
$$= Ay_c^2$$
$$\therefore \quad I_{AB} = I_{GG} + Ay_c^2$$

**Note:** The above equation cannot be applied to any two parallel axes. One of the axis ( $GG$) must be centroidal axis only.

## 6.6 MOMENT OF INERTIA FROM FIRST PRINCIPLES

For simple figures, moment of inertia can be obtained by writing the general expression for an element and then carrying out integration so as to cover the entire area. This procedure is illustrated with the following three cases:

1. Moment of inertia of a rectangle about the centroidal axis.
2. Moment of inertia of a triangle about the base.
3. Moment of inertia of a circle about the diametral axis.

*Moment of Inertia of a Rectangle about the Centroidal Axis*: Consider a rectangle of width $b$ and depth $d$ (Fig. 6.42). Moment of inertia about the centroidal axis $x$-$x$ parallel to the short side is required.

Consider an elemental strip of width $dy$ at a distance $y$ from the axis. Moment of inertia of the elemental strip about the centroidal axis $x$-$x$ is:

$$= y^2 dA$$

$$= y^2 b\, dy$$

$$\therefore \quad I_{xx} = \int_{-d/2}^{d/2} y^2 b\, dy$$

$$= b\left[\frac{y^3}{3}\right]_{-d/2}^{d/2}$$

$$= b\left[\frac{d^3}{24} + \frac{d^3}{24}\right]$$

$$I_{xx} = \frac{bd^3}{12}$$

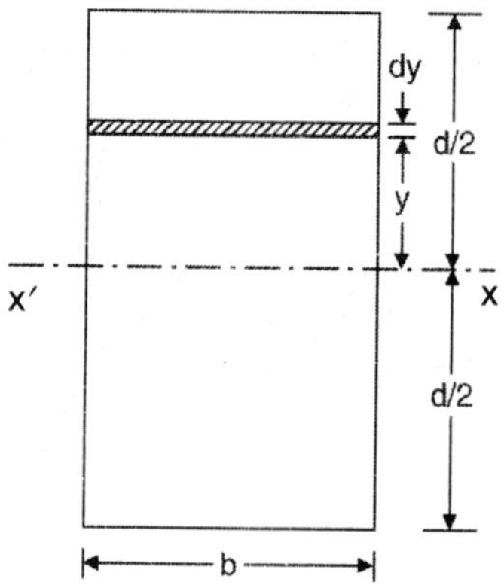


Fig. 6.42

*Moment of Inertia of a Triangle about the Base*: Moment of inertia of a triangle with base width $b$ and height $h$ is to be determined about the base $AB$ (Fig. 6.43).

Consider an elemental strip at a distance $y$ from the base $AB$. Let $dy$ be the thickness of the strip and $dA$ its area. Width of this strip is given by:

$$b_1 = \frac{(h-y)}{h} \times b$$

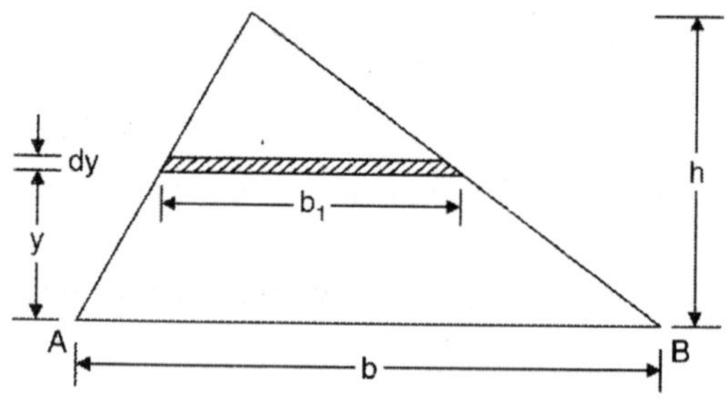


Fig. 6.43

Moment of inertia of this strip about $AB$

$$= y^2 dA$$

$$= y^2\, b_1\, dy$$

$$= y^2 \frac{(h-y)}{h} \times b \times dy$$

$\therefore$ Moment of inertia of the triangle about $AB$,

$$I_{AB} = \int_0^h \frac{y^2(h-y)bdy}{h}$$

$$= \int_0^h \left(y^2 - \frac{y^3}{h}\right) bdy$$

$$= b\left[\frac{y^3}{3} - \frac{y^4}{4h}\right]_0^h$$

$$= b\left[\frac{h^3}{3} - \frac{h^4}{4h}\right]$$

$$I_{AB} = \frac{bh^3}{12}$$

*Moment of Inertia of a Circle about the Diametral Axis*: Moment of inertia of a circle of radius $R$ is required about its diametral axis as shown in Fig. 6.44.

Consider an element of sides $rd\theta$ and $dr$ as shown in the figure. Its moment of inertia about the diametral axis $x$-$x$:

$$= y^2 dA$$

$$= (r \sin \theta)^2 \, r \, d\theta \, dr$$

$$= r^3 \sin^2 \theta \, d\theta \, dr$$

∴ Moment of inertia of the circle about axis $x$-$x$ is given by

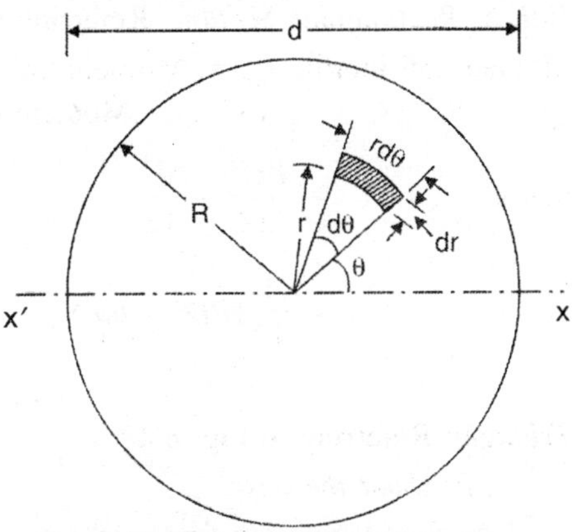


Fig. 6.44

$$I_{xx} = \int_0^R \int_0^{2\pi} r^3 \sin^2 \theta \, d\theta \, dr$$

$$= \int_0^R \int_0^{2\pi} r^3 \frac{(1-\cos 2\theta)}{2} d\theta \, dr$$

$$= \int_0^R \frac{r^3}{2}\left[\theta - \frac{\sin 2\theta}{2}\right]_0^{2\pi} dr$$

$$= \left[\frac{r^4}{8}\right]_0^R [2\pi - 0 + 0 - 0] = \frac{2\pi}{8} R^4$$

$$I_{xx} = \frac{\pi R^4}{4}$$

If $d$ is the diameter of the circle, then

$$R = \frac{d}{2}$$

$$\therefore \quad I_{xx} = \frac{\pi}{4}\left(\frac{d}{2}\right)^4$$

$$I_{xx} = \frac{\pi d^4}{64}$$

## Moment of Inertia of Standard Sections

*Rectangle*: Referring to Fig. 6.45.

(*a*) $I_{xx} = \dfrac{bd^3}{12}$ as derived from the first principles.

(*b*) $I_{yy} = \dfrac{db^3}{12}$ can be derived on the same lines.

(*c*) About the base *AB*, from parallel axis theorem,

$$I_{AB} = I_{xx} + Ay_c^2$$

$$= \frac{bd^3}{12} + bd\left(\frac{d}{2}\right)^2, \quad \text{since } y_c = \frac{d}{2}$$

$$= \frac{bd^3}{12} + \frac{bd^3}{4}$$

$$= \frac{bd^3}{3}$$

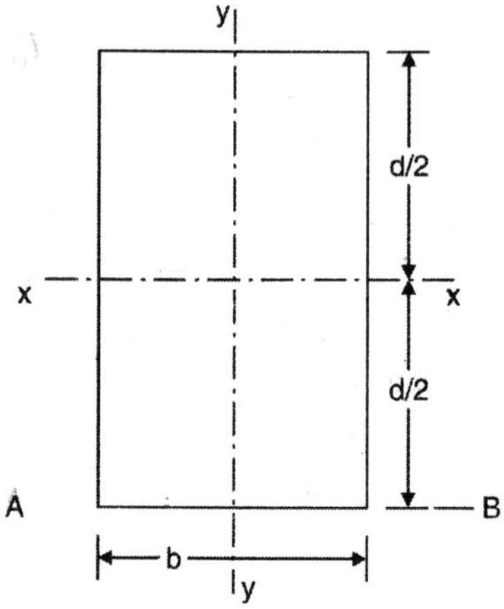


Fig. 6.45

*Hollow Rectangular Section*: Referring to Fig. 6.46,

Moment of inertia $I_{xx}$ = Moment of inertia of larger rectangle − Moment of inertia of hollow portion

that is $$= \frac{BD^3}{12} - \frac{bd^3}{12}$$

$$= \frac{1}{12}(BD^3 - bd^3)$$

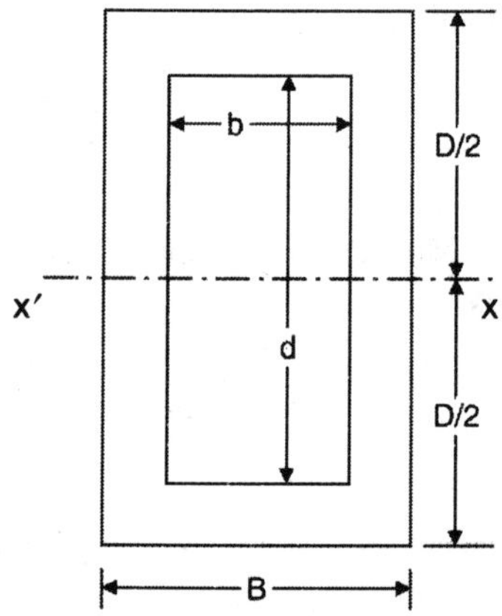


Fig. 6.46

*Triangle*: Referring to Fig. 6.47.

(*a*) *About the base:*

As found from the first principles

$$I_{AB} = \frac{bh^3}{12}$$

(*b*) *About centroidal axis x-x parallel to base:*

From parallel axis theorem,

$$I_{AB} = I_{xx} + Ay_c^2$$

Now, $y_c$, the distance between the non-centroidal axis *AB* and centroidal axis *x-x*, is equal to $\dfrac{h}{3}$.

$$\therefore \quad \frac{bh^3}{12} = I_{xx} + \frac{1}{2}bh\left(\frac{h}{3}\right)^2$$

$$= I_{xx} + \frac{bh^3}{18}$$

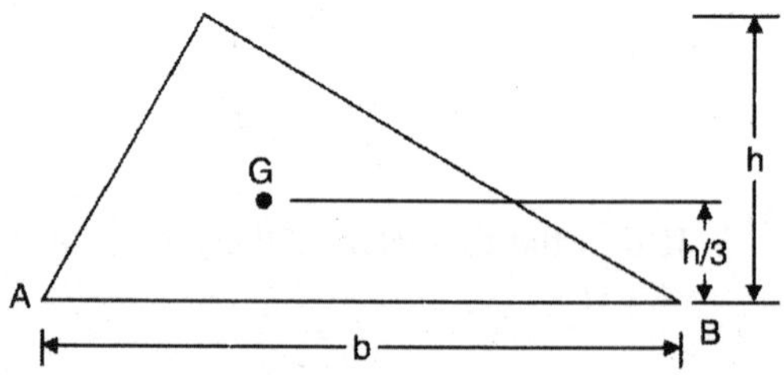


Fig. 6.47

$$\therefore \quad I_{xx} = \frac{bh^3}{12} - \frac{bh^3}{18}$$

$$= \frac{bh^3}{36}$$

*Moment of Inertia of a Circle about Any Diametral Axis*

$$= \frac{\pi d^4}{64} \quad \text{(as found from the first principles)}$$

*Moment of Inertia of a Hollow Circle*: Referring to Fig. 6.48,

$I_{AB}$ = Moment of inertia of solid circle of diameter $D$ about $AB$ – Moment of inertia of circle of diameter $d$ about $AB$. That is,

$$= \frac{\pi D^4}{64} - \frac{\pi d^4}{64}$$

$$= \frac{\pi}{64}(D^4 - d^4)$$

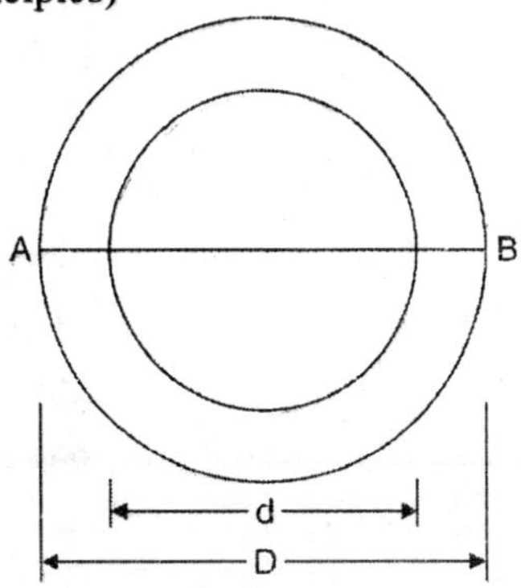


**Fig. 6.48**

*Moment of Inertia of a Semicircle*: (*a*) *About diametral axis:*

If the limit of integration is put as 0 to $\pi$ instead of 0 to $2\pi$ in the derivation for the moment of inertia of a circle about diametral axis, the moment of inertia of a semicircle is obtained. It can be observed that the moment of inertia of a semicircle (Fig. 6.49) about the diametral axis $AB$:

$$= \frac{1}{2} \times \frac{\pi d^4}{64} = \frac{\pi d^4}{128}$$

(*b*) *About centroidal axis x-x:*

Now, the distance of centroidal axis $y_c$ from the diametral axis is given by:

$$y_c = \frac{4R}{3\pi} = \frac{2d}{3\pi}$$

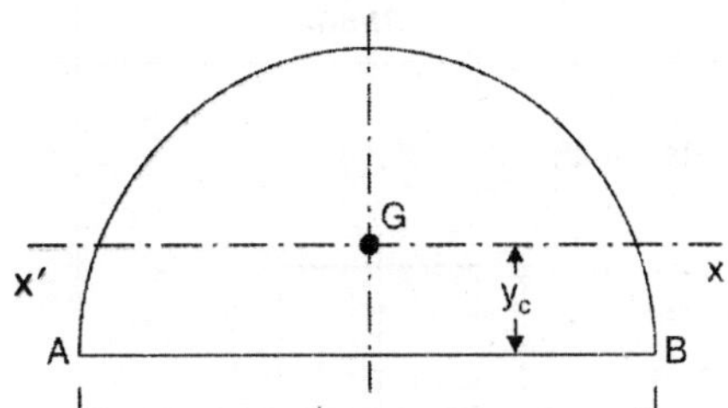


**Fig. 6.49**

and $\quad$ Area $A = \frac{1}{2} \times \frac{\pi d^2}{4} = \frac{\pi d^2}{8}$

From parallel axis theorem,

$$I_{AB} = I_{xx} + Ay_c^2$$

$$\frac{\pi d^4}{128} = I_{xx} + \frac{\pi d^2}{8} \times \left(\frac{2d}{3\pi}\right)^2$$

$$I_{xx} = \frac{\pi d^4}{128} - \frac{d^4}{18\pi}$$

$$= 0.0068598\, d^4$$

*Moment of Inertia of a Quarter of a Circle*: (*a*) *About the base:*

If the limit of integration is put as 0 to $\frac{\pi}{2}$ instead of 0 to $2\pi$ in the derivation for moment of inertia of a circle, the moment of inertia of a quarter of a circle is obtained. It can be observed that moment of inertia of the quarter of a circle about the base $AB$ is given by

$$I_{AB} = \frac{1}{4} \times \frac{\pi d^4}{64} = \frac{\pi d^4}{256}$$

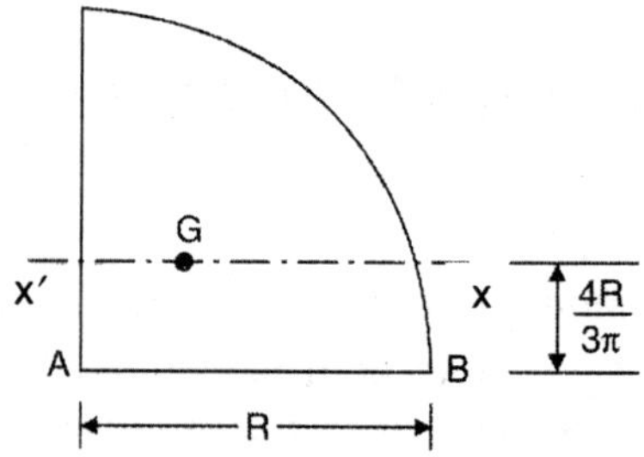


**Fig. 6.50**

(*b*) *About centroidal axis x-x:*

Now, the distance of centroidal axis $y_c$ from the base is given by:

$$y_c = \frac{4R}{3\pi} = \frac{2d}{3\pi}$$

and the area $$A = \frac{1}{4} \times \frac{\pi d^2}{4} = \frac{\pi d^2}{16}$$

From parallel axis theorem,

$$I_{AB} = I_{xx} + A y_c^2$$

$$\frac{\pi d^4}{256} = I_{xx} + \frac{\pi d^2}{16}\left(\frac{2d}{3\pi}\right)^2$$

$$I_{xx} = \frac{\pi d^4}{256} - \frac{d^4}{36\pi}$$

$$= 0.00343\ d^4$$

The moments of inertia of common standard sections are presented in Table 6.4.

**Table 6.4:** *Moments of Inertia of Standard Sections*

| ***Shape*** | ***Axis*** | ***Moment of inertia*** |
|---|---|---|
| Rectangle 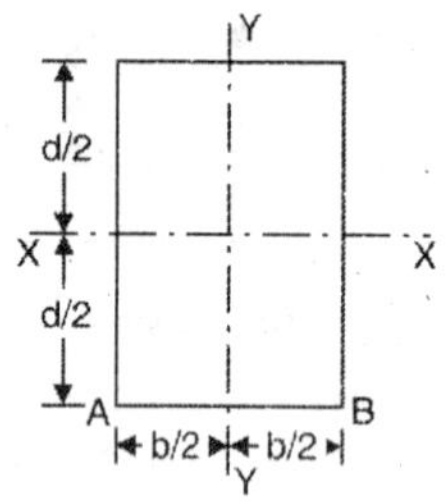  | (*a*) Centroidal axis *x-x*<br>(*b*) Centroidal axis *y-y*<br>(*c*) *A* – *B* | $I_{xx} = \frac{bd^3}{12}$<br>$I_{yy} = \frac{bd^3}{12}$<br>$I_{AB} = \frac{bd^3}{3}$ |
| Hollow rectangle 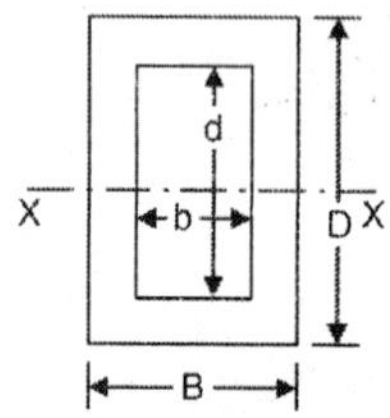  | Centroidal axis *x-x* | $I_{xx} = \frac{BD^3 - bd^3}{12}$ |

(Contd...)

| | | |
|---|---|---|
| Triangle<br>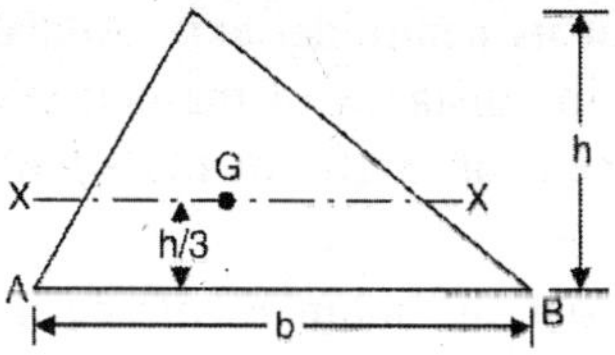 | (*a*) Centroidal axis *x-x*<br>(*b*) Base *AB* | $I_{xx} = \frac{bh^3}{36}$<br>$I_{AB} = \frac{bh^3}{12}$ |
| Circle<br>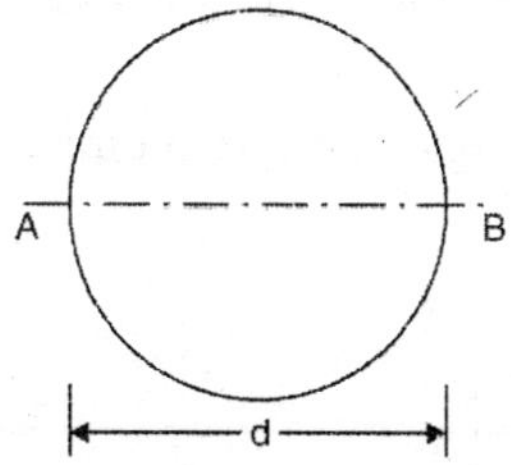 | Diametral axis | $I = \frac{\pi d^4}{64}$ |
| Hollow circle<br>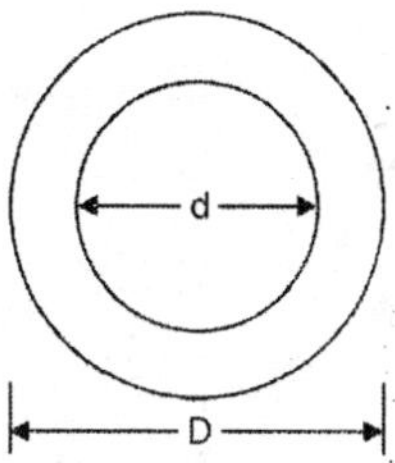 | Diametral axis | $I = \frac{\pi}{64}(D^4 - d^4)$ |
| Semicircle<br>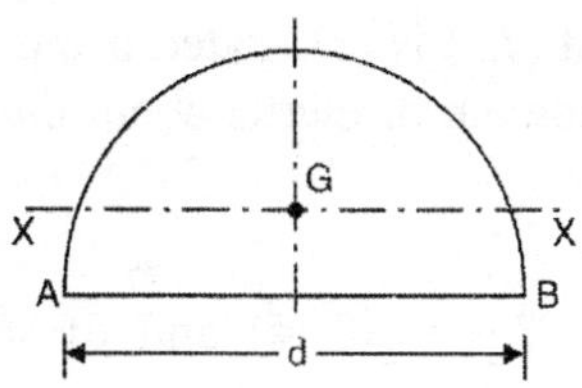 | (*a*) *A* – *B*<br>(*b*) Centroidal axis | $I_{AB} = \frac{\pi d^4}{128}$<br>$I_{xx} = 0.0068598\, d^4$ |
| Quarter of a circle<br>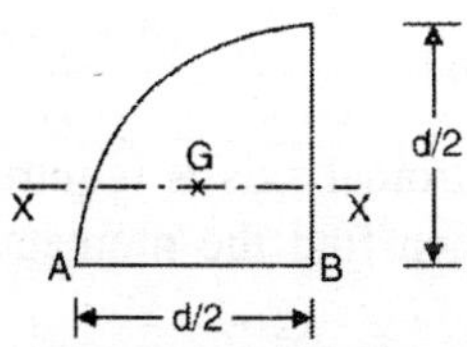 | (*a*) *A* – *B*<br>(*b*) Centroidal axis *x-x* | $I_{AB} = \frac{\pi d^4}{256}$<br>$I_{xx} = 0.00343\, d^4$ |

## 6.7 MOMENT OF INERTIA OF COMPOSITE SECTIONS

Beams and columns having composite sections are commonly used in structures. Moment of inertia of these sections about an axis can be found by the following steps:

1. Divide the given figure into a number of simple figures.
2. Locate the centroid of each simple figure by inspection or using standard expressions.
3. Find the moment of inertia of each simple figure about its centroidal axis. Add the term $Ay^2$ where $A$ is the area of the simple figure and $y$ is the distance of the centroid of the simple figure from the reference axis. This gives moment of inertia of the simple figure about the reference axis.
4. Sum up moments of inertia of all simple figures to get the moment of inertia of the composite section.

The procedure given above is illustrated below. Referring to the Fig. 6.51, it is required to find out the moment of inertia of the section about axis $A$–$B$.

1. The section in the figure is divided into a rectangle, a triangle and a semicircle. The areas of simple figures $A_1$, $A_2$ and $A_3$ are calculated.

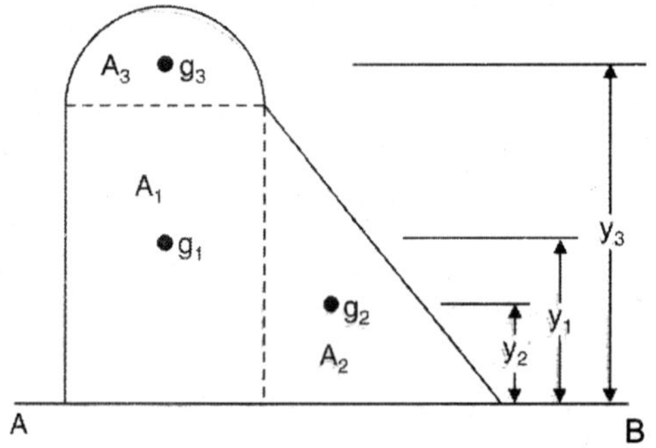


**Fig. 6.51**

2. The centroids of rectangle ($g_1$), triangle ($g_2$) and semicircle ($g_3$) are located. The distances $y_1$, $y_2$ and $y_3$ are found from the axis $AB$.
3. The moment of inertia of the rectangle about its centroid ($I_{g_1}$) is calculated using standard expression. To this, the term $A_1\,y_1^2$ is added to get the moment of inertia about the axis $AB$ as:

$$I_1 = I_{g_1} + A_1\,y_1^2$$

Similarly, the moment of inertia of the triangle ($I_2 = I_{g_2} + A_2\,y_2^2$) and of semicircle ($I_3 = I_{g_3} + A_3\,y_3^2$) about axis $AB$ are calculated.

4. The moment of inertia of the composite section about $AB$ is given by:

$$I_{AB} = I_1 + I_2 + I_3$$
$$= I_{g_1} + A_1\,y_1^2 + I_{g_2} + A_2\,y_2^2 + I_{g_3} + A_3\,y_3^2 \qquad ...(6.12)$$

In most engineering problems, moment of inertia about the centroidal axis is required. In such cases, first locate the centroidal axis as discussed in Art. 6.4 and then find the moment of inertia about this axis.

Referring to Fig. 6.52, the first moment of area about any reference axis, say $AB$ is taken and is divided by the total area of section to locate centroidal axis $x$-$x$. Then the distances of centroid of individual figures $y_{c_1}$, $y_{c_2}$ and $y_{c_3}$ from the axis $x$-$x$ are determined. The moment of inertia of the composite section about the centroidal axis $x$-$x$ is calculated using the expression:

$$I_{xx} = I_{g_1} + A_1\,y_{c1}^2 + I_{g_2} + A_2 y_{c2}^2 + I_{g_3} + A_3 y_{c3}^2 \qquad ...(6.13)$$

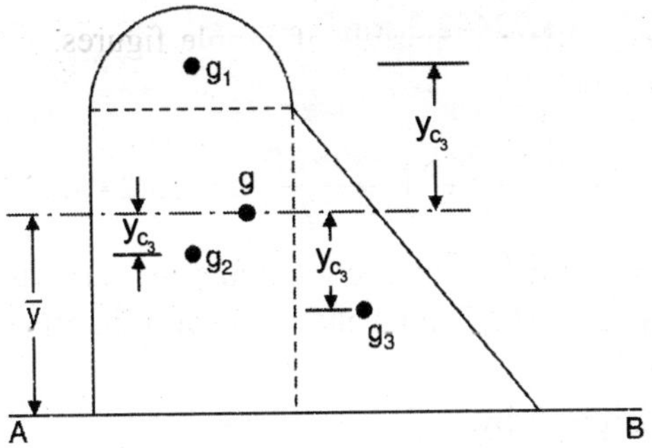


**Fig. 6.52**

Sometimes the moment of inertia is found about a convenient axis and then using parallel axis theorem, the moment of inertia about centroidal axis is found.

In the above example, the moment of inertia $I_{AB}$ is found and $y_c$, the distance of $CG$ from axis $AB$ is calculated. Then from parallel axis theorem,

$$I_{AB} = I_{xx} + Ay_c^2$$

$$I_{xx} = I_{AB} - Ay_c^2$$

where $A$ is the area of composite section.

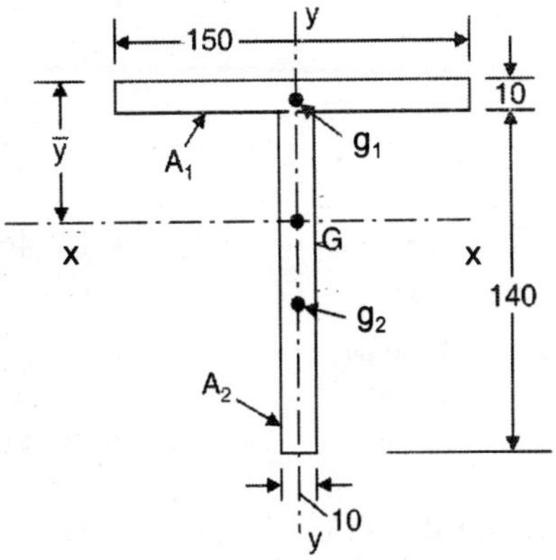


**Fig. 6.53**

**Example 6.12.** Determine the moment of inertia of the section shown in Fig. 6.53 about an axis passing through the centroid and parallel to the topmost fibre of the section.

Also determine moment of inertia about the axis of symmetry. Hence, find radii of gyration.

**Solution.** The given composite section can be divided into two rectangles as follows:

Area $\quad A_1 = 150 \times 10 = 1500 \text{ mm}^2$

Area $\quad A_2 = 140 \times 10 = 1400 \text{ mm}^2$

Total area $\quad A = A_1 + A_2 = 2900 \text{ mm}^2$

Due to symmetry, centroid lies on the symmetric axis $y$-$y'$.

The distance of the centroid from the topmost fibre is given by:

$$y_c = \frac{\text{Sum of moment of the areas about topmost fibre}}{\text{Total area}}$$

$$= \frac{1500 \times 5 + 1400(10+70)}{2900} = 41.21 \text{ mm}$$

Referring to the centroidal axes $x$-$x'$ and $y$-$y'$, the centroid of $A_1$ is $g_1$ (0.0, 36.21) and that of $A_2$ is $g_2$ (0.0, 38.79).

Moment of inertia of the section about $x$-$x'$ axis is given as

$I_{xx}$ = Moment of inertia of $A_1$ about $x$-$x'$ axis + Moment of inertia of $A_2$ about $x$-$x'$ axis.

$$\therefore \quad I_{xx} = \frac{150 \times 10^3}{12} + 1500\,(36.21)^2 + \frac{10 \times 140^3}{12} + 1400\,(38.79)^2$$

i.e., $I_{xx} = 63{,}72442.5 \text{ mm}^4$ **Ans.**

Similarly,

$$I_{yy} = \frac{10 \times 150^3}{12} + \frac{140 \times 10^3}{12} = 28{,}24166.7 \text{ mm}^4$$ **Ans.**

Hence, the moment of inertia of the section about an axis passing through the centroid and parallel to the topmost fibre is 63,72442.5 mm$^4$ and moment of inertia of the section about the axis of symmetry is 28,24166.66 mm$^4$.

The radius of gyration is given by:

$$k = \sqrt{\frac{I}{A}}$$

$$\therefore \quad k_{xx} = \sqrt{\frac{I_{xx}}{A}}$$

$$= \sqrt{\frac{63{,}72442.5}{2900}}$$

$$k_{xx} = 46.88 \text{ mm}$$ **Ans.**

Similarly, $$k_{yy} = \sqrt{\frac{28{,}24166.66}{2900}}$$

$$k_{yy} = 31.21 \text{ mm}$$ **Ans.**

**Example 6.13.** Determine the moment of inertia of the *L*-section shown in Fig. 6.54 about its centroidal axis parallel to the legs. Also find out the polar moment of inertia.

**Solution.** The given section is divided into two rectangles $A_1$ and $A_2$.

Area $A_1 = 125 \times 10 = 1250 \text{ mm}^2$

Area $A_2 = 75 \times 10 = 750 \text{ mm}^2$

Total area $= 2000 \text{ mm}^2$

First, the centroid of the given section is to be located.

The reference axes (1)–(1) and (2)–(2) are chosen as shown in Fig. 6.54.

The distance of centroid from the axis (1)–(1)

$$= \frac{\text{Sum of moment of area } A_1 \text{ and } A_2 \text{ about (1)–(1)}}{\text{Total area}}$$

i.e., $$\bar{x} = \frac{120 \times 5 + 750\left(10 + \frac{75}{2}\right)}{2000}$$

$$= 20.94 \text{ mm}$$

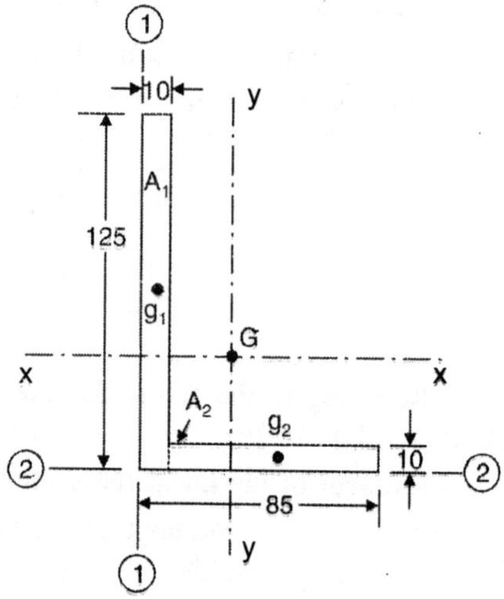


Fig. 6.54

Similarly, the distance of the centroid from the axis (2) – (2)

$$\bar{y} = \frac{1250 \times \frac{125}{2} + 750 \times 5}{2000} = 40.94 \text{ mm}$$

With respect to the centroidal axes $x$-$x'$ and $y$-$y'$, the centroid of $A_1$ is $g_1$ (15.94, 21.56) and that of $A_2$ is $g_2$ (26.56, 35.94).

$\therefore$ $I_{xx}$ = Moment of inertia of $A_1$ about $x$-$x'$ axis + Moment of inertia of $A_2$ about $x$-$x'$ axis

$$\therefore \quad I_{xx} = \frac{10 \times 125^3}{12} + 1250 \times 21.56^2 + \frac{75 \times 10^3}{12} + 750 \times 39.94^2$$

*i.e.,* $I_{xx}$ = 34,11298.9 mm$^4$ **Ans.**

Similarly,

$$I_{yy} = \frac{125 \times 10^3}{12} + 1250 \times 15.94^2 + \frac{10 \times 75^3}{12} + 750 \times 26.56^2$$

*i.e.,* $I_{yy}$ = 12,08658.9 mm$^4$ **Ans.**

Polar moment of inertia $= I_{xx} + I_{yy}$

= 34,11298.9 + 12,08658.9

$I_{zz}$ = 46,19957.8 mm$^4$ **Ans.**

**Example 6.14.** Determine the moment of inertia of the symmertic $I$-section shown in Fig. 6.55 about its centroidal axes $x$-$x'$ and $y$-$y'$. Also, determine moment of inertia of the section about a centroidal axis perpendicular to $x$-$x'$ axis and $y$-$y'$ axis.

**Solution.** The section is divided into three rectangles $A_1$, $A_2$ and $A_3$.

Area $A_1 = 200 \times 9 = 1800$ mm$^2$

Area $A_2 = (250 - 9 \times 2) \times 6.7 = 1554.4$ mm$^2$

Area $A_3 = 200 \times 9 = 1800$ mm$^2$

Total area $A = 5154.4$ mm$^2$

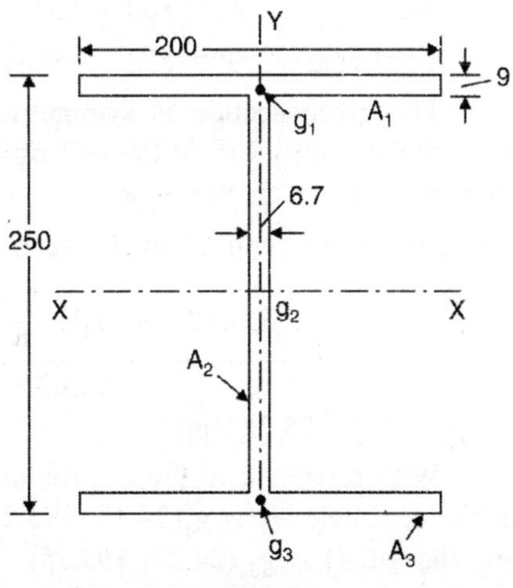

Fig. 6.55

The section is symmetrical about both $x$-$x$ and $y$-$y$ axes. Therefore, its centroid coincides with the centroid of rectangle $A_2$.

With respect to the centroidal axes $x$-$x$ and $y$-$y$, the centroid of rectangle $A_1$ is $g_1$ (0.0, 120.5), that of $A_2$ is $g_2$ (0.0, 0.0) and that of $A_3$ is $g_3$ (0.0, 120.5).

$I_{xx}$ = Moment of inertia of $A_1$ + Moment of inertia of $A_2$ + Moment of inertia of $A_3$ about $x$-$x'$ axis

$$I_{xx} = \frac{200 \times 9^3}{12} + 1800 \times 120.5^2 + \frac{6.7 \times 232^3}{12} + 0 + \frac{200 \times 9^3}{12} + 1800(120.5)^2$$

$I_{xx}$ = 5,92,69202 mm$^4$ **Ans.**

Similarly,

$$I_{yy} = \frac{9 \times 200^3}{12} + \frac{232 \times 6.7^3}{12} + \frac{9 \times 200^3}{12}$$

$I_{yy}$ = 1,20,05815 mm$^4$ **Ans.**

Moment of inertia of the section about a centroidal axis perpendicular to $x$-$x'$ and $y$-$y'$ axis is nothing but polar moment of inertia, and is given by:

$$I_{zz} = I_{xx} + I_{yy}$$
$$= 5{,}92{,}69202 + 1{,}20{,}05815$$

$I_{zz}$ = 7,12,75017 mm$^4$ **Ans.**

**Example 6.15.** Compute the second moment of area of the channel section shown in Fig. 6.56 about centroidal axes $x$-$x'$ and $y$-$y'$.

**Solution.** The section is divided into three rectangles $A_1$, $A_2$ and $A_3$.

Area $A_1 = 100 \times 13.5 = 1350$ mm$^2$

Area $A_2 = (400 - 27) \times 8.1 = 3021.3$ mm$^2$

Area $A_3 = 100 \times 13.5 = 1350.00$ mm$^2$

Total area $A = 5721.3$ mm$^2$

The given section is symmetric about horizontal axis passing through the centroid $g_2$ of the rectangle $A_2$. A reference axis (1)–(1) is chosen as shown in Fig. 6.56.

The distance of centroid of the section from (1)–(1)

$$= \frac{1350 \times 50 + 3021.3 \times \frac{8.1}{2} + 1350 \times 50}{5721.3}$$

= 25.73 mm

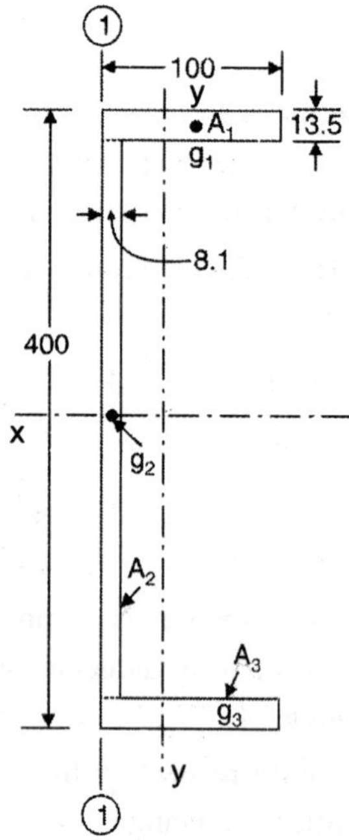

Fig. 6.56

With reference to the centroidal axes $x$-$x$ and $y$-$y$, the centroid of the rectangle $A_1$ is $g_1$(24.27, 193.25) that of $A_2$ is $g_2$ (21.68, 0.0) and that of $A_3$ is $g_3$ (24.27, 193.25).

$\therefore$ $I_{xx}$ = Moment of inertia of $A_1$, $A_2$ and $A_3$ about $x$-$x$

$$= \frac{100 \times 13.5^3}{12} + 1350 \times 193.25^2 + \frac{8.1 \times 373^3}{12} + \frac{100 \times 13.5^3}{12} + 1350 \times 193.25^2$$

$$I_{xx} = 1.359 \times 10^8 \text{ mm}^4$$ **Ans.**

Similarly,
$$I_{yy} = \frac{13.5 \times 100^3}{12} + 1350 \times 24.27^2 + \frac{273 \times 8.1^3}{12} + 3021.3 \times 21.68^2 + \frac{13.5 \times 100^3}{12} + 1350 \times 24.27^2$$

$$I_{yy} = 52{,}72557.6 \text{ mm}^4$$ **Ans.**

**Example 6.16.** Determine the polar moment of inertia of the *I*-section as shown in the Fig. 6.57. Also determine the radii of gyration with respect to *x-x* and *y-y* axes.

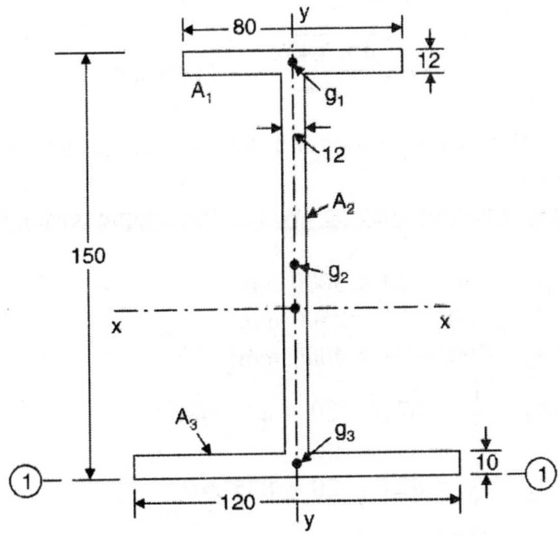


**Fig. 6.57**

**Solution.** The section is divided into three rectangles as shown in Fig. 6.57.

Area $A_1 = 80 \times 12 = 960 \text{ mm}^2$

Area $A_2 = (150 - 22) \times 12 = 1536 \text{ mm}^2$

Area $A_3 = 120 \times 10 = 1200 \text{ mm}^2$

Total area $A = 3696 \text{ mm}^2$

Due to symmetry, centroid lies on axis *y-y*. The bottom fibre (1)–(1) is chosen as reference axis to locate the centroid.

The distance of the centroid from (1)–(1)

$$= \frac{\text{Sum of moment of areas of the rectangles about (1) – (1)}}{\text{Total area of section}}$$

$$= \frac{960 \times (150 - 6) + 1536 \times \left(\frac{128}{2} + 10\right) + 1200 \times 5}{3696} = 69.78 \text{ mm}$$

With reference to the centroidal axes *x-x* and *y-y*, the centroid of the rectangles $A_1$ is $g_1$ (0.0, 74.22), that of $A_2$ is $g_2$ (0.0, 4.22) and that of $A_3$ is $g_3$ (0.0, 64.78).

$$I_{xx} = \frac{80 \times 12^3}{12} + 960 \times 74.22^2 + \frac{12 \times 128^3}{12} + 1536 \times 4.22^2 + \frac{120 \times 10^3}{12} + 1200 \times 64.78^2$$

$$I_{xx} = 1{,}24{,}70027 \text{ mm}^4$$

$$I_{yy} = \frac{12 \times 80^3}{12} + \frac{128 \times 12^3}{12} + \frac{10 \times 120^3}{12} = 19{,}70432 \text{ mm}^4$$

Polar moment of inertia $= I_{xx} + I_{yy}$

$= 1{,}24{,}70027 + 19{,}70432$

$= 1{,}44{,}40459 \text{ mm}^4$ **Ans.**

$\therefore \quad k_{xx} = \sqrt{\dfrac{I_{xx}}{A}} = \sqrt{\dfrac{1{,}24{,}70027}{3696}}$

$= 58.09$ mm **Ans.**

$k_{yy} = \sqrt{\dfrac{I_{yy}}{A}} = \sqrt{\dfrac{19{,}70432}{3696}} = 23.09$ mm. **Ans.**

**Example 6.17.** Determine the moment of inertia of the built-up section as shown in Fig. 6.58 about its centroidal axes *x-x* and *y-y*.

**Solution.** The given composite section may be divided into simple rectangles and triangles as shown in Fig. 6.58.

Area $A_1 = 100 \times 30 = 3000 \text{ mm}^2$

Area $A_2 = 100 \times 25 = 2500 \text{ mm}^2$

Area $A_3 = 200 \times 20 = 4000 \text{ mm}^2$

Area $A_4 = \dfrac{1}{2} \times 87.5 \times 20 = 875 \text{ mm}^2$

Area $A_5 = \dfrac{1}{2} \times 87.5 \times 20 = 875 \text{ mm}^2$

Total area $A = 11250 \text{ mm}^2$

Due to symmetry, centroid lies on the axis *y-y*.

A reference axis (1)–(1) is chosen as shown in the figure.

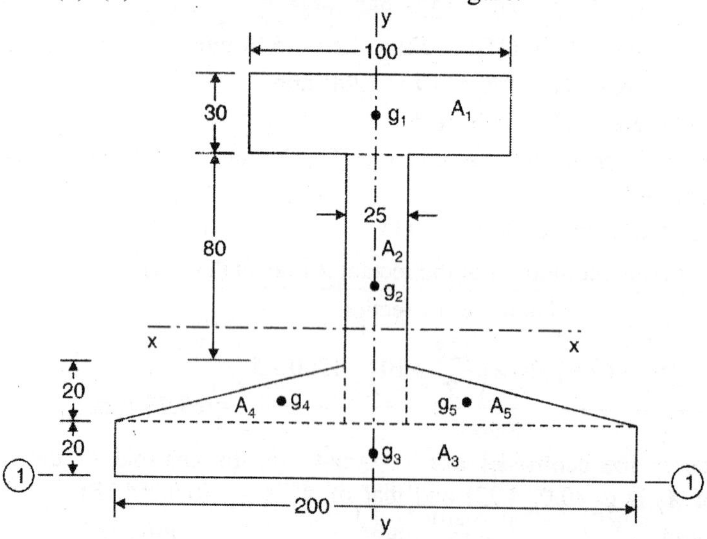


**Fig. 6.58**

The distance of the centroidal axis from (1)–(1)

$$= \frac{\text{Sum of moment of the areas about (1) - (1)}}{\text{Total area}}$$

$$\bar{y} = \frac{3000 \times 135 + 2500 \times 70 + 4000 \times 10 \times 875\left(\frac{1}{3} \times 20 + 20\right) \times 2}{11250} = 59.26 \text{ mm}$$

With reference to the centroidal axes *x-x* and *y-y*, the centroid of the rectangle $A_1$ is $g_1$ (0.0, 75.74), that of $A_2$ is $g_2$(0.0, 10.74), that of $A_3$ is $g_3$ (0.0, 49.26), the centroid of triangle $A_4$ is $g_4$ (41.66, 32.59) and that of $A_5$ is $g_5$ (41.66, 32.59).

$$I_{xx} = \frac{100 \times 30^3}{12} + 3000 \times 75.74^2 + \frac{25 \times 100^3}{12} + 2500 \times 10.74^2 + \frac{200 \times 20^3}{12} + 4000$$

$$\times 49.26^2 + \frac{87.5 \times 20^3}{36} + 875 \times 32.59^2 + \frac{87.5 \times 20^3}{36} + 875 \times 32.59^2$$

$I_{xx}$ = 3,15,43447 mm$^4$ **Ans.**

$$I_{yy} = \frac{30 \times 100^3}{12} + \frac{100 \times 25^3}{12} + \frac{20 \times 200^3}{12} + \frac{20 \times 87.5^3}{36} + 875 \times 41.66^2$$

$$+ \frac{20 \times 87.5^3}{36} + 875 \times 41.66^2$$

$I_{yy}$ = 1,97,45122 mm$^4$. **Ans.**

**Example 6.18.** Determine the moment of inertia of the built-up section as shown in Fig. 6.59 about axis *AB* passing through the topmost fibre of the section as shown.

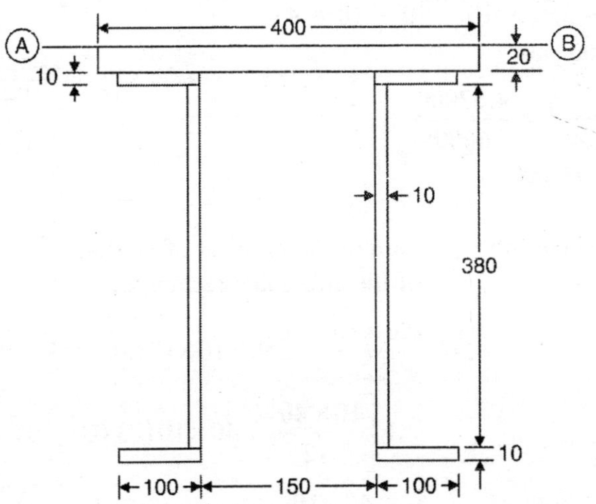


**Fig. 6.59**

**Solution.** In this problem, it is required to find out the moment of inertia of the section about axis *AB*. So there is no need to find out the position of the centroid.

The given section is split into simple rectangles as shown in Fig. 6.59.

Now,

Moment of inertia about $AB$ = Sum of moments of inertia of the rectangles about $AB$

$$= \frac{400\times20^3}{12}+400\times20\times10^2+\left[\frac{100\times10^3}{12}+100\times10\times(20+5)^2\right]\times2$$

$$+\left[\frac{100\times380^3}{12}+10\times380\times(30+190)^2\right]\times2$$

$$+\left[\frac{100\times10^3}{12}+100\times10\times(20+10+380+5)^2\right]\times2$$

$I_{AB} = 8.06093 \times 10^8 \text{ mm}^4.$ **Ans.**

**Example 6.19.** Calculate the moment of inertia of the built-up section shown in Fig. 6.60 about a centroidal axis parallel to $AB$. All members are 10 mm thick.

**Solution.** The built-up section is divided into six simple rectangles as shown in the figure.

The distance of centroidal axis from $AB$

$$= \frac{\text{Sum of moment of the areas about } AB}{\text{Total area}}$$

$$= \frac{\Sigma A_i y_i}{A}$$

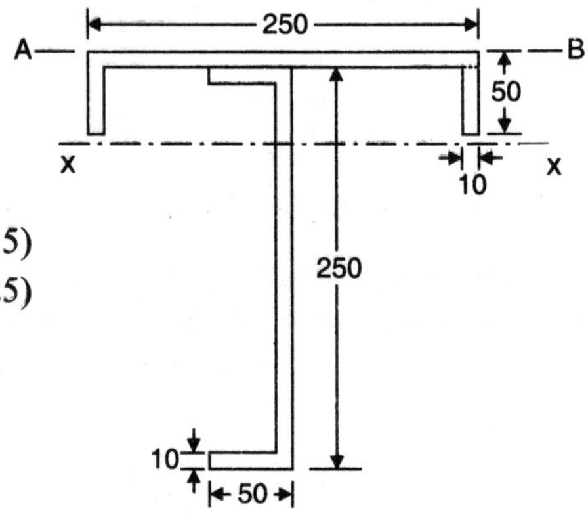


Fig. 6.60

Now, $\Sigma A_i y_i = 250 \times 10 \times 5 + 2 \times 40 \times 10 \times (10 + 20) + 40 \times 10 \times (10 + 5) + 40 \times 10 \times 255 + 250 \times 10 \times (10 + 125)$

$= 4,82000 \text{ mm}^3$

$A = 2 \times 250 \times 10 + 40 \times 10 \times 4$

$= 6600 \text{ mm}^2$

$\therefore \quad \bar{y} = \dfrac{\Sigma A_i y_i}{A} = \dfrac{4,82000}{6600}$

$= 73.03 \text{ mm}$

Now,

Moment of inertia about the centroidal axis } = { Sum of the moment of inertia of the individual rectangles

$$= \frac{250\times10^3}{12}+250\times10\times(73.03-5)^2$$

$$+\left[\frac{10\times40^3}{12}+40\times10(73.03-30)^2\right]\times2$$

$$+\frac{40\times10^3}{12}+40\times10(73.03-15)^2+\frac{10\times250^3}{12}+250$$

$$\times10(73.03-135)^2+\frac{40\times10^3}{12}+40\times10(73.03-255)^2$$

$I_{xx} = 5,03,99395 \text{ mm}^5.$ **Ans.**

**Example 6.20.** A built-up section of structural steel consists of a flange plate 400 mm × 20 mm, a web plate 600 mm × 15 mm and two angles 150 mm × 150 mm × 10 mm assembled to form a section as shown in Fig. 6.61. Determine the moment of inertia of the section about the horizontal centroidal axis.

**Solution.** Each angle is divided into two rectangles as shown in Fig. 6.61.

The distance of the centroidal axis from the bottom fibres of section

$$= \frac{\text{Sum of moment of the areas about bottom fibres}}{\text{Total area of the section}}$$

$$= \frac{\Sigma A_i y_i}{A}$$

Now, $\Sigma A_i y_i = 600 \times 15 \times \left(\frac{600}{2} + 20\right) + 140 \times 10 \times (70 + 30) \times 2 + 150 \times 10 \times (5 + 20) \times 2 + 400 \times 20 \times 10$

$= 1315000 \text{ mm}^3$

$A = 600 \times 15 + 140 \times 10 \times 2 + 150 \times 10 \times 2 + 400 \times 20$

$= 22{,}800 \text{ mm}^2$

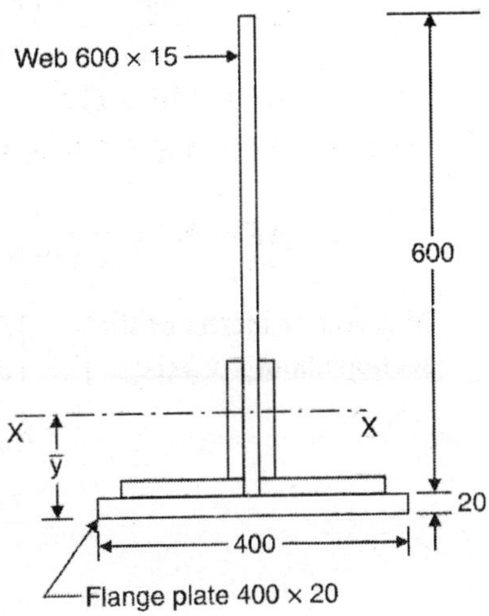

**Fig. 6.61**

$\therefore \quad \bar{y} = \frac{\Sigma A_i y_i}{A} = \frac{13{,}15000}{22{,}800}$

$= 145.39 \text{ mm}$

$$\left.\begin{array}{l}\text{Moment of inertia of the}\\ \text{section about centroidal axis}\end{array}\right\} = \left\{\begin{array}{l}\text{Sum of the moments of inertia of the}\\ \text{all simple figures about centroidal axis}\end{array}\right.$$

$$= \frac{15 \times 600^3}{12} + 600 \times 15(145.39 - 320)^2 + \left[\frac{10 \times 140^3}{12} + 1400(145.39 - 100)^2\right] \times 2 + \left[\frac{150 \times 10^3}{12} + 1500 \times (145.39 - 15)^2\right] \times 2 + \frac{400 \times 20^3}{12} + 400 \times 20 \times (145.39 - 10)^2$$

$I_{xx} = 7.45156 \times 10^8 \text{ mm}^4$. **Ans.**

**Example 6.21.** Compute the moment of inertia of the 100 mm × 150 mm rectangle as shown in Fig. 6.62 about $x$-$x$ axis to which it is inclined at an angle

$$\theta = \sin^{-1}\left(\frac{4}{5}\right)$$

**Solution.** The rectangle is divided into four triangles as shown in the figure. [The lines $AE$ and $FC$ are parallel to $x$-axis].

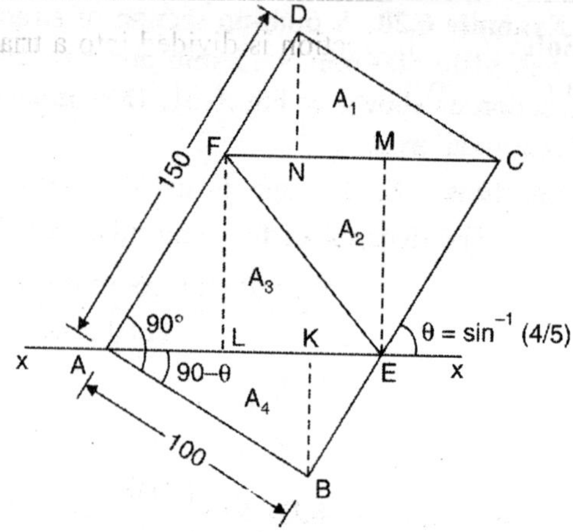


Fig. 6.62

Now $\theta = \sin^{-1}\left(\frac{4}{5}\right) = 53.13°$

From the geometry of Fig. 6.62,

$$BK = AB \sin(90° - \theta)$$

$$= 100 \sin(90° - 53.13°) = 60 \text{ mm}$$

$$ND = BK = 60 \text{ mm}$$

$$\therefore \quad FD = \frac{60}{\sin\theta} = \frac{60}{\sin 53.13°} = 75 \text{ mm}$$

$$\therefore \quad AF = 150 - FD = 75 \text{ mm}$$

Hence, $FL = ME = 75 \sin\theta = 60$ mm

$$AE = FC = \frac{AB}{\cos(90° - \theta)} = \frac{100}{0.8} = 125 \text{ mm}$$

Moment of inertia of the section about $x$-$x$ axis = Sum of the moments of inertia of individual triangular areas about $x$-$x$ axis

$$= I_{DFC} + I_{FCE} + I_{FEA} + I_{AEB}$$

$$= \frac{125 \times 60^3}{36} + \frac{1}{2} \times 125 \times 60 \times \left(60 + \frac{1}{3} \times 60\right)^2$$

$$+ \frac{125 \times 60^3}{36} + \frac{1}{2} \times 125 \times 60 \times \left(\frac{2}{3} \times 60\right)^2 + \frac{125 \times 60^3}{36} + \frac{1}{2} \times 125$$

$$\times 60 \times \left(\frac{1}{3} \times 60\right)^2 + \frac{125 \times 60^3}{36} + \frac{1}{2} \times 125 \times 60 \times \left(\frac{1}{3} \times 60\right)^2$$

$$I_{xx} = 3,60,00000 \text{ mm}^4.$$

**Ans.**

**Example 6.22.** Find the moment of inertia of the shaded area as shown in Fig. 6.63 about axis $AB$.

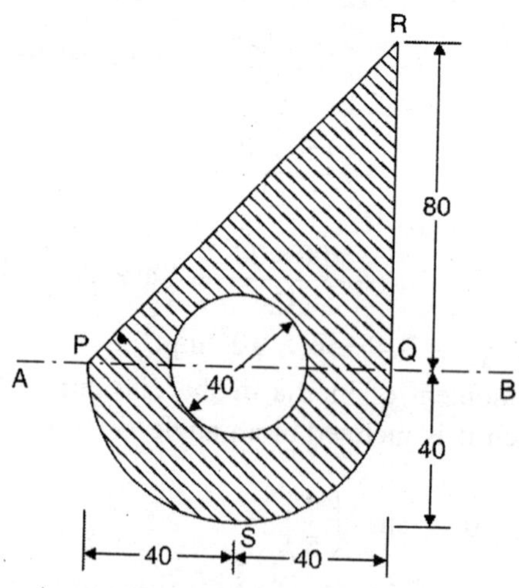


Fig. 6.63

**Solution.** The section is divided into a triangle *PQR*, a semicircle *PSQ* having base on axis *AB* and a circle having its centre on axis *AB*.

Now,

$$\left.\begin{array}{l}\text{Moment of inertia of the}\\ \text{section about axis } AB\end{array}\right\} = \left\{\begin{array}{l}\text{Moment of inertia of triangle } PQR \text{ about}\\ AB + \text{Moment of inertia of semicircle}\\ PSQ \text{ about } AB - \text{Moment of inertia of}\\ \text{circle about } AB\end{array}\right.$$

$$= \frac{80 \times 80^3}{12} + \frac{\pi}{128} \times 80^4 - \frac{\pi}{64} \times 40^4$$

$$I_{AB} = 42{,}92979 \text{ mm}^4.$$ **Ans.**

**Example 6.23.** Find the second moment of the shaded portion as shown in Fig. 6.64 about its centroidal axis.

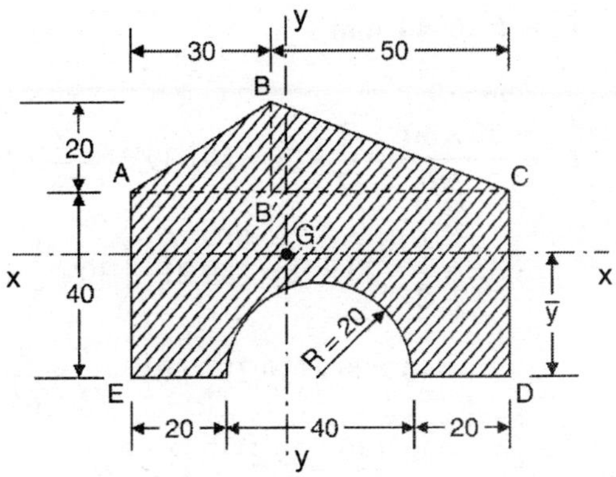


**Fig. 6.64**

**Solution.** The section is divided into three simple figures viz., a triangle *ABC*, a rectangle *ACDE* and a semicircle.

Total area = Area of triangle *ABC* + Area of rectangle *ACDE* – Area of semicircle

$$A = \frac{1}{2} \times 80 \times 20 + 40 \times 80 - \frac{1}{2} \times \pi \times 20^2$$

$$= 3371.68$$

$$A\,\bar{y} = \frac{1}{2} \times 80 \times 20\left(\frac{1}{3} \times 20 + 40\right) + 40 \times 80 \times 20 - \frac{1}{2} \times \pi\, 20^2 \times \frac{4 \times 20}{3\pi}$$

$$= 95991.77$$

$$\therefore \quad \bar{y} = \frac{95991.77}{3371.6} = 28.47 \text{ mm}$$

$$A\,\bar{x} = \frac{1}{2} \times 30 \times 20 \times \frac{2}{3} \times 30 + \frac{1}{2} \times 50 \times 20 \times \left(\frac{1}{3} \times 50 \times 30\right)$$

$$+ 40 \times 80 \times 40 - \frac{1}{2} \times \pi \times 20^2 \times 40$$

$$= 132203.6$$

$$\therefore \qquad \bar{x} = \frac{A\bar{x}}{A} = \frac{132203.6}{3371.68} = 37.21 \text{ mm}$$

$$\left.\begin{array}{l}\text{Moment of inertia about}\\ \text{centroidal } x\text{-}x \text{ axis}\end{array}\right\} = \left\{\begin{array}{l}\text{Moment of inertia of triangle } ABC \text{ about}\\ x\text{-}x \text{ axis} + \text{Moment of inertia of rectangle}\\ \text{about } x\text{-}x \text{ axis} - \text{Mioment of semicircle}\\ \text{about } x\text{-}x \text{ axis}\end{array}\right.$$

$$\therefore \qquad I_{xx} = \frac{80\times 20^3}{36} + \frac{1}{2}\times 80\times 20\left(60 - \frac{2}{3}\times 20 - 28.47\right)^2$$

$$+ \frac{80\times 40^3}{12} + 80\times 40\times(28.47-20)^2$$

$$-\left[0.0068598\times 20^4 + \frac{1}{2}\pi\times 20^2\left(28.47 - \frac{4\times 20}{3\pi}\right)^2\right]$$

$$I_{xx} = 6{,}86944 \text{ mm}^4.$$ **Ans.**

Similarly,

$$I_{yy} = \frac{20\times 30^3}{36} + \frac{1}{2}\times 20\times 30\left(39.21 - \frac{2}{3}\times 30\right)^2 + \frac{20\times 50^3}{36}$$

$$+ \frac{1}{2}\times 20\times 50\times\left[39.21 - \left(30 + \frac{1}{3}\times 50\right)\right]^2 + \frac{40\times 80^3}{12}$$

$$+ 40\times 80\times(39.21-40)^2 - \frac{1}{2}\times\frac{\pi}{64}\times 40^4 - \frac{1}{2}\times\frac{\pi}{4}$$

$$\times 40^2(40-39.21)^2$$

$$= 18{,}68392 \text{ mm}^4.$$ **Ans.**

## 6.8 PRODUCT MOMENT OF PLANE AREA

Consider the plane area as shown in Fig. 6.65. $x$-$y$ are the cartesian coordinate axes selected and $dA_i$ is an element in the area with coordinates $x_i$ and $y_i$. *The term $\Sigma x_i y_i dA_i$ to cover the entire area is termed product moment of the plane area or product moment of inertia*.

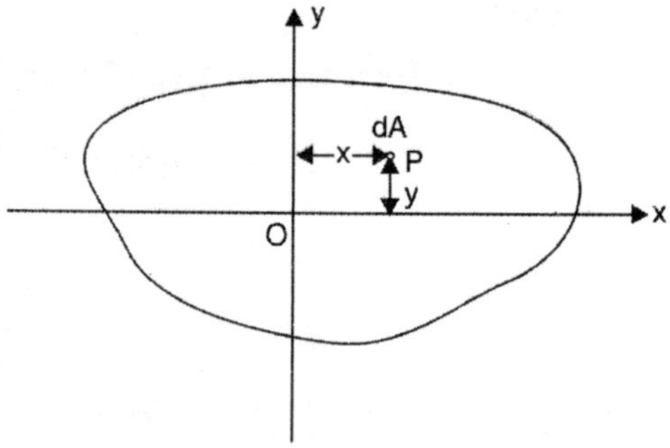

Fig. 6.65

If the term $x_i\, y_i\, dA_i$ can be expressed in general term for any element in the area and appropriate limits can be found for $x$ and $y$ to cover entire area, then $\Sigma x_i\, y_i\, dA_i$ can be replaced by $\oint xydA$. Denoting the product moment of inertia by $I_{xy}$, we have

$$\text{Product moment of area} = I_{xy} = \Sigma xydA = \oint xydA \qquad \text{...(6.14)}$$

For this term also the unit is mm$^4$ or m$^4$. We noticed $I_{xx}$ and $I_{yy}$ terms were always positive, since, they involved the terms $x^2$ and $y^2$. But the product moment of area $I_{xy}$ can be negative also, since one of the coordinates may be positive and the other may be negative.

For composite sections, product moment of inertia found by $I_{xy} = A_1x_1y_1 + A_2x_2y_2 + \ldots$ where $x_i\, y_i$ are the coordinates of centroidal axis of $A_i$.

The term product moment of inertia will be very useful when the bending stresses in a structure with unsymmetric cross-sections are to be determined.

## 6.9 PRINCIPAL MOMENT OF INERTIA

The axis about which product moment of inertia is zero is termed as principal axis. Incidently the moments of inertia are maximum and minimum about these axes, and they are known as principal moments of inertia. If the axes considered are through the centroid of the area, then the axes are known as centroidal principal axes.

Consider the product moment of inertia about a symmetric axis shown in Fig. 6.66. If we consider the two elemental areas at a time, which are equal but one having the coordinates $(x, y)$ and the other $(-x, y)$, we find their contribution to the term $I_{xy}$ is zero. Since, $y$-axis is symmetric axis, we can find one to one match for $xydA$ and $-xydA$ terms over entire area and hence $I_{xy}$ is zero. Hence, a symmetric axis is a principal axis. If we consider $I_{xy}$ term about an axis at right angles to the symmetric axis, then also we find $I_{xy}$ term will be zero. Hence, the symmetric axis and axis at right angle to symmetric axis are principal axes.

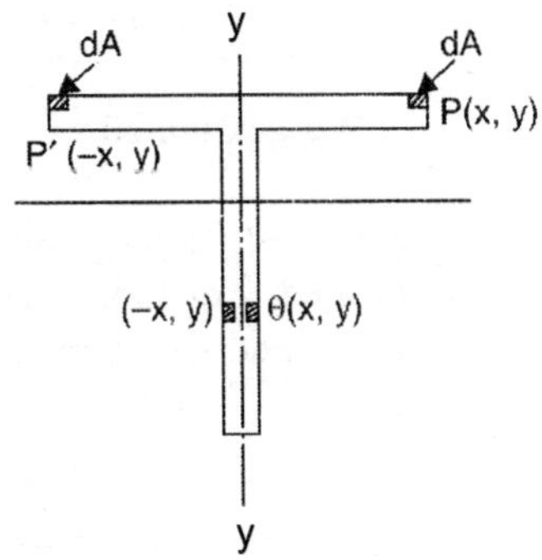

Fig. 6.66

If there is no axis of symmetry in the plane area considered, any convenient cartesian axis $x$-$y$ is selected and $I_{xx}$, $I_{yy}$ and $I_{xy}$ are computed. If $u$, $v$ are the principal axes, making angle θ with $x$, $y$ axes (Ref. Fig. 6.67), then it is possible to find the relations between $I_{uv}$ and $I_{xy}$, $I_{xx}$, $I_{yy}$. Then make use of the definition that $I_{uv}$ should be zero, since, $u$-$v$ are principal axes to get angle θ. Then the principal moments of inertias $I_{uu}$ and $I_{vv}$ can be found.

Consider the elemental area $dA$ at point $P$. Its coordinates are $x$ and $y$ with respect to $x$-$y$ axis and $u$ and $v$ with respect to $u$-$v$ axes. $PM$ and $PN$ are perpendicular to $x$-axis and $u$-axis. Let $M'$ be the perpendicular to $u$-axis. $N'$ is the intersection of $PN$ with the line $MN'$ which is parallel to $x$-axis.

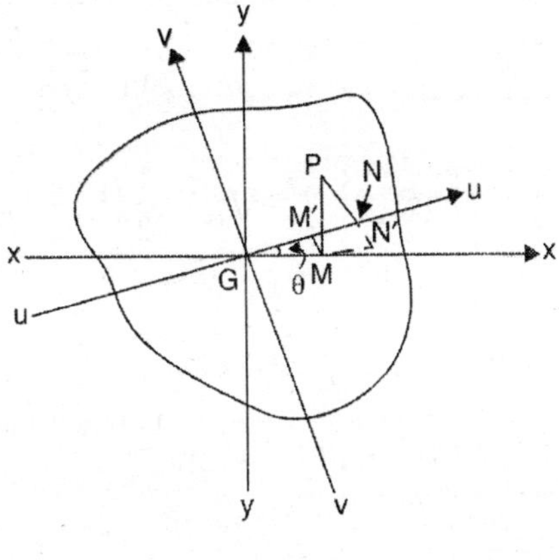

Fig. 6.67

Now, $u = GN = GM' + M'N = GM' + MN'$

$= GM \cos\theta + PM \sin\theta$

$= x \cos\theta + y \sin\theta$ ...(6.15)

$v = PN = PN' - NN' = PN' - MM'$

$= PM \cos\theta - GM \sin\theta$

$= y \cos\theta - x \sin\theta$ ...(6.16)

According to definitions

$$I_{xx} = \Sigma y^2 dA,\ I_{yy} = \Sigma x^2 dA,\ I_{xy} = \Sigma xydA$$

and

$$I_{uu} = \Sigma v^2 dA,\ I_{vv} = \Sigma u^2 dA,\ I_{uv} = \Sigma uvdA$$

Consider the term $I_{uv}$,

$$I_{uv} = \Sigma uvdA$$

$$= \Sigma(x \cos\theta + y \sin\theta)(y \cos\theta - x \sin\theta)\, dA$$

$$= \Sigma(xy \cos^2\theta - x^2 \sin\theta \cos\theta + y^2 \sin\theta \cos\theta - xy \sin^2\theta)dA$$

$$= \cos^2\theta\, \Sigma xydA - \sin\theta \cos\theta\, \Sigma x^2 dA + \sin\theta \cos\theta\, \Sigma y^2 dA - \sin^2\theta\, \Sigma xydA$$

$$= I_{xy} \cos^2\theta - I_{yy} \sin\theta \cos\theta + I_{xx} \sin\theta \cos\theta - I_{xy} \sin^2\theta$$

$$= I_{xy}(\cos^2\theta - \sin^2\theta) - \sin\theta\cos\theta\,(I_{yy} - I_{xx})$$

$$= I_{xy}\cos 2\theta - \frac{(I_{yy} - I_{xx})}{2}\sin 2\theta = I_{xy}\cos 2\theta + \frac{(I_{xx} - I_{yy})}{2}\sin 2\theta \quad \text{...(6.17)}$$

Since $u$-$v$ are principal axes,

$$I_{uv} = 0$$

$$\therefore \quad \text{We get, } \tan 2\theta = \frac{2I_{xy}}{I_{yy} - I_{xx}} = \frac{-2I_{xy}}{I_{xx} - I_{yy}} \quad \text{...(6.18)}$$

Using $I_{xx}$, $I_{yy}$ and $I_{xy}$ values, θ can be found from Eqn. (6.18), in other words, the principal axis can be located. Equation (6.18) gives two values for θ which differ by 90°, one being minimum principal axis and the other maximum principal axis.

Now,

$$I_{uu} = \Sigma v^2 dA$$

$$= \Sigma(y\cos\theta - x\sin\theta)^2 dA$$

$$= \Sigma(y^2\cos^2\theta - 2xy\sin\theta\cos\theta + x^2\sin^2\theta)dA$$

$$= \Sigma y^2\cos^2\theta\, dA - \Sigma 2xy\sin\theta\cos\theta\, dA + \Sigma x^2\sin^2\theta\, dA$$

$$= I_{xx}\cos^2\theta - I_{xy}\sin 2\theta + I_{yy}\sin^2\theta$$

$$= I_{xx}\frac{1+\cos 2\theta}{2} + I_{yy}\frac{1-\cos 2\theta}{2} - I_{xy}\sin 2\theta$$

$$= \frac{1}{2}(I_{xx} + I_{yy}) + \frac{(I_{xx} - I_{yy})}{2}\cos 2\theta - I_{xy}\sin 2\theta \quad \text{...(6.19)}$$

Representing Eqn. (6.18) for the direction of principal axis '2θ' by the sides of the triangles as shown in Fig. 6.68, we get

$$\sin 2\theta = \frac{I_{xy}}{\sqrt{\left(\frac{I_{yy} - I_{xx}}{2}\right)^2 + I_{xy}^2}} \quad \text{and} \quad \cos 2\theta = \frac{\frac{I_{yy} - I_{xx}}{2}}{\sqrt{\left(\frac{I_{yy} - I_{xx}}{2}\right)^2 + I_{xy}^2}}$$

Thus,

$$I_1 = I_{uu} = \frac{1}{2}(I_{xx} + I_{yy}) + \frac{I_{xx} - I_{yy}}{2}\,\frac{\frac{I_{yy} - I_{xx}}{2}}{\sqrt{\left(\frac{I_{yy} - I_{xx}}{2}\right)^2 + I_{xy}^2}} - I_{xy}\frac{I_{xy}}{\sqrt{\left(\frac{I_{yy} - I_{xx}}{2}\right)^2 + I_{xy}^2}}$$

$$= \frac{1}{2}(I_{xx} + I_{yy}) - \sqrt{\left(\frac{I_{yy} - I_{xx}}{2}\right)^2 + I_{xy}^2} \quad \text{...(6.20)}$$

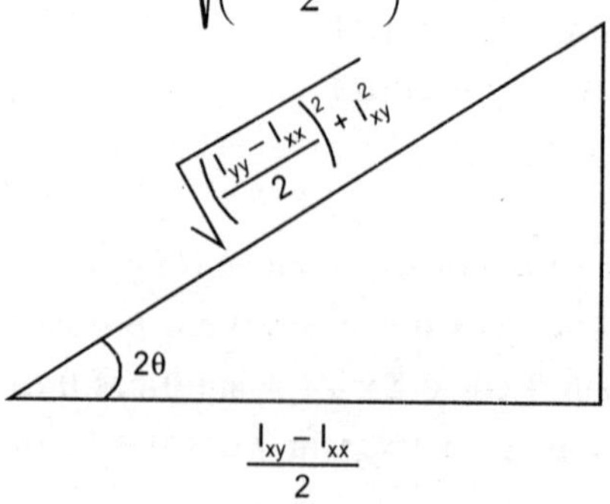


**Fig. 6.68**

Similarly,

$$I_2 = I_{vv} = \Sigma u^2 dA = \Sigma(x \cos\theta + y \sin\theta)^2 dA$$
$$= \Sigma x^2 \cos^2\theta\, dA + \Sigma 2xy \sin\theta \cos\theta\, dA + \Sigma y^2 \sin^2\theta\, dA$$
$$= I_{yy} \cos^2\theta\, dA + I_{xy}\, 2 \sin\theta \cos\theta + I_{xx} \sin^2\theta$$
$$= I_{yy} \frac{1+\cos 2\theta}{2} dA + I_{xx} \frac{1-\cos 2\theta}{2} + I_{xy} \sin 2\theta$$
$$= \frac{I_{yy}+I_{xx}}{2} + \frac{I_{yy}-I_{xx}}{2} \cos 2\theta + I_{xy} \sin 2\theta$$
$$= \frac{I_{xx}+I_{yy}}{2} + \frac{I_{yy}-I_{xx}}{2} \frac{\frac{I_{yy}-I_{xx}}{2}}{\sqrt{\left(\frac{I_{yy}-I_{xx}}{2}\right)^2 + I_{xy}^2}} + I_{xy} \frac{I_{xy}}{\sqrt{\left(\frac{I_{yy}-I_{xx}}{2}\right)^2 + I_{xy}^2}}$$
$$= \frac{I_{xx}+I_{yy}}{2} + \sqrt{\left(\frac{I_{yy}-I_{xx}}{2}\right)^2 + I_{xy}^2} \qquad \text{...(6.21)}$$

Thus, Eqn. (6.18) gives the direction of principal axis and Eqns. (6.20) and (6.21) give the principal moment of inertia.

**Example 6.24.** Determine the centroidal principal moment of inertia of the equal angle section 50 mm × 50 mm × 10 mm as shown in Fig. 6.69.

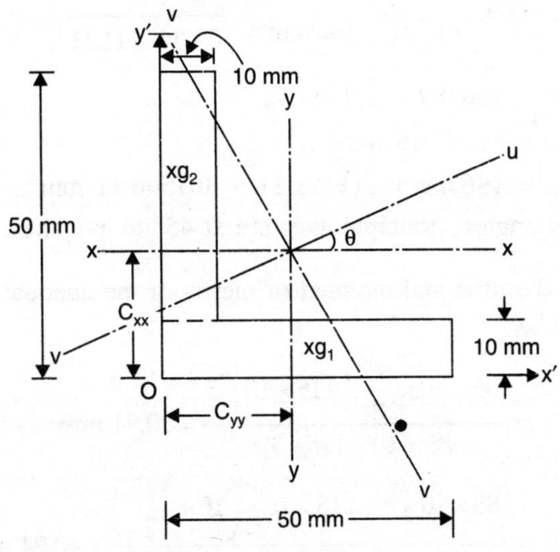


**Fig. 6.69**

**Solution.** Let $C_{xx}$ and $C_{yy}$ be the distances of centroid from $x'$ and $y'$ axes as shown in Fig. 6.69.

$$\therefore \qquad C_{xx} = \frac{50\times10\times5 + 40\times10\times(10+20)}{50\times10+40\times10} = 16.11 \text{ mm}$$

Similarly, $\qquad C_{yy} = 16.11$ mm

$$\therefore \qquad I_{xx} = \frac{1}{12} \times 50 \times 10^3 + 50 \times 10 \times (16.11 - 5.0)^2 + \frac{1}{12} \times 40^3 \times 10 + 40 \times 10 \times (30 - 16.11)^2$$

$$= 196389 \text{ mm}^4$$

Similarly, $I_{yy} = 196389 \text{ mm}^4$ [since it is equal angle].

Coordinates of the two simple rectangles w.r.t. centroidal axes are,

$$g_1[25 - 16.11, -(16.11 - 5)], \; g_2[-(16.11 - 5), 30 - 16.11]$$

*i.e.*, $g_1(8.89, -11.11)$ and $g_2(-11.11, 13.89)$

$$\therefore \qquad I_{xy} = A_1x_1y_1 + A_2x_2y_2$$

$$= 50 \times 10 \times 8.89 \times (-11.11) + 40 \times 10 \times (-11.11) \times 13.89$$

$$= -111111.11 \text{ mm}^4$$

Direction of principal axis is given by $\theta$, where

$$\tan 2\theta = \frac{2I_{xy}}{I_{yy} - I_{xx}} = \infty, \text{ since } I_{yy} = I_{xx}$$

$$\therefore \qquad 2\theta = 90° \quad \text{or} \quad \theta_1 = 45° \quad \text{and} \quad \theta_2 = 135°$$

$$\therefore \qquad I_{uu} = \frac{1}{2}\left(I_{xx} + I_{yy}\right) - \sqrt{\left(\frac{I_{yy} - I_{xx}}{2}\right)^2 + I_{xy}^2}$$

$$= \frac{1}{2}(196389 + 196389) - \sqrt{0 + 111111.11^2}$$

$$= 196389 - 111111.11$$

$$I_{uu} = 85277.89 \text{ mm}^4$$ **Ans.**

Similarly, $I_{vv} = 196389 + 111111.11 = 307500.11 \text{ mm}^4$. **Ans.**

Thus, in case of equal angles, principal axes are at 45° to $x$-$y$ axes.

**Example 6.25.** Determine the principal moment of inertia of the unequal angle section 125 × 85 × 10 mm as shown in Fig. 6.70.

**Solution.**

$$C_{xx} = \frac{85 \times 10 \times \frac{85}{2} + 115 \times 10 \times 5}{85 \times 10 + 115 \times 10} = 20.94 \text{ mm}$$

$$C_{yy} = \frac{85 \times 10 \times 5 + 115 \times 10 \times \left(10 + \frac{115}{2}\right)}{85 \times 10 + 115 \times 10} = 40.94 \text{ mm}$$

The coordinates $g_1$ and $g_2$ w.r.t. centroidal $x$, $y$ axes are

$$g_1 \left[\frac{85}{2} - 20.4, -(20.94 - 5)\right]; \left[-(20.94 - 5), 10 + \frac{115}{2} - 40.94\right]$$

*i.e.*, $g_1[22.466, -35.94]$; $g_2[-15.94, 26.56]$

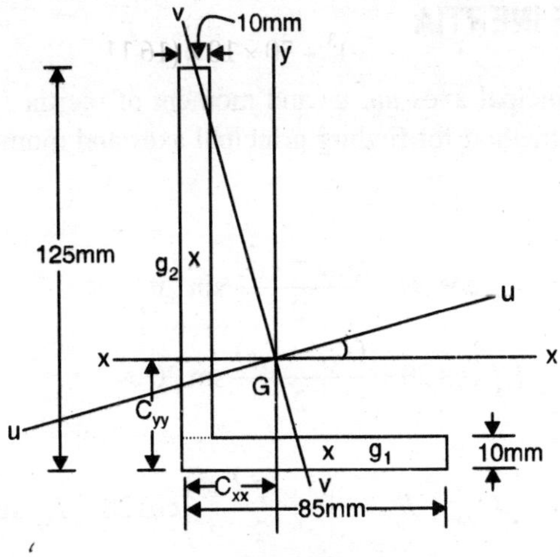


**Fig. 6.70**

$$I_{xx} = \frac{1}{12} \times 85 \times 10^3 + 85 \times 10 \times (-35.94)^2 + \frac{1}{12} \times 10 \times 115^3$$

$$+ 10 \times 115\,(26.56)^2$$

$$= 3183659 \text{ mm}^4$$

$$I_{yy} = \frac{1}{12} \times 10 \times 85^3 + 10 \times 85 \times 22.46^2 + \frac{1}{12} \times 115 \times 10^3$$

$$+ 115 \times 10\,(-15.95)^2$$

$$= 1242700 \text{ mm}^4$$

$$I_{xy} = A_1 x_1 y_1 + A_2 x_2 y_2$$

$$= 85 \times 10 \times 22.46\,(-35.94) + 115 \times 10 \times (-15.94) \times 26.54$$

$$= -\,1172635 \text{ mm}^4$$

$$\tan 2\theta = \frac{-1172635}{\dfrac{1242700 - 3183659}{2}} = 1.2083$$

$\therefore$ $$2\theta = 50.39$$

$\therefore$ $$\theta_1 = 25.19°;\ \theta_2 = 115.19°$$

$$I_1 = I_{uu} = \frac{I_{xx} + I_{yy}}{2} - \sqrt{\left(\frac{1242700 - 3183659}{2}\right)^2 + (1172635)^2}$$

$$= \frac{3183659 + 1242700}{2} - \sqrt{\left(\frac{1242700 - 3183659}{2}\right)^2 + (1172635)^2}$$

$$= 2213179 - 1522138$$

$\therefore$ $$I_1 = I_{uu} = 691041 \text{ mm}^4$$ **Ans.**

$$I_2 = I_{vv} = 2213179 + 1522138$$

$$I_{vv} = 3735317 \text{ mm}^4$$ **Ans.**

## 6.10 MOHR'S CIRCLE OF INERTIA

By drawing Mohr's circle, principal axes *uu*, *vv* and moment of inertia $I_{uu}$ and $I_{vv}$ can be easily determined. Thus, it is a graphical method for finding principal axes and moment of inertia about principal axes.

From Eqn. (6.17),

$$I_{uv} = I_{xy}\cos 2\theta - \frac{(I_{yy} - I_{xx})}{2}\sin 2\theta$$

$$= I_{xy}\cos 2\theta + \frac{(I_{xx} - I_{yy})}{2}\sin 2\theta \qquad \ldots(a)$$

From Eqn. (6.19),

$$I_{uu} = \frac{1}{2}(I_{xx} + I_{yy}) + \left(\frac{I_{xx} - I_{yy}}{2}\right)\cos 2\theta - I_{xy}\sin 2\theta$$

i.e.,

$$I_{uu} - \frac{1}{2}(I_{xx} + I_{yy}) = \frac{I_{xx} - I_{yy}}{2}\cos 2\theta - I_{xy}\sin 2\theta \qquad \ldots(b)$$

Squaring and adding Eqns. (*a*) and (*b*) and noting that $\cos^2 2\theta + \sin^2 2\theta = 1$, we get

$$\left(I_{uu} - \frac{(I_{xx} + I_{yy})}{2}\right)^2 + I_{uv}^2 = \left(\frac{I_{xx} - I_{yy}}{2}\right)^2 + I_{xy}^2$$

This is the equation of the circle of the form

$$(I_{uu} - a)^2 + I_{uv}^2 = R^2$$

where $a = \dfrac{I_{xx} + I_{yy}}{2}$ is the centre of circle

and $R = \sqrt{\left(\dfrac{I_{xx} - I_{yy}}{2}\right)^2 + I_{xy}^2}$ is the radius of the circle.

*A circle with centre at* $a = \dfrac{I_{xx} + I_{yy}}{2}$ *and radius* $R = \sqrt{\left(\dfrac{I_{xx} - I_{yy}}{2}\right)^2 + I_{xy}^2}$ *is a cartesian system with moment of inertia* ($I_{uu}$ and $I_{uv}$) *along one axis and product of inertia along another axis is known as Mohr's circle.*

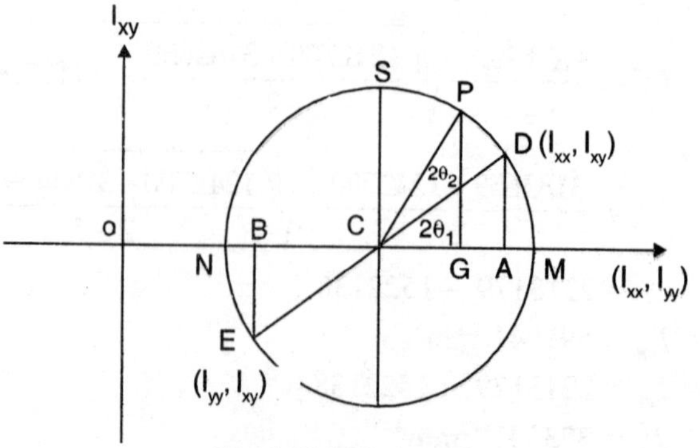


**Fig. 6.71** *Mohr's circle*

Figure 6.71 shows Mohr's circle for moment of inertias $I_{xx}$ and $I_{yy}$ and product moment of inertia $I_{xy}$.

To construct Mohr's circle:

- Select cartesian coordinate system with origin at $O$.
- Let $x$-axis represent moment of inertia and $y$-axis represent product moment of inertia.

Let $P$ be a point such that $\angle DCP = 2\theta$. To show that if coordinates represent the normal and tangential stresses on a plane which is at an angle $\theta$ to the plane of $I_{xx}$, drop perpendicular $PG$ on moment of inertia axis. Then

$$CG = CP \cos(2\theta_1 + 2\theta)$$
$$= R(\cos 2\theta_1 . \cos 2\theta - \sin 2\theta_1 \sin 2\theta)$$
$$= R\left(\frac{CA}{R}\cos 2\theta - \frac{AD}{R}\sin 2\theta\right)$$
$$= CA \cos 2\theta - AD \sin 2\theta$$
$$= \frac{I_{xx} - I_{yy}}{2}\cos 2\theta - I_{xy}\sin 2\theta$$

∴ Abscissa of $P$

$$OG = OC + CG$$
$$= \frac{I_{xx} + I_{yy}}{2} + \frac{I_{xx} - I_{yy}}{2}\cos 2\theta - I_{xy}\sin 2\theta$$
$$= I_{uu}$$

- Locate point $D$ to represent $I_{xx}$ and $I_{xy}$.
- Locate point $E$ to represent $I_{yy}$ and $I_{xy}$.
- Join $DE$, let the point of intersection with $x$-axis be $C$.
- $C$ as centre and radius $R = CD = CE$, draw a circle. This is Mohr's circle.

Let $DA$ and $EB$ be perpendiculars to $x$-axis (Moment of inertia axis) and $\angle DCA = 2\theta_1$

Then $$CE = CD = R$$
$$BC = CA = R \cos 2\theta_1$$

*i.e.,* $C$ is midpoint of $AB$.

∴ $$OC = OB + BC = OB + \frac{AB}{2}$$
$$= \frac{1}{2}(2OB + AB) = \frac{1}{2}(OB + OA)$$
$$= \frac{1}{2}\left(I_{yy} + I_{xx}\right)$$

Now, $$CA = \frac{1}{2}AB = \frac{1}{2}(OA - OB) = \frac{1}{2}(I_{xx} - I_{yy})$$

∴ $$R = \sqrt{CA^2 + AD^2} = \sqrt{\left(\frac{I_{xx} - I_{yy}}{2}\right)^2 + I_{xy}^2}$$

Thus, circle has centre at $\frac{I_{xx} - I_{yy}}{2}$ and Radius $= \sqrt{\left(\frac{I_{xx} - I_{yy}}{2}\right)^2 + I_{xy}^2}$

Ordinate of $P = PG$

$= R \sin (2\theta_1 + 2\theta)$

$= R \sin 2\theta_1 \,.\, \cos 2\theta + R \cos 2\theta_1 \,.\, \sin 2\theta$

$= AD. \cos 2\theta + CA \sin 2\theta$

$$= I_{xy} \cos 2\theta + \frac{I_{xx} - I_{yy}}{2} \sin 2\theta$$

$= I_{uv}$

Thus the coordinates of $P$ represent values of moment of inertia and product moment of inertia, which makes angle $\theta$ with the $x$-axis.

Hence the point $M$, $N$ and $S$ represent the values of maximum principal moment of inertia, minimum principal moment of inertia and maximum product moment of inertia. It may be observed on the axis of maximum product moment of inertia, moment of inertia is $\frac{I_{xx} + I_{yy}}{2}$.

It may also be noted that principal axis is at $\frac{1}{2}$ $\angle DCM$ to $x$-$x$ axis.

Using Mohr's circle method, Examples 6.25 and 6.26 may be solved.

## 6.11 CENTRE OF GRAVITY OF SOLIDS

Centre of gravity of solids may be found using Eqn. (6.1) which will be same as those found from Eqns. (6.2) and (6.3) if the mass is uniform. Hence centre of gravity of solids, centre of gravity of mass or centroid of volumes is the same for all solids with uniform mass. For standard solids, the centre of gravity may be found from first principles and the results obtained for standard solids may be used to find centre of gravity of composite solids. The procedure is illustrated with Examples 6.26 to 6.29.

**Example 6.26.** Locate the centre of gravity of the right circular cone of base radius $r$ and height $h$ as shown in Fig. 6.72.

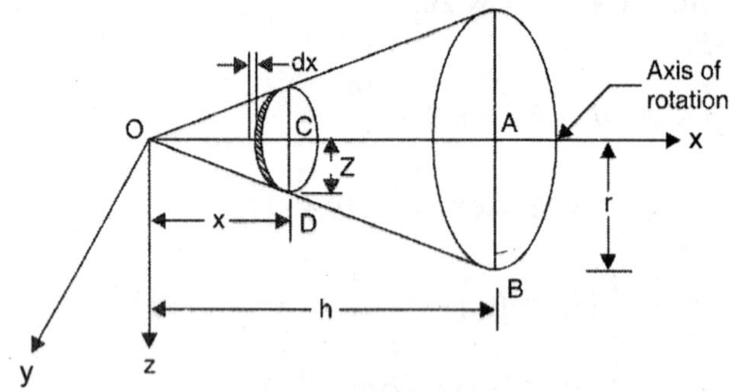


**Fig. 6.72**

**Solution.** Taking origin at the vertex of the cone and selecting the axis as shown in Fig. 6.72, it can be observed that due to symmetry the coordinates of centre of gravity $\bar{y}$ and $\bar{z}$ are equal to zero, *i.e.*, the centre of gravity lies on the axis of rotation of the cone. To find its distance $\bar{x}$ from the vertex, consider an elemental plate at a distance $x$. Let the thickness of the elemental plate be $dx$. From the similar triangles $OAB$ and $OCD$, the radius of elemental plate $z$ is given by

$$z = \frac{x}{h} r$$

∴ Volume of the elemental plate $dv$

$$dv = \pi z^2 dx = \pi x^2 \frac{r^2}{h^2} dx$$

If $\gamma$ is the unit weight of the material of the cone, then weight of the elemental plate is given by:

$$dW = \gamma \pi x^2 \frac{r^2}{h^2} dx \quad ...(i)$$

$$W = \int_0^h \gamma \frac{\pi r^2}{h^2} x^2 \, dx$$

$$= \gamma \frac{\pi r^2}{h^2} \left[ \frac{x^3}{3} \right]_0^h$$

$$= \gamma \pi \frac{r^2 h}{3} \quad ...(ii)$$

$$\left[ \textbf{Note: } \frac{\pi r^2 h}{3} \text{ is volume of cone.} \right]$$

Now, moment of the plate about $z$-axis

$$\int x \cdot dW = \int_0^h \gamma \frac{\pi r^2}{h^2} x^2 \cdot x \cdot dx$$

$$= \gamma \frac{\pi r^2}{h^2} \left[ \frac{x^4}{4} \right]_0^h$$

$$= \gamma \frac{\pi r^2 h^2}{4} \quad ...(iii)$$

If $\bar{x}$ is the distance of c.g. from Eqn. 6.1, we get

$$W\bar{x} = \int x \, dW$$

*i.e.,*

$$\frac{\pi r^2 h}{3} \bar{x} = \frac{\gamma \pi r^2 h^2}{4}$$

∴

$$\bar{x} = \frac{3}{4} h$$

Thus, in a right circular cone, centre of gravity lies at a distance $\frac{3}{4}h$ from vertex along the axis of rotation, *i.e.*, at a distance $\frac{h}{4}$ from the base.

**Example 6.27.** Determine the centre of gravity of a solid hemisphere of radius $r$ from its diametral axis.

**Solution.** Due to symmetry, centre of gravity lies on the axis of rotation. To find its distance $\bar{x}$ from the base along the axis of rotation, consider an elemental plate at a distance $x$ as shown in Fig. 6.73.

Now, $\qquad x^2 + z^2 = r^2$

$$z^2 = r^2 - x^2 \qquad (i)$$

Volume of elemental plate

$$dv = \pi z^2\, dx = \pi(r^2 - x^2)dx \qquad (ii)$$

∴ Weight of elemental plate

$$dw = \gamma dv = \gamma\pi(r^2 - x^2)dx \qquad (iii)$$

∴ Weight of hemisphere

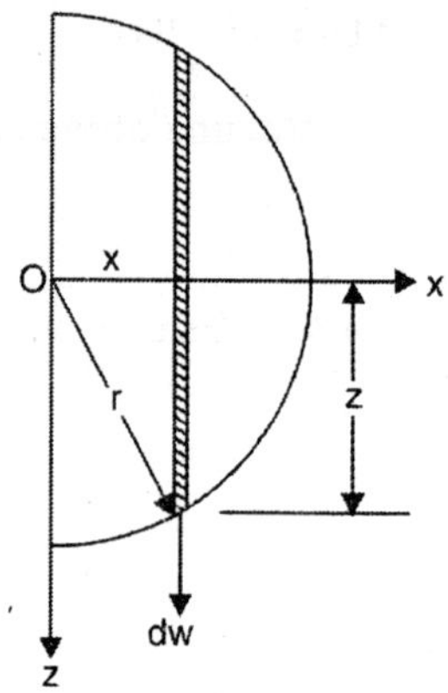


Fig. 6.73

$$W = \int dw = \int_0^r \gamma\pi\left(r^2 - x^2\right)dx$$

$$= \gamma\pi\left[r^2 x - \frac{x^3}{3}\right]_0^r$$

$$= \frac{2\gamma\pi r^3}{3} \qquad \ldots(iv)$$

Moment of weight about $z$-axis

$$= \int_0^r x\,dw$$

$$= \int_0^r x\,\pi\left(r^2 - x^2\right)dx$$

$$= \pi\left[r^2\frac{x^2}{2} - \frac{x^4}{4}\right]_0^r$$

$$= \frac{\pi r^4}{4} \qquad \ldots(v)$$

∴ $\bar{x}$, the distance of centre of gravity from base is given by:

$$W\bar{x} = \int_0^r x\,dw$$

*i.e.*, From Eqns. (*iv*) and (*v*) above, we get

$$\frac{2\gamma\pi r^3}{3}\bar{x} = \frac{\gamma\pi r^4}{4} \qquad \bar{x} = \frac{3}{8}r$$

Thus, the centre of gravity of a solid hemisphere of radius $r$ is at a distance $\frac{3}{8}r$ from its diametral axis.

**Example 6.28.** Determine the maximum height $h$ of the cylindrical portion of the body with hemispherical base as shown in Fig. 6.74, so that it is in stable equilibrium on its base.

**Solution.** The body will be stable on its base as long as its centre of gravity is in hemispherical base. The limiting case is when it is on the plane $x$-$x$ as shown in the figure.

Centroid lies on the axis of rotation.

Mass of cylindrical portion

$$m_1 = \pi r^2 h\rho, \text{ where } \rho \text{ is unit mass of material.}$$

Its centre of gravity $g_1$ is at a height

$$z_1 = \frac{h}{2} \text{ from } x\text{-axis.}$$

Mass of hemispherical portion

$$m_2 = \rho \frac{2\pi r^3}{3}$$

and its $CG$ is at a distance

$$z_2 = \frac{3r}{8} \text{ from } x\text{-}x \text{ plane.}$$

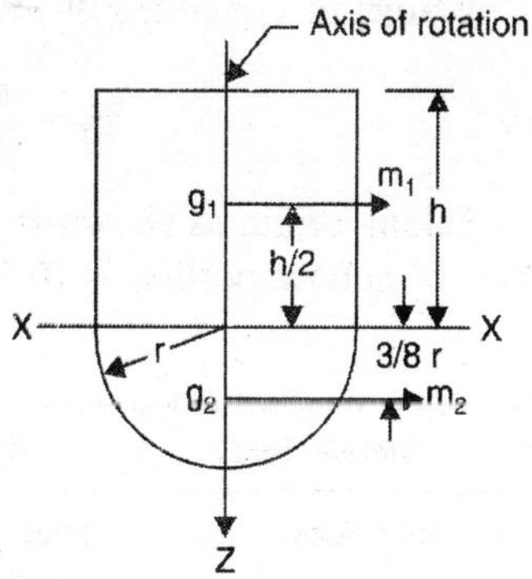


Fig. 6.74

Since centroid is to be on $x$-$x$ plane, $\bar{z} = 0$

*i.e.,* $$\Sigma m_i z_i = 0$$

$$\therefore \quad \frac{m_1 h}{2} - m_2 \frac{3}{8} r = 0$$

$$\pi r^2 h\rho \frac{h}{2} = \rho \frac{2\pi r^3}{3} \frac{3}{8} r$$

$$\therefore \quad h^2 = \frac{1}{2} r^2$$

$$\therefore \quad h = \frac{r}{\sqrt{2}} = 0.707 r$$ **Ans.**

**Example 6.29.** A concrete block of size 0.60 m × 0.75 m × 0.5 m is cast with a hole of diameter 0.2 m and depth 0.3 m as shown in Fig. 6.75. The hole is completely filled with steel balls weighing 2500 N. Locate the centre of gravity of the body. Take the weight of concrete = 25000 N/m$^3$.

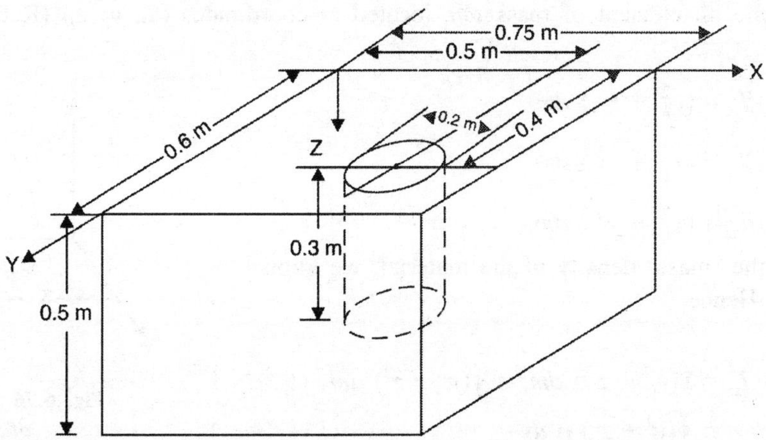


Fig. 6.75

**Solution.** Weight of solid concrete block:

$$W_1 = 0.6 \times 0.75 \times 0.5 \times 25000 = 5625 \text{ N}$$

Weight of concrete ($W_2$) removed for making hole:

$$W_2 = \frac{\pi}{4} \times 0.2^2 \times 0.3 \times 25000 = 235.62 \text{ N}$$

Taking origin as shown in the figure, the centre of gravity of solid block is (0.375, 0.3, 0.25) and that of hollow portion is (0.5, 0.4, 0.15). The following Table 6.5 may be prepared now.

**Table 6.5**

| Simple body | $W_i$ | $x_i$ | $W_i x_i$ | $y_i$ | $W_i y_i$ | $z_i$ | $W_i z_i$ |
|---|---|---|---|---|---|---|---|
| 1. Solid block | 5625 | 0.375 | 2109.38 | 0.3 | 1687.5 | 0.25 | 1406.25 |
| 2. Hole in concrete block | –235.62 | 0.5 | –117.81 | 0.4 | –94.25 | 0.15 | –35.34 |
| 3. Steel balls | 2500 | 0.5 | 1250.0 | 0.4 | 1000.0 | 0.15 | 375.0 |

$$\Sigma W_i = 7889.38 \quad \Sigma W_i x_i = 3241.57 \quad \Sigma W_i y_i = 2593.25 \quad \Sigma W_i z_i = 1745.91$$

$$\therefore \quad \bar{x} = \frac{\sum W_i x_i}{W} = \frac{\sum W_i x_i}{\sum W_i} \qquad \bar{x} = \frac{3241.57}{7889.38} = 0.411 \text{ m} \qquad \textbf{Ans.}$$

Similarly,

$$\bar{y} = \frac{2593.25}{7887.38} = 0.329 \text{ m} \qquad \textbf{Ans.}$$

$$\bar{z} = \frac{1745.91}{7889.38} = 0.221 \text{ m} \qquad \textbf{Ans.}$$

## 6.12 MASS MOMENT OF INERTIA

Mass moment of inertia of a body about an axis is defined as the sum total of product of its elemental masses and square of their distance from the axis.

Thus for the *i*th element of mass $dm_i$ located at coordinates ($x_i$, $y_i$, $z_i$) (Ref. Fig. 6.76), we have

$$dI_{xx} = (y_i^2 + z_i^2)\, dm_i$$

$$dI_{yy} = (x_i^2 + z_i^2)\, dm_i$$

and

$$dI_{zz} = (x_i^2 + y_i^2)\, dm_i$$

If ρ is the mass density of the material, we know that $dm = \rho dv$. Hence

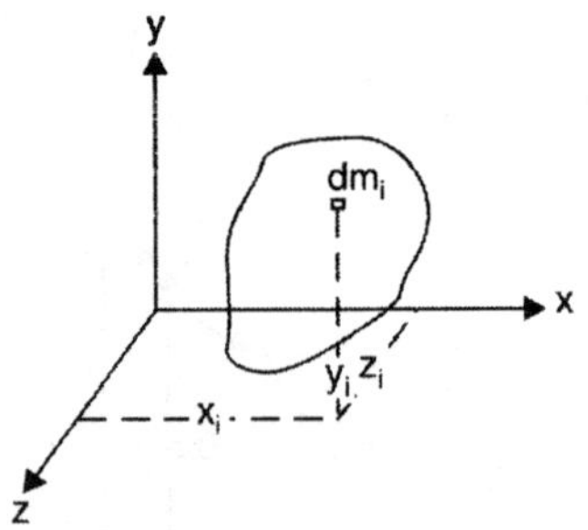


Fig. 6.76 *Mass moment of inertia*

$$I_{xx} = \Sigma(y_i^2 + z_i^2)\, dm_i = \oint (y^2 + z^2)\, dm$$
$$= \oint (y^2 + z^2)\, \rho\, dv$$
$$I_{yy} = \Sigma(x_i^2 + z_i^2) dm_i = \oint (x^2 + z^2)\, dm$$
$$= \oint (x^2 + z^2)\, \rho\, dv$$
$$I_{zz} = \Sigma(x_i^2 + y_i^2)\, dm_i = \oint (x^2 + y^2)\, dm$$

$$= \oint (x^2 + y^2)\, \rho dv \qquad ...(6.22)$$

The product inertia of mass of the body are similarly defined as,

$$I_{xy} = I_{yx} = \oint xy\ \rho\ dv$$

$$I_{xz} = I_{zx} = \oint xz\ \rho\ dv$$

and

$$I_{yz} = I_{zy} = \oint yz\ \rho\ dv$$

Thus for a body, there are nine components of inertia matrix as shown below:

$$I = \begin{bmatrix} I_{xx} & I_{xy} & I_{xz} \\ I_{yx} & I_{yy} & I_{yz} \\ I_{zx} & I_{zy} & I_{zz} \end{bmatrix}$$

In the above matrix as already shown,

$$I_{xy} = I_{yx};\ I_{xz} = I_{zx} \text{ and } I_{yz} = I_{zy}$$

We can also see that,

$$I_{xx} + I_{yy} + I_{zz} = \oint 2(x^2 + y^2 + z^2)\ \rho\ dv = \oint 2r^2 \rho\ dv$$

where $r$ is the position vector $x\mathbf{i} + y\mathbf{j} + z\mathbf{k}$. Since whatever be the orientation of the coordinate system $x, y, z$ at the origin, sum of the inertia at that origin is constant.

**Radius of Gyration**

*Radius of Gyration* is that distance which when squared and multiplied with total mass of the body gives the mass moment of inertia of the body. Thus if $I$ is moment of inertia of a body of mass $M$ about an axis, then its radius of gyration about that axis is given by the relation

$$I = Mk^2 \text{ or } k = \sqrt{\frac{I}{M}} \qquad ...(6.23)$$

A physical meaning may be assigned to the radius of gyration as the distance at which the entire mass can be assumed to be concentrated such that the moment of inertia of the actual body and the concentrated mass are the same.

## 6.13 DETERMINATION OF MASS MOMENT OF INERTIA FROM FIRST PRINCIPLES

Mass moment of inertia of simple bodies can be determined from its definition. The following steps may be followed:

1. Take a general element.
2. Write the expression for mass of the element, $dm$, and its distance, $r$, from the axis.
3. Integrate the term $r^2 dm$ between suitable limits such that the entire mass of the body is covered. The procedure is illustrated with the examples given below.

**Example 6.30.** Determine the mass moment of inertia of a uniform rod of length $L$ about its: (*a*) centroidal axis normal to rod and (*b*) axis at the end of the rod and normal to it.

**Solution.** *(a) About Centroidal Axis Normal to Rod:*

Consider an elemental length $dx$ at a distance $x$ from centroidal axis $y$-$y$ as shown in Fig. 6.77. Let the mass of rod be $m$ per unit length. Then mass of the element $dm = m\ dx$.

$$\therefore \qquad I = \int_{-\frac{L}{2}}^{\frac{L}{2}} x^2 dm = \int_{-\frac{L}{2}}^{\frac{L}{2}} x^2 \times m\, dx$$

$$= m\left[\frac{x^3}{3}\right]_{-\frac{L}{2}}^{\frac{L}{2}} = \frac{mL^3}{12}$$

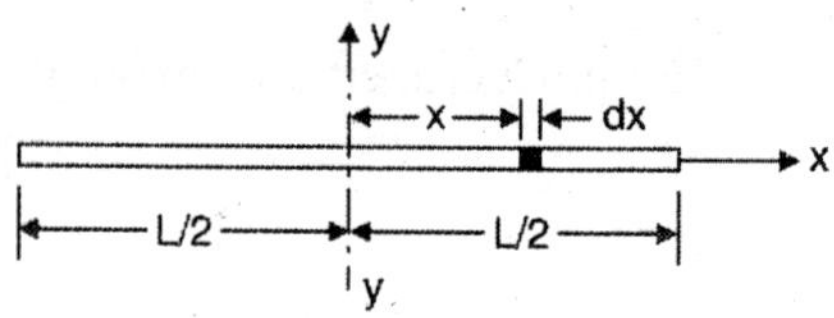


**Fig. 6.77**

i.e.,
$$I = \frac{ML^2}{12}$$

where $M = mL$ is total mass of the rod.

*(b) About Axis at the End of the Rod and Normal to it:*

Consider an element of length $dx$ at a distance $x$ from end as shown in Fig. 6.78.

Moment of inertia of the rod about $y$-axis:

$$I = \int_0^L x^2 m\, dx$$

$$= m\left[\frac{x^3}{3}\right]_0^L = \frac{mL^3}{3}$$

$$I = \frac{ML^2}{3}$$

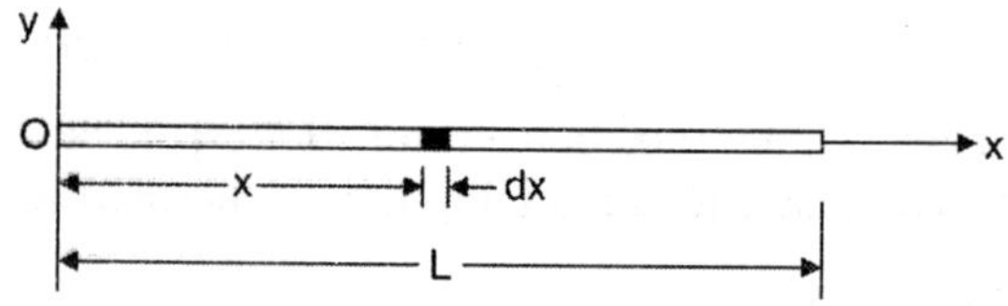


**Fig. 6.78**

**Example 6.31.** Determine the moment of inertia of a rectangular plate of size $a \times b$ and thickness $t$ about its centroidal axis.

**Solution.** *To find $I_{xx}$:*

Consider an elemental strip of width $dy$ at a distance $y$ from $x$-axis as shown in Fig. 6.79 (*a*). Mass of the element:

$$dm = \rho\, b \times t \times dy \qquad (\rho\text{-unit mass of the material})$$

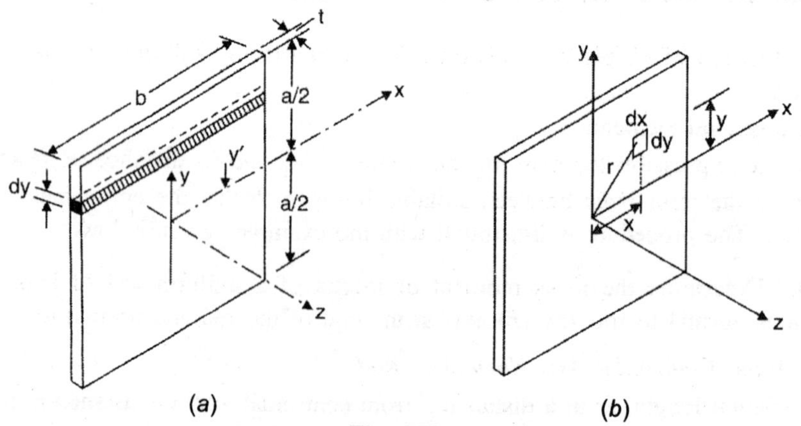


**Fig. 6.79**

$$\therefore \quad I_{xx} = \int_{-a/2}^{a/2} y^2 dm = \int_{-a/2}^{a/2} y^2 \rho\, b\, t\, dy$$

$$= \rho\, bt \left[\frac{y^3}{3}\right]_{-a/2}^{a/2}$$

$$= \frac{\rho bta^3}{12}$$

but Mass of the plate $\quad M = \rho\, bta$

$$\therefore \quad I_{xx} = \frac{Ma^2}{12}$$

*To find $I_{yy}$:*

Taking an elemental strip parallel to $y$-axis, it can be easily shown that:

$$I_{yy} = \frac{Mb^2}{12}$$

*To find $I_{zz}$:*

Consider an element of size $dx\ dy$ and thickness $t$ as shown in Fig. 6.79 (*b*). Now

$$r^2 = x^2 + y^2$$

$$I_{zz} = \int r^2 dm$$

$$= \int (x^2 + y^2)\, dm$$

$$= \int x^2\, dm + \int y^2\, dm$$

$$= I_{xx} + I_{yy}$$

$$= \frac{Ma^2}{12} + \frac{Mb^2}{12}$$

$$I_{zz} = \frac{1}{12} M\left(a^2 + b^2\right)$$

**Example 6.32.** Find the moment of inertia of circular plate of radius $R$ and thickness $t$ about its centroidal axis.

**Solution.** Consider an elemental area $r\ d\theta\ dr$ and thickness $t$ as shown in Fig. 6.80.

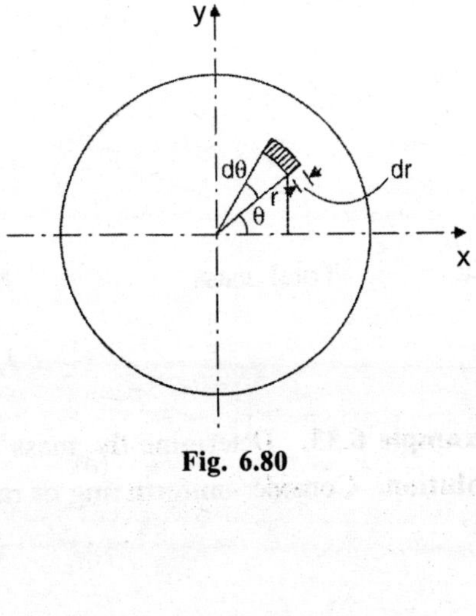


**Fig. 6.80**

$dm$ = Mass of the element = $\rho r\ d\theta\ dr\ t = \rho\ t\ r\ d\theta\ dr$

Its distance from $x$-axis = $r \sin\theta$

$$\therefore \quad I_{xx} = \oint (r \sin\theta)^2 dm$$

$$= \int_0^R \int_0^{2\pi} r^2 \sin^2\theta \rho\ t\ r\ d\theta\ dr$$

$$= \rho t \int_0^R \int_0^{2\pi} r^3 \left( \frac{1-\cos 2\theta}{2} \right) dr d\theta$$

$$= \rho t \int_0^R \frac{r^3}{2} \left[ \theta - \frac{\sin 2\theta}{2} \right]_0^{2\pi} dr$$

$$= \rho t \int_0^R \frac{r^3}{2} \times 2\pi \, dr$$

$$= \rho t \pi \left[ \frac{r^4}{4} \right]_0^R$$

$$= \rho t \frac{\pi R^4}{4}$$

Mass of the plate $M = \rho \times \pi R^2 t$

$\therefore$ $$I_{xx} = \frac{MR^2}{4}$$

Similarly, $$I_{yy} = \frac{MR^2}{4}$$

Actually $I = \dfrac{MR^2}{4}$ is moment of inertia of circular plate about any diametral axis in the plate.

To find $I_{zz}$, consider the same element.

$$I_{zz} = \oint r^2 dm = \int_0^R \int_0^{2\pi} r^2 \, \rho \, t \, r \, dr \, d\theta$$

$$= \rho t \int_0^R r^3 [\theta]_0^{2\pi} \, dr$$

$$= \rho t \int_0^R 2\pi r^3 \, dr$$

$$= \rho t \, 2\pi \left[ \frac{r^4}{4} \right]_0^R$$

$$= \rho t \, 2\pi \frac{R^4}{4} = \rho t \frac{\pi R^4}{2}$$

but Total mass $$M = \rho t \pi R^2$$

$\therefore$ $$I_{zz} = \frac{MR^2}{2}$$

**Example 6.33.** Determine the mass moment of inertia of a circular ring of uniform cross-section.

**Solution.** Consider uniform ring of radius $R$ as shown in Fig. 6.81. Let its mass per unit length be $m$.

Hence, Total mass $M = 2\pi R\ m$.

Consider an elemental length $ds = Rd\theta$ at an angle $\theta$ to the diametral axis $x$-$x$. The distance of the element from $x$-axis is $R \sin \theta$ and mass of element is $m\ r\ d\theta$.

$$\therefore \quad I = \int_0^{2\pi} (R\sin\theta)^2\ mRd\theta$$

$$= R^3 m \int_0^{2\pi} \sin^2\theta\ d\theta$$

$$= mR^3 \int_0^{2\pi} \left(\frac{1-\cos 2\theta}{2}\right) d\theta$$

$$= \frac{mR^3}{2}\left[\theta - \frac{\sin 2\theta}{2}\right]_0^{2\pi} = mR^3\pi$$

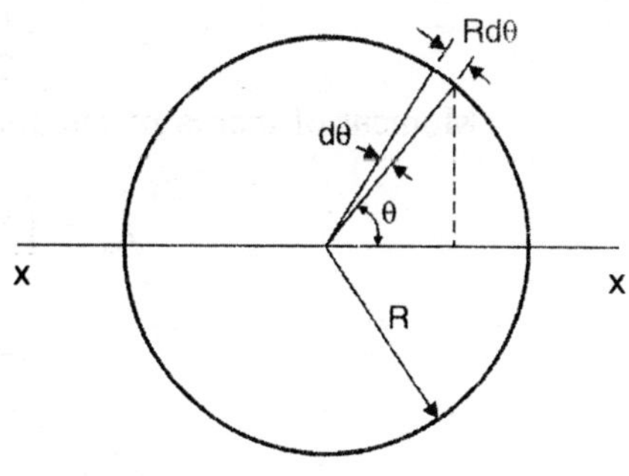


Fig. 6.81

but $\quad M = 2\pi Rm$

$$\therefore \quad I = \frac{MR^2}{2}$$

**Example 6.34.** Find the mass moment of inertia of the solid cone of height $h$ and base radius $R$ about:

1. its axis of rotation, and
2. an axis through vertex normal to the axis of rotation.

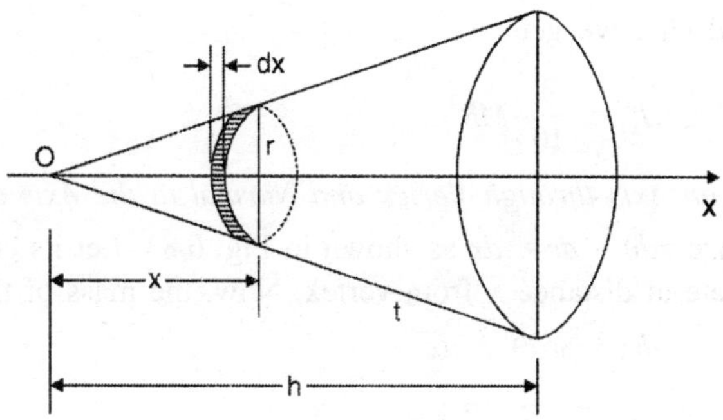


**Fig. 6.82**

**Solution.** *Moment of Inertia about its Axis of Rotation:*

Consider an elemental plate at distance $x$. Let its radius be $r$ and thickness $dx$.

Mass of the elemental plate = $\rho\pi r^2\ dx$

But, the moment of inertia of circular plate about normal axis through its centre is

$$= \frac{1}{2} \times \text{Mass} \times \text{Square of radius}$$

$$= \frac{1}{2} \times \rho\pi r^2\ dx \cdot r^2 = \rho \frac{\pi r^4 dx}{2} \quad ...(i)$$

But now, $$r = \left(\frac{x}{h}\right) R$$

$\therefore$ Moment of inertia of the elemental plate about $x$-axis

$$= \rho \frac{\pi}{2} R^4 \frac{x^4}{h^4} dx \qquad ...(ii)$$

$\therefore$ Moment of inertia of the cone about $x$-axis,

$$I_{xx} = \int_0^h \frac{\rho\pi}{2} R^4 \frac{x^4}{h^4} dx$$

$$= \frac{\rho\pi}{2} \frac{x^4}{h^4} \left[\frac{x^5}{5}\right]_0^h$$

$$I_{xx} = \frac{\pi R^4 h}{10} \qquad ...(iii)$$

but Mass of the cone $M = \int_0^h \pi r^2\, dx = \int_0^h \pi R^2 \frac{x^2}{h^2} dx$

$$= \frac{\pi R^2}{h^4} \left[\frac{x^3}{3}\right]_0^h$$

$$= \frac{\pi R^2 h}{3} \qquad ...(iv)$$

From Eqns. (*iii*) and (*iv*), we get

$$I_{xx} = \frac{3}{10} MR^2$$

*Moment of Inertia about an Axis through Vertex and Normal to the Axis of Rotation:*

Consider an element of size $rd\theta \times dr \times dx$ as shown in Fig. 6.83. Let its coordinates be $x$, $y$, $z$ and $z_1$ be the radius of the plate at distance $x$ from vertex. Now, the mass of this element

$$dm = \rho r d\theta\, dr\, dx$$

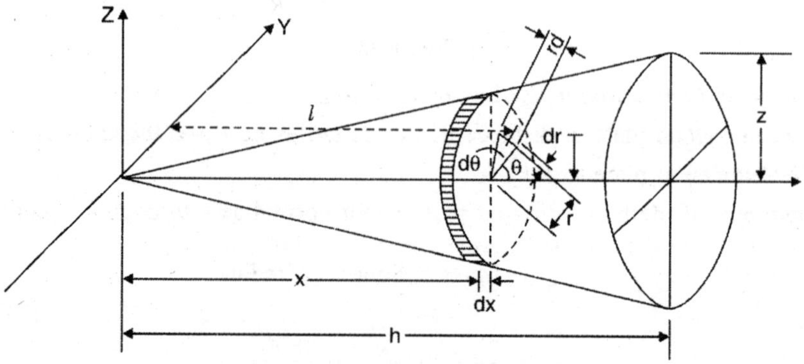


**Fig. 6.83**

and its distance from $y$-axis is given by:

$$l = \sqrt{x^2 + z^2}$$

$\therefore$ Moment of inertia of the cone about $y$-axis:

$$I_y = \oint l^2\, dm$$

$$= \int_0^h \int_0^{z_1} \int_0^{2\pi} \left(x^2 + z^2\right) \rho r\, d\theta\, dr\, dx$$

but $z = r \sin\theta$

$$\therefore \quad I_{yy} = \int_0^h \int_0^{z_1} \int_0^{2\pi} \left(x^2 + r^2 \sin^2\theta\right) \rho\, r\, d\theta\, dr\, dx$$

$$= \int_0^h \int_0^{z_1} \int_0^{2\pi} \rho\left(x^2 r + r^3 \frac{1 - \cos 2\theta}{2}\right) d\theta\, dr\, dx$$

$$= \int_0^h \int_0^{z_1} \rho\left[x^2 r\theta + \frac{r^3}{2}\left(\theta - \frac{\sin 2\theta}{2}\right)\right]_0^{2\pi} dr\, dx$$

$$= \int_0^h \int_0^{z_1} \rho\left[2\pi x^2 r + \frac{r^3}{2} 2\pi\right] dr\, dx$$

$$= \int_0^h \rho\left[\pi x^2 r^2 + \pi \frac{r^4}{4}\right]_0^{z_1} dx$$

$$= \int_0^h \rho\left[\pi x^2 z_l^2 + \frac{\pi z_l^4}{4}\right]_0^{z_1} dx$$

but $z_l = \dfrac{x}{h} R$

$$\therefore \quad I_{yy} = \int_0^h \rho\left[x^2 \frac{x^2}{h^2} R^2 + \frac{\pi}{4} \frac{x^4}{h^4} R^4\right] dx$$

$$= \rho\left[\pi \frac{x^5}{5h^2} R^2 + \frac{\pi x^5}{20h^4} R^4\right]_0^h$$

$$= \rho\left[\frac{\pi}{5} h^3 R^2 + \frac{\pi}{20} h R^4\right] = \frac{\pi\rho R^2 h}{5}\left(h^2 + \frac{R^2}{4}\right)$$

but Mass of cone $M = \dfrac{1}{3} \rho\pi R^2 h$

$$\therefore \quad I_{yy} = \frac{3M}{5}\left(h^2 + \frac{R^2}{4}\right)$$

**Example 6.35.** Determine the moment of inertia of a solid sphere of radius R about its diametral axis.

**Solution.** Consider an elemental plate of thickness $dy$ at distance $y$ from the diametral axis as shown in Fig. 6.84. Radius of this elemental circular plate $x$ is given by the relation:

$$x^2 = R^2 - y^2 \quad \text{...(i)}$$

$\therefore$ Mass of the elemental plate $dm = \rho\, \pi x^2 dy$

$$= \rho\, \pi\, (R^2 - y^2) dy \quad \text{...(ii)}$$

Moment of inertia of this circular plate element about $y$-axis is:

$$= \frac{1}{2} \times \text{Mass} \times \text{Square of radius}$$

$$= \frac{1}{2} \times \rho\, \pi x^2 dy \times x^2$$

$$= \rho\, \frac{\pi}{2}\, x^4\, dy$$

$$= \rho \frac{\pi}{2}\, (R^2 - y^2)^2\, dy$$

$$= \rho\, \frac{\pi}{2}\, (R^4 - 2R^2y^2 + y^4)\, dy$$

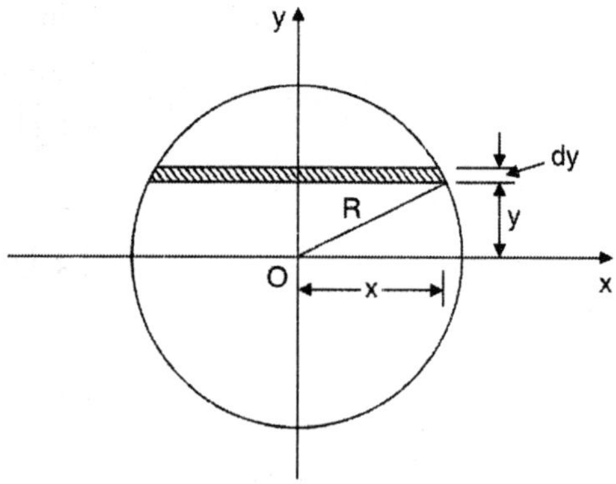


Fig. 6.84

$\therefore$
$$I_{yy} = 2\int_0^R \rho \frac{\pi}{2}\, (R^4 - 2R^2y^2 + y^4) dy$$

$$= \rho\pi \left[ R^4 y - \frac{2R^2y^3}{3} + \frac{y^5}{5} \right]_0^R$$

$$= \rho\pi R^5 \left[ 1 - \frac{2}{3} + \frac{1}{5} \right]$$

$$= \frac{8}{15} \rho\pi R^5 \quad \text{...(iii)}$$

but Mass of sphere, $M = 2\int_0^R dm = 2\int_0^R \rho\pi\, x^2 dy$

$$= 2\int_0^R \rho\pi \left( R^2 - y^2 \right) dy$$

$$= 2\rho\pi \left( R^2 y - \frac{y^3}{3} \right)_0^R = 2\rho\pi \left[ R^3 - \frac{R^3}{3} \right]$$

$$M = \frac{4\rho R^3}{3} \quad \text{...(iv)}$$

From Eqns. (*iii*) and (*iv*) above, we get

$$I_{yy} = \frac{2}{5}\, MR^2$$

## QUESTIONS

**6.1** Distinguish between centre of gravity and centroid.

**6.2** Prove that centroid of an area lies on the axis of symmetry, if it exists.

**6.3** Determine the centroid of the following figures from first principles:

(*a*) A triangle from its base.

(*b*) A semi-circle from its diametral axis.

(*c*) A sector of a circle from its centre along symmetric axis.

**6.4** Define the terms: Polar moment of inertia and radius of gyration of a cross-section.

**6.5** State and prove perpendicular axis and parallel axis theorem of moment of inertia.

**6.6** Determine the moment of inertia of the following standard sections from first principles:

(*a*) A triangle about its base.

(*b*) A circle about its diametral axis.

(*c*) A semicircle about its diametral axis.

**6.7** Explain the term product moment and principal moment of inertia of a plane area.

**6.8** Derive the expressions for principal moment of inertia of a plane area in terms of known values of moment of inertia and product moment of inertia about *x-x* and *y-y* axes.

**6.9** Explain how Mohr's circle is to be drawn for finding principal moment of inertia.

**6.10** Determine the centre of gravity of the following solids from first principles:

(*a*) Right circular cone of base radius $r$ and height $h$ from its apex.

(*b*) Solid hemisphere of radius $r$ from its centroid axis.

**6.11** Explain the term mass moment of inertia and radius of gyration of a solid.

**6.12** Determine the mass moment of inertia of the following:

(*a*) A rectangular plate of size $a \times b$ and thickness '$t$' about its centroidal axes.

(*b*) A circular plate of radius $R$ and thickness '$t$' about its centroidal axes.

## EXERCISES

**6.1** Determine the centroid of the built-up section in Fig. 6.85. Express the coordinates of centroid with respect to $x$ and $y$ axes shown. [**Ans.** $\bar{x}$ = 48.91 mm; $\bar{y}$ = 61.30 mm]

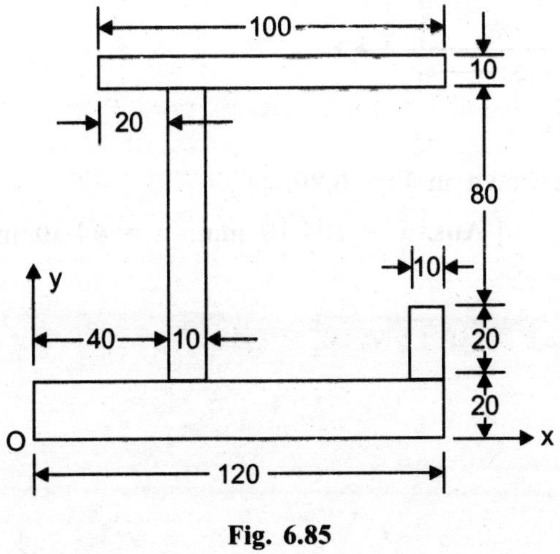


Fig. 6.85

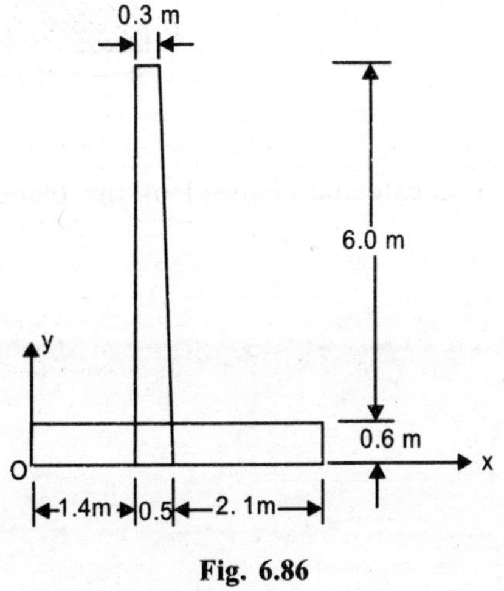


Fig. 6.86

**6.2** Determine the centroid of the reinforced concrete retaining wall section as shown in Fig. 6.86. **[Ans.** $\overline{x}$ = 1.848 m; $\overline{y}$ = 1.825 m]

**6.3** Find the coordinates of the centroid of the shaded area with respect to the axes shown in Fig. 6.87. **[Ans.** $\overline{x}$ = 43.98 mm; $\overline{y}$ = 70.15 mm]

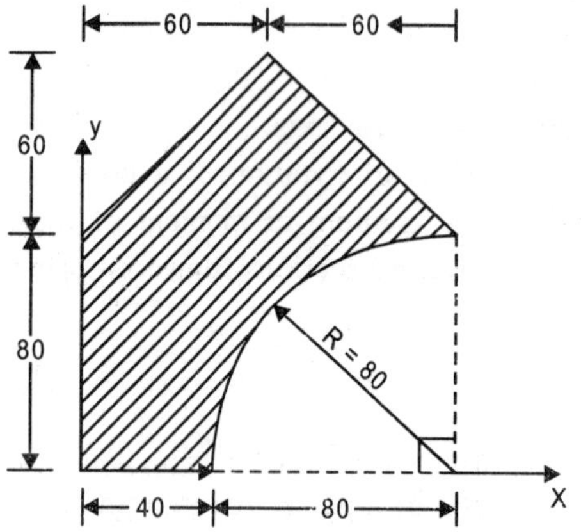


**Fig. 6.87**

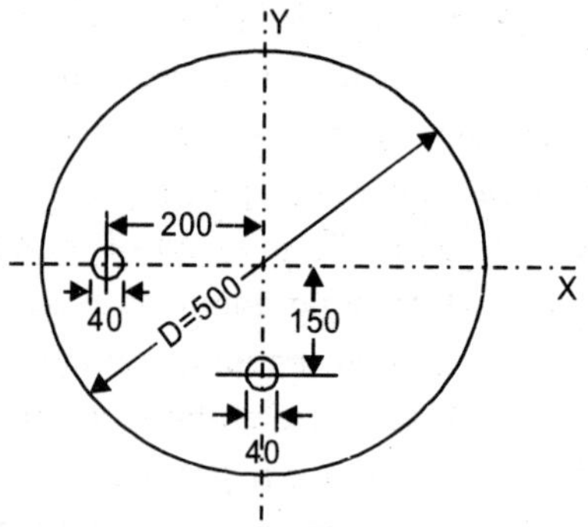


**Fig. 6.88**

**6.4** A circular plate of uniform thickness and of diameter 500 mm as shown in Fig. 6.88 has two circular holes of 40 mm diameter each. Where should a 80 mm diameter hole be drilled so that the centre of gravity of the plate will be at the geometric centre.

**[Ans.** $\overline{x}$ = 50 mm; $\overline{y}$ = 37.5 mm]

**6.5** With respect to the coordinate axes *x* and *y*, locate the centroid of the shaded area shown in Fig. 6.89. **[Ans.** $\overline{x}$ = 97.47 mm; $\overline{y}$ = 70.69 mm]

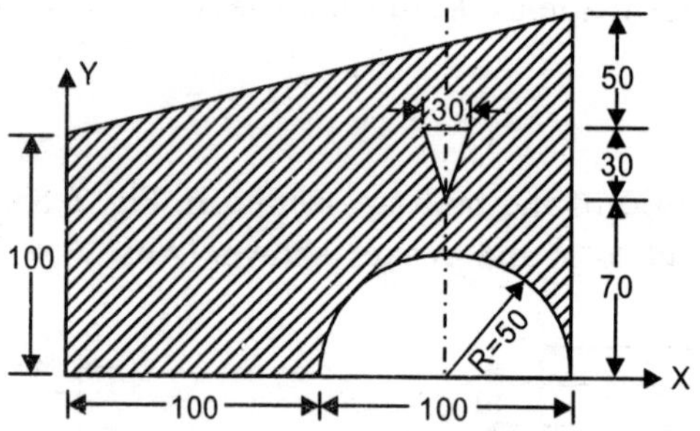


**Fig. 6.89**

**6.6** Locate the centroid of the plane area as shown in Fig. 6.90.

[ **Ans.** $\overline{x}$ = 104.10 mm; $\overline{y}$ = 44.30 mm]

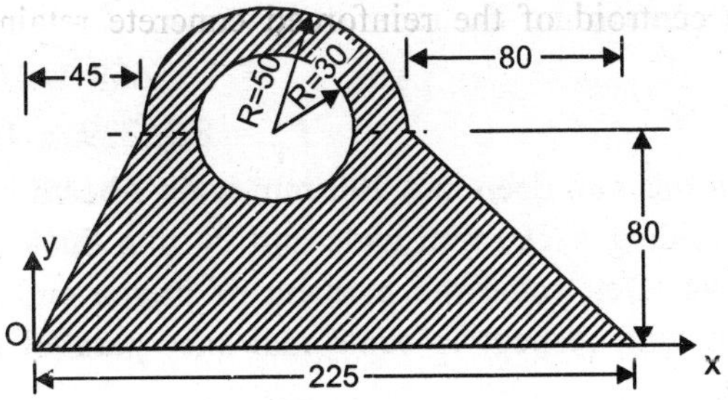


Fig. 6.90

**6.7** Determine the coordinates of the centroid of shaded area shown in Fig. 6.91 with respect to the corner point $O$. Take $x$ = 40 mm. [**Ans.** $\overline{x}$ = 2600 mm; $\overline{y}$ = 113.95 mm]

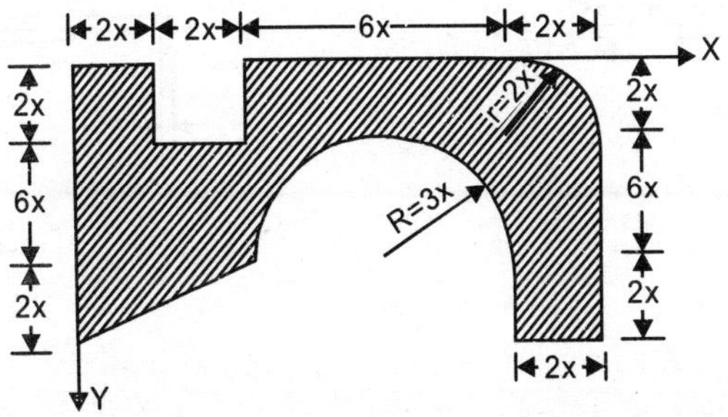


Fig. 6.91

**6.8** $ABCD$ is a square section of sides 100 mm. Determine the ratio of moment of inertia of the section about centroidal axis parallel to a side to that about diagonal $AC$.

[**Ans.** 1]

**6.9** The cross-section of a rectangular hollow beam is as shown in Fig. 6.92. Determine the polar moment of inertia of the section about centroidal axes.

[**Ans.** $I_{xx}$ = 1,05,38667 mm$^4$; $I_{yy}$ = 49,06667 mm$^4$; $I_{zz}$ = 1,54,45334 mm$^4$]

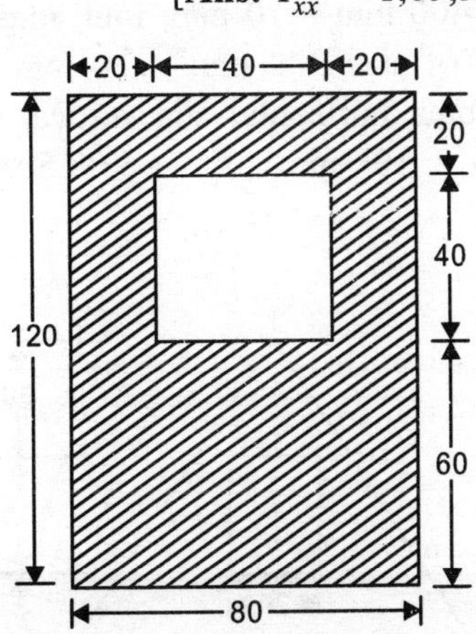


Fig. 6.92

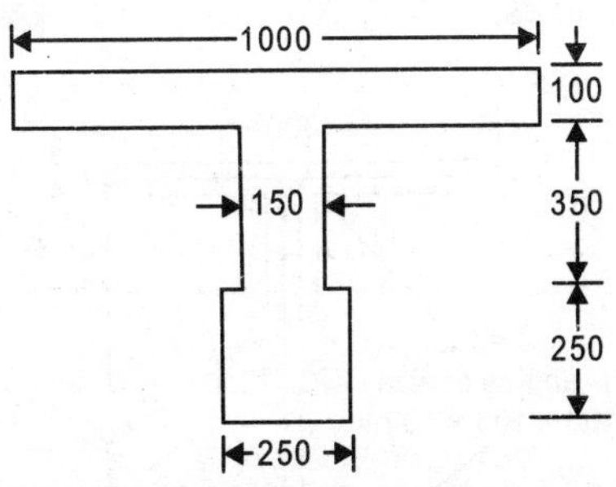


Fig. 6.93

**6.10** The cross-section of a prestressed concrete beam is shown in Fig. 6.93. Calculate the moment of inertia of this section about the centroidal axes parallel to and perpendicular to top edge. Also determine the radii of gyration.

[**Ans.** $I_{xx}$ = 1.15668 × $10^{10}$ mm$^4$; $k_{xx}$ = 231.95 mm; $I_{yy}$ = 8.75729 × $10^9$ mm$^4$; $k_{yy}$ = 201.82 mm]

**6.11** The strength of a 400 mm deep and 200 mm wide *I*-beam of uniform thickness 10 mm, is increased by welding a 250 mm wide and 20 mm thick plate to its upper flanges as shown in Fig. 6.94. Determine the moment of inertia and the radii of gyration of the composite section with respect to centroidal axes parallel to and perpendicular to the bottom edge *AB*.

[**Ans.** $I_{xx}$ = 3.32393 × $10^8$ mm$^4$; $k_{xx}$ = 161.15 mm; $I_{yy}$ = 3,94,06667 mm$^4$; $k_{yy}$ = 55.49 mm]

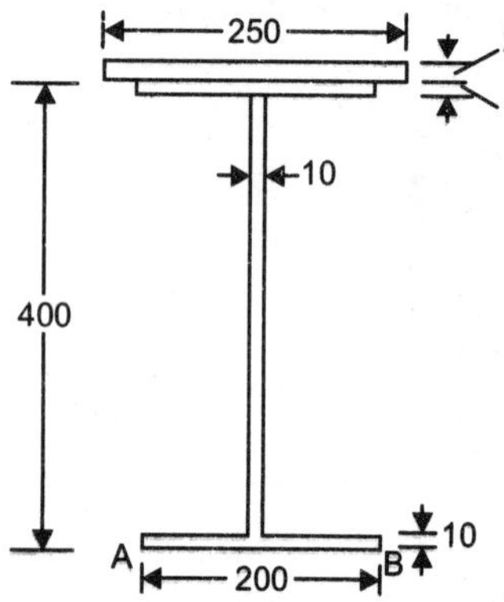


**Fig. 6.94**

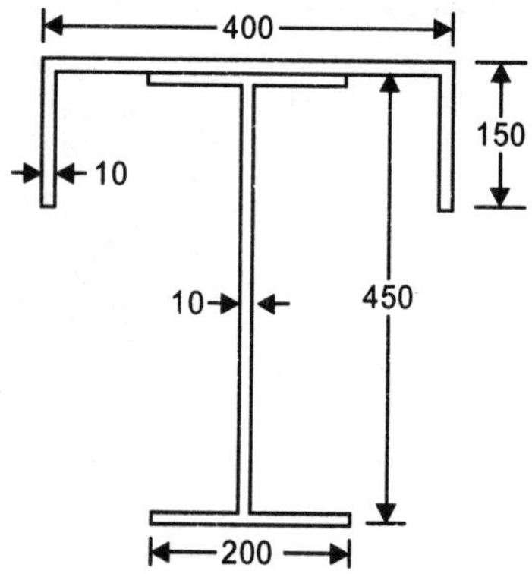


**Fig. 6.95**

**6.12** The cross-section of a gantry girder is as shown in Fig. 6.95. It is made up of an *I*-section of depth 450 mm, flange width 200 mm and a channel of size 400 mm × 150 mm. Thickness of all members is 10 mm. Find the moment of inertia of the section about the horizontal centroidal axis. [**Ans.** $I_{xx}$ = 4.2198 × $10^8$ mm$^4$]

**6.13** A plate girder is made up of a web plate of size 400 mm × 10 mm, four angles of size 100 mm × 100 mm × 10 mm and cover plates of size 300 mm × 10 mm as shown in Fig. 6.96. Determine the moment of inertia about horizontal and vertical centroidal axes. [**Ans.** $I_{xx}$ = 5.35786 × $10^8$ mm$^4$; $I_{yy}$ = 6,08,50,667 mm$^4$]

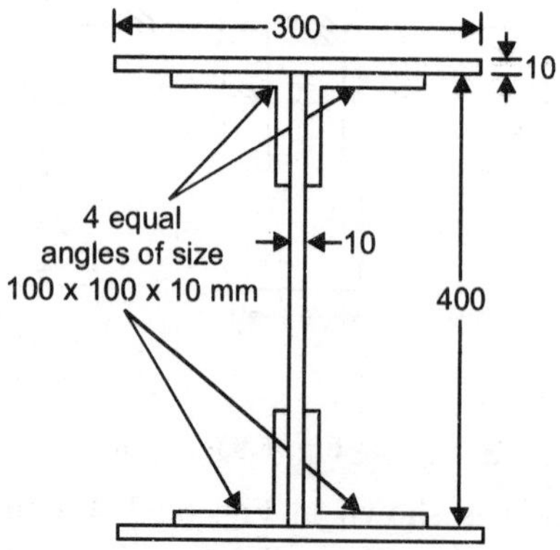


**Fig. 6.96**

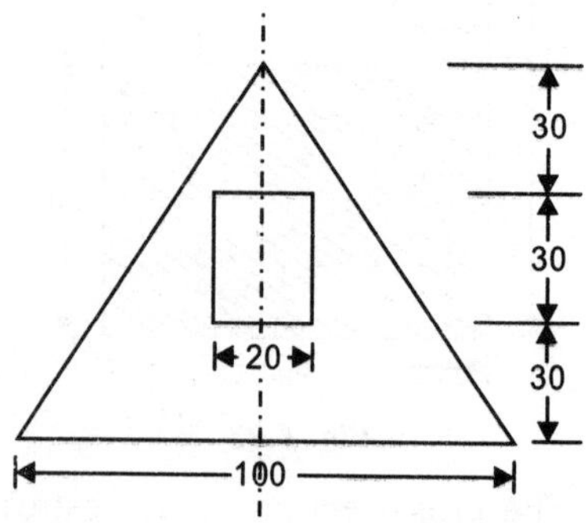


**Fig. 6.97**

**6.14** Determine the moment of inertia and radii of gyration of the area shown in Fig. 6.97 about the base *A—B* and the centroidal axis parallel to *AB*.

[**Ans.** $I_{AB}$ = 48,15000 mm$^4$; $I_{xx}$ = 18,24231 mm$^4$; $k_{AB}$ = 35.14 mm; $k_{xx}$ = 21.63 mm]

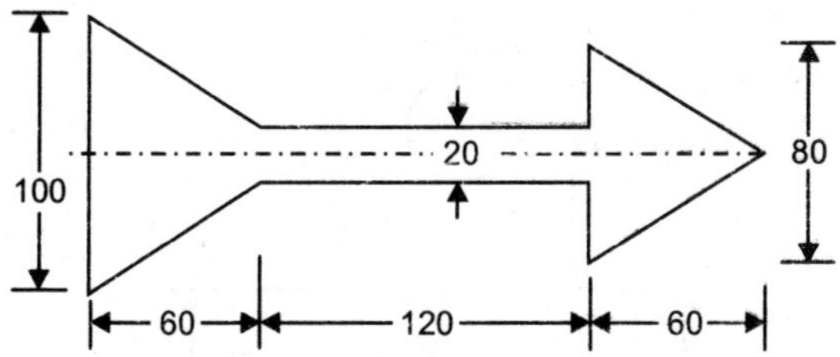


**Fig. 6.98**

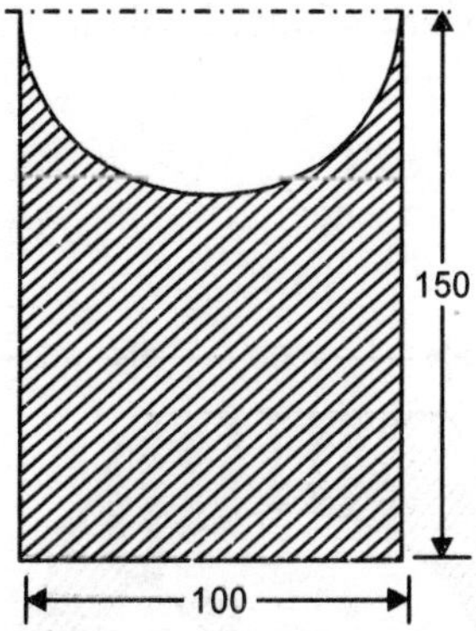


**Fig. 6.99**

**6.15** Determine the moment of inertia of the section shown in Fig. 6.98 about the vertical centroidal axis. [**Ans.** $I_{yy}$ = 5,03,82857 mm$^4$]

**6.16** A semi-circular cut is made in rectangular wooden beam as shown in Fig. 6.99. Determine the polar moment of inertia of the section about the centroidal axis.

[**Ans.** $I_{xx}$ = 3,35,81456 mm$^4$; $I_{yy}$ = 1,00,45631 mm$^4$; $I_{zz}$ = 2,20,98980 mm$^4$]

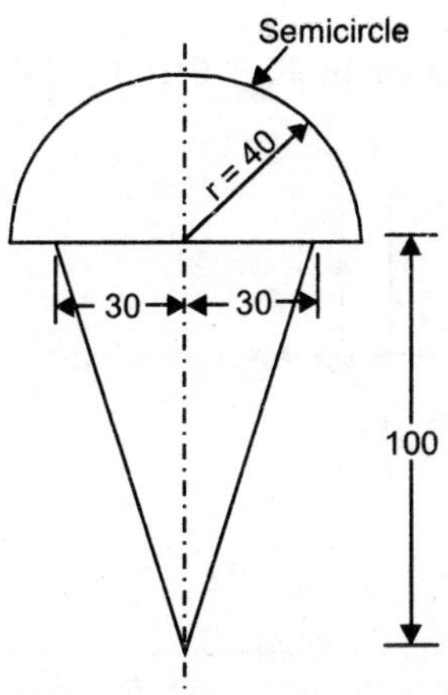


**Fig. 6.100**

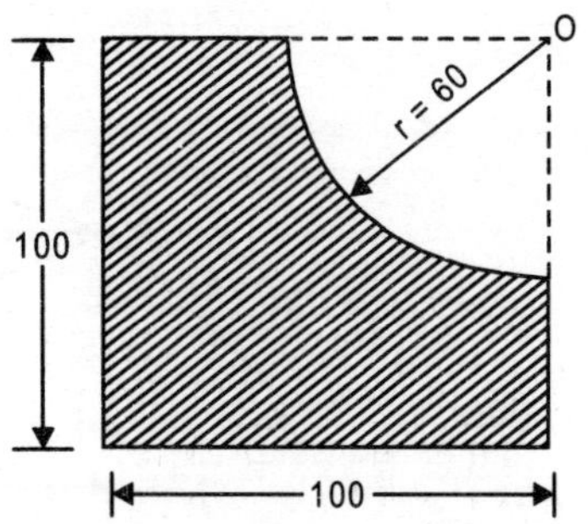


**Fig. 6.101**

**6.17** Determine the moment of inertia of the section shown in Fig. 6.100 about the horizontal centroidal axis. Also find the moment of inertia of the section about the symmetrical axis. Hence find the polar moment of inertia.

[**Ans.** $I_{xx}$ = 54,09046 mm$^4$; $I_{yy}$ = 14,55310 mm$^4$; $I_{zz}$ = 68,64356 mm$^4$]

**6.18** The cross-section of a machine part is as shown in Fig. 6.101. Determine its moment of inertia and radius of gyration about the horizontal centroidal axis.

[**Ans.** $I_{xx}$ = 52,49090.85 mm$^4$; $k_{xx}$ = 27.05 mm]

**6.19** The cross-section of a plain concrete culvert is as shown in Fig. 6.102. Determine the moment of inertia about the horizontal centroidal axis.

[**Ans.** $I_{xx} = 5.45865 \times 10^{10}$ mm$^4$]

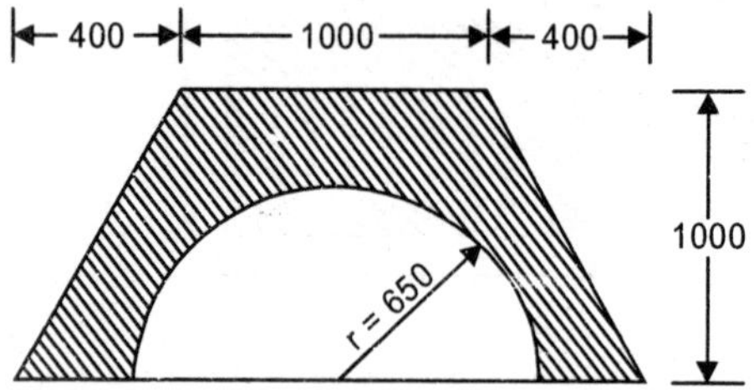


**Fig. 6.102**

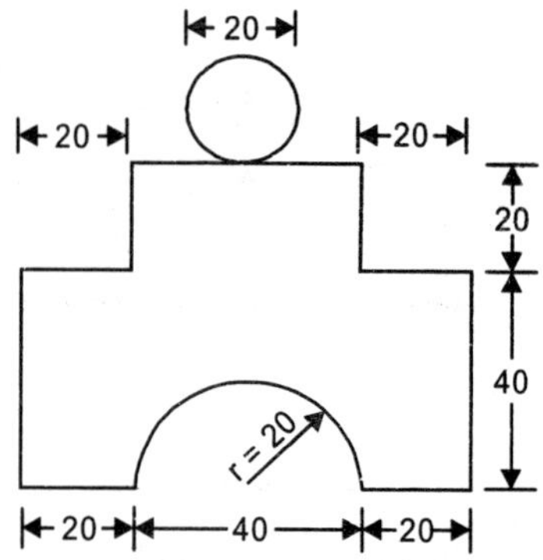


**Fig. 6.103**

**6.20** Determine the centroid of the built-up section shown in Fig. 6.103 and find the moment of inertia and radius of gyration about the horizontal centroidal axis.

[**Ans.** $I_{xx}$ = 12,67942 mm$^4$; $k_{xx}$ = 18.55 mm]

**6.21** Determine the centre of gravity of the pyramid shown in Fig. 6.104. $\left[\textbf{Ans. } x = \frac{3}{4}h\right]$

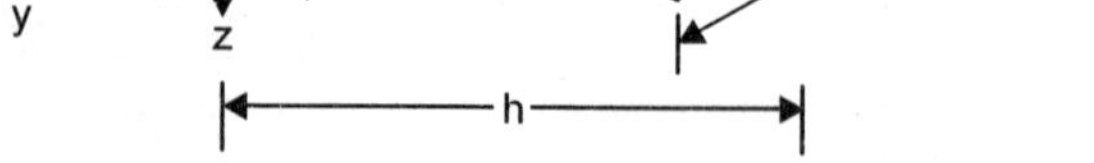


**Fig. 6.104**

**6.22** A steel ball of diameter 150 mm rests centrally over a concrete cube of size 150 mm. Determine the centre of gravity of the system, taking weight of concrete = 25000 N/m$^3$ and that of steel 80000 N/m$^3$. [**Ans.** 168.94 mm from base]

**6.23** Locate the centre of gravity of the wire as shown in Fig. 6.105. Portion *BC* is in *x-y* plane and semicircle *CD* is parallel to *x-z* plane.

[**Ans.** $\overline{x}$ = 124.02 mm; $\overline{y}$ = 110.41 mm; $\overline{z}$ = 11.28 mm]

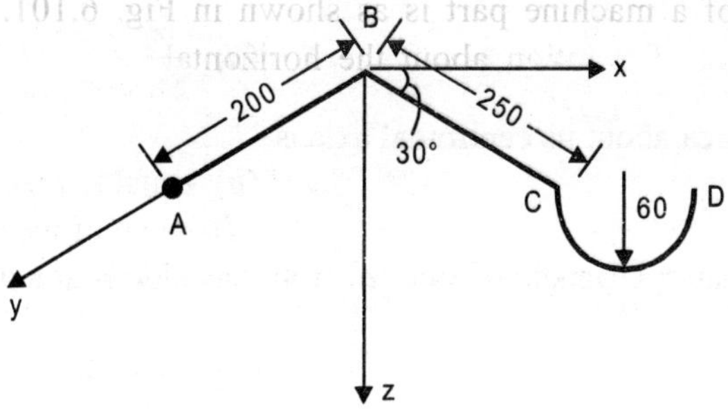


Fig. 6.105

**6.24** Determine the moment of inertia of the link shown in Fig. 6.106 about *x*-axis.

[**Ans.** 17.23 × $10^8$ ρ units]

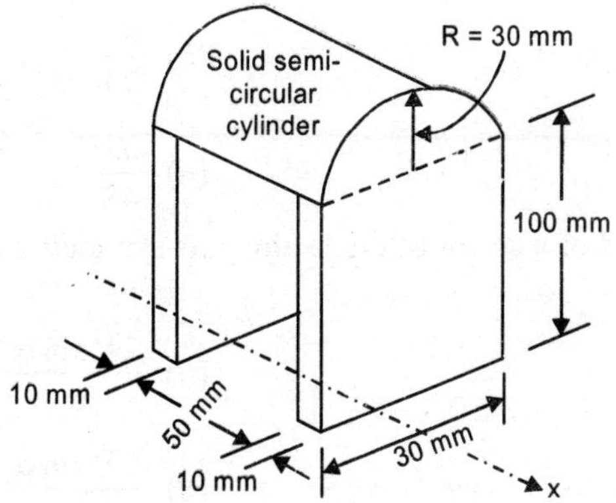


Fig. 6.106

## OBJECTIVE TYPE QUESTIONS

**Select the correct option in the following:**

**6.1** Centroid is the point through which:

(*a*) resultant of self weight of body acts.
(*b*) the moment of area about any axis through that point is zero.
(*c*) second moment of area is maximum.
(*d*) None of the above.

**6.2** Centroid of a triangle of base width *b* and height *h* from its apex is

(*a*) $\frac{2h}{3}$  (*b*) $\frac{h}{3}$

(*c*) $\frac{h}{6}$  (*d*) $\frac{h}{12}$

**6.3** The centroid of a semicircle of radius *r* from its diametral axis is

(*a*) $\frac{3r}{4\pi}$  (*b*) $\frac{4r}{3\pi}$

(c) $\frac{4r}{\pi}$ (d) $\frac{3r}{8\pi}$

**6.4** Moment of a given area about its centroidal axis is

(a) twice the area (b) equal to area

(c) zero (d) None of the above

**6.5** The centroid of equilateral triangle of side '$a$' from any side is at a distance

(a) $\frac{a\sqrt{3}}{2}$ (b) $\frac{a\sqrt{2}}{3}$

(c) $\frac{a}{2\sqrt{3}}$ (d) $\frac{a}{3\sqrt{2}}$

**6.6** The first moment of a semicircle about its diametral axis is

(a) $\frac{d^3}{12}$ (b) $\frac{d^3}{24}$

(c) $\frac{d^3}{36}$ (d) $\frac{d^3}{48}$

**6.7** The distance of centroid of a sector of circle along its symmetric axis measured from its centre $O$ is

(a) $\frac{R\sin\alpha}{\alpha}$ (b) $\frac{2R\sin\alpha}{\alpha}$

(c) $\frac{R\sin\alpha}{2\alpha}$ (d) $\frac{2R\sin\alpha}{3\alpha}$

**6.8** A is a triangle of base width $b$ and height $h$. $B$ is a triangle of base width $2b$ and height $2h$. If distance of centroid of $A$ from base is 30 mm. Then the distance of centroid of $B$ from its base is

(a) 30 mm (b) 60 mm

(c) 90 mm (d) 120 mm

**6.9** If an area is symmetric about $y$-axis and origin is on this axis, then coordinate of the centroid is

(a) $\bar{x} = 0$ (b) $\bar{y} = 0$

(c) $\bar{x} = 0, \bar{y} = 0$ (d) None of the above

**6.10** The moment of inertia about $x$-$x$ axis is

(a) $\Sigma x^2 dA$ (b) $\Sigma y^2 dA$

(c) $\Sigma x dA$ (d) $\Sigma y dA$

**6.11** If $I_G$ is the moment of inertia of a rectangle about its centroidal axis parallel to the base and $I_{AB}$ is the moment of inertia about its base,

(a) $I_G < I_{AB}$ (b) $I_G = I_{AB}$

(c) $I_G > I_{AB}$ (d) None of the above

**6.12** The unit of moment of inertia of an area is

(a) $m^2$ (b) $m^3$

(c) $m^4$ (d) kg-$m^4$.

**6.13** The cross-section of block $A$ has width $b$ and depth $d$; and that of block $B$ has width $b$ and depth $2d$. Then the ratio of moment of inertia of block $B$ to block $A$ about their centroidal axis parallel to base is

(a) 2 (b) 4
(c) 8 (d) 16

**6.14** Block $A$ has cross-sectional area $b \times d$ while that of block $B$ are $2b \times 2d$. Then the ratio of centroidal moment of inertia parallel to the base is

(a) 4 (b) 8
(c) 16 (d) 32

**6.15** Area of cross-section of block $A$ is $2b \times 2d$. Block $B$ has cross-section of size $2b \times 2d$, but a central portion of size $b \times d$ removed. The ratio of moment of inertia of block $B$ to block $A$ about centroidal axis parallel to base is

(a) $\frac{1}{4}$ (b) $\frac{1}{2}$
(c) $\frac{3}{4}$ (d) $\frac{15}{16}$

**6.16** The moment of inertia of a rectangle of size $a \times b$ about its base is

(a) $\frac{bd^3}{12}$ (b) $\frac{bd^3}{8}$
(c) $\frac{bd^3}{4}$ (d) $\frac{bd^3}{3}$

**6.17** Moment of inertia of a triangle about its centroidal axis parallel to base is

(a) $\frac{bh^3}{12}$ (b) $\frac{bh^3}{24}$
(c) $\frac{bh^3}{36}$ (d) $\frac{bh^3}{64}$

**6.18** Moment of inertia of a quadrant of a circle about its diametral axis is

(a) $\frac{\pi d^4}{64}$ (b) $\frac{\pi d^4}{128}$
(c) $\frac{\pi d^4}{256}$ (d) None of the above

**6.19** The radius of gyration of a circle of radius $r$ is

(a) $\frac{r}{2}$ (b) $\frac{r}{4}$
(c) $\frac{4r}{3\pi}$ (d) $\frac{2r}{3\pi}$

**6.20** Unit of radius of gyration is

(a) m (b) $m^2$
(c) $m^3$ (d) $m^4$

**6.21** The moment of inertia of a square of side '*a*' about its centroidal axis parallel to a side is

(*a*) $\frac{a^4}{4}$ (*b*) $\frac{a^4}{8}$

(*c*) $\frac{a^4}{12}$ (*d*) $\frac{a^4}{36}$

**6.22** The moment of inertia of an equilateral triangle of side '*a*' about its centroidal axis parallel to the base is

(*a*) $\frac{a^4}{32}$ (*b*) $\frac{a^4}{64}$

(*c*) $\frac{a^4}{96}$ (*d*) $\frac{a^4\sqrt{3}}{96\sqrt{2}}$

**6.23** Polar moment of inertia of a square of side '*a*' about its centroid is

(*a*) $\frac{a^4}{12}$ (*b*) $\frac{a^4}{8}$

(*c*) $\frac{a^4}{6}$ (*d*) None of the above

**6.24** The unit of mass moment of inertia is

(*a*) kg/m (*b*) $m^4$

(*c*) kg-$m^2$ (*d*) None of the above

**6.25** Moment of inertia of an area about an axis, which is not a centroidal axis can be found with the help of

(*a*) Pappus theorem (*b*) Parallel axis theorem

(*c*) Perpendicular axis theorem (*d*) None of the above

**6.26** Mass moment of inertia of a cone about its axis of rotation is

(*a*) $\frac{3}{10}MR^2$ (*b*) $\frac{MR^2}{2}$

(*c*) $\frac{MR^3}{12}$ (*d*) None of the above

**6.27** Moment of inertia of a solid sphere of radius *R* about its diametral axis is

(*a*) $\frac{MR^2}{2}$ (*b*) $\frac{2}{5}MR^2$

(*c*) $\frac{3}{10}MR^2$ (*d*) None of the above

**6.28** Mass moment of inertia of a cylinder about its axis of rotation is

(*a*) $\frac{MR^2}{2}$ (*b*) $\frac{2MR^2}{5}$

(*c*) $\frac{3}{10}MR^2$ (*d*) None of the above

CHAPTER

# 7

# VIRTUAL WORK METHOD

So far in this book, we saw application of Newtonian mechanics to many engineering mechanics problems. There were many scientists before and after Newton who found solutions to mechanics problems by completely different approach. Jean Bernoulli (350 BC) and Leonardo da Vinci (1435) are the names associated with this school of thought. They imagined a small displacement of the bodies and found out the work done by various forces in the system and said if the body is in equilibrium, then the total work done by the system should be zero. This approach is called 'virtual work method.' The word *virtual* is used since in reality there is no actual displacement. The term *virtual* is defined by Webster as "being in essence of effect but not in fact."

The work done is virtual if the displacements are virtual or forces acting are virtual. Hence in virtual work approach, we may have the following two approaches:

(*i*) Principles of virtual displacements,

(*ii*) Principles of virtual forces.

In this chapter, we consider the principle of virtual displacements.

## 7.1 WORK DONE BY FORCES AND MOMENTS

Work done by a force = Force × Distance moved in the direction of force.

In Fig. 7.1, various forces acting on a particle are shown. If the particle moves a distance $s$ in $x$ direction, from $A$ to $B$, then the work done by various forces are as given below:

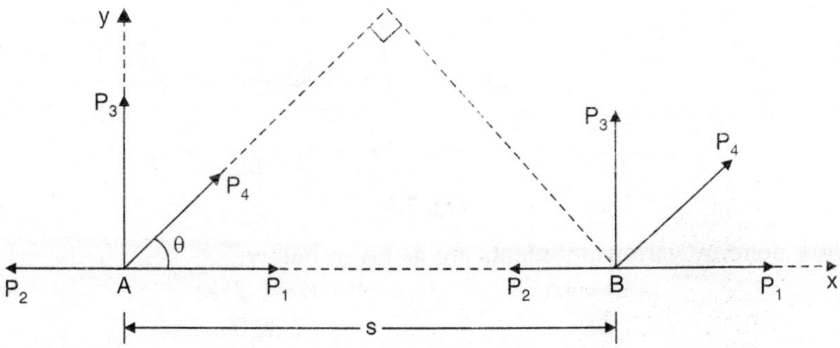


**Fig. 7.1**

| *Forces* | *Work done* |
|---|---|
| $\mathbf{P}_1$ | $P_1 s$ |
| $\mathbf{P}_2$ | $P_2(-s) = -P_2 s$ |
| $\mathbf{P}_3$ | $P_3 \times 0 = 0$ |
| $\mathbf{P}_4$ | $P_4 s \cos\theta$ |

Similarly if the particle moves from $A$ to $C$ in vertical direction (Fig. 7.2), the work done by various forces is as given below:

| *Forces* | *Work done* |
|---|---|
| $\mathbf{P}_1$ | $P_1 \times 0 = 0$ |
| $\mathbf{P}_2$ | $P_2 \times 0 = 0$ |
| $\mathbf{P}_3$ | $P_3 \times s' = P_3 s'$ |
| $\mathbf{P}_4$ | $P_4 \times s' \sin\theta$ |

If the displacement is at a particular angle, say $\theta$ to the direction of the force, as is the case of force $\mathbf{P}_4$ in the above cases, it will be convenient to find the perpendicular and parallel components of the forces to displacement and then find the work done. The component of $\mathbf{P}_4$ in $x$ and $y$ directions are:

$$P_{4x} = P_4 \cos\theta$$
$$P_{4y} = P_4 \sin\theta$$

In Fig. 7.1

Displacement in $x$ direction $= s$

Displacement in $y$ direction $= 0$

$\therefore$ The work done by force $\mathbf{P}_4$, is

$$= P_{4x}\, s + P_{4y}\, 0 = P_{4x}\, s = Ps \cos\theta$$

Similarly in Fig. 7.2,

Displacement in $x$ direction $= 0$

Displacement in $y$ direction $= s'$

$\therefore$ Hence, Work done by $\mathbf{P}_4 = P_{4x} \times 0 + P_{4y}\, s'$

$$= P_4 \sin\theta\, s' = P_4\, s' \sin\theta$$

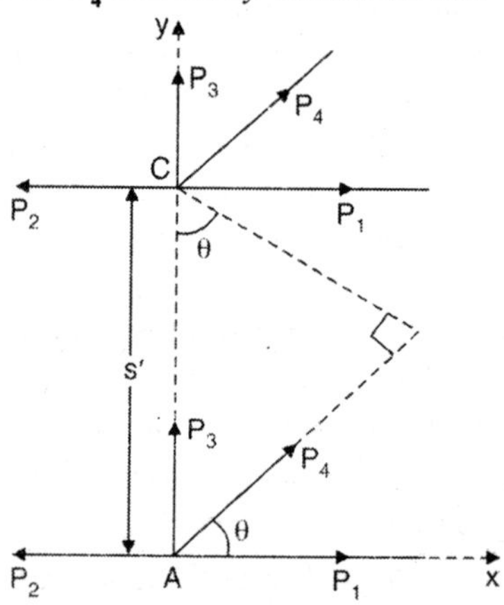

Fig. 7.2

On the same lines, we can say that work done by a moment is equal to the product of the moment and the angular rotation in the direction of the moment. In Fig. 7.3($a$), the moments $\mathbf{M}_1$, $\mathbf{M}_2$ and $\mathbf{M}_3$ are acting on a body which rotates by angle $\theta$ as shown in Fig. 7.3($b$).

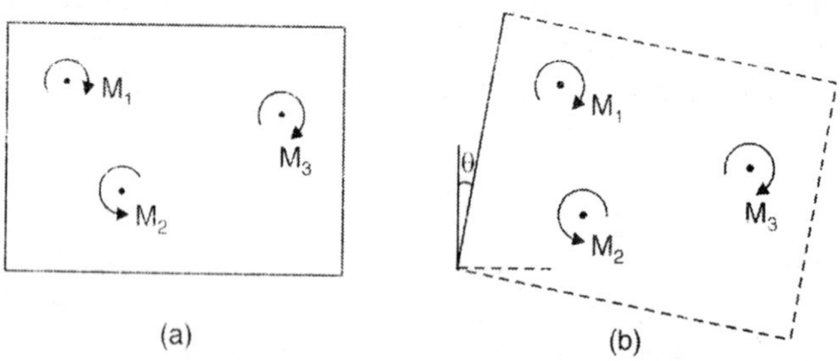

Fig. 7.3

Then the work done by various moments are as given below:

| *Moments* | *Work done* |
|---|---|
| $\mathbf{M}_1$ | $M_1\theta$ |
| $\mathbf{M}_2$ | $-M_2\theta$ |
| $\mathbf{M}_3$ | $M_3\theta$ |

1. No work is done by a force if displacement is at right angles to the line of action of force.
2. Work done is –ve if the displacement is in the direction opposite to force.
3. Work done by a moment is –ve if rotation is opposite to the direction of moment.

## 7.2 THE METHOD

The method consists of the following steps:

1. Give an elemental virtual displacement to the body in equilibrium in suitable direction.
2. Write the work done by each force in the system.
3. Sum up to get total work done by system of forces and equate it to zero to get required equations. Many times such equations are identical to equations of equilibrium. However, the method of getting these equations is completely different in this method.

A number of problems are solved to explain and to illustrate the virtual work method.

## 7.3 SIGN CONVENTION

In this chapter, the positive senses of moments and displacements are as given below:

1. Forces and displacements are +ve if they are in positive directions of coordinates as shown in Fig. 7.4.
2. Moments and rotations are positive if they are in clockwise directions.

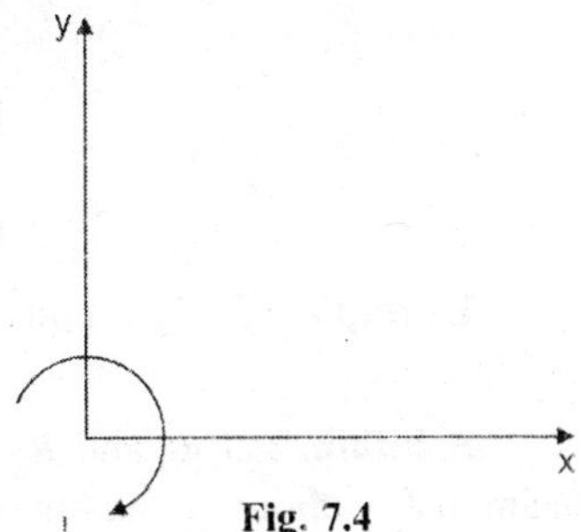


Fig. 7.4

**Example 7.1.** Determine the resultant of a system of concurrent forces acting on a body as shown in Fig. 7.5(*a*).

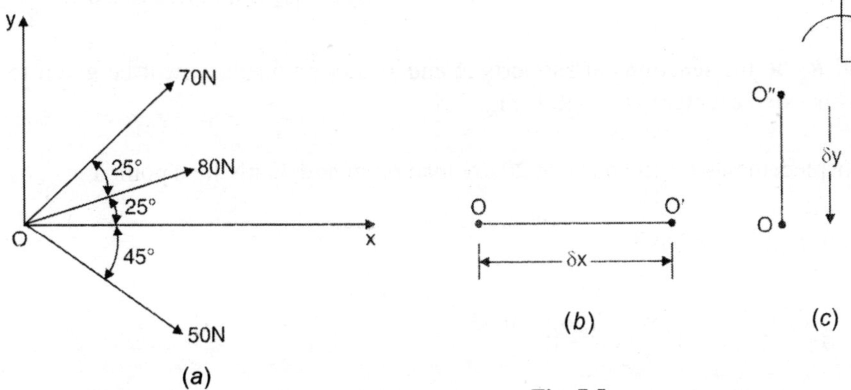


Fig. 7.5

**Solution.** Let $E$ be the equilibriant of the system which makes anticlockwise angle $\theta$ with $x$-axis. Then the system is in equilibrium with force **F**. Let us give a virtual displacement of $\delta x$ in $x$-direction as shown in Fig. 7.5(*b*).

Then total virtual work done by the system of given forces is

$$50 \cos 45° \,\delta x + 80 \cos 25° \,\delta x + 70 \cos 50° \,\delta x + E \cos \theta \,\delta x$$

Equating it to zero, we get

$$(50 \cos 45° + 80 \cos 25° + 70 \cos 50° + E \cos \theta)\, \delta x = 0$$

Since $\delta x \neq 0$,

$$50 \cos 45° + 80 \cos 25° + 70 \cos 50° + E \cos \theta = 0$$

or $$-E \cos \theta = 50 \cos 45° + 80 \cos 25° + 70 \cos 50° = 152.86 \text{ N} \quad ...(7.1)$$

Let us now give virtual displacement $\delta y$ to point $O$ as shown in Fig. 7.5($c$). Then virtual work done by the system of forces and the equilibriant are:

$$-50 \sin 45° \, \delta y + 80 \sin 25° \, \delta y + 70 \sin 50° \, \delta y + E \sin \theta \, \delta y$$

Since the system is in equilibrium with $E$, we get

$$(-50 \sin 45° + 80 \sin 25° + 70 \sin 50° + E \sin \theta) \, \delta y = 0$$

Since $\delta y \neq 0$,

$$-50 \sin 45° + 80 \sin 25° + 70 \sin 50° + E \sin \theta = 0$$

or $$-E \sin \theta = -50 \sin 45° + 80 \sin 25° + 70 \sin 50° = 52.07 \sin 25° \quad ...(7.2)$$

Resultant is exactly equal but opposite of equilibriant. Hence, Eqns. (7.1) and (7.2) can be written as

$$R \cos \theta = 152.86 \text{ N}$$

and $$R \sin \theta = 52.07 \text{ N}$$

$\therefore$ $$R = \sqrt{152.86^2 + 52.07^2} = 161.48 \text{ N}$$

and $$\theta = \tan^{-1} \frac{52.07}{152.86}$$

$\therefore$ $$\theta = 18.81°$$

**Example 7.2.** Determine the reactions $R_A$ and $R_B$ developed in the simply supported beam shown in Fig 7.6($a$).

**Solution.** Let $R_A$ and $R_B$ be the reactions at supports $A$ and $B$. Let $\delta y$ displacement be given to beam at $B$ without giving any displacement at $A$. (Ref. Fig. 7.6.)

The corresponding displacements for the beam at 20 kN load point and 40 kN load point are $\frac{1}{3} \delta y$ and $\frac{2}{3} \delta y$. By virtual work principle, we have

$$R_A \times 0 - 20 \times \frac{1}{3} \delta y - 40 \times \frac{2}{3} \delta y + R_B \, \delta y = 0$$

or $$\left( \frac{-20}{3} - \frac{80}{3} + R_B \right) \delta y = 0$$

Since $\delta y \neq 0$ $$\frac{-20}{3} - \frac{80}{3} + R_B = 0$$

or $$R_B = \frac{20}{3} + \frac{80}{3} = \frac{100}{3} = \mathbf{33.33 \text{ kN}}$$ **Ans.**

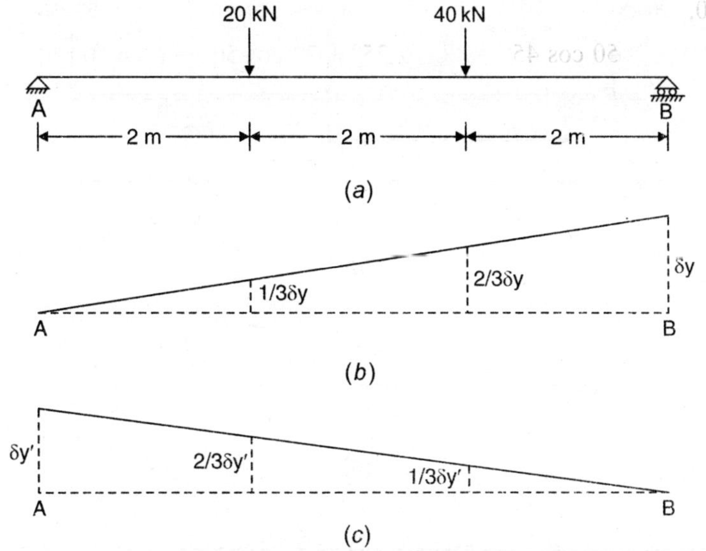


**Fig. 7.6**

Now let us give virtual displacement to beam as shown in Fig. 7.6*c*. Since the body is in equilibrium, virtual work equation is given as

$$R_A\,\delta y' - 20 \times \frac{1}{3}\delta y - 40 \times \frac{2}{3}\delta y + R_B \times 0 = 0$$

$$\therefore \qquad \left(R_A - \frac{40}{3} - \frac{40}{3}\right)\delta y' = 0$$

Since $\delta y \neq 0$,

$$R_A - \frac{40}{3} - \frac{40}{3} = 0$$

or

$$R_A = \frac{80}{3} = \mathbf{26.667\ kN} \qquad \textbf{Ans.}$$

**Example 7.3.** Determine the reactions in the overhanging beam as shown in Fig. 7.7(*a*).

**Solution.** Let $\mathbf{R_A}$ and $\mathbf{R_B}$ be the reactions as shown in Fig. 7.7(*a*). Let the virtual displacement $\delta y$ be given to the beam as shown in Fig. 7.7(*b*). Then the virtual displacement of various loads are as given in the figure. Applying virtual work principle for the beam which is in equilibrium, we get

$$20\left(\frac{\delta y}{8}\right) + R_A \times 0 - 60\left(\frac{4}{8}\delta y\right) + R_B\frac{6}{8}\delta y - 30 \times \delta y = 0$$

or

$$\left(\frac{20}{8} - \frac{60 \times 4}{8} + R_B\frac{6}{8} - 30\right)\delta y = 0$$

Since $\delta y \neq 0$,

$$\frac{20}{8} - \frac{60 \times 4}{8} + R_B\frac{6}{8} - 30 = 0$$

$$\therefore \qquad \mathbf{R_B = 76.667\ kN} \qquad \textbf{Ans.}$$

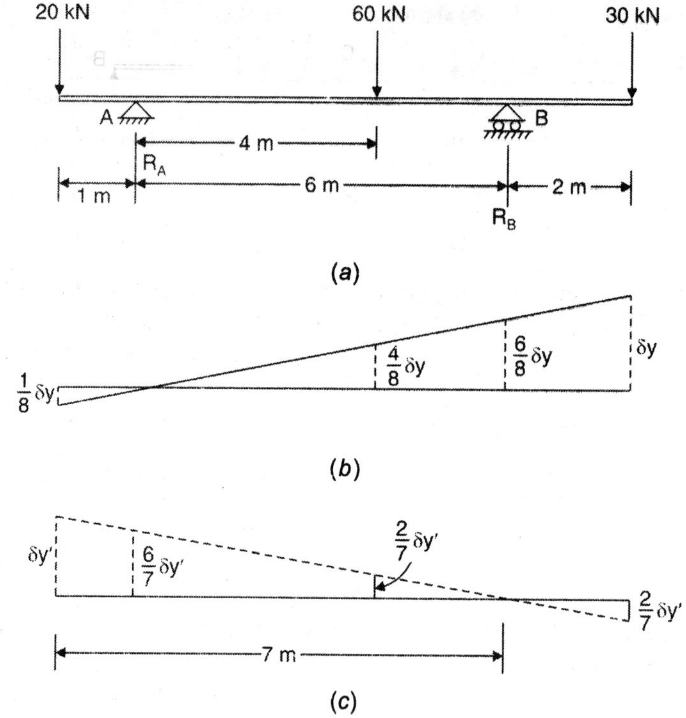


**Fig. 7.7**

Next, let us give virtual displacement $\delta y'$ to the beam as shown in Fig. 7.7(*c*).

The displacements of various load/reaction points are as shown in the figure. From virtual work principle,

$$-20(\delta y') + R_A \frac{6}{7}\delta y' - 60\left(\frac{2}{7}\delta y'\right) + R_B \times 0 - 30\left(-\frac{2}{7}\delta y'\right) = 0$$

$$\left(-20 + R_A \frac{6}{7} - \frac{60 \times 2}{7} + 30 \times \frac{2}{7}\right)\delta y' = 0$$

Since $\delta y' \neq 0$,
$$-20 + R_A \frac{6}{7} - \frac{60 \times 2}{7} + 30 \times \frac{2}{7} = 0$$

$\mathbf{R_A = 33.333\ kN}$ **Ans.**

**Example 7.4.** Determine the reaction at $A$ in simply supported beam shown in Fig. 7.8(*a*).

**Solution.** Let the virtual displacement $\delta y$ be given at point $A$ as shown in Fig. 7.8(*b*).

Total uniformly distributed load = 40 × 2 = 80 kN

Its centre of gravity is at midpoint of $AC$.

Displacement of the load at $A = \delta y$

Displacement of the load at $C = \dfrac{\delta y}{2}$

∴ Displacement of 80 kN load at its centroid

$$= \text{Average of } \delta y \text{ and } \frac{\delta y}{2} = \frac{\delta y + \delta y / 2}{2} = 0.75\ \delta y$$

Hence, Work done by UDL = – 40 × 2 (0.75 $\delta y$) = – 60 $\delta y$

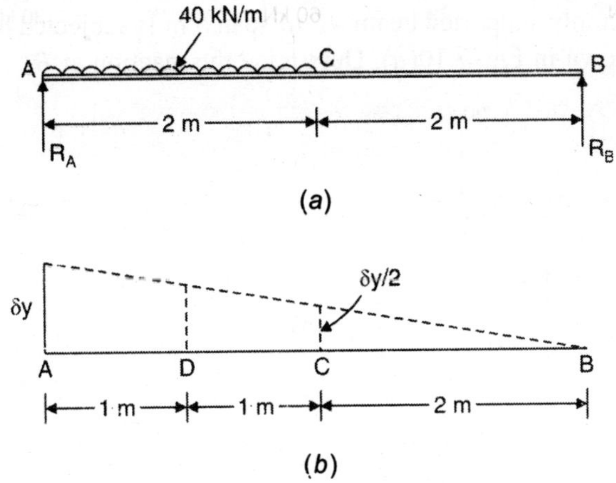


**Fig. 7.8**

Writing virtual work equation for the system of forces, we get

$$R_A\ \delta y - 60\ \delta y + R_B \times 0 = 0$$

$$(R_A - 60)\ \delta y = 0$$

Since, $\delta y \neq 0$ $\qquad R_A = \mathbf{60\ kN}$ **Ans.**

**Example 7.5.** Determine the reaction at $A$ in the overhanging beam $AB$ as shown in Fig. 7.9 ($a$).

**Solution.** Let the virtual displacement $\delta y$ be given at point $C$ keeping displacement at $B$ zero as shown in Fig. 7.9($b$). Let $\mathbf{R_A}$ and $\mathbf{R_B}$ be the reactions at $A$ and $B$, respectively.

The centroid of UDL moves down by $\dfrac{0 + 2\delta y}{2} = \delta y$. Total UDL = 30 × 6 kN; downward. Hence virtual work equation is

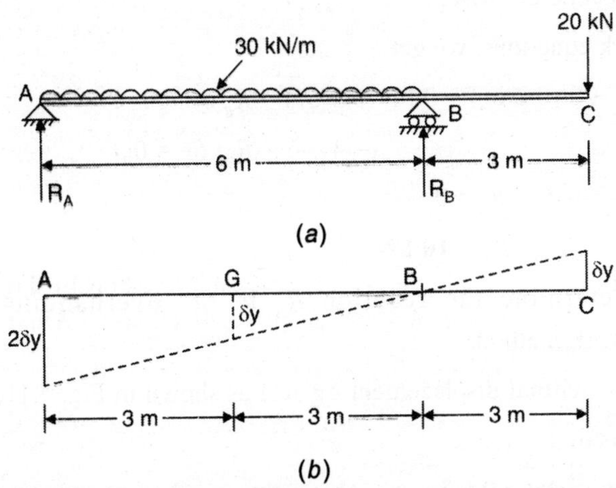


**Fig. 7.9**

$$R_A(-2\delta y) - 30 \times 6\,(-\delta y) + R_B \times 0 - 20\,(\delta y) = 0$$

$$(-2R_A + 180 - 20)\ \delta y = 0$$

Since $\delta y \neq 0$, $\qquad -2R_A + 180 - 20 = 0$

$\therefore \qquad R_A = \mathbf{80\ kN}$ **Ans.**

**Example 7.6.** The simply supported beam $AB$ of span 5 m is subjected to a concentrated load and an external moment as shown in Fig. 7.10($a$). Determine the reaction at $B$.

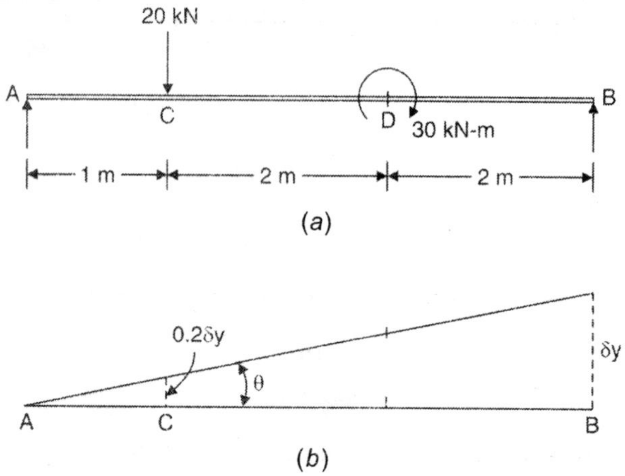


**Fig. 7.10**

**Solution.** Let a virtual displacement $\delta y$ be given at $B$ as shown in Fig. 7.10($b$).

Displacement at $A = 0$

Displacement at $C = 0.2\ \delta y$

Rotation of moment 30 kNm is,

$$\theta = \frac{-\delta y}{5} = -0.2\ \delta y$$

–ve sign, since it is anticlockwise.

Writing virtual work equations, we get

$$R_A \times 0 - 20\ (0.2\ \delta y) + 30\ (-\ 0.2\ \delta y) + R_B\ \delta y = 0$$

$$(-\ 4 - 6 + R_B)\ \delta y = 0$$

Since $\delta y \neq 0$, $\quad -\ 4 - 6 + R_B = 0$

or $\quad R_B = \mathbf{10\ kN}$ **Ans.**

**Example 7.7.** Determine the reaction $R_A$ in the overhanging beam, as shown in Fig. 7.11($a$), by virtual work method.

**Solution.** Let us give virtual displacement $\delta y$ at $A$ as shown in Fig. 7.11($b$).

Displacement at $A$ is $\delta y$

Displacement at $D$ is $\frac{3}{5}\ \delta y = 0.6\ \delta y$.

$\therefore$ Displacement at centroid of UDL $= \dfrac{\delta y + 0.6\delta y}{2} = 0.8\ \delta y$

Displacement at $B = 0$

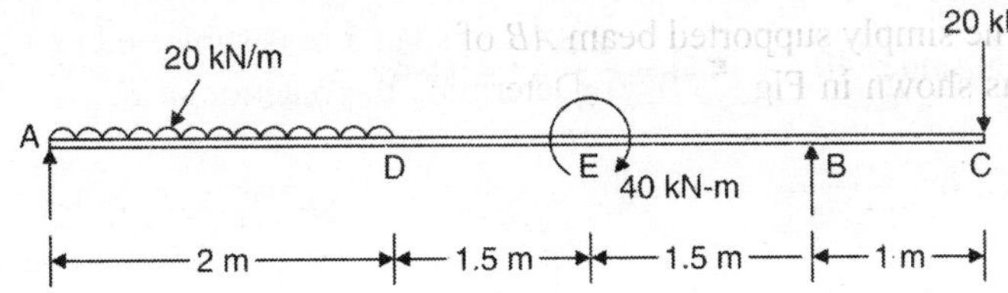


(a)

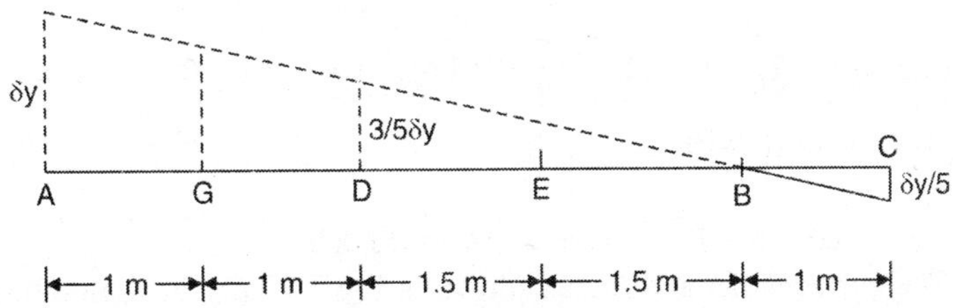


(b)

**Fig. 7.11**

Displacement at $C$ is $-\dfrac{1}{5}\ \delta y = -\ 0.2\ \delta y$

Rotation of moment $= \dfrac{\delta y}{5} = 0.2\delta y$

Writing the virtual work equation for the system, we get

$$R_A\ \delta y - 20 \times 2\ (0.8\ \delta y) + 40\ (0.2\ \delta y) + R_B \times 0 - 20\ (-\ 0.2\ \delta y) = 0$$

$$(R_A - 32 + 8 + 0 + 4)\ \delta y = 0$$

Since $\delta y \neq 0$

$$R_A - 32 + 8 + 4 = 0 \quad \text{or} \quad R_A = \mathbf{20\ kN}$$ **Ans.**

**Example 7.8.** Determine the reaction $R_A$ in the beam as shown in Fig 7.12($a$).

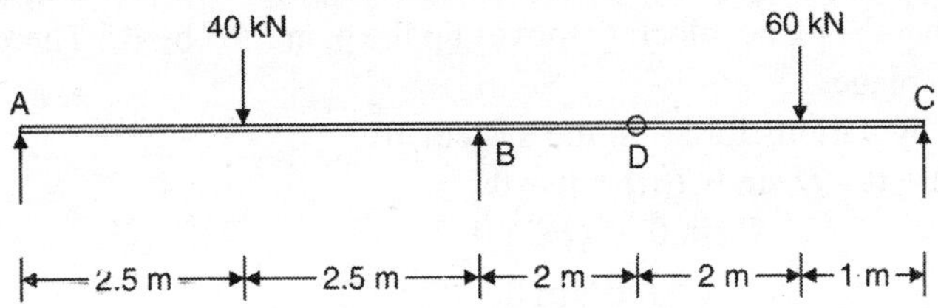


(a)

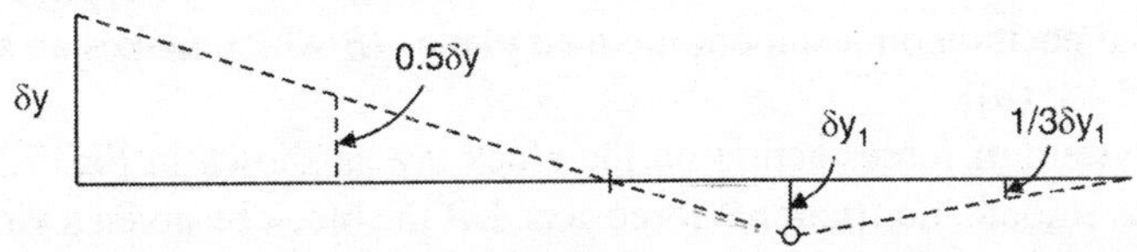


(b)

**Fig. 7.12**

**Solution.** Let us give virtual displacement δy at $A$ and no displacements at $B$ and $C$. Since $D$ is hinged, the shape of displaced structure is as shown in Fig. 7.12($b$).

Now, displacement under 40 kN load $= \frac{2.5}{5}\delta y = 0.5\,\delta y$

Displacement at $D$, $\delta y_1 = -\frac{2}{5}\delta y = -0.4\,\delta y$

Displacement of 60 kN load $= -\frac{1}{3}\delta y_1 = -\frac{1}{3} \times 0.4\,\delta y$

$\therefore$ Virtual work equation is

$$R_A\,\delta y - 40\,(0.5\,\delta y) + R_B \times 0 - 60\left(-\frac{1}{3} \times 0.4\,\delta y\right) + R_C \times 0 = 0$$

$\therefore$ $(R_A - 20 + 8)\,\delta y = 0$

Since $\delta y \neq 0$,

$R_A - 20 + 8 = 0$ or $R_A =$ **12 kN** **Ans.**

**Example 7.9.** The surfaces *AB* and *BC* as shown in Fig 7.13 (*a*), are smooth. Determine the ratio of loads *P* and *Q* for which the system is in equilibrium.

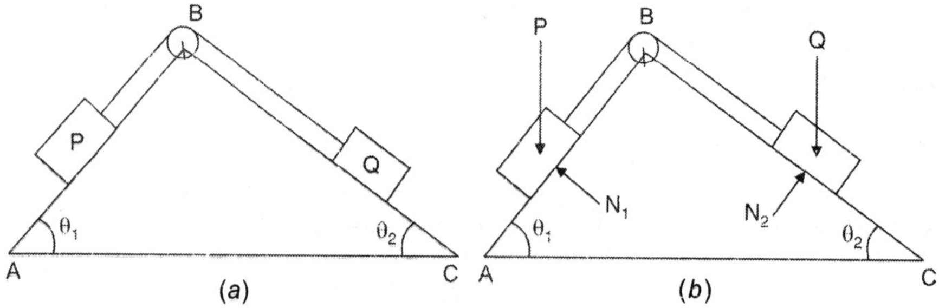


**Fig. 7.13**

**Solution.** Since the surfaces *AB* and *BC* are smooth frictional forces do not exist. There are normal reactions $N_1$ and $N_2$ acting on the blocks as shown in Fig. 7.13 (*b*). If displacement $\delta s$ is given to block of weight *P* down the plane, block *Q* moves up the plane *BC* by $\delta s$. There is no movement in the direction of normal to planes.

The work done by various forces in the system are:

$$P \sin\theta_1\,(\delta s) + 0 - Q \sin\theta_2\,(\delta s) + 0 = 0$$

*i.e.*,
$$P \sin\theta_1 = Q \sin\theta_2$$

or
$$\frac{P}{Q} = \frac{\sin\theta_2}{\sin\theta_1}$$ **Ans.**

**Example 7.10.** Determine the magnitude of horizontal force **P** to be applied to a block of weight 1500 N to hold it in the position on a smooth inclined plane *AB* which makes an angle of 30° with the horizontal [Ref. Fig. 7.14. (*a*)].

**Solution.** The system of forces acting on the block are as shown in Fig. 7.14(*b*).

Since the plane is smooth, no frictional force acts. Let the block be given a virtual displacement $\delta s$ up the plane.

Then, Displacement in vertical direction $= \delta s \sin\theta$

Then, Displacement in horizontal direction $= \delta s \cos\theta$

Then, Displacement in normal to plane $= 0$

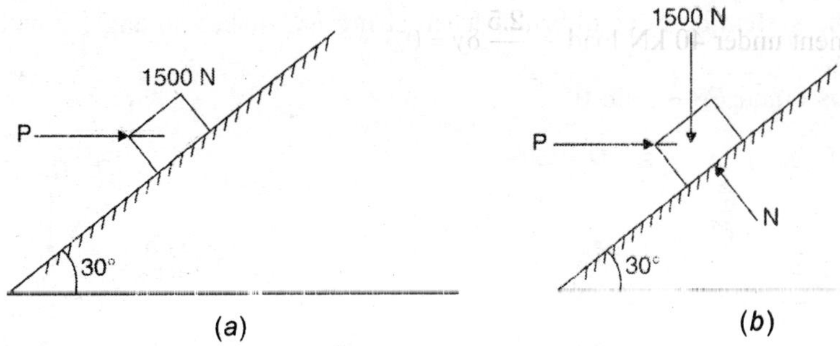


**Fig. 7.14**

∴ From principle of virtual work,

$$-1500\ (\delta s \sin 30°) + P\ \delta s \cos 30° + 0 = 0.$$

or $$P \cos 30° = 1500 \sin 30°$$

$$P = 1500 \tan 30° = \mathbf{866.03\ N}$$ **Ans.**

**Example 7.11.** A block weighing 1000N is to be held on an inclined plane *AB* which makes angle of 70° with horizontal. For what range of force **P** it can be kept in equilibrium. Coefficient of friction between the plane and the block is 0.3 [Ref. Fig. 7.15(*a*)].

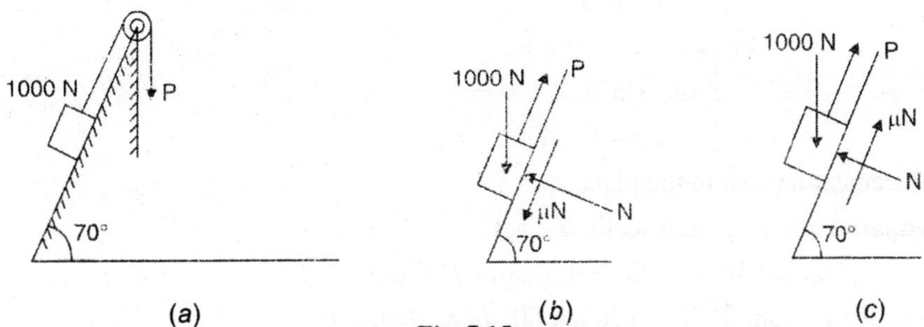


**Fig. 7.15**

**Solution.** (*a*) When motion is impending up the plane. The system of forces acting in this situation is shown in Fig. 7.15(*b*). Frictional force acts down the plane

$$F = \mu N = \mu\ 1000 \cos 70° = 0.3 \times 1000 \cos 70° = 300 \cos 70°$$

Let us give virtual displacement $\delta s$ up the plane. Let the direction up the plane be taken +ve.

Then from principle of virtual work, we get

$$-1000 \sin 70°\ (\delta s) - 300 \cos 70°\ (\delta s) + P\ \delta s = 0$$

∴ $$-1000 \sin 70° - 300 \cos 70° + P = 0$$

or $$P = 1000 \sin 70° + 300 \cos 70° = 1042.3\ \text{N}$$

(*b*) When motion is impending down the plane. The frictional force acts up the plane as shown in Fig. 7.15(*c*). Let $\delta s$ be virtual displacement up the plane. Then virtual work equation is,

$$-1000 \sin 70°\ \delta s + F\ \delta s + P\ \delta s = 0$$

∴ $$-1000 \sin 70° + 300 \cos 70° + P = 0$$

or $$P = 1000 \sin 70° - 300 \cos 70° = 837.1\ \text{N}$$

Hence, the block is in equilibrium for $P = 837.1$ N to 1042.3 N.

**Example 7.12.** Ball 1, Shown in Fig. 7.16 weighs 200 N and ball 2 weighs 150 N. They are connected by an inextensible string of length l. Determine the position of the balls if the system is left on the inclined plane as shown in the figure. Assume the planes are smooth.

**Solution.** Let the system be in equilibrium when string $DE$ makes an angle $\theta$ with plane $AB$ as shown in Fig. 7.16($b$).

Now, $DB = l \cos\theta$ and $EB = l \sin\theta$

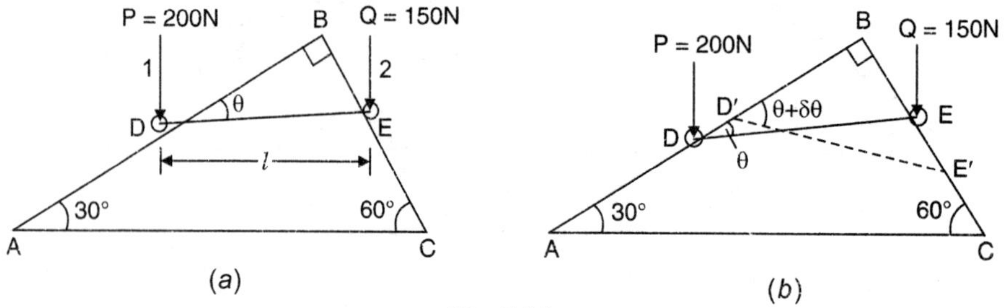


**Fig. 7.16**

Let a virtual displacement be given to the system of balls so that they move to $D'$ $E'$, *i.e.*, by an angle $\delta\theta$.

Component of $P$ down the plane $= P \sin 30° = 200 \sin 30°$

Component of $Q$ down the plane $= Q \sin 60° = 150 \sin 60°$

Since
$$D_B = l \cos\theta$$
$$\delta(D_B) = -l \sin\theta\, \delta\theta$$
and
$$E_B = l \sin\theta$$
$$\delta(E_B) = l \cos\theta \cdot \delta\theta$$

No displacement normal to the plane.

∴ From principle of virtual work, we get

$$200 \sin 30°\, \delta\,(DB) + 150 \sin 60°\, \delta\,(EB) = 0$$
$$200 \sin 30° (-l \sin\theta)\, \delta\theta + 150 \sin 60°\, l \cos\theta\, \delta\theta = 0$$

or
$$\tan\theta = \frac{150 \sin 60°}{200 \sin 30°} = 1.299$$
$$\theta = \mathbf{52.41°}$$ **Ans.**

**Example 7.13.** For the system of connected bodies, as shown in Fig. 7.17($a$), determine force **P** required to make the motion impending to the left. Coefficient of friction for all contacting surfaces is 0.25. Pulleys are smooth.

**Solution.** Since motion is impending to the left, frictional forces are acting in the directions shown in Fig. 7.17($b$).

$$F_1 = 0.25\, N_1 = 0.25 \times 250 = 62.5 \text{ N}$$
$$F_2 = 0.25\, N_2 = 0.25 \times 1000 \cos 45° = 176.78 \text{ N}$$
$$F_3 = 0.25\, N_3 = 0.25 \times 500 = 12.5 \text{ N}$$

Let us give a virtual displacement towards left. Taking the displacement of the system to left as +ve, virtual work equation is

$$P\delta s - F_1\, \delta s - 1000 \sin 45°\, \delta s - F_2\, \delta s - F_3\, \delta s = 0$$
∴
$$(P - F_1 - 1000 \sin 45° - F_2 - F_3)\, \delta s = 0$$

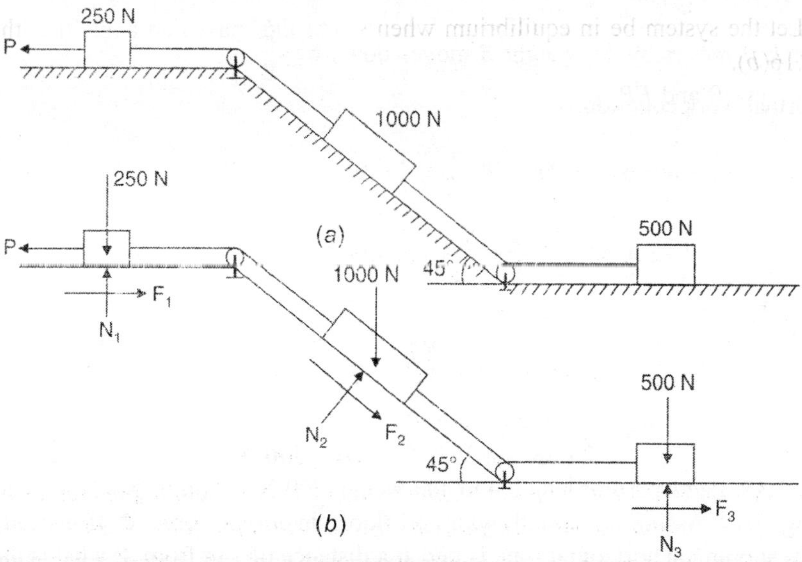


Fig. 7.17

Since $\delta s \neq 0$,

$$\mathbf{P} = F_1 + 1000 \sin 45° + F_2 + F_3$$
$$= 62.5 + 1000 \sin 45° + 176.78 + 125$$
$$= \mathbf{1071.38\ N}$$ **Ans.**

**Example 7.14.** Using the principle of virtual work, determine weight $\mathbf{W_B}$ for which the motion of block $A$ is impending up the plane. Given: coefficient of friction between block $A$ and the plane = 0.3 and pulleys are smooth [Ref. Fig. 7.18(*a*)].

**Solution.**
$$\tan \theta = \frac{3}{4}$$
$$\theta = 36.87°$$
$$F = \mu N = \mu W_A \cos \theta$$
$$= 0.3 \times 200 \cos 36.87° = 48 \text{ N}.$$

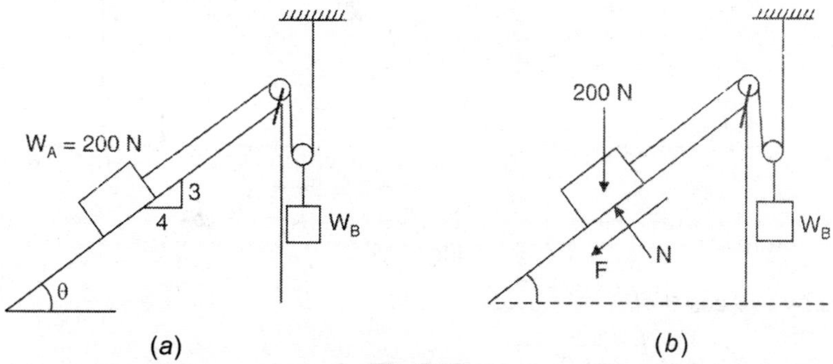


Fig. 7.18

Let us give virtual displacement of $\delta s$ up the plane to block $A$.

Component of block $A$ down the plane = $W_A \sin \theta$

Component normal to the plane do not do any work, since there is no displacement normal to the plane.

When block $A$ moves by $\delta s$, weight $B$ moves down by $\frac{\delta s}{2}$.

Hence, virtual work equation is

$$-200 \sin \theta \, \delta s - F \, \delta s + W_B \left(\frac{\delta s}{2}\right) = 0$$

or
$$\left(-200 \sin \theta - F + \frac{W_B}{2}\right) \delta s = 0$$

Since $\delta s \neq 0$

$$-200 \sin \theta - F + \frac{W_B}{2} = 0$$

or
$$W_B = 2\,(200 \sin \theta + F)$$
$$= 2\,(200 \sin 36.87° + 48) = \mathbf{336\ N}$$ **Ans.**

**Example 7.15.** Ladder $AB$ of length 4 m and weight 200 N is held in position by applying force $P$ as shown in Fig. 7.19. Assuming smooth wall and floor, determine force **P**. If instead of force **P** for preventing from slipping, a horizontal rope is tied at a distance of 1 m from $A$, what is the tension in the rope ?

**Solution.** (*a*) **P** at floor level

$$AB = l = 4\text{m}$$

Let ladder rotate by angle $\delta\theta$ as shown in Fig. 7.19(*b*).

Taking $C$ as origin.

Coordinate of $A = x = -4 \sin \theta$

$$\therefore \quad \delta x = -4 \cos \theta \, \delta\theta$$

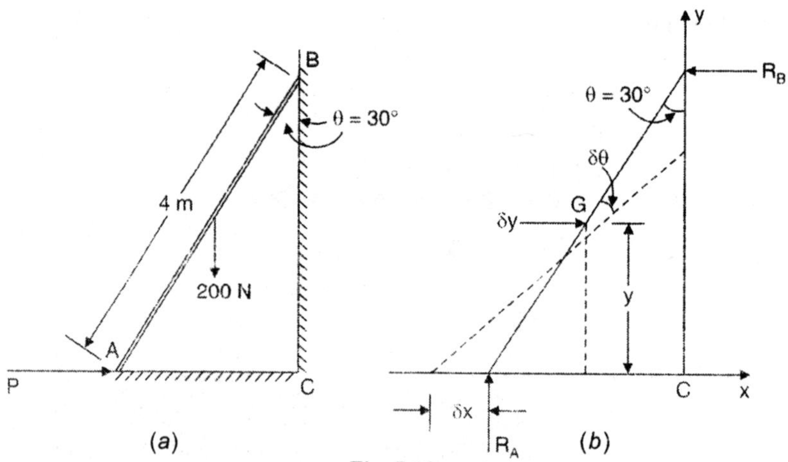


Fig. 7.19

**P** moves by $-4 \cos \theta \, \delta\theta$

Now,
$$y = \frac{l}{2} \cos \theta$$

$$\delta y = -\frac{l}{2} \sin \theta \, \delta\theta = -2 \sin \theta \, \delta\theta, \text{ since } l = 4 \text{ m}.$$

The reactions $\mathbf{R_A}$ and $\mathbf{R_B}$ are not doing any work, since there is no movement of their point of applications in the direction they act. Applying principle of virtual work, we get

$$P\,\delta x - 200\,\delta y = 0$$

$$P(-4\cos\theta)\,\delta\theta - 200(-2\sin\theta)\,\delta\theta = 0$$

$$P(-4\cos\theta) + 400\sin\theta = 0$$

or $$P = \frac{400\sin\theta}{4\cos\theta} = 100\tan 30° = \mathbf{57.735\ N}$$ **Ans.**

(*b*) If a rope is used instead of force **P**, the system of forces is as shown in Fig. 7.20(*a*).

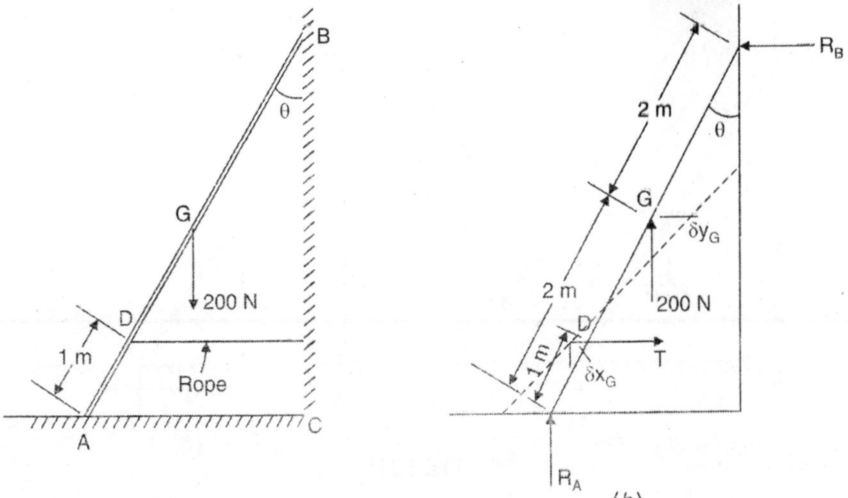


**Fig. 7.20**

The system of forces in this case are as shown in Fig. 7.20(*b*). Taking *C* as origin,

$$x_D = -3\sin\theta$$

$$\delta x_D = -3\cos\theta\,\delta\theta$$

$$y_G = 2\cos\theta$$

$$\delta y_G = -2\sin\theta\,\delta\theta$$

There is no displacement in the direction of $R_B$ and $R_A$. Hence from the principle of virtual work,

$$T(\delta x_D) + 200(\delta y_G) = 0$$

$$T(-3\cos\theta\,\delta\theta) - 200(-2\sin\theta\,\delta\theta) = 0$$

$$(-3T\cos\theta + 400\sin\theta)\,\delta\theta = 0$$

Since $\delta\theta \neq 0$

$$3T\cos\theta = 400\sin\theta$$

$$T = \frac{400}{3}\tan\theta = \frac{400}{3}\tan 30° = \mathbf{76.98\ N}$$ **Ans.**

**Example 7.16.** The ladder shown in Fig. 7.21(*a*) is 6 m long and is supported by a horizontal floor and vertical wall. The coefficient of friction between the floor and the ladder is 0.4 and between the wall and the ladder is 0.25. The weight of the ladder is 200 N. The ladder also supports a vertical load of 900 N at *C* which is at a distance of 1 m from *B*. Determine the least value of θ for which the ladder may be placed without slipping. Determine the reaction at this stage.

**Solution.** The various forces acting on the ladder when it is on the verge of slipping are as shown in Fig. 7.21(*b*). Note that frictional forces always oppose impending motions and they are equal to μ × normal reaction, where μ is coefficient of friction. To calculate virtual works, we need the displacement in the direction of forces. Normal reactions $\mathbf{N_A}$ and $\mathbf{N_B}$ are not going to do any work since displacements in direction of those forces is always zero. Taking $O$ as origin,

now, $x_A = 6 \sin\theta$ $\qquad \delta x_A = 6 \cos\theta\, \delta\theta$

$y_G = 3 \cos\theta$ $\qquad \delta y_G = -3 \sin\theta\, \delta\theta$

$y_C = 5 \cos\theta$ $\qquad \delta y_C = -5 \sin\theta\, \delta\theta$

and $y_B = 6 \cos\theta$ $\qquad \delta y_B = -6 \sin\theta\, \delta\theta$

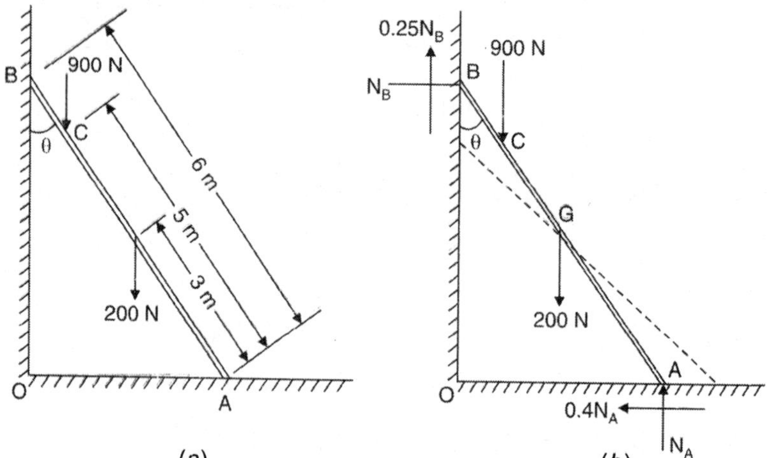


**Fig. 7.21**

Applying principles of virtual work, we get

$-0.4\, N_A\, 6 \cos\theta\, \delta\theta - 200\,(-3 \sin\theta\, \delta\theta) - 900\,(-5 \sin\theta\, \delta\theta) + 0.25\, N_B\,(-6 \sin\theta\, \delta\theta) = 0$

Since $\delta\theta \neq 0$,

$$-2.4\, N_A \cos\theta + 600 \sin\theta + 4500 \sin\theta - 1.5\, N_B \sin\theta = 0 \qquad ...(7.3)$$

From the equilibrium conditions,

$\Sigma F_x = 0$, gives $0.4\, N_A = N_B$

and $\Sigma F_y = 0$, gives $N_A + 0.25\, N_B = 900 + 200 = 1100$

$N_A + 0.25 \times 0.4\, N_A = 1100$

$\therefore \qquad N_A = 1000$ N

Hence $\qquad N_B = 0.4 \times 1000 = 400$ N

Substituting these values in Eqn. (7.3), we get

$-2.4 \times 1000 \cos\theta + 600 \sin\theta + 4500 \sin\theta - 1.5 \times 400 \sin\theta = 0.$

*i.e.,* $\qquad 4500 \sin\theta = 2400 \cos\theta$

or $\qquad \tan\theta = \dfrac{2400}{4500}$

$\theta = \mathbf{28.073}$ **Ans.**

**Example 7.17.** Determine the force in the member FH of the truss as shown in Fig 7.22. Each load is 10 kN and all triangles are equilaterals with sides 4 m.

**Solution.** Due to symmetry

$$R_A = R_B = \frac{1}{2} \text{ Total load} = \frac{1}{2} \times 10 \times 7 = 35 \text{ kN}$$

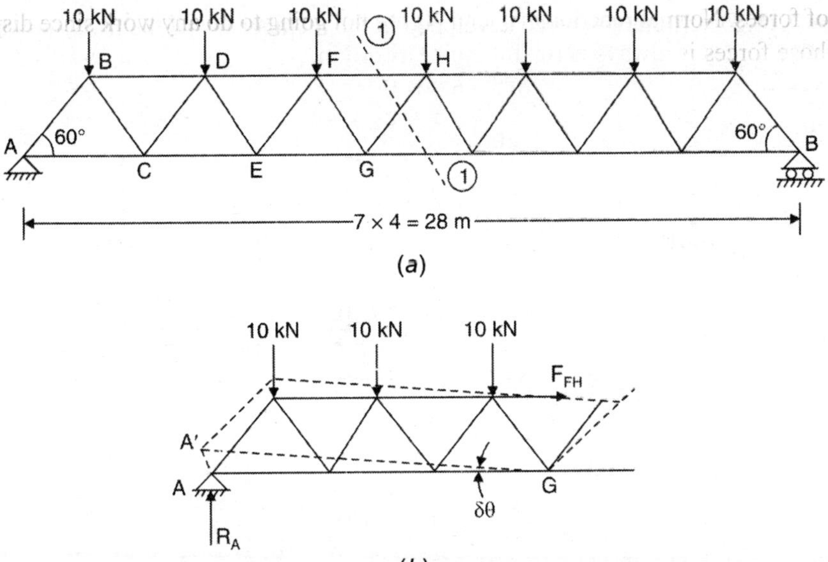


**Fig. 7.22**

Take section 1-1, as shown in Fig. 7.22 (*a*), and consider left side portion as shown in Fig. 7.22 (*b*). Imagine a rotation of δθ about point *G*.

Vertical displacement of $A = 12\ \delta\theta$

Vertical displacement of $B = 10\ \delta\theta$

Vertical displacement of $D = 6\ \delta\theta$

and Vertical displacement of $F = 2\ \delta\theta$

Horizontal displacement of $F_{FH} = 2 \tan 60°$ towards right

∴ From the principle of virtual displacement, we have

$$R_A\ 12\ \delta\theta - 10 \times 10\ \delta\theta - 10 \times 6\ \delta\theta - 10 \times 2\ \delta\theta + F_{FH}\ 2 \tan 60° = 0$$

$$2\ F_{FH} \tan 60° = -R_A\ 12 + 100 + 60 + 20$$

$$= -35 \times 12 + 180 = -240$$

$$F_{FH} = -\frac{240}{2 \tan 60°} = -69.28 \text{ kN}$$

*i.e.*, $F_{FH}$ direction is reversed.

*i.e.*, $F_{FH}$ is compressive force of **69.28 kN** **Ans.**

**Note:** The difference between the displacement in this case and the ladder problem. Here the rotation is free rotation about a point. Hence

Vertical displacement = Horizontal coordinate × δθ

and Horizontal displacement = Vertical distance × δθ

In ladder problem, the two edges are constrained to move along the specified two edges. In such cases:

$$\text{Vertical displacement } (-y \text{ direction}) \text{ of point } D = \frac{d(y_D)}{dy}$$

$$\text{and} \quad \text{Horizontal displacement} = \frac{d(x_D)}{dx}$$

**Example 7.18.** Determine the force developed in the member DF of the truss shown in Fig. 7.23.

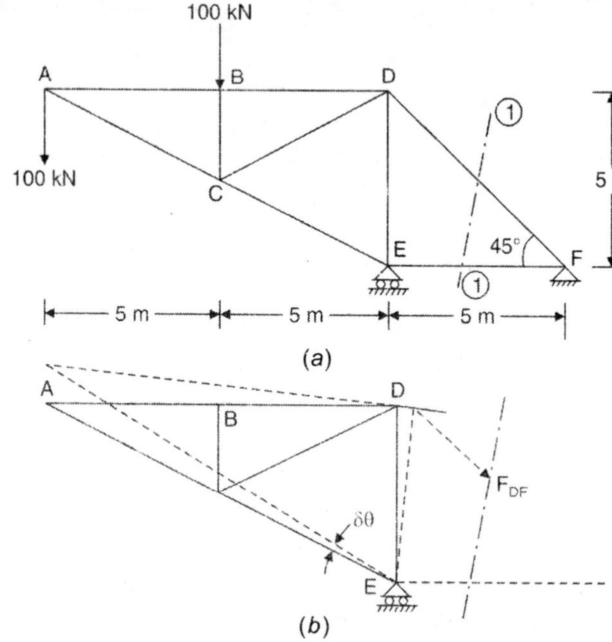


**Fig. 7.23**

**Solution.** Take the section 1-1, as shown in Fig. 7.23($a$), and consider left side portion. Imagine a virtual rotation $\delta\theta$ at support $E$. Then

Vertical displacement of point $A = 10\ \delta\theta$

Vertical displacement of point $B = 5\ \delta\theta$

Vertical displacement of point $D = 0$

Horizontal displacement of point $D = 5\ \delta\theta$

$\therefore$ From virtual work principle,

$$-100 \times 10\ \delta\theta - 100 \times 5\ \delta\theta + F_{DF} \cos 45° \times 5\ \delta\theta = 0$$

$$\therefore \quad F_{DF}\ 5 \cos 45° = 100 \times 10 + 100 \times 5 = 1500$$

$$\therefore \quad F_{DF} = \frac{1500}{5 \cos 45°} = 300\sqrt{2} = 424.26 \text{ kN (tensile).}$$

**Example 7.19.** Neglecting the friction, determine the effort required to lift a load by the screw jack shown in Fig. 7.24. Take pitch of the jack = $p$.

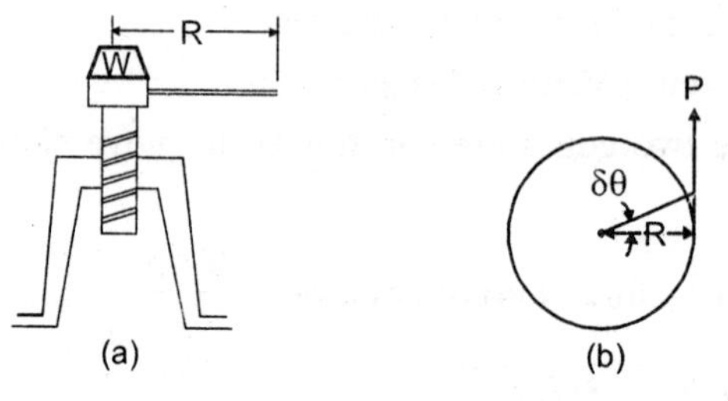


**Fig. 7.24**

**Solution.** Let **P** be the effort required. Let a virtual rotation $\delta\theta$ be given to effort.

$\therefore$ Virtual work done by **P** = $PR\delta\theta$

For one full rotation of effort, the load moves by distance $p$.

$\therefore$ For $\delta\theta$ rotation distance through which weight **W** moves = $\frac{p}{2\pi}\,\delta\theta$

From virtual work principle, total work done is zero.

$$\therefore \quad PR\delta\theta = W\frac{p}{2\pi}\,\delta\theta \quad \text{or} \quad \mathbf{P} = \frac{Wp}{2\pi R} \qquad \textbf{Ans.}$$

**Example 7.20.** Neglecting friction, determine the pressure applied by screw press when effort **P** is applied [Ref. Fig. 7.25(*a*)].

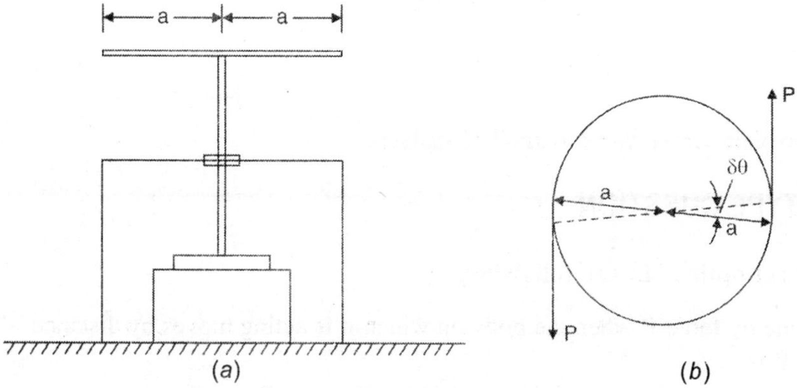


**Fig. 7.25**

**Solution.** Let a rotation $\delta\theta$ be given to the screw press by applying efforts **P** as shown in Fig. 7.25(*b*).

Work done by effort = $2Pa\,\delta\theta = 2Pa\,\delta\theta$

If $p$ is the pitch of the screw, work done by load **W** is

$$= W \times \frac{p}{2\pi}\,\delta\theta$$

Hence from principle of virtual work, we have

$$2\,Pa\,\delta\theta = W\frac{p}{2\pi}\,\delta\theta$$

or

$$P = \frac{Wp}{4\pi\,a} \qquad \textbf{Ans.}$$

## EXERCISES

**7.1** Determine the vertical reaction developed at support $B$ in the beam as shown in Fig. 7.26.

[**Ans.** $\mathbf{V_B}$ = 35 kN use virtual work method].

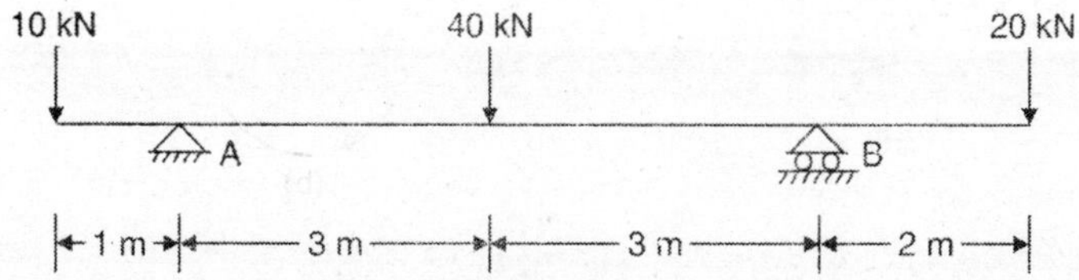


**Fig. 7.26**

**7.2** Determine the reaction $\mathbf{V_B}$ in the overhanging beam shown in Fig 7.27 by virtual work method.

[**Ans.** $\mathbf{V_B}$ = 38.33 kN]

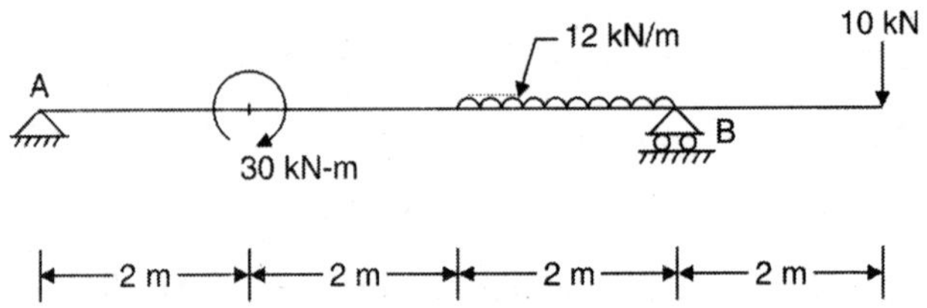


**Fig. 7.27**

## QUESTION

**7.1.** Briefly explain virtual work method of analysis.

## OBJECTIVE TYPE QUESTION

**Select the correct options in the following:**

**7.1.** Work done by force **P** when the body on which it is acting moves by distance '*s*' at right angles to force **P** is

(*a*) **P × S**
(*b*) **P × (–S)**
(*c*) Zero
(*d*) None of these

# CHAPTER 8

# D'ALEMBERT'S PRINCIPLE

The dynamics problems which need the consideration of forces causing the motion are are called kinetic problems.

In this chapter, analysis of kinetic problems of a body in plane motion using D' Alembert's principle are explained. Dynamic problem can be converted into a static equilibrium problem by applying D' Alembert's principle. This is done by introducing an additional force, and then, the usual equations of static equilibrium are used to get the solution.

## 8.1 NEWTON'S SECOND LAW OF MOTION

D' Alembert's principle is in fact an application of Newton's second law to a moving body and looking at it from a different angle. Observing the motion of falling bodies, Galileo discovered first two laws of motion which are now commonly known as Newton's laws of motion. However, Newton generalised the laws and demonstrated their validity by astronomical predictions. These laws are presented in the first chapter. According to Newton's second law, the rate of change of momentum is directly proportional to the impressed force and takes place in the direction, in which the force acts. The above definition leads to the statement that *force is directly proportional to product of mass and acceleration* and unit of force is so selected that constant of proportionality reduces to unity. Hence, finally Newton's second law reduces to the statement, Force = Mass × Acceleration.

Instead of single force, if a system of forces acts on a particle, the above statement reduces to the resultant force equal to *the product of mass and acceleration in the direction of the resultant force.* Mathematically,

$$R = m\,a \qquad \text{...(8.1)}$$

where, $R$ is the resultant of forces acting on the particle. Hence many times Newton's second law is stated as: *A particle acted upon by an unbalanced system of forces has an acceleration directly proportional and in line with the resultant force.*

## 8.2 D'ALEMBERT'S PRINCIPLE

French mathematician Jean le Rond d' Alembert proved in 1743 that the Newton's second law of motion is applicable not only to the motion of a particle but also to the motion of a body and looked at Eqn. (8.1) from different angle. The equation $R = ma$ may be written as

$$R - ma = 0 \qquad \ldots(8.2)$$

The term '$-ma$' may be looked as a force of magnitude $m \times a$, applied in the opposite direction of motion and is termed as the inertia force or reverse effective force. D' Alembert looks at Eqn. (8.2) as an equation of equilibrium and states that *the system of forces acting on a body in motion is in dynamic equilibrium with the inertia force of the body.* This is known as D' Alembert's principle.

Let the body as shown in Fig. 8.1, be subjected to a system of forces causing the body to move with an acceleration $a$ in the direction of the resultant. Then apply a force equal to '$m\ a$' in the reversed direction of acceleration as shown in Fig. 8.1. Now according to D' Alembert's principle, the equations of equilibrium $\Sigma F_x = 0$ and $\Sigma F_y = 0$ may be used for the system of forces as shown in Fig. 8.2.

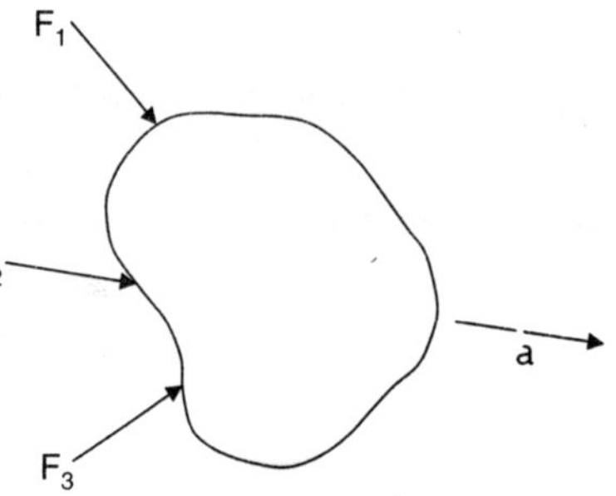


**Fig. 8.1**

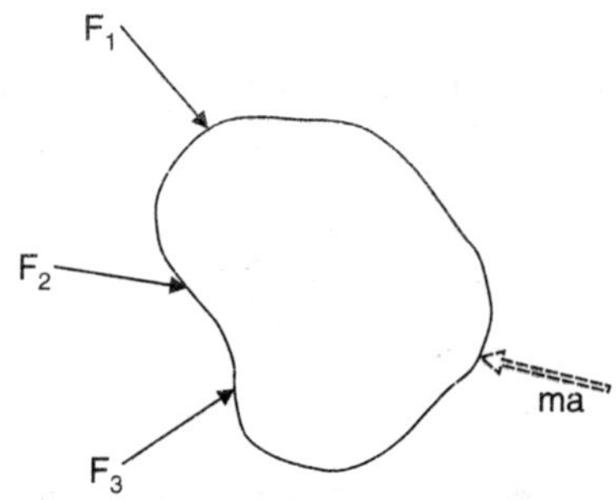


**Fig. 8.2**

The inertia force $-ma$ has a physical meaning. According to Newton's first law of motion, a body continues to be in the state of rest or of uniform motion in a straight line unless acted by an external force. That means every body has a tendency to continue in its state of rest or of uniform motion. This tendency is called inertia. Hence inertia force is the resistance offered by a body to the change in its state of rest or of uniform motion.

Many scientists are critical of D' Alembert's principle. Usually equilibrium equations are applied to a system of forces acting on a body. Inertia force is not acting on the moving body. Actually this is the force exerted by the moving body to resist the change in its state; hence, D' Alembert is criticized for messing-up the concept of equations of equilibrium.

However, many engineers prefer to use D' Alembert's principle, since just by applying a reverse effective force, the moving body can be treated as a body in equilibrium and can be analysed using equations of static equilibrium.

**Example 8.1:** A man weighing W Newton entered a lift which moves with an acceleration of a m/s$^2$. Find the force exerted by the man on the floor of lift when

(*a*) lift is moving downward.

(*b*) lift is moving upward.

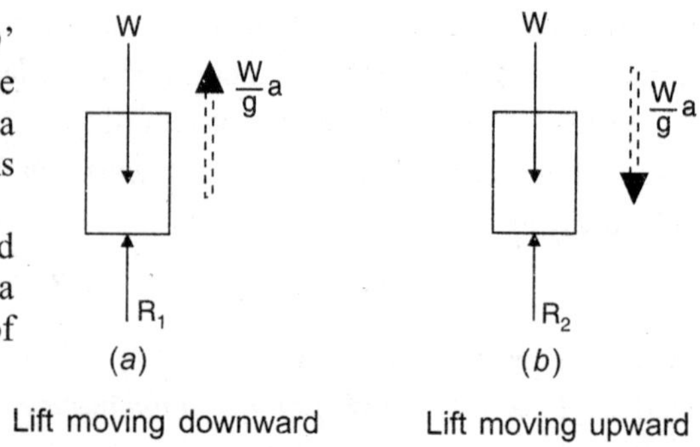


**Fig. 8.3**

**Solution.** (*a*) **When the lift is moving down-ward,** the inertia force $ma = \dfrac{W}{g}a$, should be applied in upward direction as shown in Fig.8.3.(*a*).

Sum of forces in vertical direction = 0 for the lift, gives

$$R_1 - W + \frac{W}{g}a = 0$$

$$R_1 = W\left(1 - \frac{a}{g}\right) \qquad \ldots(8.3)$$

(*b*) **When lift is moving upwards,** inertia force $\frac{W}{g}a$ should be applied in the downward direction as shown in Fig. 8.3(*b*). Applying the equation of equilibrium, we get

$$R_2 - W - \frac{W}{g}a = 0$$

$$R_2 = W\left(1 + \frac{a}{g}\right) \qquad \ldots(8.4)$$

Thus when lift is moving with acceleration downward, the man exerts less force on the floor of the lift, and while moving upward, he exerts more force. From Eqns. (8.3) and (8.4) it may be observed that if the lift moves with uniform velocity, then the acceleration $a$ is zero and hence the man exerts force equal to his own weight on the lift.

**Example 8.2.** An elevator cage of a mine shaft weighing 8 kN, when empty, is lifted or lowered by means of a wire rope. Once a man weighing 600 N, entered into it and lowered with uniform acceleration such that when a distance of 187.5 m was covered, the velocity of the cage was 25 m/s. Determine the tension in the rope and the force exerted by the man on the floor of the cage.

**Solution.** In this problem,

Initial velocity $u = 0$

Final velocity $v = 25$ m/s

and Distance covered $s = 187.5$ m

Using the equation of motion, we get

$$v^2 - u^2 = 2as$$

$$25^2 - 0 = 2a\ 187.5$$

or $$a = 1.667 \text{ m/s}^2$$

Figure 8.4 (*a*) shows free body diagram of elevator cage and the man with inertia force $\frac{W}{g}a = \frac{8000+600}{9.81}$ $a$ applied in upward direction (since the motion is downward).

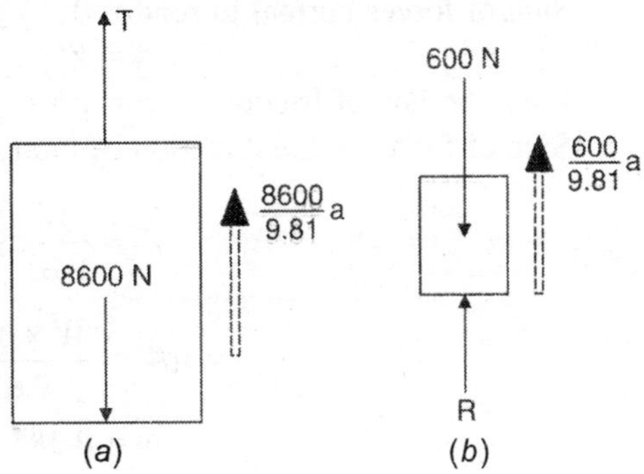


**Fig. 8.4**

Summing up the forces in vertical direction, we have

$$T + \frac{8600}{9.81}1.667 - 8600 = 0$$

$\therefore$ $$\mathbf{T = 7138.90\ N}$$ **Ans.**

Figure 8.4 (*b*) shows the free body diagram of the man along with inertia force $\frac{600}{9.81}a$ applied in upward direction. Sum of the vertical forces should be zero. That is,

$$\Sigma V = 0, \text{ gives}$$

$$R + \frac{600}{9.81}a - 600 = 0$$

*i.e.*, $$R + \frac{600}{9.81} \times 1.667 - 600 = 0$$

or $$\mathbf{R = 498.06\ N}$$ **Ans.**

**Example 8.3.** A motorist travelling at a speed of 70 kmph suddenly applies brakes and halts after skidding 50 m. Determine:

1. The time required to stop the car.
2. The coefficient of friction between the tyres and the road.

**Solution.** Initial velocity $u = 70$ kmph

$$= \frac{70 \times 1000}{60 \times 60} = 19.44 \text{ m/s}$$

Final velocity $v = 0$

Displacement $s = 50$ m

Using the equation of linear motion $v^2 = u^2 + 2as$, we get

$$0 = (19.44)^2 + 2a \times 50$$

$$a = -3.78 \text{ m/s}^2$$

*i.e.*, the retardation is 3.78 m/s$^2$.

Using the relation $v = u + at$, we get

$$0 = 19.44 - 3.78\ t$$

$\therefore$ $$\mathbf{t = 5.14\ s}$$ **Ans.**

Inertia force must be applied in the opposite direction of acceleration, which means, it should be applied in the direction of motion while retarding. Figure 8.5 shows the free body of the motor along with inertia force.

Sum of forces normal to road = 0

$$N = W$$

From the law of friction, $F = \mu N = \mu W$

Sum of forces in the direction of motion = 0, gives

$$F = \frac{W}{9.81} 3.78$$

$$\mu W = \frac{W \times 3.78}{9.81}$$

$\therefore$ $$\mu = \mathbf{0.385}$$ **Ans.**

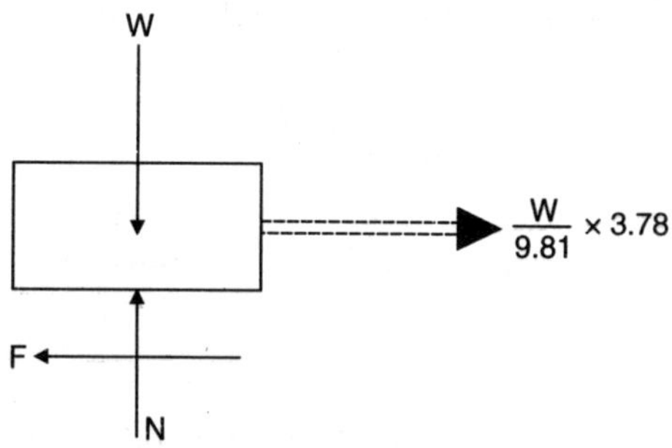


**Fig. 8.5**

**Example 8.4.** A block weighing 1 kN rests on a horizontal plane as shown in Fig. 8.6 (a). Find the magnitude of force **P** required to give the block an acceleration of 3 m/s$^2$ to the right. The coefficient of friction between the block and the plane is 0.25.

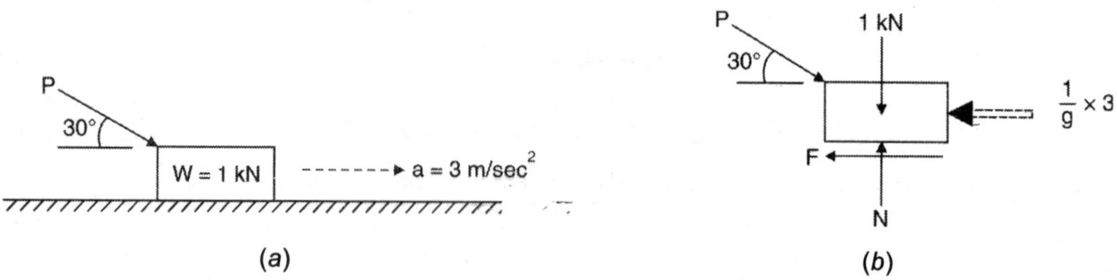


**Fig. 8.6**

**Solution.** Free body diagram of the block along with inertia force $ma = \frac{1}{g} \times 3 = \frac{3}{9.81}$ kN is shown in Fig. 8.6(*b*). In the figure, **N** is the normal reaction and **F** is the frictional force.

$$\Sigma V = 0, \text{ gives}$$

$$N - 1 - P \sin 30° = 0$$

or

$$N = 1 + \frac{P}{2} \qquad \text{...(8.5)}$$

From the law of friction, we have

$$F = \mu N = 0.25 \left(1 + \frac{P}{2}\right) \qquad \text{...(8.6)}$$

$$\Sigma H = 0, \text{ gives}$$

$$P \cos 30° - F - \frac{3}{9.81} = 0$$

$$P \cos 30° - 0.25 \left(1 + \frac{P}{2}\right) - \frac{3}{9.81} = 0$$

$\therefore$ **P = 0.561 kN** **Ans.**

**Example 8.5.** A 750 N crate rests on a 500 N cart. The coefficient of friction between the crate and the cart is 0.3 and between cart and the road is 0.2. If the cart is to be pulled by a force **P** [Ref. Fig. 8.7 (*a*)] such that the crate does not slip, determine : (*a*) the maximum allowable magnitude of **P** and (*b*) the corresponding acceleration of the cart.

**Solution.** Let the maximum acceleration be *a*, at which 750 N crate is about to slip. Hence frictional force will have limiting value (= μ N). Consider the free body diagram of crate along with inertia force, as shown in Fig. 8.7(*b*).

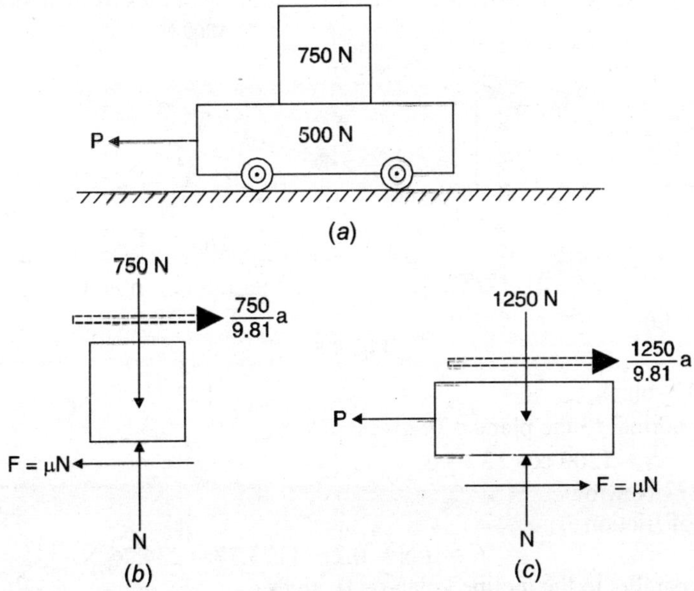


**Fig. 8.7**

$$\Sigma V = 0, \text{ gives}$$
$$\mathbf{N} = \mathbf{W} = 750 \text{ Newton}$$

Frictional force,
$$\mathbf{F} = \mu \text{ N} = 0.3 \times 750$$
$$= 225 \text{ N}$$
$$\Sigma H = 0, \text{ gives}$$
$$225 = \frac{750}{9.81} a$$

$\therefore$
$$\mathbf{a} = \mathbf{2.943 \text{ m/s}^2}$$
**Ans.**

Consider now dynamic equilibrium of crate and the cart shown as a single body in Fig. 8.7(*c*).

$$\Sigma V = 0, \text{ gives}$$
$$N = 1250 \text{ Newton}$$
$$\text{Frictional force} = \mu \text{ N} = 0.2 \times 1250 = 250 \text{ Newton}$$
$$\Sigma H = 0, \text{ gives}$$
$$P - 250 - \frac{1250}{9.81} \times 2.943 = 0$$
$$\mathbf{P} = \mathbf{625 \text{ N}}$$
**Ans.**

**Example 8.6.** A body weighing 1200 N rests on a rough plane inclined at 12° to the horizontal. It is pulled up the plane by means of a light flexible rope running parallel to the plane and passing over a light frictionless pulley at the top of the plane as shown in Fig. 8.8 (*a*). The portion of the rope beyond the pulley hangs vertically down and carries a weight of 800 N at its end. If the coefficient of friction for the plane and the body is 0.2, find:

(*a*) tension in the rope.

(*b*) acceleration with which the body moves up the plane.

(*c*) the distance moved by the body in 3 seconds after starting from rest.

**Solution.** Let *a* be the acceleration of the system.

Free body diagrams of 1200 N block and 800 N block are shown in Figs. 8.8(*b*) and 8.8(*c*) along with inertia forces. According to D' Alembert, we can treat these bodies as in static equilibrium.

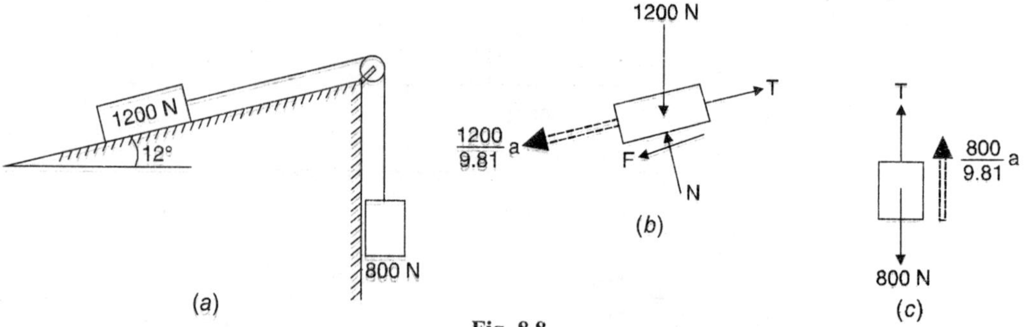


**Fig. 8.8**

Consider 1200 N block.

Sum of forces normal to the plane = 0, gives

$$N - 1200 \cos 12° = 0$$
$$N = 1173.77 \text{ Newton}$$

From the law of friction

$$F = \mu \text{ N} = 0.2 \times 1173.77 = 234.76 \text{ N}$$

Sum of forces parallel to the inclined plane = 0, gives

$$\frac{1200}{9.81} a + 1200 \sin 12° + F - T = 0$$

*i.e.,* $122.32\,a - T = -484.25$ ...(8.7)

Consider the free body diagram of 800 N block shown in Fig. 8.8(*c*).

$$T + \frac{800}{9.81}a = 800 \quad ...(8.8)$$

Adding Eqns. (8.7) and (8.8), we get

$$\left(122.32 + \frac{800}{9.81}\right) a = 800 - 484.25$$

$\therefore$ **a = 1.549 m/s²** **Ans.**

Substituting it in (8.8), we get

$$T = 800 - \frac{800}{9.81} \times 1.549$$

*i.e.,* **T = 673.68 N** **Ans.**

Initial velocity = 0

$a$ = 1.549 m/s²

$t$ = 3 s

Using the equation $s = ut + \frac{1}{2}at^2$, we get **distance moved in 3 seconds** as

$$= 0 \times 3 + \frac{1}{2} \times 1.549 \times 3^2$$

= **6.971 m** **Ans.**

**Example. 8.7.** Two weights 800 N and 200 N are connected by a thread and they move along a rough horizontal plane under the action of a force of 400 N applied to the 800 N weight as shown in Fig. 8.9(*a*). The coefficient of friction between the sliding surface of the weights and the plane is 0.3. Using D' Alembert's principle, determine the acceleration of the weight and tension in the thread.

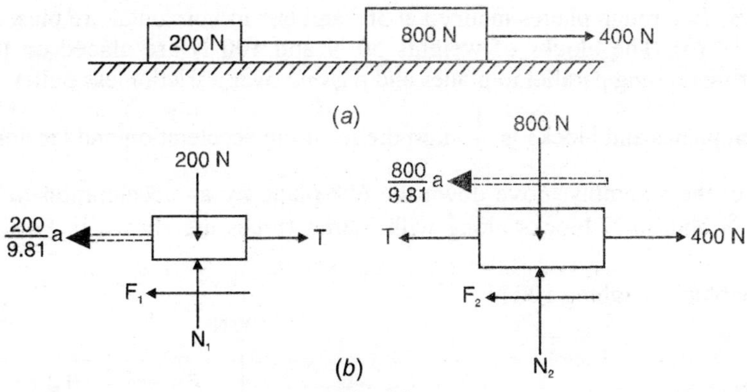


**Fig. 8.9**

**Solution.** Free body diagrams of 200 N and 800 N blocks along with inertia forces are shown in Fig. 8.9(*b*), in which *a* is acceleration of the system.

Consider the dynamic equilibrium of 200 N weight

$\Sigma V = 0$, gives

$N_1 = 200\ N.$ ...(8.9)

From law of friction, $F_1 = \mu\, N_1 = 0.3 \times 200 = 60$ N ...(8.10)

$\Sigma H = 0$, gives

$$T - F_1 - \frac{200}{9.81}a = 0$$

i.e., $$T - \frac{200}{9.81}a = 60 \quad \text{...(8.11)}$$

Since $F_1 = 60$ N

Consider 800 N body.

$\Sigma V = 0$, gives

$N_2 = 800$ N ...(8.12)

From law of friction, $F_2 = \mu N_2 = 0.3 \times 800$

$= 240$ N ...(8.13)

$\Sigma H = 0$, gives

$$-T - \frac{800}{9.81}a - F_2 + 400 = 0$$

or $$T + \frac{800}{9.81}a = 160 \quad \text{...(8.14)}$$

Since $F_2 = 240\ N$

Subtracting Eqns. (8.11) from (8.14), we get

$$\left(\frac{200}{9.81} + \frac{800}{9.81}\right) a = 160 - 60$$

$\therefore$ **a = 0.981 m/s²** **Ans.**

Substituting it in Eqn. (8.14), we get

$$T = 160 - \frac{800}{9.81} \times 0.981$$

i.e., **T = 80 N.** **Ans.**

**Example 8.8.** Two rough planes inclined at 30° and 60° to horizontal are placed back to back as shown in Fig. 8.10 (*a*). The blocks of weights 50 N and 100 N are placed on the faces and are connected by a string running parallel to planes and passing over a frictionless pulley. If the coefficient of friction between planes and blocks is $\frac{1}{3}$, find the resulting acceleration and tension in the string.

**Solution.** Let the assembly move down the 60° plane by an acceleration '*a*' m/s². Free body diagrams of 100 N and 50 N blocks along with inertia forces are shown in Figs. 8.10(*b*) and (*c*), respectively.

Consider the block weighing 100 N:

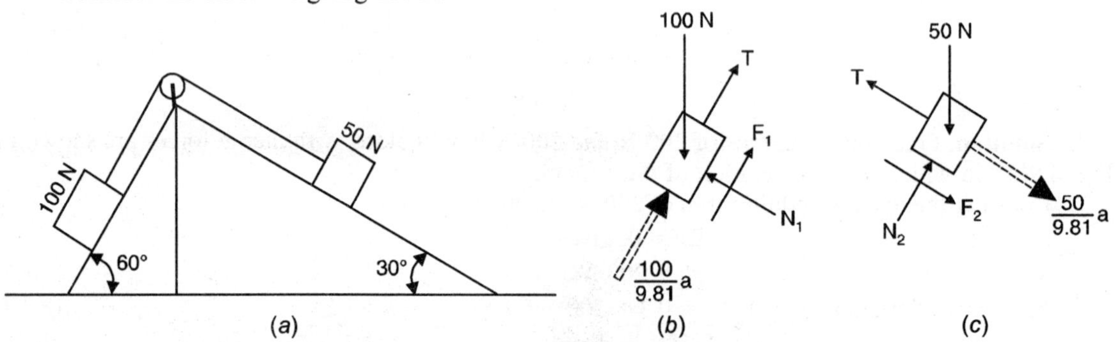


**Fig. 8.10**

Sum of forces normal to the plane = 0, gives

$$N_1 = 100 \cos 60° = 50 \text{ N} \quad ...(8.15)$$

From the law of friction,

$$F_1 = \mu \text{N} = \frac{1}{3} \times 50 = 16.67 \text{ N} \quad ...(8.16)$$

Sum of forces parallel to the plane = 0, gives

$$T + \frac{100}{9.81} a - 100 \sin 60° + F_1 = 0$$

$$T + \frac{100}{9.81} a = 69.93 \quad ...(8.17)$$

since $F_1 = 16.67$

Now consider 50 N block:

Sum of forces normal to plane = 0, gives

$$N_2 = 50 \cos 30° = 43.30 \text{ N} \quad ...(8.18)$$

From the law of friction, $F_2 = \mu N_2$

$$= \frac{1}{3} \times 43.3 = 14.43 \text{ N} \quad ...(8.19)$$

Sum of forces parallel to 30° plane = 0, gives

$$\frac{50}{9.81} a + F_2 + 50 \sin 30° - T = 0$$

$$\frac{50}{9.81} a - T = -39.43 \quad ...(8.20)$$

since $F_2 = 14.43$

Adding Eqns. (8.17) and (8.20), we get

$$\left(\frac{100}{9.81} + \frac{50}{9.81}\right) a = 69.93 - 39.43$$

or $\mathbf{a = 1.9947 \ m/s^2}$ **Ans.**

From Eqn. (8.17),

$\mathbf{T} = 69.93 - 100/9.81 \times 1.9947 = \mathbf{49.6 \ N}$ **Ans.**

**Example 8.9.** Two blocks *A* and *B* are released from rest on a 30° incline, when they are 18 m apart. The coefficient of friction under the upper block *A* is 0.2 and that under the lower block *B* is 0.4 [Ref. Fig. 8.11(*a*)]. In what time block *A* reaches the block *B*? After they touch and move as a single unit, what will be the contact force between them? Weights of the blocks *A* and *B* are 100 *N* and 80 *N*, respectively.

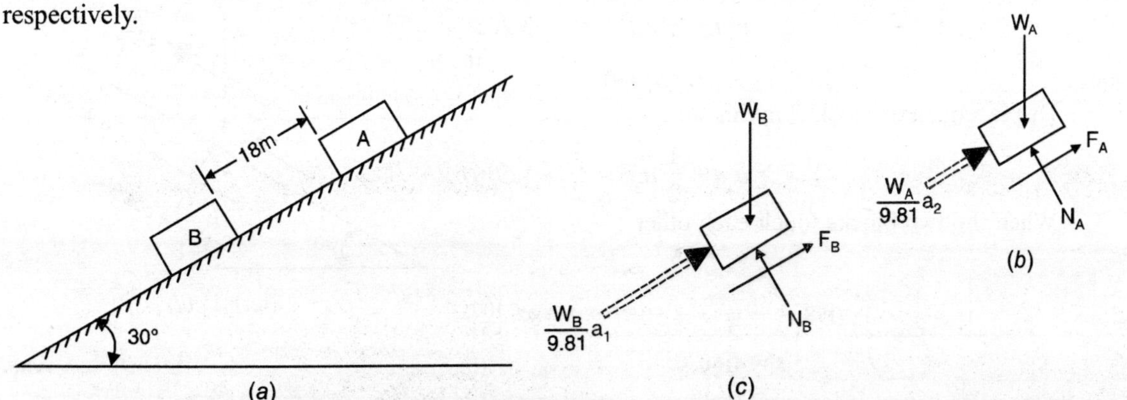


**Fig. 8.11**

**Solution.** Let block $A$ moves with an acceleration $a_1$ and block $B$ with an acceleration $a_2$. The free body diagrams of the blocks $A$ and $B$ along with inertia forces are shown in Figs. 8.11($b$) and 8.11($c$), respectively.

Consider block $A$:

Sum of forces normal to the plane = 0, gives

$$N_A = W_A \cos\theta = W_A \cos 30° \quad \text{...(8.21)}$$

From the law of friction $F_A = \mu N_A$

$$= 0.2\, W_A \cos 30° \quad \text{...(8.22)}$$

Sum of forces parallel to the plane = 0, gives

$$\frac{W_A}{9.81} a_1 + F_A - W_A \sin 30° = 0$$

$$\therefore \quad \frac{W_A}{9.81} a_1 + 0.2\, W_A \cos 30° - W_A \sin 30° = 0$$

$$\frac{a_1}{9.81} + 0.2 \cos 30° - \sin 30° = 0$$

$$a_1 = 3.2058 \text{ m/s}^2 \quad \text{...(8.23)}$$

Consider block $B$:

Sum of forces normal to the plane = 0, gives

$$N_B = W_B \cos 30° \quad \text{...(8.24)}$$

From the law of friction $F_B = \mu N_B$

*i.e.,* $F_B = 0.4\, W_B \cos 30°$ ...(8.25)

Sum of forces parallel to the plane = 0, gives

$$\frac{W_B}{9.81} a_2 + F_B - W_B \sin 30° = 0$$

$$\frac{W_B}{9.81} a_2 + 0.4\, W_B \cos 30° - W_B \sin 30° = 0$$

$$\therefore \quad a_2 = 1.5067 \text{ m/s}^2$$

Let $t$ be the time elapsed until the blocks touch each other.

Displacement of block $A$ in this period

$$s_1 = u_1 t + \frac{1}{2} a_1 t^2 = \frac{1}{2} \times 3.2058\, t^2$$

since Initial velocity $u_1 = 0$

Displacement of block $B$ in this time

$$s_2 = u_2 t + \frac{1}{2} a_2 t^2 = \frac{1}{2} \times 1.5067\, t^2$$

When the two blocks touch each other

$$s_1 = s_2 + 18$$

$$\frac{1}{2} \times 3.2058\, t^2 = \frac{1}{2} \times 1.5067\, t^2 + 18$$

$$\therefore \quad \mathbf{t = 4.60\ s} \qquad \textbf{Ans.}$$

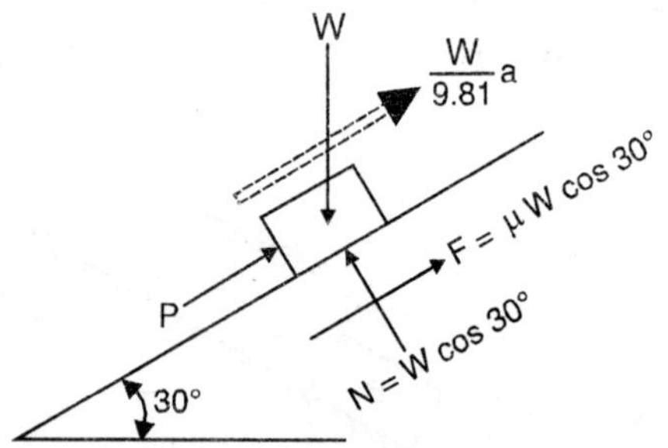


**Fig. 8.11 (*d*)**

After the blocks touch each other, let the common acceleration be $a$. Summing up the forces including inertia forces along the inclined plane.

$$\frac{100}{9.81}a + 0.2 \times 100 \cos 30° - 100 \sin 30 + \frac{80}{9.81}a + 0.4 \times 80 \cos 30° - 80 \sin 30° = 0$$

$$\mathbf{a} = 2.45 \text{ m/s}^2$$

Considering the free body diagram of any one of the blocks, contact force **P** can be obtained. Free body diagram of block $A$ along with inertia force is shown in Fig. 8.11($d$), in which W = 100 N.

Now, Sum of forces parallel to plane – 0, gives

$$P - 100 \sin 30° + F + \frac{100}{9.81}a = 0$$

$$P - 100 \sin 30° + 0.2 \times 100 \cos 30° + \frac{100}{9.81} \times 2.45 = 0$$

$$\mathbf{P = 7.7\ N}$$ **Ans.**

**Example 8.10.** Two bodies weighing 300 $N$ and 450 $N$ are hung to the ends of a rope passing over an ideal pulley as shown in Fig. 8.12($a$). With what acceleration the heavier body comes down ? What is the tension in the string ?

**Solution.** Let $a$ be the acceleration with which the system moves and $T$ be the tension in the string.

Free body diagrams of 300 N block and 450 N block along with inertia forces are shown in Figs. 8.12($b$) and 8.12($c$).

Consider the body weighing 300 N,

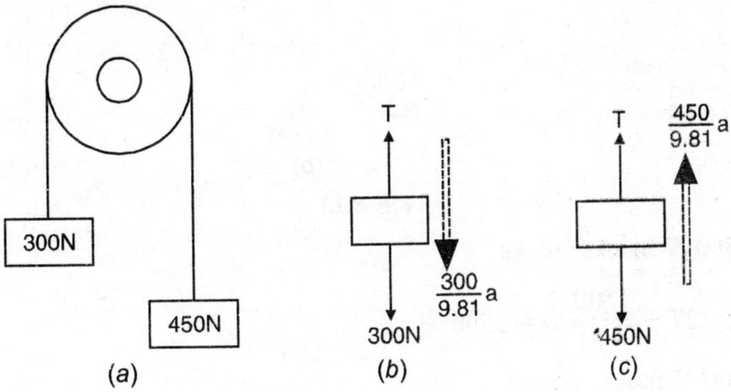


**Fig. 8.12**

$$\Sigma V = 0, \text{ gives}$$

$$T - \frac{300}{9.81}a - 300 = 0$$

*i.e.*, $$T - \frac{300}{9.81}a = 300 \qquad \ldots(8.26)$$

Consider the body weighing 450 N, we get

$$T + \frac{450}{9.81}a = 450 \qquad \ldots(8.27)$$

Subtracting Eqn. (8.26) from Eqn. (8.27), we get

$$\frac{450}{9.81}a + \frac{300}{9.81}a = 450 - 300$$

$$\mathbf{a = 1.962\ m/s^2}$$ **Ans.**

Substituting this value in Eqn. (8.26), we get

$$T = 300 + \frac{300}{9.81} \times 1.962$$

*i.e.,* $$\mathbf{T = 360\ N}$$ **Ans.**

**Example 8.11.** Determine the tension in the string and accelerations of blocks *A* and *B* weighing 1500 N and 500 N connected by an inextensible string as shown in Fig. 8.13(*a*). Assume pulleys as frictionless and weightless.

**Solution.** In this pulley system, it may be observed that if 1500 N block moves downward by distance $x$, 500 N block moves up by $2x$. Hence if acceleration of 1500 N block is $a$, that of 500 N block is $2a$. The free body diagrams of 1500 N block and 500 N blocks are shown in Figs. 8.13(*b*) and (*c*), along with inertia forces. According to D' Alembert's principle, the system of forces shown in Figs. 8.13(*b*) and (*c*) may be treated to be in equilibrium.

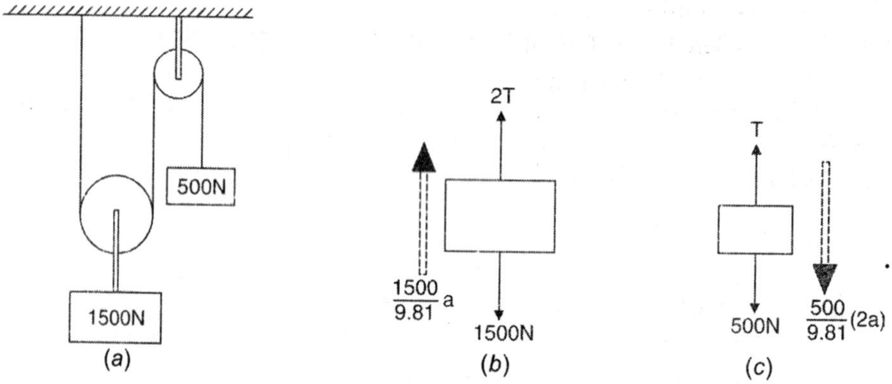


**Fig. 8.13**

Considering 1500 N block, we get

$$2T + \frac{1500}{9.81}a = 1500 \qquad \text{...(8.28)}$$

Considering 500 N block, we get

$$T - \frac{500}{9.81}(2a) = 500 \qquad \text{...(8.29)}$$

From Eqns. (8.28) and (8.29), we get

$$\left(\frac{1500}{9.81} + \frac{2000}{9.81}\right)a = 500$$

$\therefore$ $$\mathbf{a = 1.401\ m/s^2}$$ **Ans.**

Substituting it in Eqn. (8.28), we get

$$2T = 1500 - \frac{1500}{9.81} \times 1.401$$

$$\mathbf{T = 642.89\ N}$$ **Ans.**

**Example 8.12.** An engine of weight 500 kN pulls a carriage weighing 1500 kN up an incline of 1 in 100. The train starts from rest and moves with a constant acceleration against a resistance of 5 N/kN. It attains a maximum speed of 36 kmph in 1 km distance. Determine the tension in the coupling between the carriage and the engine and the tractive force developed by the engine.

**Solution.** Initial velocity $u = 0$

$$\text{Final velocity } v = 36 \text{ kmph} = \frac{36 \times 1000}{60 \times 60} = 10 \text{ m/s}^2$$

Displacement $s = 1 \text{ km} = 1000 \text{ m}$

From the kinematic equation $v^2 = u^2 + 2as$, we get

$$10^2 = 0 + 2 \times a \times 1000$$

$$a = 0.05 \text{ m/s}^2$$

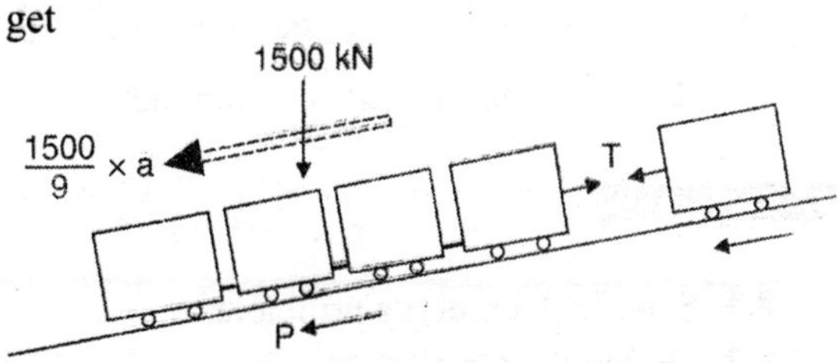


**Fig. 8.14**

To find the tension in the coupling between engine and train, dynamic equilibrium of the coaches only may be considered. Let **T** be the tension in coupling. Various forces acting parallel to the track are shown in Fig. 8.14.

Tractive resistance = 5 × 1500 = 7500 N = 7.5 kN down the plane.

$$\text{Component of weight of train} = 1500 \times \frac{1}{100} = 15 \text{kN} \text{ down the plane.}$$

$\left(\textbf{Note: } \text{Since slope of track is very small, } \sin\theta = \tan\theta = \frac{1}{100}.\right)$

$$\text{Inertia force} = \frac{W}{g} a$$

$$= \frac{1500}{9.81} \times 0.05 = 7.645 \text{ kN (down the plane)}$$

Dynamic equilibrium equation is

$$T - 7.5 - 15 - 7.645 = 0$$

or $\mathbf{T = 30.145}$ **kN** **Ans.**

To find tractive force developed by the engine, consider dynamic equilibrium of entire train.

Figure 8.15 shows various forces in the direction of track acting on the entire train along with inertia force. Let **P** be tractive force developed.

Total tractive resistance = 5 × 2000 = 10,000 N

= 10 kN, down the plane.

$$\text{Inertia force} = \frac{W}{g} a$$

$$= \frac{2000}{9.81} \times 0.05 = 10.194 \text{ kN, down the plane.}$$

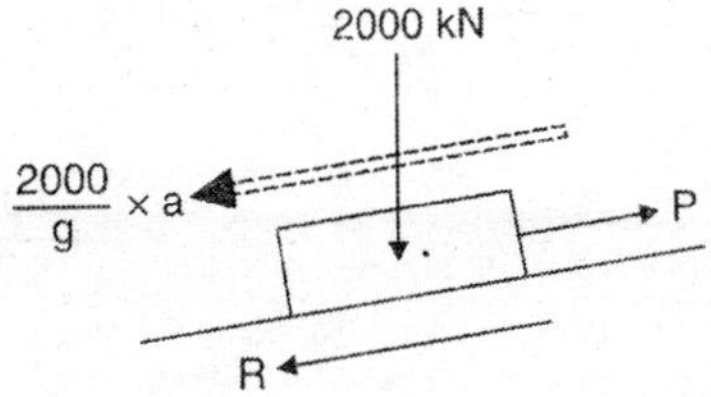


**Fig. 8.15**

Component of weight down the plane

$$= 2000 \times \frac{1}{100} = 20\,\text{kN}$$

Dynamic equilibrium equation along the track is

$$P - 10 - 10.194 - 20 = 0$$

$\therefore$ **P = 40.194 kN** **Ans.**

## IMPORTANT DEFINITION

1. **D'Alembert's principle** states that the system of forces acting on a body in motion is in dynamic equilibrium with the inertia force of the body.

## QUESTIONS

**8.1** State D'Alembert's principle.

**8.2** Define the term 'inertia force'.

## EXERCISES

**8.1** A mine cage weighs 12 kN and can carry a maximum load of 20 kN. The average frictional resistance of the slide guys is 500 N. What constant cable tension is required to give a loaded cage an upward velocity of 3 m/s, from rest in a distance of 3 m? [**Ans.** 37.393 kN]

**8.2** A train weighing 3000 kN is moving up a slope 2 in 100 with an acceleration of 0.04 m/s$^2$. Tractive resistance is 6 N/kN. Determine the acceleration of the train if it moves with the same tractive force:

(*a*) on a level track.

(*b*) down the plane inclined at 2 in 100. [**Ans.** 0.2362 m/s$^2$; 0.4324 m/s$^2$]

**8.3** The coefficient of friction between the crate and the cart shown in Fig. 8.16 is 0.25 and that between cart and the road is 0.15. If the cart is pulled by a force of 500 N, determine the acceleration of the cart. Check whether there is slipping of the crate at this stage.

[**Ans. a** = 1.4138 m/s$^2$; No slipping]

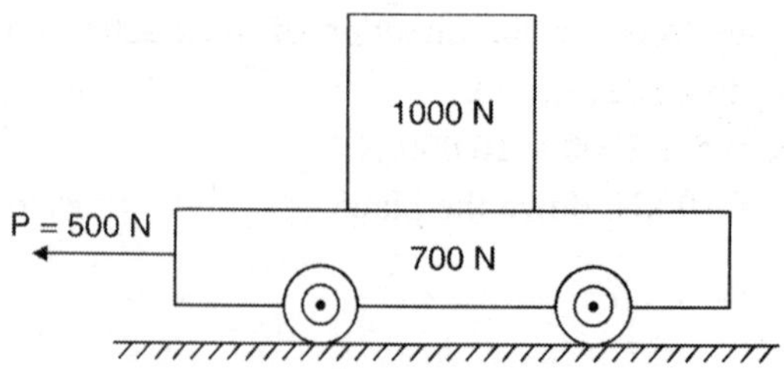


**Fig. 8.16**

**8.4** A block weighing 2500 N rests on a level horizontal plane which has a coefficient of friction 0.20. This block is pulled by a force of 1000 N, which is acting at an angle of 30° to the horizontal. Find the velocity of the block, after it moves 30 m, starting from rest. If the force of 1000 N is then removed, how much further will it move?

[**Ans.** $v$ = 10.4748 m/s; $s$ = 27.961 m]

**8.5** In what distance will body 1 shown in Fig. 8.17 attain a velocity of a 3 m/s starting from rest? Take coefficient of friction between the blocks and plane as 0.2. Assume pulley is smooth. What is the tension in the chord? [**Ans.** $s$ = 12.953 m; **T** = 159.08 N]

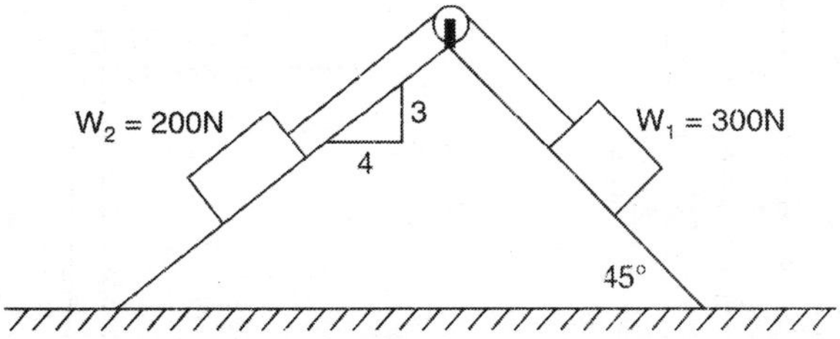


**Fig. 8.17**

**8.6** Determine the constant force **P** that will give the system of bodies shown in Fig. 8.18 a velocity of 3 m/s after moving a distance of 4.5 m from the position of rest. Coefficient of friction at all contact points is 0.2. Assume pulleys as frictionless. [**Ans.** 448.39 N]

**8.7** Two blocks of weight $\mathbf{W_1}$ and $\mathbf{W_2}$ are connected by inextensible wire passing over a smooth pulley as shown in Fig. 8.19. If $\mathbf{W_1}$ is greater than $\mathbf{W_2}$, find the tension in the string and the acceleration of the system.

$$\left[\textbf{Ans. } \mathbf{a} = g\left(\frac{W_1 - W_2}{W_1 + W_2}\right); \mathbf{T} = \frac{2\,W_1 W_2}{W_1 + W_2}\right]$$

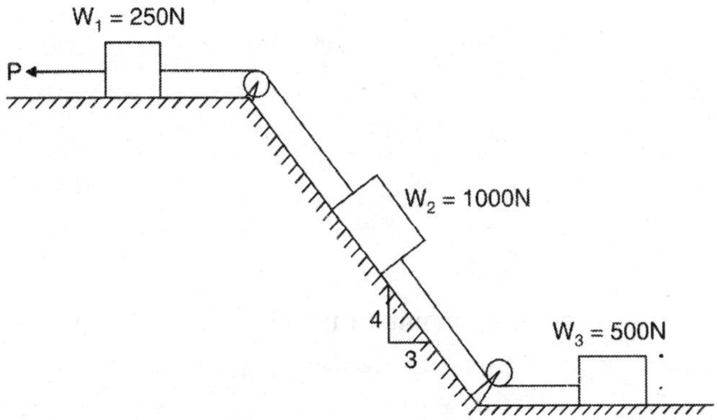


**Fig. 8.18**

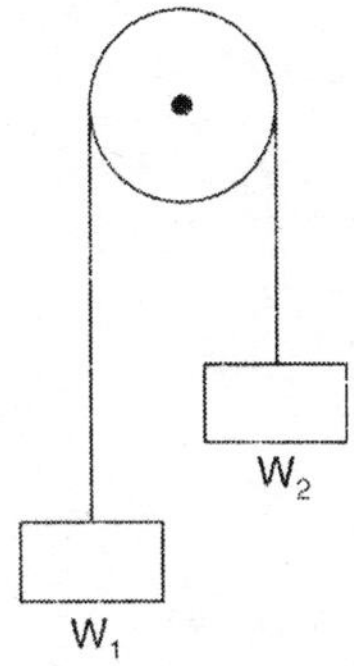


**Fig. 8.19**

**8.8** Find the acceleration of the three weights shown in Fig. 8.20. Assume pulleys are frictionless. Find also the tension in the cables.

$$\left[\textbf{Ans. } \text{Acceleration of 200 N block} = \frac{13}{37}g;\ 40\text{ N block} = \frac{23}{37}g;\ 40\text{ N block} = \frac{3}{37}g\right.$$

$$\left.\text{Tension in upper cable} = 129.73\text{ N; Tension in lower cable} = 64.85\text{ N}\right]$$

**8.9** The weights of the three blocks shown in Fig. 8.21 are $\mathbf{W_A}$ = 100 N, $\mathbf{W_B}$ = 200 N and $\mathbf{W_C}$ = 200 N. Coefficient of friction between block *A* and the floor is 0.2, that between floor and block *C* is 0.25. Assuming pulleys as weightless and smooth, find the acceleration of each block.

$\left[\textbf{Hint: } a_B = \left(\frac{a_A + a_C}{2}\right)\right]$ **[Ans.** $a_A$ = 0.5 g, $a_B$ = 0.3 g, $a_C$ = 0.1 g]

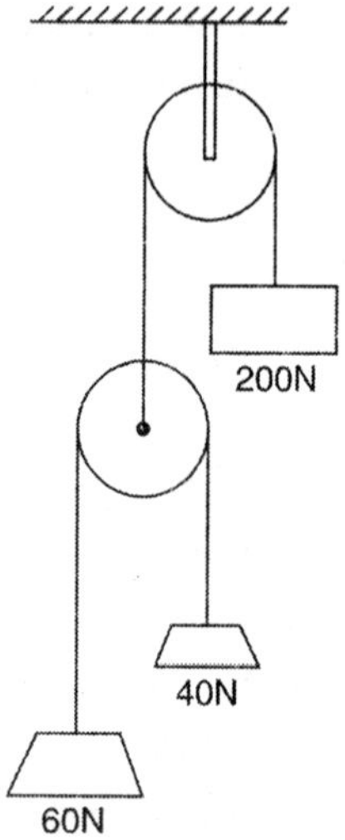


Fig. 8.20

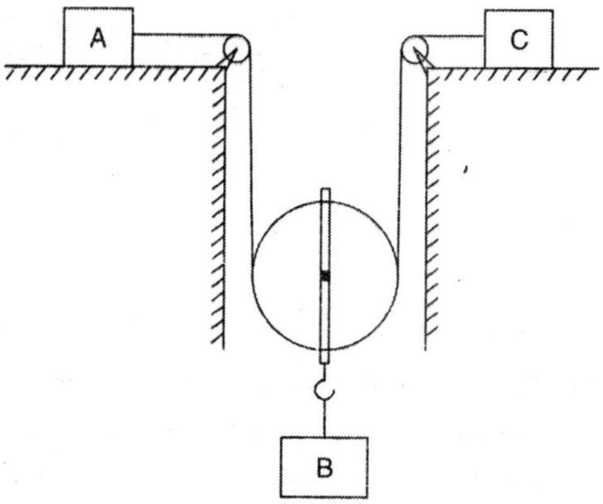


Fig. 8.21

CHAPTER

# 9

# WORK-ENERGY METHOD

In the last chapter, kinetic problems were solved using D' Alembert's dynamic equilibrium condition. In this chapter, another approach, called Work-Energy approach, is used to solve kinetic problems. This method is advantageous over D'Alembert's method when the problem involves velocities, rather than acceleration. The terms work, energy and power are explained first, then the work-energy equation is derived. Using this equation, a number of kinetic problems are solved.

## 9.1 WORK

The *work done* by a force on a moving body is defined as *the product of the force and the distance moved in the direction of the force.* In Fig. 9.1, various forces acting on a particle are shown. If the particle moves a distance $s$ in $x$ direction, from $A$ to $B$, then the work done by various forces are as given below:

| *Force* | *Work done* |
|---|---|
| $P_1$ | $P_1 s$ |
| $P_2$ | $+P_2(-s) = -P_2 s$ |
| $P_3$ | $P_3 \times 0 = 0$ |
| $P_4$ | $P_4 s \cos\theta$ |

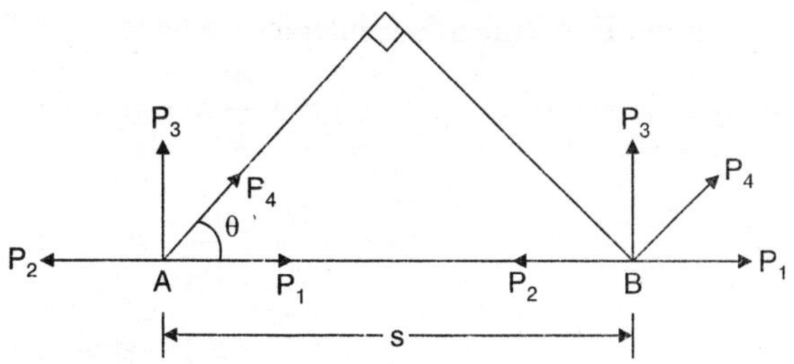


**Fig. 9.1**

The general expression for work done by a force **P** is $Ps \cos\theta$, where $\theta$ is the angle between the force and the direction of motion. This expression may be rearranged as $P \cos\theta \times s$, giving the new definition of work done. Thus work done by a force may be defined as *the product of component of force in the direction of motion and the distance moved.*

From the definition of work, it is obvious that unit of work is obtained by multiplying unit of force by unit of length. Hence, if newton is unit of force and metre is unit of displacement, unit of work will be N-m. One N-m of work is denoted by the term *Joule* (*J*). Hence one Joule may be defined as *the amount of work done by one Newton force when the particle moves* 1 *metre in the direction of that force.*

The other commonly used units are: *kilo joules kJ* (*i.e.*, kN-m) or *milli Joules mJ* (N-mm), etc.

## 9.2 WORK DONE BY A VARYING FORCE

Let the varying force acting at any instance on the particle be $P$. Now if the particle moves a small distance $\delta s$, then the work done by the force is $P \times \delta s$. Work done by the force in moving the body by a distance $s$ is $\Sigma Ps$. Thus if a force versus displacement curve is drawn (Fig. 9.2), the area under the curve gives the work done by the force. If the variation of $P$ is in a regular fashion, then

$$\Sigma Ps = \int P\, ds \qquad ...(9.1)$$

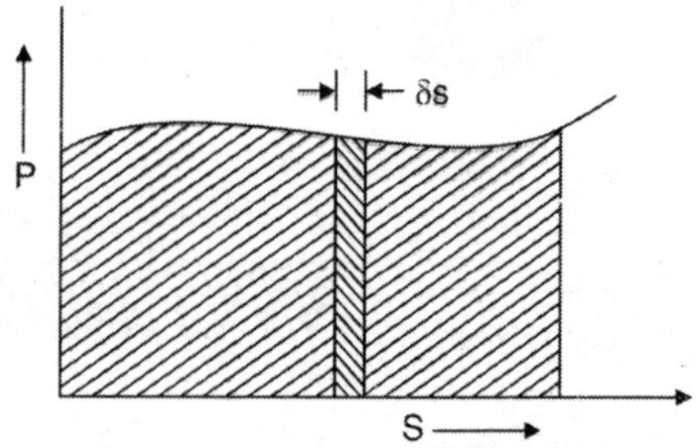


**Fig. 9.2**

## 9.3 ENERGY

*Energy* is defined as the *capacity to do work.* There are many forms of energy such as heat energy, mechanical energy, electrical energy and chemical energy. In engineering mechanics, we are interested in mechanical energy. This energy may be classified into potential energy and kinetic energy.

Potential energy is the capacity to do work *due to the position of the body.* A body of weight '$W$' held at a height $h$ possesses an energy $Wh$.

*Kinetic energy* is *the capacity to do work due to motion of the body.* Consider a car moving with a velocity $v$ m/s. (Fig. 9.3). If the engine is stopped, it still moves forward, doing work against frictional resistance and stops at a certain distance $s$. From the kinematic of the motion, we have

$$0 - u^2 = 2as, \qquad a = -\frac{u^2}{2s}$$

From D'Alembert's principle, we have

$$F + \frac{W}{g}a = 0$$

i.e.,

$$F - \frac{W}{g} \times \frac{u^2}{2s} = 0$$

or

$$F = \frac{Wu^2}{2gs}$$

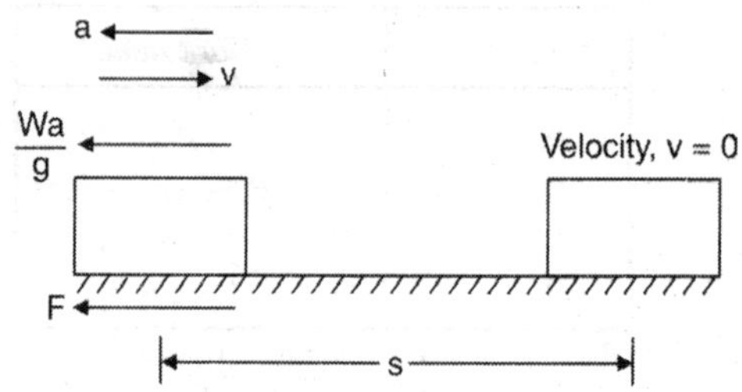


**Fig. 9.3**

$$\therefore \qquad \text{Work done} = F \times s = \frac{Wu^2}{2g} \qquad ...(9.2)$$

This work is done by the energy stored initially in the body.

$$\therefore \qquad \text{Kinetic energy} = \frac{1}{2} \times \frac{W}{g} v^2; \text{ where } v \text{ is the velocity of the body.} \qquad ...(9.3)$$

Unit of energy is same as that of work, since it is nothing but capacity to do work. It is measured in *Joules* J (N-m) or *kilo Joules* kJ (*i.e.*, kN-m).

## 9.4 POWER

*Power* is defined as *time rate of doing work.* Unit of power is *Watt* (*W*) and is defined as one Joule of work done in one second. In practice, kilowatt is the commonly used unit which is equal to 1000 watts. Horse power is the unit used in MKS and FPS systems.

1 metric H.P. = 735.75 watts

1 British H.P. = 745.8 watts

**Example 9.1.** A pump lifts 40 $m^3$ of water to a height of 50 m and delivers it with a velocity of 5 m/s. What is the amount of energy spent during this process? If the job is done in half an hour, what is the input power of the pump which has an overall efficiency of 70%?

**Solution.** Output energy of the pump is spent in lifting 40 $m^3$ of water to a height of 50 m and delivering it with the given kinetic energy of delivery.

Work done in lifting 40 $m^3$ of water to a height of 50 m is,

$$= Wh, \text{ where } W \text{ is weight of 40 m}^3 \text{ of water}$$
$$= 40 \times 9810 \times 50$$
$$= 1{,}96{,}20000 \text{ N-m}$$

[**Note:** 1 $m^3$ of water weighs = 9810 Newtons]

$$\text{Kinetic energy at delivery} = \frac{1}{2} \times \frac{W}{g} v^2 = \frac{1}{2} \times \frac{40 \times 9810}{9.81} \times 5^2$$
$$= 5{,}00000 \text{ N-m}$$
$$\text{Total energy spent} = 1{,}96{,}20000 + 5{,}00000$$

***Energy spent* = 2,01,20000 N-m Ans.**

This energy is spent by the pump in half an hour, *i.e.*, in 30 × 60 = 1800 s.

$$\therefore \quad \text{Output power of pump} = \text{Output energy spent per second}$$
$$= \frac{2{,}01{,}20000}{1800} = 11177.8 \text{ watts}$$
$$= 11.1778 \text{ kW.}$$

$$\textbf{Input power} = \frac{\text{Output power}}{\text{Efficiency}}$$
$$= \frac{11.1778}{0.7} = \mathbf{15{,}9683 \text{ kW}}$$ **Ans.**

**Example 9.2.** A man wishes to move wooden box of 1 metre cube to a distance of 5 m with the least amount of work. If the block weighs 1 kN and the coefficient of friction is 0.3, find whether he should tip it or slide it.

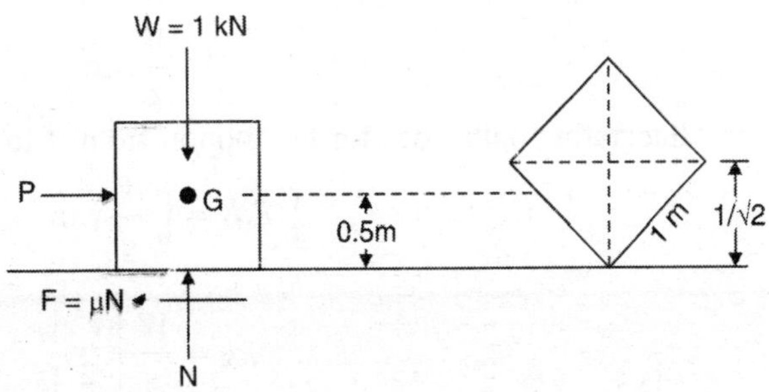


**Fig. 9.4**

**Solution.** *Work done in sliding*

Normal reaction **N** = **W** = 1 kN

∴ Friction force = **F** = $\mu N$ = 0.3 kN

Applied force **P** = **F** = 0.3 kN

Work to be done in sliding to a distance of 5 m = $P \times 5$

$$= 0.3 \times 5 = 1.5 \text{ kN-m}$$
$$= 1.5 \text{ kJ} \qquad ...(i)$$

*Work to be done in tipping*:

In one tipping, the centre of gravity of box is to be raised to a height (Ref. Fig. 9.4)

$$= \frac{1}{\sqrt{2}} - 0.5 = 0.207 \text{ m}$$

∴ Work done in one tipping $= Wh$

$$= 1 \times 0.207$$
$$= 0.207 \text{ kJ}$$

To move a distance of 5 m, five tippings are required. Hence work to be done in moving it by 5 metres, by tipping

$$= 5 \times 0.207 = 1.035 \text{ kJ} \qquad ...(ii)$$

Since the man needs to spend only 1.035 kJ while tipping and it is less than 1.5 kJ to be spent in sliding, the man should move the box by tipping.

## 9.5 WORK–ENERGY EQUATION FOR TRANSLATION

Consider the body as shown in Fig. 9.5, subject to a system of forces $F_1$, $F_2$, ... and moving with an acceleration $a$ in $x$-direction. Let its initial velocity at $A$ be $u$ and final velocity when it moves distance $AB = s$ be $v$. Then the resultant of system of the forces must be in $x$-direction. Let

$$R = \Sigma F_x \qquad ...(9.4)$$

From Newton's second law of motion,

$$R = \frac{W}{g} a$$

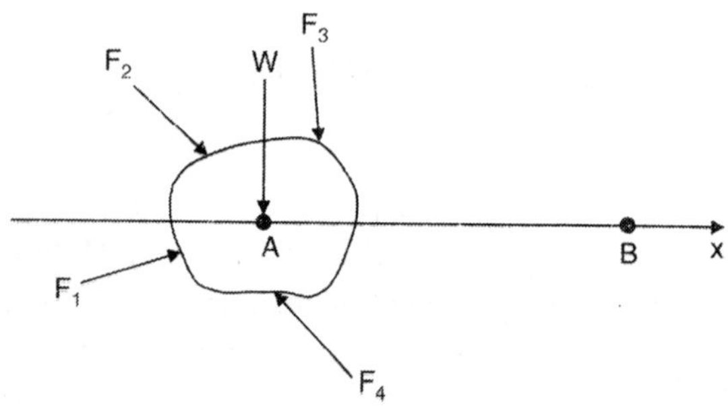


Fig. 9.5

Multiplying both sides by elementary distance $ds$, we get

$$Rds = \frac{W}{g} a \, ds$$
$$= \frac{W}{g} v \frac{dv}{ds} ds \quad \text{since } a = v \frac{dv}{ds}$$
$$= \frac{W}{g} v dv$$

Integrating both sides for the motion from $A$ to $B$, we get

$$\int_0^s R ds = \int_u^v \frac{W}{g} v \, dv$$

$$Rs = \frac{W}{g} \left[ \frac{v^2}{2} \right]_u^v$$

$$= \frac{W}{2g} \left( v^2 - u^2 \right) \qquad ...(9.5)$$

Now $Rs$ is the work done by the forces acting on the body. $\frac{W}{2g}v^2$ is final kinetic energy and $\frac{Wu^2}{2g}$ is initial kinetic energy. Hence we can say, work done in a motion is equal to change in kinetic energy. That is,

$$\text{Work done} = \text{Final kinetic energy} - \text{Initial kinetic energy} \qquad ...(9.6)$$

and it is called *Work-Energy Equation.*

This work energy principle may be stated as *the work done by a system of forces acting on a body during a displacement is equal to the change in kinetic energy of the body during the same displacement.*

Using this work energy equation a number of kinetic problems can be solved. This will be found more useful than D' Alembert's principle when we are not interested in finding acceleration in the problem, but mainly interested in velocity and distance.

**Example 9.3.** A body weighing 300 N is pushed up a 30° plane by a 400 N force acting parallel to the plane. If the initial velocity of the body is 1.5 m/s and coefficient of kinetic friction is $\mu = 0.2$, what velocity will the body have after moving 6 m?

**Solution.** Consider the free body diagram of the body as shown in Fig. 9.6.

Sum of forces normal to plane = 0, gives $N = 300 \times \cos 30° = 259.81$ Newton

$\therefore$ Frictional force $F = \mu N$

$= 0.2 \times 259.81 = 51.96$ N

Initial velocity $u = 1.5$ m/s

Displacement $s = 6$ m

Let the final velocity be = $v$ m/s

Equating the work done by forces along the plane to change in kinetic energy, we get

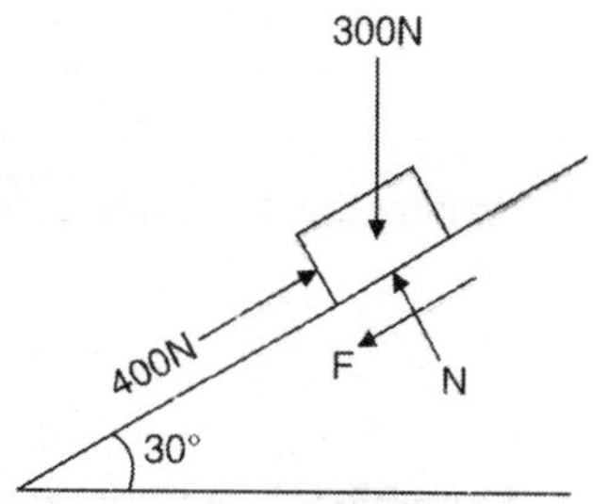

Fig. 9.6

$$(400 - F - W\sin\theta)s = \left(\frac{1}{2}\right)\frac{W}{g}\left(v^2 - u^2\right)$$

$$\left(400 - 51.96 - 300 \times \frac{1}{2}\right)6 = \frac{1}{2} \times \left(\frac{300}{9.81}\right)\left(v^2 - 1.5^2\right)$$

$$77.71 = v^2 - 2.25$$

$\therefore$ $\quad v = \mathbf{8.942\ m/s}$ **Ans.**

**Example 9.4.** Find the power of a locomotive, drawing a train whose weight including that of engine is 420 kN up an incline 1 in 120 at a steady speed of 56 kmph, the frictional resistance being 5 N/kN.

While the train is ascending the incline, the steam is shut off. Find how far it will move before coming to rest, assuming that the resistance to motion remains the same.

**Solution.** Figure 9.7(*a*) shows the system of forces acting on the locomotive while moving up the incline with steady speed.

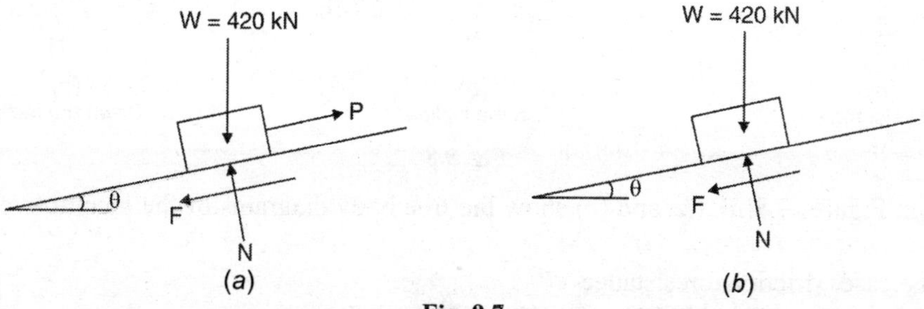

Fig. 9.7

$$v = 56 \text{ kmph} = \frac{56 \times 1000}{60 \times 60} = 15.556 \text{ m/s}$$

$$F = 5 \times 420 = 2100 \text{ Newton}$$
$$= 2.1 \text{ kN}$$

$$P = F + W \sin \theta = 2.1 + 420 \times \frac{1}{120}$$
$$= 5.6 \text{ kN}$$

∴ **Power of locomotive** = Work done by $P$ per second

$= P \times$ Distance moved per second

$= P \times v = 5.6 \times 15.556$

$=$ **87.11 kN** **Ans.**

When steam is put off, let it move a distance $s$ before coming to rest

Initial velocity $u = 15.556$ m/s

Final velocity $v = 0$

Figure 9.7($b$) shows the system of forces acting in this motion. Resultant force parallel to the plane is

$$= F + W \sin \theta$$

$$= 2.1 + 420 \times \left(\frac{1}{120}\right) = 5.6 \text{ kN (down the plane.)}$$

Writing the work-energy equation for motion up the plane, we have

$$-5.6 \times s = \frac{1}{2} \times \frac{420}{9.81} (0 - 15.556^2)$$

$s =$ **924.98 m** **Ans.**

**Example 9.5.** A tram car weighs 120 kN, the tractive resistance being 5 N/kN. What power will be required to propel the car at a uniform speed of 20 kmph?

1. On level surface.
2. Up an incline of 1 in 300.
3. Down an inclination of 1 in 300.

Take efficiency of motor as 80%.

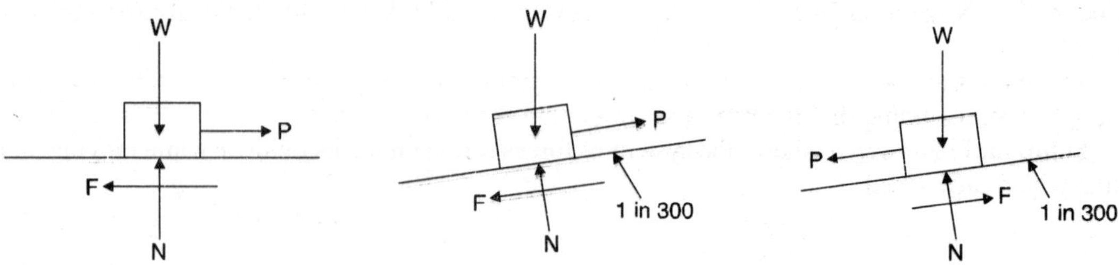


($a$) On level track ($b$) Up the incline ($c$) Down the incline

**Fig. 9.8**

**Solution.** Figures 9.8($a$), ($b$) and ($c$) show the free body diagrams of the locomotive in the three cases given.

In all the cases, frictional resistance

$$F = 5 \text{ N/kN} = 5 \times 120 = 600\ N = 0.6 \text{ kN}$$

The locomotive is moving with uniform velocity. Hence it is in equilibrium.

Now, $v = 20 \text{ kmph} = \dfrac{20 \times 1000}{60 \times 60} = 5.556 \text{ m/s}$

**1. On level track:**

$$P = F = 0.6 \text{ kN}$$

$\therefore$ $$\text{Output power} = Pv = 0.6 \times 5.556 = 3.3333 \text{ kW}$$

$$\eta = 80\% = 0.8$$

$\therefore$ $$\textbf{Input power} = \frac{\text{Output power}}{\eta} = \frac{3.3333}{0.8} = \mathbf{4.167 \text{ kW}}$$ **Ans.**

**2. Up the plane:**

The component of weight $W \sin \theta$ acts down the plane and $\sin \theta \approx \tan \theta = \dfrac{1}{300}$

$\therefore$ $$P = F + W \sin \theta = 0.6 + 120 \times \frac{1}{300} = 1 \text{ kN}$$

$$\text{Output power required} = P \times v = 1 \times 5.5556 = \mathbf{5.5556 \text{ kW}}$$

$\therefore$ $$\textbf{Input power of engine} = \frac{\text{Output power}}{\eta} = \frac{5.5556}{0.8} = \mathbf{6.94 \text{ kW}}$$ **Ans.**

**3. Down the incline plane:**

Referring to Fig. 9.8(*c*), we have

$$P = F - W \sin \theta = 0.6 - 120 \times \frac{1}{300} = 0.2$$

$\therefore$ $$\text{Output power} = 0.2 \times 5.5556 = 1.1111 \text{ kW}$$

$$\textbf{Input power} = \frac{1.1111}{0.8} = \mathbf{1.389 \text{ kW}}$$ **Ans.**

**Example 9.6.** In a police investigation of tyre marks, it was concluded that a car while in motion along a straight level road skidded for a total of 60 metres after the brakes were applied. If the coefficient of friction between the tyres and the pavement is estimated as 0.5, what was the probable speed of the car just before the brakes were applied?

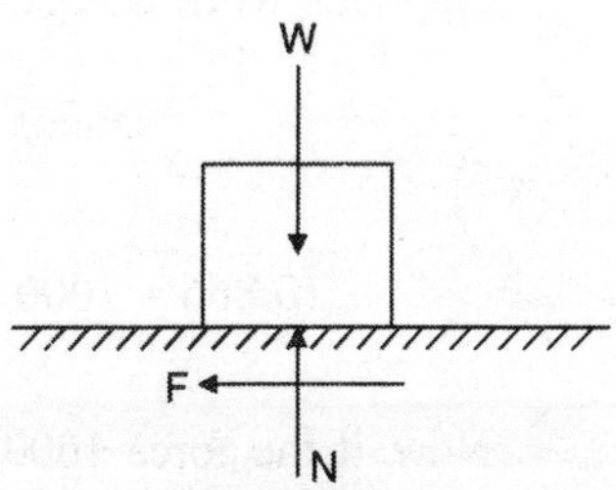

**Fig. 9.9**

**Solution.** Let the probable speed of the car just before brakes were applied be *u* m/s. Free body diagram of the car is shown in Fig. 9.9.

Now, $\Sigma F_v = 0$, gives

$$N = W$$

$\therefore$ $$F = \mu N = \mu W = 0.5 \text{ W} \quad \ldots(1)$$

Only force in the direction of motion is *F.*

Now, Final velocity = 0

Displacement, $s = 60$ m

Applying work-energy equation, we have

$$-F \times s = \frac{W}{2g}\left(v^2 - u^2\right)$$

$$-0.5\, W \times 60 = \frac{W}{2 \times 9.81}\left(0 - u^2\right)$$

$\therefore$ $u = 24.261$ m/s

$$= \frac{24.261 \times 60 \times 60}{1000} = \mathbf{87.34\ kmph}$$ **Ans.**

**Example 9.7.** A block weighing 2500 N rests on a level horizontal plane for which coefficient of friction is 0.20. This block is pulled by a force of 1000 N acting at an angle of 30° to the horizontal. Find the velocity of the block after it moves 30 m starting from rest. If the force of 1000 N is then removed, how much further will it move? Use work-energy method.

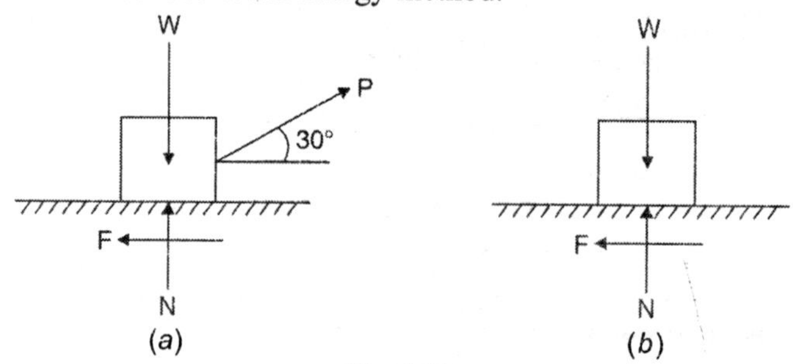


**Fig. 9.10**

**Solution.** Free body diagrams of the block for the two cases are shown in Figs. 9.10(*a*) and (*b*).

When pull *P* is acting:

$$\mathbf{N} = W - P \sin 30°$$
$$= 2500 - 1000 \sin 30° = 2000 \text{ Newton}$$
$$\mathbf{F} = \mu N = 0.2 \times 2000$$
$$= 400 \text{ Newton}$$

Initial velocity = 0

Let final velocity be *v*.

Displacement $(s) = 30$ m

Applying work-energy equation for the horizontal motion, we have

$$(P \cos 30° - F)s = \frac{W}{2g}\left(v^2 - u^2\right)$$

$$(0.866 \times 1000 - 400)\, 30 = \frac{2500}{2 \times 9.81}\left(v^2 - 0\right)$$

$$\mathbf{v = 10.4745\ m/s}$$ **Ans.**

Now, if the force 1000 N is removed (Ref. Fig. 9.10), let the distance moved be '*s*' before the body comes to rest.

$\therefore$ Initial velocity = 10.4745 m/s

Final velocity = 0

Applying work-energy equation for the motion in horizontal direction, we get

$$-F \times s = \frac{W}{2g}\left(v^2 - u^2\right)$$

$$-400 \times s = \frac{2500}{2 \times 9.81}\left(0 - 10.4745^2\right)$$

$$s = \mathbf{27.96\ m}$$

**Ans.**

**Example 9.8.** A small block starts from rest at point *A* and slides down the inclined plane as shown in Fig. 9.11(*a*). What distance along the horizontal plane will it travel before coming to rest? The coefficient of kinetic friction between the block of either plane is 0.3. Assume that the initial velocity with which it starts to move along *BC* is of the same magnitude as that gained in sliding from *A* to *B*.

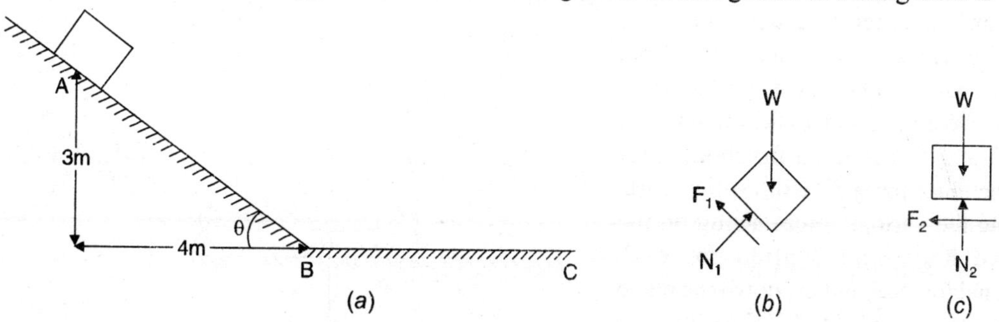


**Fig. 9.11**

**Solution.** Length $AB = \sqrt{3^2 + 4^2} = 5$ m

$\therefore$ $\sin\theta = 0.6$ and $\cos\theta = 0.8$

Consider the free body diagram of the block on inclined plane at *A* [Fig. 9.11(*b*)].

It moves down the plane. Hence

Sum of forces normal to plane = 0,. gives

$$N_1 = W\cos\theta = W \times 0.8$$

$\therefore$ $$F_1 = \mu N_1 = 0.3\ W \times 0.8 = 0.24\ W$$

Applying work-energy equation for the motion from *A* to *B*, velocity at *B*, $v_B$ is given by

$$(W\sin\theta - F_1)s = \frac{W}{2g}\left(v_B^2 - 0\right)$$

$$(0.6\ W - 0.24\ W) \times 5 = \frac{W}{2 \times 9.81}\left(v_B^2\right)$$

$$v_B = 5.943 \text{ m/s}$$

For the motion on horizontal plane, free body diagram of the block is shown in Fig. 9.11(*c*).

Initial velocity = $v_B$ = 5.943 m/s

Final velocity = 0

Writing work-energy equation for the motion along *BC*, we have

$$-F_2 s = \frac{W}{2g}\left(0 - v_B^2\right)$$

but $$F_2 = \mu N = \mu W = 0.3W$$

$\therefore$ $$-0.3Ws = -\frac{W}{2 \times 9.81}(5.943)^2$$

$\therefore$ $$s = \mathbf{6\ m}$$

**Ans.**

## 9.6 MOTION OF CONNECTED BODIES

Work-energy equation may be applied to the connected bodies also. There is no need to separate connected bodies and work out for forces in the connecting member. Note the various forces acting on connected bodies. Equate summation of work done by forces acting on bodies to the summation of change in kinetic energy of the bodies. While writing work done note that force components in the direction of motion are to be multiplied by the distance moved.

Consider the two connected bodies as shown in Fig. 9.12.

Under the pull $P$, both bodies move the same distance and with the same velocity. Hence, initial velocity, final velocity and displacement are the same for the two bodies. Out of the three forces $W_1$, $N_1$ and $F_1$ acting on first body, only the frictional force $F_1$ will do the work. Among the various forces acting on the second body, the applied force $P$, frictional force $F_2$ and the down the plane component of weight, $W_2 \sin\theta$ will do the work. Hence the work-energy equation for that system will be

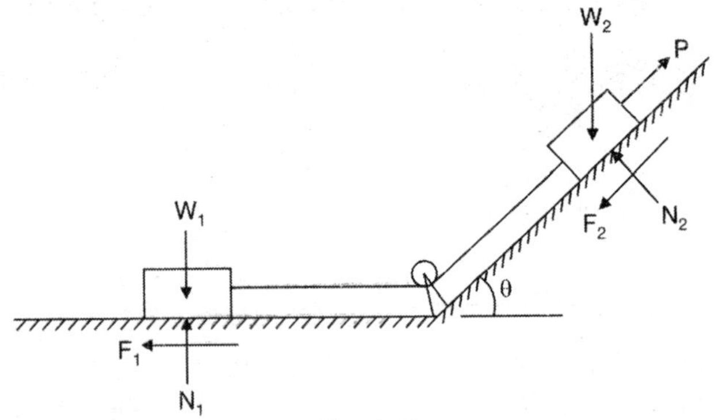


**Fig. 9.12**

$$-F_1 s + (P - F_2 - W_2 \sin\theta)s = \frac{W_1}{2g}\left(v^2 - u^2\right) + \frac{W_2}{2g}\left(v^2 - u^2\right)$$

*i.e.*,
$$(-F_1 + P - F_2 - W_2 \sin\theta)s = \frac{W_1 + W_2}{2g}\left(v^2 - u^2\right)$$

**Example 9.9.** Determine the constant force $P$ that will give the system of bodies shown in Fig. 9.13($a$) a velocity of 3 m/s after moving 4.5 m from rest. Coefficient of friction between the blocks and the plane is 0.3. Pulleys are smooth.

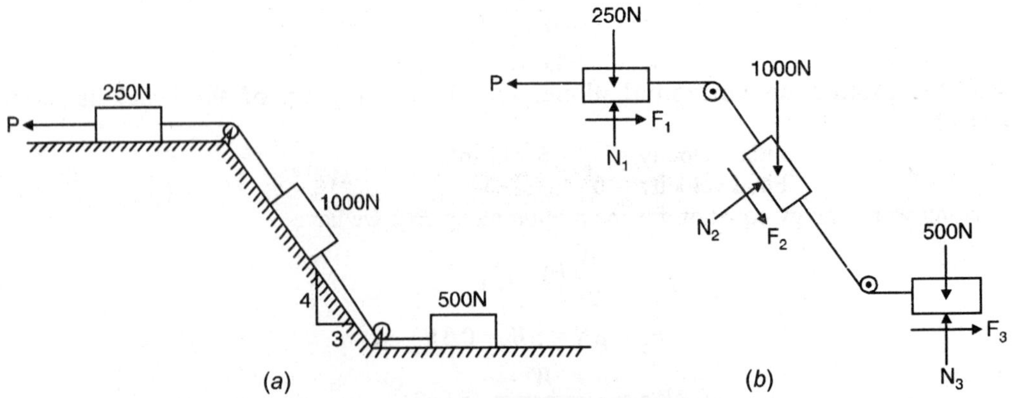


**Fig. 9.13**

**Solution.** The system of forces acting on connecting bodies is shown in Fig. 9.13(*b*).

$$N_1 = 250 \text{ N}$$

$$\therefore \quad F_1 = \mu N_1 = 0.3 \times 250 = 75 \text{ N}$$

$$N_2 = 1000 \cos \theta = 1000 \times \frac{3}{5} = 600 \text{ N}$$

$$\therefore \quad F_2 = 0.3 \times N_2 = 0.3 \times 600 = 180 \text{ N}$$

$$N_3 = 500 \text{ N}$$

$$\therefore \quad F_3 = \mu N_3 = 0.3 \times 500 = 150 \text{ N}$$

Let the constant force be *P*. Writing work-energy equation, we have

$$(P - F_1 - F_2 - 1000 \sin \theta - F_3)s = \frac{W_1 + W_2 + W_3}{2g}(v^2 - u^2)$$

$$(P - 75 - 180 - 1000 \times 0.8 - 150)\, 4.5 = \frac{250 + 1000 + 500}{2 \times 9.81}(3^2 - 0)$$

$$P = 1383.39 \text{ N}$$ **Ans.**

[**Note:** Work done is force × distance moved in the direction of force. Hence, $N_1$, $N_2$, $N_3$, 250, 1000 cos θ and 500 forces are not contributing to the work done.]

**Example 9.10.** In what distance will body *A* of Fig. 9.14 attain a velocity of 3 m/s starting from rest?

**Solution.** Take μ = 0.2. Pulleys are frictionless and weightless.

Let $\theta_1$ and $\theta_2$ be the slopes of inclined planes.

Then
$$\sin \theta_1 = \frac{4}{5} = 0.8, \quad \cos \theta_1 = 0.6$$

$$\sin \theta_2 = \frac{3}{5} = 0.6, \quad \cos \theta_2 = 0.8$$

By observing pulley system, it may be concluded that if 1500 N body moves a distance '*s*', 2000 *N* body moves a distance 0.5 *s*, and if velocity of 1500 N block is *v* that of 2000 N block will be 0.5 v. Assuming 2000 N body moves up the plane and 1500 N body moves down the plane, the forces acting on 1500 N that will do work are:

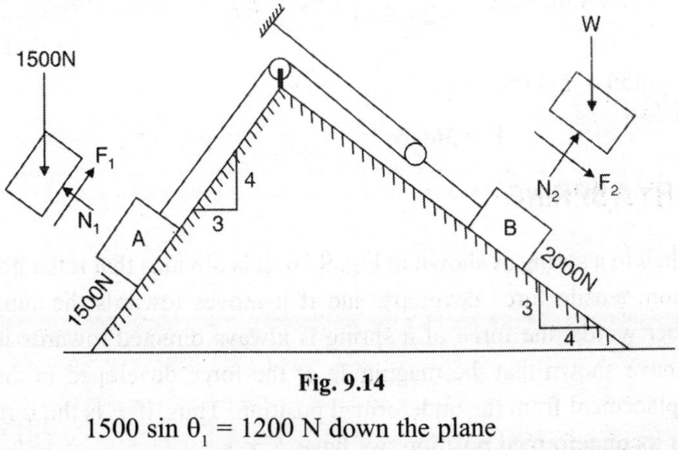


**Fig. 9.14**

$$1500 \sin \theta_1 = 1200 \text{ N down the plane}$$

and $F_1 = \mu \times 1500 \cos\theta_1 = 0.2 \times 1500 \times 0.6 = 180$ N up the plane

The forces acting on 2000 N block that do work when it slides up are:

$$2000 \sin\theta_2 = 2000 \times 0.6 = 1200 \text{ N down the plane}$$

and
$$F_2 = 0.2 \times 2000 \cos\theta_2 = 0.2 \times 2000 \times 0.8$$
$$= 320 \text{ N down the plane.}$$

Equating work done by various forces to change in the kinetic energy of the system, we get

$$(1200 - 180)s - (1200 + 320)\, 0.5s = \left\{\frac{1500}{2 \times 9.81}\right\}(v^2 - 0) + \frac{2000}{2 \times 9.81}\left\{(0.5v)^2 - 0\right\}$$

In the present case, $v = 3$ m/s

$$260s = \frac{1500}{2 \times 9.81} \times 3^2 + \frac{2000}{2 \times 9.81} \times 1.5^2$$

$$s = \mathbf{3.529 \text{ m}}$$ **Ans.**

Since $s$ is positive, the assumed direction of motion is correct. If it comes out negative, note that recalculations are to be made since the frictional force changes the sign, if motion is reversed.

**Example 9.11.** Two bodies weighing 300 N and 450 N are hung to the ends of a rope passing over an ideal pulley as shown in Fig. 9.15(*a*). How much distance the blocks will move in increasing the velocity of system from 2 m/s to 4 m/s? How much is the tension in the string? Use work-energy method.

**Solution.** 450 N block moves down and 300 N block moves up. The arrangement is such that both bodies will be having same velocity and both will move by the same distance. Let '$s$' be the distance moved. Writing work-energy equation for the system, we get

$$450s - 300s = \left[\frac{450}{2 \times 9.81}\right](v^2 - u^2) + \left[\frac{300}{2 \times 9.81}\right](v^2 - u^2)$$

$$150s = \left[\frac{450}{2 \times 9.81}\right] \times (4^2 - 2^2) + \left[\frac{300}{2 \times 9.81}\right](4^2 - 2^2)$$

$$s = \mathbf{3.058 \text{ m}}$$ **Ans.**

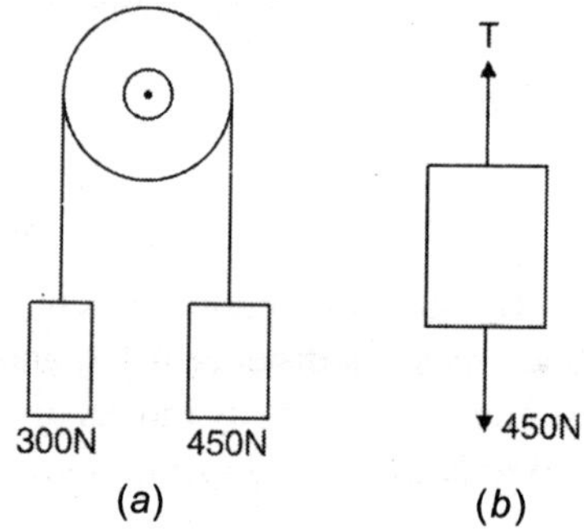

**Fig. 9.15**

Let $T$ be the tension in the string. Consider work-energy equation for any one body, say 450 N body as shown in Fig. 9.15(*b*).

$$450s - Ts = \left[\frac{450}{2 \times 9.81}\right](4^2 - 2^2)$$

$$(450 - T)3.058 = \left[\frac{450}{2 \times 9.81}\right] \times 12$$

$$\mathbf{T = 360 \text{ N}}$$ **Ans.**

## 9.7 WORK DONE BY A SPRING

Consider a body attached to a spring as shown in Fig. 9.16. It is obvious that if the body moves out from its undeformed position, tensile force develops, and if it moves towards the supports, compressive force develops. In other words, the force of a spring is always directed towards its normal position. Experimental results have shown that the magnitude of the force developed in the spring is directly proportional to its displacement from the undeformed position. Thus, if $F$ is the force in the spring due to deformation $x$ from its undeformed position, we have $F \propto x$.

*i.e.,* $F = kx$ ...(9.7)

where the constant of proportionality $k$ is called spring constant and is defined as a force required for unit deformation of the spring. Hence the unit of spring constant is N/m or kN/m.

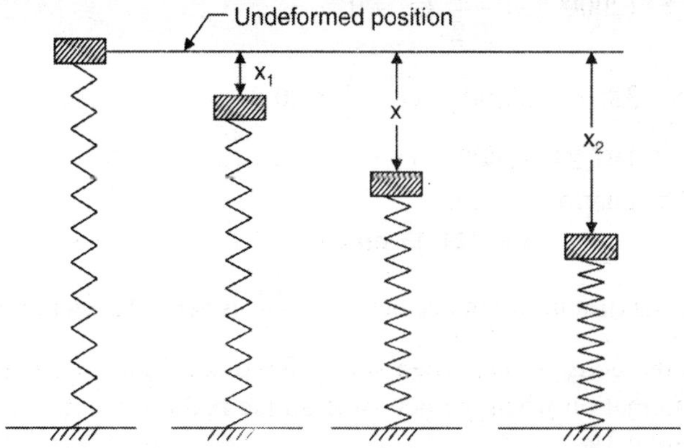


**Fig. 9.16**

At any instant if the displacement is $dx$, the work done by spring force $dU$ is given by

$$dU = -Fdx = -kxdx$$

**Note:** The negative sign is used since the force of spring is in the opposite direction of displacement.

∴ Work done in displacement of the body from $x_1$ to $x_2$ is given by

$$U = \int_{x_1}^{x_2} -kx\,dx = -k\left[\frac{x^2}{2}\right]_{x_1}^{x_2}$$

$$= -\frac{1}{2}k\left(x_2^2 - x_1^2\right)$$

If the work done is to be found in moving from undeformed position to displacement $x$, then

$$U = -\frac{1}{2}k\left(x^2 - 0^2\right)$$

$$= -\frac{1}{2}kx^2 \quad ...(9.8)$$

**Note:** The negative sign is used with the expression for work done by the spring, since whenever spring is deformed, the force of spring is in the opposite direction of deformation. However, if a deformed spring is allowed to move towards its normal position, work done will be positive, since the movement and the force of spring are in the same direction.

**Example 9.12.** A 3000 N block starting from rest as shown in Fig. 9.17 slides down a 50° incline. After moving 2 m, it strikes a spring whose modulus is 20 N/mm. If the coefficient of friction between the block and the incline is 0.2, determine the maximum deformation of the spring and the maximum velocity of the block.

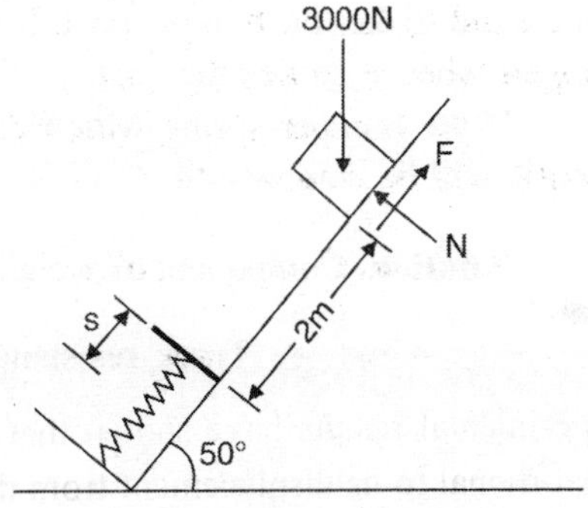


**Fig. 9.17**

**Solution.** Normal reaction $N = 3000 \cos 50°$

∴ Frictional force $F = \mu N = 0.2 \times 3000 \cos 50°$

$= 385.67$ N.

Let the maximum deformation of spring be $s$ mm. The body was at rest and is again at rest, when it moves a distance (2000 + $s$) millimetres. Applying work-energy equation, we have

$$(3000 \sin 50° - F)(2000 + s) - \frac{1}{2} ks^2 = 0$$

$$(3000 \sin 50° - 385.67)(2000 + s) = \frac{1}{2} \times 20 \times s^2$$

$$191.246\,(2000 + s) = s^2$$

Solving the quadratic equation, we get

$$s = \mathbf{721.43\ mm}$$ **Ans.**

The velocity will be maximum, when acceleration $\frac{dv}{dt}$ is zero. Acceleration is zero, when force is zero. Net force acting on the body is zero when spring force developed balances the force exerted by the body. Let $x$ be the deformation when the net force on the body in the direction of motion is zero.

Then referring to Fig. 9.18,

$$kx = W \sin \theta - F$$

$$20x = 3000 \sin 50° - 385.67$$

$$x = 95.62 \text{ mm}$$

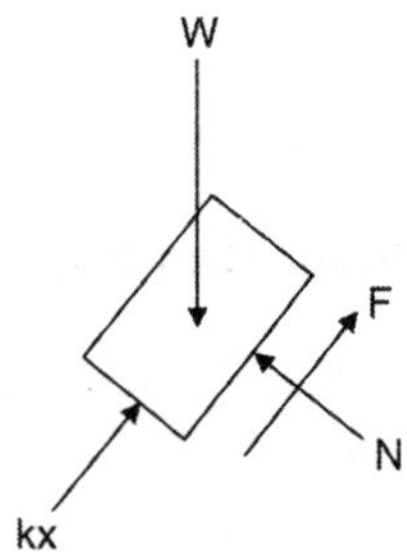

Fig. 9.18

Now applying work-energy equation, we have

$$(3000 \sin 50° - 385.67)(2000 + x) - \frac{1}{2}kx^2 = \frac{3000}{2 \times 9810}(v^2 - 0)$$

$$(3000 \sin 50° - 385.67)(2000 + 95.62) - \frac{1}{2} \times 20 \times 95.67^2$$

$$= \left[\frac{3000}{2 \times 9810}\right] v^2$$

$$v = 5060.9 \text{ mm/s}$$

*i.e.,* $$v = \mathbf{5.061\ m/s}$$ **Ans.**

**Note:** Maintain the consistency of units

$$g = 9.81 \text{ m/s}$$
$$= 9810 \text{ mm/s}$$

**Example 9.13.** A wagon weighing 500 kN starts from rest, runs 30 metre down one per cent grade and strikes the bumper post. If the rolling resistance of the track is 5 N/kN, find the velocity of the wagon when it strikes the post.

If the bumper spring which compresses 1 mm for every 15 kN, determine by how much this spring will be compressed.

**Solution.** Component of weight down the plane $= W \sin \theta = 500 \times \frac{1}{100} = 5$ kN

Track resistance = 5 N/kN

$= 5 \times 500 = 2500$ N

$= 2.5$ kN

The wagon starts from rest ($u = 0$), and moves a distance $s = 30$ m before striking the bumper.

Let the velocity of wagon while striking be $v$ m/s.
Applying work-energy equation (Ref. Fig. 9.19), we get

$$(W \sin \theta - F)\, s = \frac{W}{2g}\left(v^2 - u^2\right)$$

$$(5 - 2.5) \times 30 = \frac{500}{2 \times 9.81}\left(v^2 - 0\right)$$

$$v = 1.716 \text{ m/s}$$

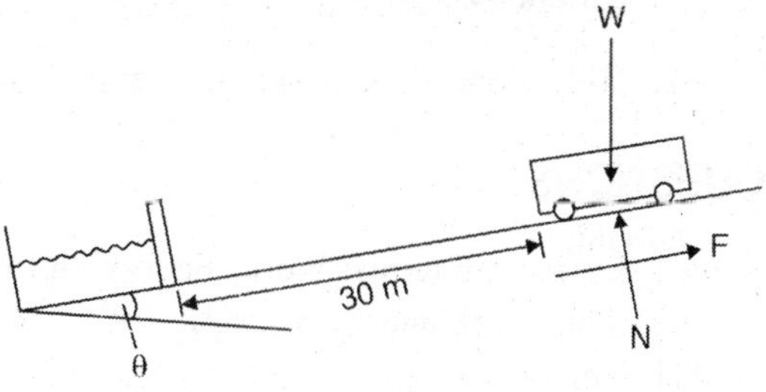


Fig. 9.19

Let spring compression be $x$.
The spring constant is

$$k = 15 \text{ kN/mm}$$
$$= 15000 \text{ kN/m}$$

and velocity is zero.
Applying work-energy equation, we get

$$(5.0 - 2.5)(30 + x) - \frac{1}{2} kx^2 = \frac{W}{2g}(0 - 0)$$

$$2.5(30 + x) - \frac{1}{2} \times 15000\, x^2 = 0$$

$$3000x^2 - x - 30 = 0$$

$$x = \frac{+1 \pm \sqrt{1 + 4 \times 3000 \times 30}}{2 \times 3000}$$

$$= 0.1002 \text{ m}$$

*i.e.,* **$x$ = 100.2 mm** **Ans.**

**Note:** Compared to work done by wagon in moving 30 m, if the work done in moving through $x$ is neglected, then work-energy equation reduces to

$$(5 - 2.5)\, 30 - \frac{1}{2} kx^2 = 0$$

$$x = \sqrt{\frac{2.5 \times 30 \times 2}{15000}} = 0.1 \text{ m}$$

*i.e.,* $x = 100$ mm.

## IMPORTANT DEFINITIONS

1. **Work done** by a force on a moving body is defined as the product of the force and the distance moved in the direction of the force.
2. The amount of work done by one newton force when the particle moves 1 metre in the direction of that force is termed as one **Joule** work.
3. **Energy** is defined as the capacity to do the work. **Potential energy** is the capacity to do work due to the position of the body and **kinetic energy** is the capacity to do the work due to the motion of the body.
4. **Power** is defined as time rate of doing work.

## IMPORTANT FORMULAE

1. Kinetic energy $= \frac{1}{2}\frac{W}{g}v^2$
2. Work done = Final kinetic energy – Initial kinetic energy.

## QUESTIONS

**9.1** Define the terms: Work, Energy and Power.
**9.2** State work-energy principle.
**9.3** Derive the expression for work done by a spring of spring constant '$k$'.
**9.4** Distinguish between potential energy and kinetic energy.

## EXERCISES

**9.1** A 10,000 kN train is accelerated at a constant rate up a 2% grade. The track resistance is constant at 9 N/kN. The velocity increases from 9 m/s to 18 m/s in a distance of 600 metres. Determine the maximum power developed by the locomotive.

[**Ans.** $P$ = 496.422 kN; power = 8935.6 kW]

**9.2** Find the power required to pull a train up an incline of 1 in 200 at a speed of 36 kmph, if the weight of the train is 3000 kN and the track resistance is 5 N/kN. Also determine the maximum speed with which the train moves up an incline of 1 in 100 with the same power.

[**Ans.** Power = 300 kW; $v$ = 24 kmph]

**9.3** On seeing a child on the road, the driver of a car applied brakes instantaneously. The car just ran over the child after skidding a total distance of 28.2 metres in the direction of motion before coming to a stop. The traffic police sued the driver on the ground of overspeeding. The speed limit of the section was 40 kmph. Would you justify the police action if the coefficient of friction between the tyres and the road is 0.5? The weight of the car is 20 kN and it was travelling on a level surface. [**Ans.** $v$ = 59.877 kmph, police action is justified]

**9.4** A mine cage weighs 12 kN and can carry a maximum load of 20 kN. The average frictional resistance of the side guides is 500 N. What constant cable tension is required to give a loaded cage an upward velocity of 3 m/s from rest in a distance of 3 m? [**Ans.** $T$ = 37.47 kN]

**9.5** A 500 N body moves along the two inclines for which the coefficient of friction is 0.2 (Fig. 9.20). If the body starts from rest at $A$ and slides 60 m down the 30° incline, how far will it then move along the other incline? What will be its velocity when it returns to $B$?

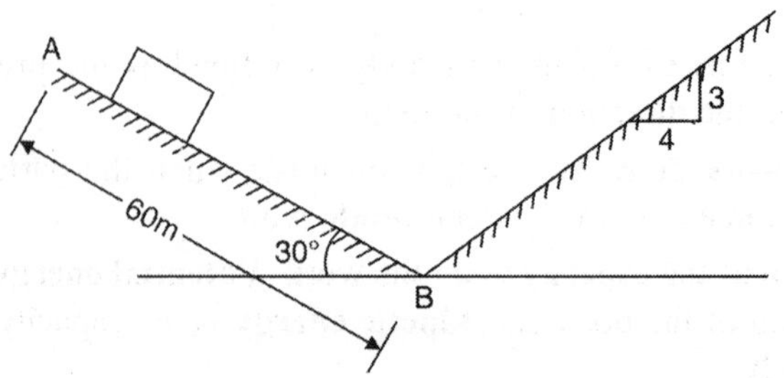


**Fig. 9.20**

[**Ans.** $s$ = 25.8 m; $v$ = 14.92 m/s]

**9.6** In what distance will body $A$ shown in Fig. 9.21 attain a velocity of 3 m/s starting from rest? Take coefficient of friction between the blocks and the plane as 0.2. Assume the pulley is smooth. [**Ans.** $s$ = 6.541 m]

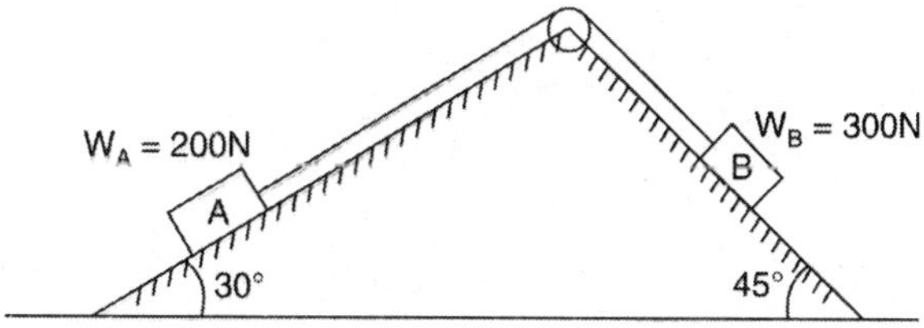


**Fig. 9.21**

**9.7** Two blocks of weights 400 N and 500 N are connected by inextensible flexible wire running around a smooth pulley as shown in Fig. 9.22. Find what will be the velocity of the system if the distance moved by the blocks is 3 m starting from rest. [**Ans.** $v$ = 2.557 m/s]

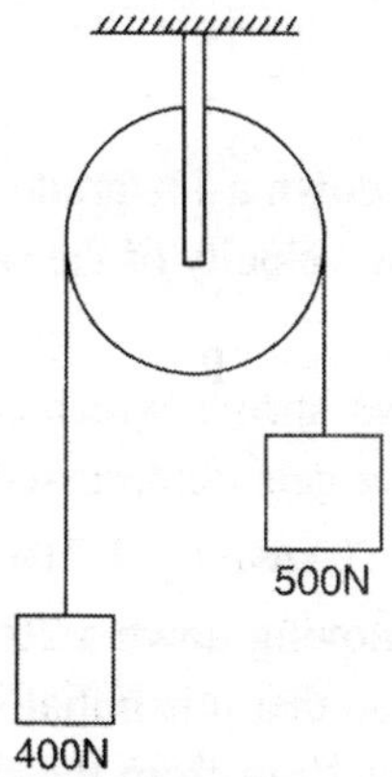


**Fig. 9.22**

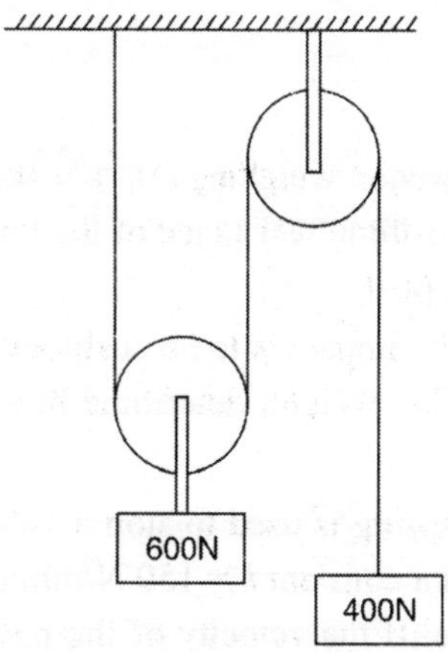


**Fig. 9.23**

**9.8** Two blocks are connected by inextensible wires as shown in Fig. 9.23. Find by how much distance block 400 N will move in increasing its velocity to 5 m/s from 2 m/s. Assume pulleys are frictionless and weightless. [**Ans.** $s$ = 5.89 m]

**9.9** By using work-energy equation, calculate the velocity and acceleration of the block $A$ shown in Fig. 9.24 after it has moved 6 m from rest. The coefficient of kinetic friction is 0.3 and the pulleys are considered to be frictionless and weightless. Also calculate the tension in the string attached to $A$.

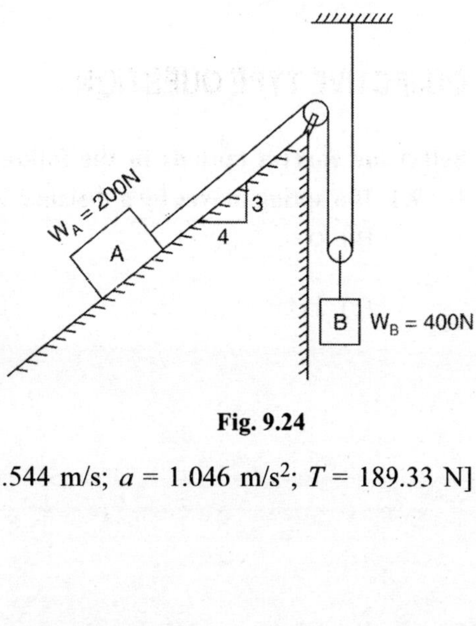


**Fig. 9.24**

[**Ans.** $v$ = 3.544 m/s; $a$ = 1.046 m/s$^2$; $T$ = 189.33 N]

**9.10** How far block $A$ shown in Fig. 9.25 will move when its velocity increases from 3 m/s to 8 m/s? Assume pulleys are weightless and frictionless. Radius of larger pulley is 0.3 m and that of smaller pulley is 0.2 m. **[Ans.** $s$ = 15.885 m]

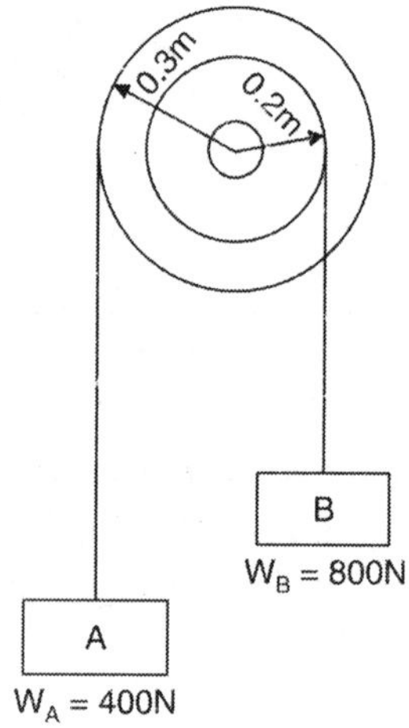


**Fig. 9.25**

**9.11** A wagon weighing 600 kN starts from rest, runs 30 m down a 1% grade and strikes a post. If the rolling resistance of the track is 5 N per kN, find the velocity of the wagon when it strikes the post.

If the impact is to be cushioned by means of one bumper spring, which compresses 1 mm per 25 kN weight, determine how much the bumper spring gets compressed.

**[Ans.** $v$ = 1.716 m/s; $s$ = 84.85 mm]

**9.12** A spring is used to stop a 1000 N package, which is moving down a 20° incline. The spring has a constant $k$ = 150 N/mm and is held by the cables so that it is initially compressed by 100 mm. If the velocity of the package is 6 m/s, when it is 10 m from the spring, determine the maximum additional of the spring in bringing the package to rest. Take coefficient of friction = 0.25. **[Ans.** $s$ = 195.6 mm]

## OBJECTIVE TYPE QUESTION

**Select the correct options in the following:**

**9.1** If a spring moves by a distance '$x$' from its underformed position, work done is equal to

(*a*) $kx$ (*b*) $-kx^2$

(*c*) $kx^2$ (*d*) $-\frac{1}{2}kx^2$

# CHAPTER 10

# MOMENTUM METHOD

It is clear from the discussion of the previous chapters that for solving kinetic problems, involving force and acceleration, D'Alembert's principle is useful and that for the problems involving force, velocity and displacement, the work-energy method is useful. In this chapter, the **Impulse Momentum method** is dealt which is useful for solving the problems involving force, time and velocity.

## 10.1 LINEAR IMPULSE AND MOMENTUM

If $R$ is the resultant force acting on a body of mass $m$, then from Newton's second law,

$$R = ma$$

But acceleration $a = dv/dt$

$$\therefore \quad R = m\frac{dv}{dt}$$

*i.e.,* $Rdt = mdv$

$$\therefore \quad \int Rdt = \int mdv$$

If initial velocity is $u$ and after time interval $t$ it becomes $v$, then

$$\int_0^t Rdt = m\Big[v\Big]_u^v = mv - mu \qquad ...(10.1)$$

The term $\int_0^t R \cdot dt$ is called *impulse*. If the resultant force is in newton and time is in second, the unit of impulse will be N-s.

If $R$ is constant during time interval $t$, then impulse is equal to $R \times t$.

The term mass × velocity is called *momentum*.

Now, $mv = \frac{W}{g}v$

Substituting dimensional equivalence, we get

$$= \frac{\text{N}}{\text{m/s}^2}\ \text{m/s} = \text{N-s}$$

Thus the momentum has also unit N-s. Equation (10.1) satisfies the requirement of dimensional homogenity. Equation (10.1) can now be expressed as

$$\text{Impulse} = \text{Final momentum} - \text{Initial momentum} \quad ...(10.2)$$

Since the velocity is a vector, impulse is also a vector. The impulse momentum equation (Eqn. 10.1 or 10.2) holds good when the directions of $R$, $u$ and $v$ are the same. Impulse momentum equation can be stated as follows:

*The component of the resultant linear impulse along any direction is equal to change in the component of momentum in that direction.*

The impulse momentum equation can be applied in any convenient direction and the kinetic problems involving force, velocity and time can be solved easily.

**Example 10.1.** A glass marble, whose weight is 0.2 N, falls from a height of 10 m and rebounds to a height of 8 metres. Find the impulse and the average force between the marble and the floor, if the time during which they are in contact is 1/10 of a second.

**Solution.** Applying kinematic equations, for the freely falling body (Ref. Fig. 10.1), the velocity with which marble strikes the floor is

$$= \sqrt{2gh} = \sqrt{2 \times 9.81 \times 10}$$

$$= 14.007 \text{ m/s (downward)} \quad ...(i)$$

Similarly applying kinematic equations for the marble moving up, we get the velocity of rebound

$$= \sqrt{2g \times 8} = \sqrt{2 \times 9.81 \times 8}$$

$$= 12.528 \text{ m/s} \quad \text{(upward)} \quad ...(ii)$$

Taking upward direction as positive and applying impulse momentum equation, we get

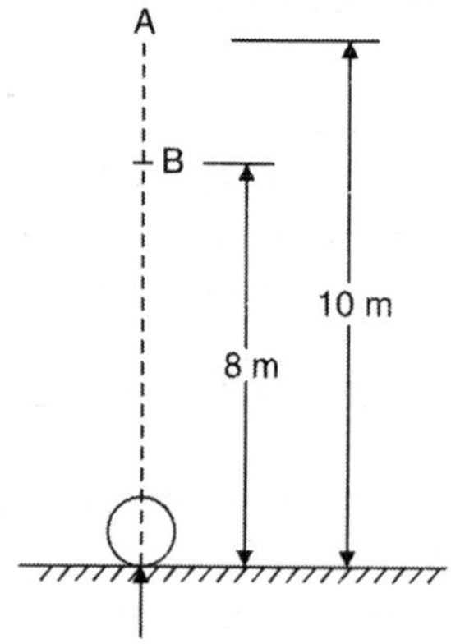


**Fig. 10.1**

$$\text{Impulse} = \frac{W}{g}(v - u) = \frac{0.2}{9.81}[12.52 - (-14.007)]$$

$$= \mathbf{0.541 \text{ N-s}}$$ **Ans.**

If $F$ is the average force, then

$$Ft = 0.541$$

$$F \times 1/10 = 0.541 \text{ N}$$

$$\mathbf{F = 5.41 \text{ N}}$$ **Ans.**

**Example 10.2.** A 1 N ball is bowled to a batsman. The velocity of ball was 20 m/s horizontally just before batsman hit it. After hitting, it went away with a velocity of 48 m/s at an inclination of 30° to horizontal as shown in Fig. 10.2($a$). Find the average force exerted on the ball by the bat if the impact lasts for 0.02 s.

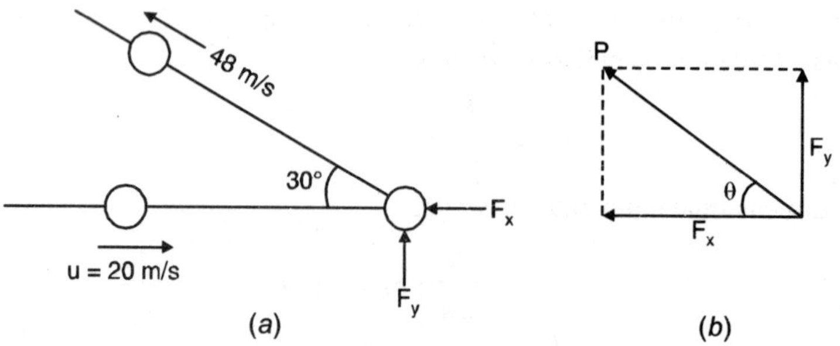


**Fig. 10.2**

**Solution.** Let $F_x$ be the horizontal component and $F_y$ be the vertical component of the force. Applying impulse moment condition in horizontal direction, we get

$$F_x \times 0.02 = \frac{1}{9.81} \ [48 \cos 30° - (-20)]$$

$$F_x = 313.81 \text{ N}$$

Applying impulse momentum equation in vertical direction, we get

$$F_y \times 0.02 = \frac{1}{9.81} \ (48 \sin 30° - 0)$$

$\therefore$ $$F_y = 122.32 \text{ N}$$

$\therefore$ Resultant force

$$F = \sqrt{P_x^2 + P_y^2} = \sqrt{313.81^2 + 122.32^2}$$

$$\mathbf{F = 336.81 \text{ N}}$$ **Ans.**

$$\theta = \tan^{-1}\left(\frac{P_y}{P_x}\right) = \tan^{-1}\left(\frac{122.32}{313.81}\right)$$

*i.e.,* **$\theta$ = 21.30° to horizontal as shown in Fig. 10.2(*b*)** **Ans.**

**Example 10.3.** A 1500 N block is in contact with a level plane, the coefficient of friction between two contact surfaces being 0.1. If the block is acted upon by a horizontal force of 300 N, what time will elapse before the block reaches a velocity of 16 m/s starting from rest ? If 300 N force is then removed, how much longer will the block continue to move? Solve the problem using impulse momentum equation.

**Solution.** Consider the FBD of the block as shown in Fig. 10.3.

Normal reaction = 1500 N

$\therefore$ Frictional force $\mathbf{F} = \mu\mathbf{N} = 0.1 \times 1500 = 150$ N

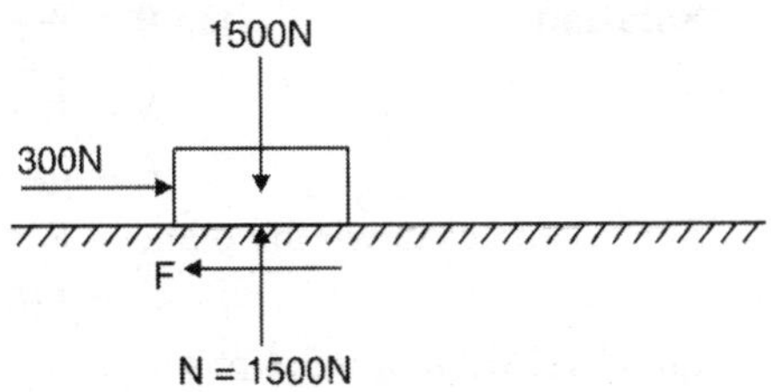


**Fig. 10.3**

Applying impulse momentum equation in the horizontal direction, we have

$$(300 - 150)\ t = \left[\frac{1500}{9.81}\right] (v - u)$$

$$= \left[\frac{1500}{9.81}\right] (16 - 0)$$

$$\mathbf{t = 16.31 \text{ s}}$$ **Ans.**

If force is then removed, the only horizontal force is $F$ = 150 N. Applying impulse momentum equation for the motion towards right, we have

$$-150\ t = \frac{1500}{9.81} (0 - 16)$$

$$\mathbf{t = 16.31 \text{ s}}$$

**The block takes another 16.31 s before it comes to rest.** **Ans.**

**Example 10.4.** A 20 kN-automobile is moving at a speed of 70 kmph when the brakes are fully applied causing all four wheels to skid. Determine the time required to stop the automobile (*a*) on concrete road for which $\mu$ = 0.75, (*b*) on ice for which $\mu$ = 0.08.

**Solution.** Initial velocity of the vehicle

$$u = 70 \text{ kmph} = \frac{70 \times 1000}{60 \times 60} = 19.44 \text{ m/s}$$

Final velocity $v = 0$

Free body diagram is shown in Fig. 10.4

$$F = \mu N = \mu W = 20\ \mu$$

Applying impulse momentum equation, we have

$$-Ft = \frac{W}{g}(v - u)$$

$$-20\ \mu t = \left[\frac{20}{9.81}\right](0 - 19.44)$$

$$\therefore \quad t = \frac{1.982}{\mu}$$

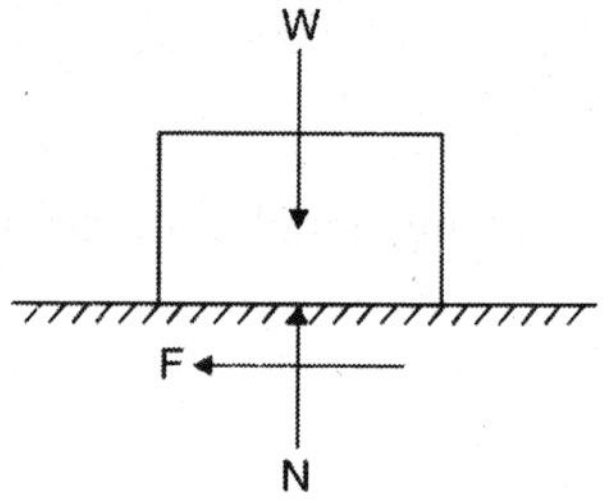

**Fig. 10.4**

(*i*) **On concrete road,** $\mu = 0.75$

$$\therefore \quad t = \frac{1.982}{0.75} = \mathbf{2.64\ s}$$ **Ans.**

(*ii*) **On ice,** $\mu = 0.08$

$$\therefore \quad t = \frac{1.982}{0.08} = \mathbf{24.78\ s}$$ **Ans.**

**Example 10.5.** A block weighing 130 N is on an incline, whose slope is 5 vertical to 12 horizontal. Its initial velocity down the incline is 2.4 m/s. What will be its velocity 5 s. later? Take coefficient of friction at contact surface = 0.3.

**Solution.** $\tan\theta = \frac{5}{12} \quad \therefore \quad \theta = 22.62°$

$$N = W\cos\theta = 130\cos 22.62° = 120 \text{ newton}$$

$$F = \mu N = 0.3 \times 120 = 36 \text{ newton}$$

Sum of forces down the plane $= R = W\sin\theta - F$

$$= 130\sin 22.26° - 36 = 14.0 \text{ N}$$

Initial velocity $u = 2.4$ m/s

Let final velocity be $v$ m/s

Time interval $t = 5$ s

Applying impulse momentum equation,

$$Rt = \frac{W}{g}(v - u), \text{ we get}$$

$$14 \times 5 = \left[\frac{130}{9.81}\right](v - 2.4)$$

$$\therefore \quad \mathbf{v = 7.68\ m/s}$$ **Ans.**

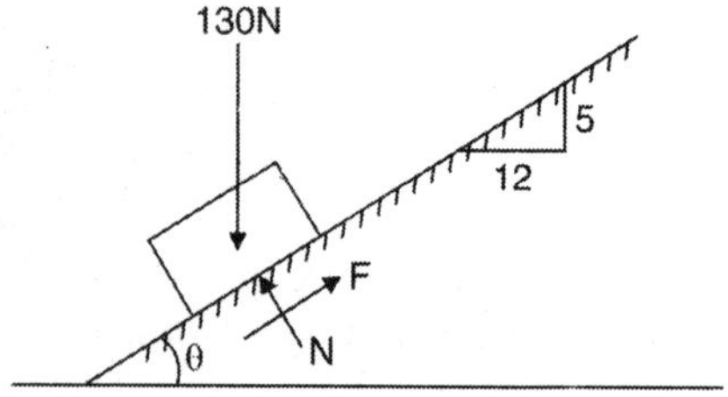

**Fig. 10.5**

## 10.2 CONNECTED BODIES

The problems involving connected bodies may be solved by any one of the following two methods:

*First Method:* Free body diagrams of each body is drawn separately. Impulse momentum equation for each body in the direction of its motion is written, and then the equations are solved to get the required values.

*Second Method:* If the connected bodies have same displacement in the same time, the impulse of internal tension in connecting chords will get cancelled. Hence free body diagram of combined bodies may be considered and impulse momentum equation applied in the direction of motion of combined bodies. *This method is applicable only if displacement of each body is the same in given time.*

**Example 10.6.** Determine the time required for the weights shown in Fig. 10.6(*a*) to attain a velocity of 9.81 m/s. What is the tension in the chord? Take $\mu = 0.2$ for both planes. Assume the pulleys as frictionless.

**Solution.** ***First Method:*** Free body diagram of the two blocks are as shown in Fig. 10.6(*b*).

For 2000 N block:

$$N_1 = W_1 \cos 30°$$
$$= 2000 \cos 30°$$
$$= 1732.05 \text{ N}$$
$$F_1 = \mu N_1 = 0.2 \times 1732.05$$
$$= 346.41 \text{ N}$$

For 1800 N block:

$$N_2 = W_2 \cos 60°$$
$$= 1800 \cos 60° = 900 \text{ N}$$
$$F_2 = \mu N_2 = 0.2 \times 900$$
$$= 180 \text{ N.}$$

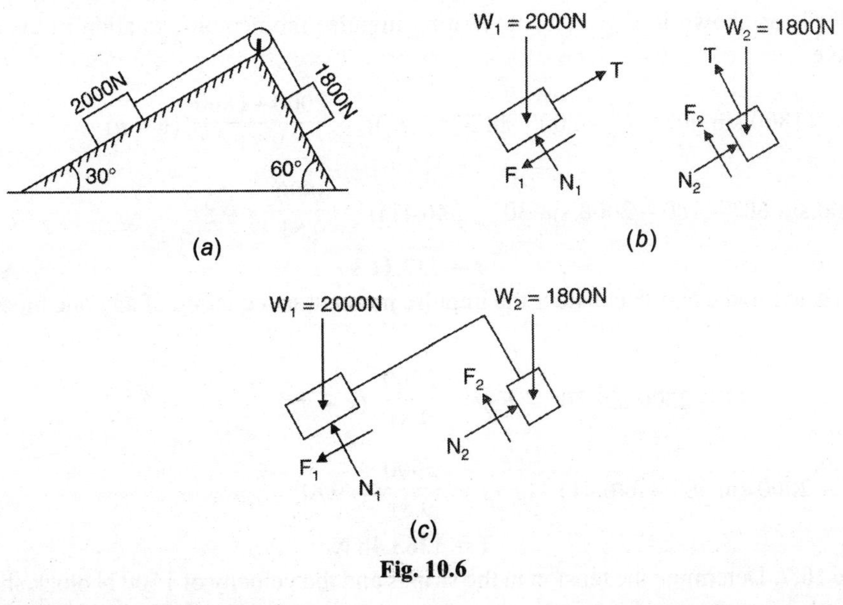


**Fig. 10.6**

Let $T$ be the tension in the chord.

Initial velocity $u = 0$

Final velocity $v = 9.81$ m/s.

Applying impulse momentum equation for the 2000 N block in upward direction parallel to the plane, we get

$$(T - 2000 \sin 30° - F_1)\ t = \left[\frac{2000}{9.81}\right]\ (v - u)$$

$$(T - 2000 \sin\ 30° - 346.41)\ t = 2000, \text{ since } v - u = 9.81$$

$$(T - 1346.41)\ t = 2000 \qquad ...(i)$$

Applying impulse momentum equation for 1800 N block in the direction parallel to 60° inclined plane, we get

$$(1800 \sin 60° - T - F_2)\ t = \left[\frac{1800}{9.81}\right]\ (v - u)$$

$$(1800 \sin 60° - T - 180)t = \left[\frac{1800}{9.81}\right]\ (9.81 - 0)$$

$$(1378.85 - T)\ t = 1800 \qquad ...(ii)$$

Dividing Eqn. (*i*) by Eqn. (*ii*), we get

$$\frac{(T - 1346.41)}{(1378.85 - T)} = \frac{2000}{1800}$$

$\therefore$ **T = 1363.48 N** **Ans.**

Substituting it in Eqn. (*ii*), we get

$$(1378.85 - 1363.48)\ t = 1800$$

$\therefore$ ***t* = 117.11 s** **Ans.**

***Second Method:*** Since the displacement of both bodies are same in given time, consider combined FBD of the blocks as shown in Fig. 10.6(*c*). Writing impulse momentum equation in the direction of motion, we have

$$(1800 \sin 60° - F_2 - 2000 \sin 30° - F_1)t = \frac{2000 + 1800}{9.81}\ (v - u)$$

$$(1800 \sin 60° - 180 - 2000 \sin 30° - 346.41)t = \frac{3800}{9.81} \times 9.81$$

$\therefore$ ***t* = 117.11 s** **Ans.**

To find tension in the chord, consider the impulse momentum equation of any one block, say 2000 N block.

$$(T - 2000 \sin 30° - F_1)t = \frac{2000}{9.81}\ (v - u)$$

$$(T - 2000 \sin 30° - 346.41)\ 117.11 = \frac{2000}{9.81} \times 9.81$$

$\therefore$ **T = 1363.48 N** **Ans.**

**Example 10.7.** Determine the tension in the strings and the velocity of 1500 N block shown in Fig. 10.7(*a*) 5 seconds after starting from

(*a*) rest.

(*b*) starting with a downward velocity of 3 m/s.

Assume pulleys as weightless and frictionless.

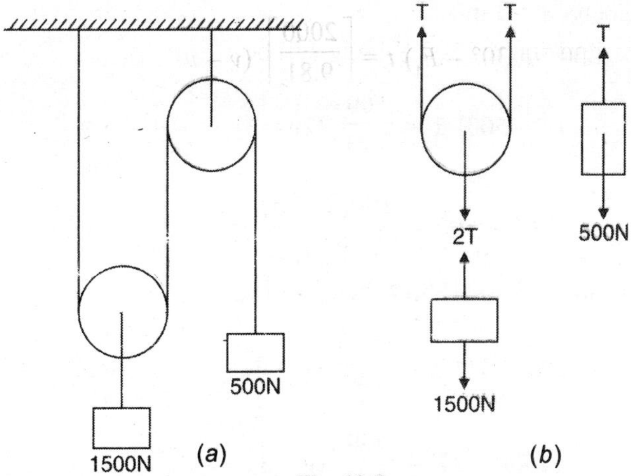


**Fig. 10.7**

**Solution.** When 1500 N block moves a distance $s$, in the same time 500 N block moves a distance $2s$. Hence if velocity of 1500 N block is $v$ m/s that of 500 N block will be $2v$ m/s. Let $T$ be the tension in the chord connecting 500 N block. Hence tension in the wire connecting 1500 N block will be $2T$ [see Fig. 10.7($b$)]. Since the velocities of two blocks are different only first method is to be used.

*Case* ($a$): Initial velocity $u = 0$, $t = 5$ s.

Writing impulse momentum equation for 500 N block, we have

$$(T - 500)t = \frac{500}{9.81}(2v - u)$$

$$(T - 500)5 = \frac{500}{9.81} \times (2v - 0)$$

$$T - 500 = \frac{200v}{9.81} \qquad \ldots(i)$$

Applying impulse momentum equation to 1500 N block,

$$(1500 - 2T)\,t = \frac{1500}{9.81}(v - 0)$$

$$(1500 - 2T)\,5 = \frac{1500}{9.81}v$$

$$1500 - 2T = \frac{300v}{9.81} \qquad \ldots(ii)$$

Adding Eqn. ($ii$) in 2 times Eqn. ($i$), we get

$$1500 - 2 \times 500 = \frac{400\,v}{9.81} + \frac{300\,v}{9.81} = \frac{700\,v}{9.81}$$

$\therefore$ $\quad \mathbf{v = 7.007\ m/s}$ **Ans.**

Substituting it in Eqn. ($i$), we get

$$T - 500 = \frac{200}{9.81} \times 7.007$$

$\therefore$ $\quad \mathbf{T = 642.86\ N}$ **Ans.**

*Case* (*b*): Initial velocity $u = 3$ m/s.

Impulse momentum equation for 500 N body will be

$$(T - 500)\,5 = \frac{500}{9.81}\,(2v - 3)$$

i.e.,
$$T - 500 = \frac{100(2v - 3)}{9.81} \qquad ...(iii)$$

Impulse momentum equation for 1500 N body will be

$$(1500 - 2T)\,5 = \frac{1500}{9.81}\,(v - 3)$$

i.e.,
$$1500 - 2T = \frac{300}{9.81}\,(v - 3) \qquad ...(iv)$$

Adding Eqn. (*iv*) and 2 times Eqn. (*i*), we get

$$1500 - 1000 = \frac{100}{9.81}\,(7v - 15)$$

$$\mathbf{v = 9.15\ m/s} \qquad \textbf{Ans.}$$

Substituting it in Eqn. (*iii*), we get

$$T - 500 = \frac{100}{9.81}\,(2 \times 9.15 - 3)$$

$$\therefore \qquad \mathbf{T = 655.96\ N} \qquad \textbf{Ans.}$$

**Example 10.8.** The system shown in Fig. 10.8(*a*) has a rightward velocity of 3 m/s. Determine its velocity after 5 seconds. Take $\mu = 0.2$ for the surfaces in contact. Assume pulleys to be frictionless.

**Solution.** Since all bodies have same displacement in given time, consider the combined FBD of the system.

$$N_1 = 500 \text{ N}$$
$$F_1 = 0.2 \times 500 = 100 \text{ N}$$
$$N_2 = 1000 \cos 30° = 866.03$$
$$F_2 = 0.2 \times N_2 = 173.2 \text{ N}$$

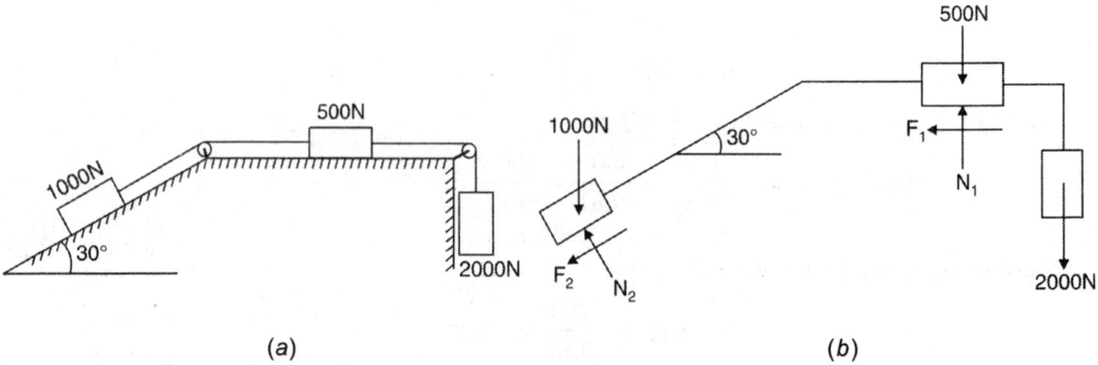


**Fig. 10.8**

Writing impulse momentum equation for whole system, we get

$$(2000 - F_1 - 1000 \sin 30° - F_2)t = \frac{2000 + 500 + 1000}{9.81} (v - u)$$

$$(2000 - 100 - 1000 \sin 30° - 173.2)\ 5 = \left[\frac{3500}{9.81}\right] (v - 3)$$

$$v = \mathbf{20.19\ m/s}$$ **Ans.**

**Example 10.9.** The system shown in Fig. 10.9(*a*) has a rightward velocity of 4 m/s, just before a force **P** is applied. Determine the value of *P* that will give a leftward velocity of 6 m/s in a time interval of 20 s. Take coefficient of friction = 0.2 and assume ideal pulley.

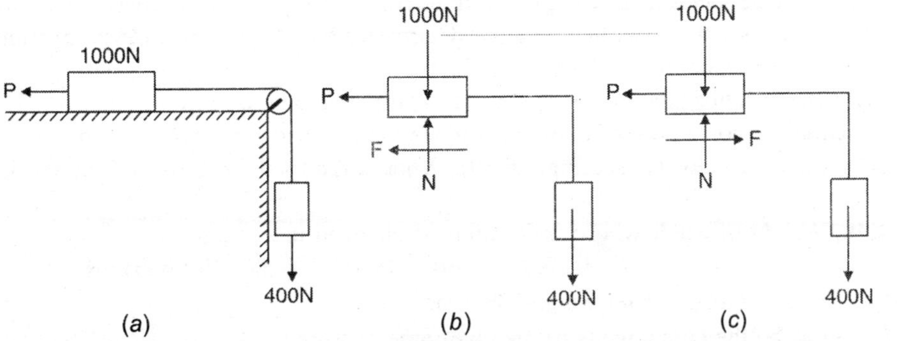


**Fig. 10.9**

**Solution.** When the system is moving rightward, frictional force acts leftward as shown in Fig. 10.9(*b*). Force **P** first brings the system to stationary position, then starts moving leftward. At this stage, the frictional force acts rightward as shown in Fig. 10.9(*c*). Let $t_1$ be the time required to bring the system to the stationary condition.

$$\mathbf{N} = 1000 \text{ newton}$$

$$\mathbf{F} = 0.2 \times 1000 = 200 \text{ N}$$

Applying impulse momentum equation for the motion upto stationary condition, we have

$$(400 - 200 - P)\ t_1 = \frac{400 + 1000}{9.81} (0 - 4)$$

$$(P - 200)\ t_1 = \frac{5600}{9.81} \qquad ...(i)$$

Applying impulse momentum equation for the motion from stationary position to leftward motion, after total time of 20 s, we have

$$(P - F - 400)\ (20 - t_1) = \frac{1000 + 400}{9.81} (v - 0)$$

$$(P - 600)\ (20 - t_1) = \frac{1400}{9.81} \times 6 \qquad ...(ii)$$

Simultaneous Eqns. (*i*) and (*ii*) may be solved to get $t_1$ and *P*. Trial and error method may be advantageously used here. Looking at Eqn. (*ii*), the value of *P* should be more than 600. Let us take a trial value of *P* as 700. From Eqn. (*i*)

$$t_1 = 1.142 \text{ s}$$

Substituting it in Eqn. (*ii*), we get

$$P = 645.41 \text{ N}$$

Substituting this value of *P* in Eqn. (*i*), we get

$$t_1 = 1.282 \text{ s}$$

Substituting it in Eqn. (*ii*), we get

$$P = 645.74 \text{ N}$$

This value is almost same as trial value 646.41 N

Hence, **P = 645.74 N.** **Ans.**

## 10.3 FORCE OF JET ON A VANE

In hydroelectricity generating stations, a jet of water is made to impinge on the vanes of turbines and get deflected by a certain angle. During this process, a force is exerted by the jet on the vane and that causes rotation of turbine. This mechanical energy is further converted into electric energy. The force exerted by the jet on the vane, moving or stationary, can be determined by applying impulse momentum equations. This is illustrated with Examples 10.10 and 10.11.

**Example 10.10.** A nozzle issues a jet of water 50 mm in diameter, with a velocity of 30 m/s which impinges tangentially upon a perfectly smooth and stationary vane, and deflects it through an angle of 30° without any loss of velocity (see Fig. 10.10). What is the total force exerted by the jet upon the vane?

**Solution.** Weight of water whose momentum is changed in *t* second

$$= (\pi/4)\,(0.05)^2 \times 30 \times 9810 \times t = 577.86\,t \text{ newtons.}$$

(**Note:** 1 cubic metre of water weighs 9810 newton)

Let $P_x$ and $P_y$ be the components of reactive force of vane.

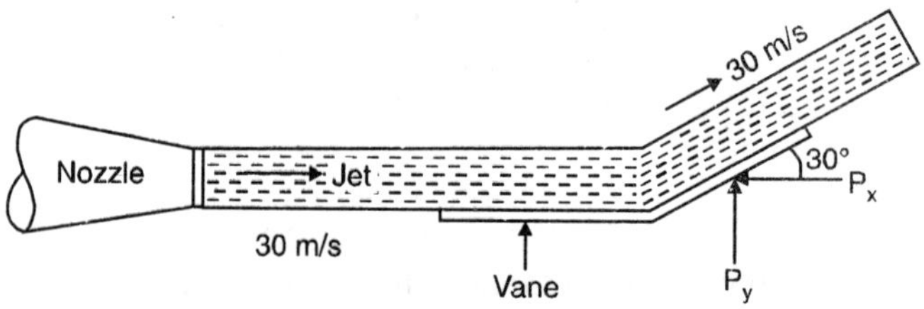


**Fig. 10.10**

Applying the impulse momentum equation in *x*-direction, we get

$$-P_x t = \frac{577.86t}{9.81}\,(30 \cos 30° - 30)$$

$$P_x = 236.75 \text{ N}$$

Applying the impulse momentum equation in *y*-direction, we have

$$P_y t = \frac{577.86t}{9.81}\,(30 \sin 30° - 0)$$

$$P_y = 883.58 \text{ N}$$

$$P = \sqrt{P_x^2 + P_y^2} = \sqrt{236.75^2 + 883.58^2} = 914.75 \text{ N}$$

Inclination with horizontal,

$$\theta = \tan^{-1}\left(\frac{P_y}{P_x}\right) = \tan^{-1}\frac{883.58}{236.75}$$

$$\theta = 75.0°$$

Force **P**, as shown in Fig. 10.11(*a*), is the reactive force of the vane. The force of jet is equal and opposite to this force as shown in Fig. 10.11(*b*).

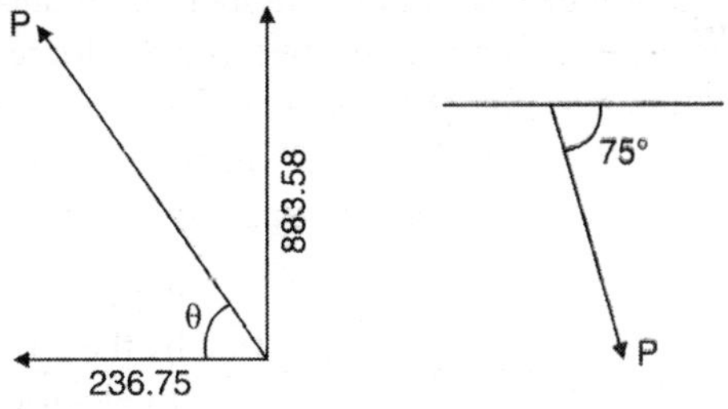


**Fig. 10.11**

**Example 10.11.** In the previous example if the vane is moving with a velocity of 10 m/s towards right, what will be the pressure exerted by the jet.

**Solution.** Velocity of approach

$= 30 - 10 = 20$ m/s

Weight of water impinging in $t$ second

$= (\pi/4) \times (0.05)^2 \times 20 \times t \times 9810$

$= 385.24\ t$

Velocity of departure = Velocity of approach

$= 30 - 10 = 20$ m/s

Applying impulse moment equation in $x$-direction, we have

$$-P_x t = \frac{385.24\ t}{9.81}(20 \cos 30° - 20)$$

$$P_x = 105.22 \text{ N}$$

Applying impulse moment equation in $y$-direction, we get

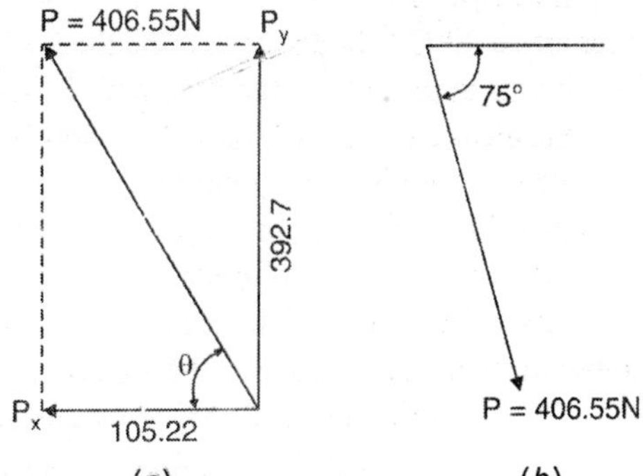


**Fig. 10.12**

$$P_y t = \frac{385.24\ t}{9.81}(20 \sin 30° - 0)$$

$$P_y = 392.70 \text{ N}$$

$$P = \sqrt{P_x^2 + P_y^2} = \sqrt{105.22^2 + 392.70^2}$$

*i.e.*, **P = 406.55 N** **Ans.**

Its inclination to the horizontal

$$\theta = \tan^{-1}\left(\frac{P_y}{P_x}\right) = \tan^{-1}\left(\frac{392.70}{105.22}\right) = 75°$$

Reaction of vane $P$ is as shown in Fig. 10.12(*a*). The pressure exerted by the jet is equal and opposite to it and is as shown in Fig. 10.12(*b*).

## 10.4 CONSERVATION OF MOMENTUM

In a system, if the resultant force $R$ is zero, the impulse momentum reduces to final momentum equal to initial momentum. Such situation arises in many cases because the force system consists of only action and reaction on the elements of the system. The resultant force is zero, only when entire system is considered, but not when the free body of each element of the system is considered. When a person jumps off a boat, the action of the person is equal and opposite to the reaction of the boat. Hence, the resultant force is zero in the system. If $W_1$ is the weight of the person and $W_2$ that of the boat, $v$ is velocity of the person and the boat before the person jumps out of the boat and $v_1$, $v_2$ are the velocities of person and the boat after jumping, according to principle of conservation of momentum,

$$\frac{W_1 + W_2}{g} v = \frac{W_1}{g} v_1 + \frac{W_2}{g} v_2$$

Similar equation holds good when we consider the system of a gun and shell. The *principle of conservation of momentum* may be stated as, *the momentum is conserved in a system in which resultant force is zero.* In other words, *in a system if the resultant force is zero, initial momentum will remain equal to final momentum.*

It must be noted that conservation of momentum applies to entire system and not to individual elements of the system.

**Example 10.12.** A 800 N man, moving horizontally with a velocity of 3 m/s, jumps off the end of a pier into a 3200 N boat. Determine the horizontal velocity of the boat (*a*) if it had no initial velocity and (*b*) if it was approaching the pier with an initial velocity of 0.9 m/s.

**Solution.** Weight of man $W_1 = 800$ N

Velocity with which man is running $v = 3$ m/s.

Weight of the system after man jumps into boat = 800 + 3200 = 4000 N

(*a*) Initial velocity of boat = 0

Since the action of the man is equal to the reaction of the boat, the principle of conservation of momentum can be applied to the system consisting of the man and the boat.

Initial momentum = Final momentum

$$\frac{800}{9.81} \times 3 + \frac{3200}{9.81} \times 0 = \frac{4000}{9.81} v$$

$$\mathbf{v = 0.6 \text{ m/s}}$$ **Ans.**

(*b*) Initial velocity of boat = 0.9 m/s towards the pier

= –0.9 m/s

Applying principle of conservation of momentum, we get

$$\frac{800}{9.81} \times 3 + \frac{3200}{9.81} \times (-0.9) = \frac{4000}{9.81} v$$

$$\therefore \quad \mathbf{v = -0.12 \text{ m/s}}$$

*i.e.,* **velocity of boat and man will be 0.12 m/s towards the pier.** **Ans.**

**Example 10.13.** A car weighing 11,000 N and running at 10 m/s holds three men, each weighing 700 N. The men jump off from the back end gaining a relative velocity of 5 m/s with the car. Find the speed of the car if the three men jump off: (*i*) in succession, (*ii*) all together.

**Solution.** (*i*) *When three men jump off in succession*

Initial velocity $u = 10$ m/s.

Let the velocity when

(*a*) first man jumps be $v_1$ m/s.

(*b*) second man jumps be $v_2$ m/s.

(*c*) third man jumps be $v_3$ m/s.

Velocity of the first man w.r.t. fixed point when he jumps = $v_1 - 5$

Applying principle of conservation of momentum when the first man jumps, we get

$$(11{,}000 + 3 \times 700)\ 10 = (11{,}000 + 2 \times 700)\ v_1 + 700\ (v_1 - 5)$$
$$= (11{,}000 + 3 \times 700)\ v_1 - 700 \times 5$$

$$v_1 = 10 + \frac{700 \times 5}{11{,}000 + 3 \times 700} \qquad ...(i)$$

When the second man jumps:

$$(11{,}000 + 2 \times 700)v_1 = (11{,}000 + 700)v_2 + 700(v_2 - 5)$$

$$v_2 = v_1 + \frac{700 \times 5}{11{,}000 + 2 \times 700} \qquad ...(ii)$$

When the third man jumps:

$$(11{,}000 + 700)\ v_2 = 11{,}000v_3 + 700\ (v_3 - 5)$$

$$v_3 = v_2 + \frac{700 \times 5}{11{,}000 + 700} \qquad ...(iii)$$

From Eqns. (*i*), (*ii*) and (*iii*), we get

$$v_3 = 10 + 700 \times 5 \left(\frac{1}{11{,}000 + 3 \times 700} + \frac{1}{11{,}000 + 2 \times 700} + \frac{1}{11{,}000 + 700}\right) = \mathbf{10.849\ m/s}$$ **Ans.**

(*ii*) *When three men jump together*

Let the velocity of the car be $v$ when three men jump together. Applying principle of conservation of momentum, we get

$$(11{,}000 + 3 \times 700)\ 10 = 11{,}000v + 3 \times 700(v - 5)$$

where, $(v - 5)$ is the relative velocity of the men when they jump,

$$v = 10 + \frac{3 \times 700 \times 5}{11{,}000 + 3 \times 700}$$

*i.e.*, $\mathbf{v = 10.802\ m/s}$ **Ans.**

**Example 10.14.** A car weighing 50 kN and moving at 54 kmph along the main road collides with a lorry of weight 100 kN which emerges at 18 kmph from a cross road at right angles to main road. If the two vehicles lock after collision, what will be the magnitude and direction of the resulting velocity ?

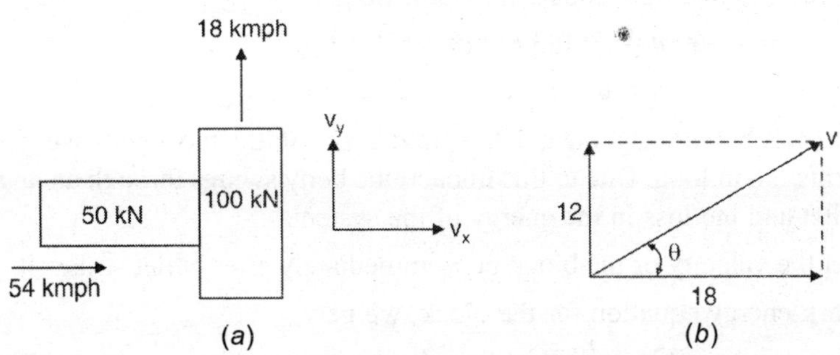


**Fig. 10.13**

**Solution.** Let the velocity of the vehicles after collision be $v_x$ in $x$-direction (along main road) and $v_y$ in $y$-direction (along cross road) as shown in Fig. 10.13($a$). Applying impulse momentum equation along $x$-direction, we get

$$\frac{50 \times 54}{9.81} + 0 = \frac{(50+100)}{9.81} v_x$$

$$\therefore \qquad v_x = 18 \text{ kmph}$$

Applying impulse momentum equation along $y$-direction, we get

$$0 + \frac{100 \times 18}{9.81} = \frac{(50+100)}{9.81} v_y$$

$$v_y = 12 \text{ kmph}$$

$\therefore$ Resultant velocity $v = \sqrt{v_x^2 + v_y^2} = \sqrt{18^2 + 12^2} = \mathbf{21.63\ kmph}$ **Ans.**

Its inclination to main road

$$\theta = \tan^{-1}\left(\frac{v_y}{v_x}\right) = \tan^{-1} 12/18 = \mathbf{33.69°}$$ **Ans.**

as shown in Fig. 10.13($b$).

**Example 10.15.** A gun weighing 300 kN fires a 5 kN projectile with a velocity of 300 m/s. With what velocity will the gun recoil ? If the recoil is overcome by an average force of 600 kN how far will the gun travel? How long will it take?

**Solution.** Applying principle of conservation of momentum to the system of gun and the projectile, we get

$$0 = 300 \times v + 5 \times 300$$

$$v = -5 \text{ m/s}$$

*i.e.*, Gun will have a velocity of 5 m/s in the direction opposite to that of bullet.

**Ans.**

Let the gun recoil for a distance $s$.

Using work-energy equation, we have

$$-600 \times s = \frac{300}{2 \times 9.81} (0 - 5^2)$$

$$\mathbf{s = 0.637\ m}$$ **Ans.**

Applying impulse momentum equation to gun, we get

$$-600t = (300/9.81)(0 - 5)$$

$$\mathbf{t = 0.255\ s}$$ **Ans.**

**Example 10.16.** A bullet weighing 0.3 N is fired horizontally into a body weighing 100 N which is suspended by a string 0.8 m long. Due to this impact, the body swings through an angle of 30°. Find the velocity of the bullet and the loss in the energy of the system.

**Solution.** Let the velocity of the block be $u$ immediately after bullet strikes it.

Applying work-energy equation for the block, we get

$$-Wh = (W/2g)(v^2 - u^2)$$

$$-100.3 \times 0.8\,(1 - \cos 30°) = \left[\frac{100.3}{2 \times 9.81}\right](0 - u^2)$$

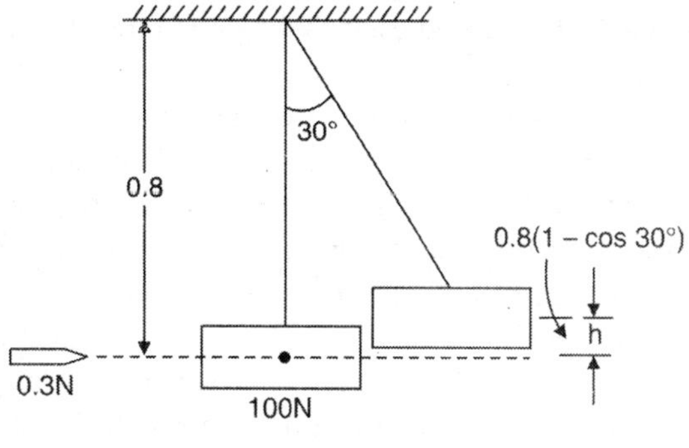


Fig. 10.14

(**Note:** Work done by the weight is negative since weight is a force acting downwards whereas body has moved upwards.)

$$u = 1.025 \text{ m/s}$$

Let $v$ be the velocity of bullet before striking the block. Applying principle of conservation of momentum to the bullet and block system, we get

$$\frac{0.3}{9.81}v + 0 = \frac{100 + 0.3}{9.81}u$$

$$0.3\,v = 100.3 \times 1.025$$

$$\mathbf{v = 342.69 \ m/s}$$ **Ans.**

$$\text{Initial energy of bullet} = \frac{0.3}{2 \times 9.81}(342.69)^2 = 1795.68 \text{ J}$$

Energy of the block and bullet system

$$= \frac{1}{2} \times \frac{100 + 0.3}{9.81} \times 1.025^2 = 5.37 \text{ J}$$

$$\text{Loss of energy} = 1795.68 - 5.37$$

$$= \mathbf{1790.31 \ J.}$$ **Ans.**

**Example 10.17.** A bullet weighs 0.5 N and moving with a velocity of 400 m/s hits centrally a 30 N block of wood moving away at 15 m/s and gets embedded in it. Find the velocity of the bullet after the impact and the amount of kinetic energy lost.

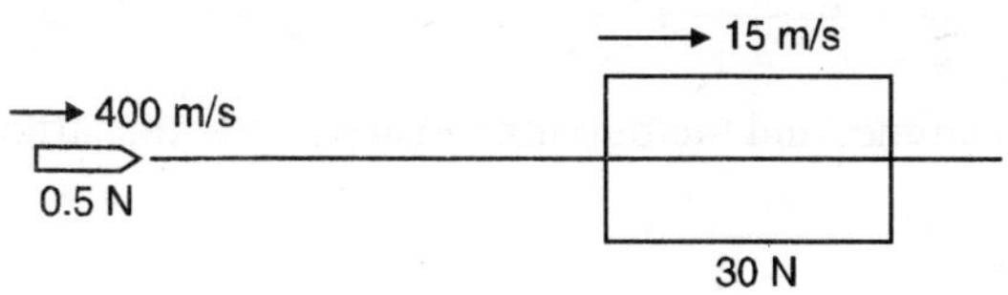


Fig. 10.15

**Solution.** Initial momentum of the system = Final momentum

$$\frac{0.5}{9.81} \times 400 + \frac{30}{9.81} \times 15 = \frac{(30 + 0.5)}{9.81}v$$

$$v = 21.31 \text{ m/s}$$

Kinetic Energy lost = Initial K.E. – Final K.E.

$$= \left(\frac{1}{2} \times \frac{0.5}{9.81} \times 400^2 + \frac{1}{2} \times \frac{30}{9.81} \times 15^2\right) - \frac{1}{2} \times \frac{30.5}{9.81} \times 21.31^2$$

$$= \mathbf{3715.47 \ J}$$ **Ans.**

## 10.5 PILE AND PILE HAMMER

If the safe bearing capacity of the soil is too less, a set of reinforced concrete or steel poles are driven in the soil. Such poles are known as piles. Over the group of piles, concrete cap is cast and on it the structure is built.

The piles are driven by pile hammer. It consists of a movable weight called the hammer (see Fig. 10.16). The hammer is raised to a convenient height $h$ and freely dropped. It is guided to fall over the pile. After the hammer strikes the pile, the hammer and the pile move downwards together. The kinetic energies of the pile and the hammer are utilised in doing the work against resistance of the ground and pile gets driven by a distance $s$. By repeated hammering, the pile is driven to required depth. If the distance moved per blow is known, earth resistance can be calculated. A general equation is derived in Ex. 10.18, and then two specific examples are solved.

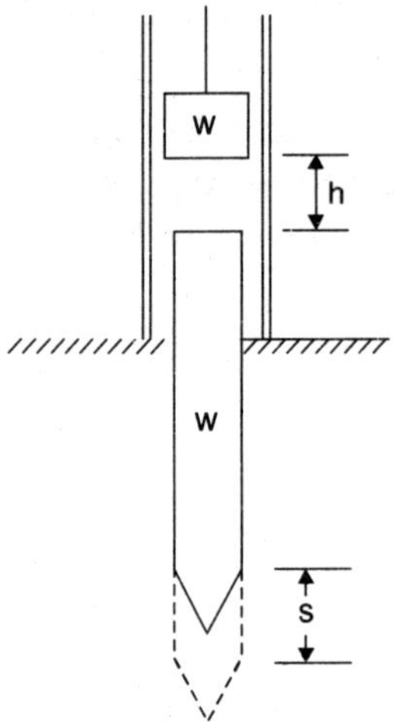


**Fig. 10.16**

**Example 10.18.** A pile of weight **W** is driven vertically through a distance $s$ when a hammer of weight **w** is dropped from a height $h$. Calculate the average resistance of the ground, the loss of kinetic energy during the impact and the time during which the pile is in motion.

**Solution.** Initial velocity of hammer = 0

Distance moved before striking pile = $h$ m

Gravitational acceleration = $g$ m/s$^2$

$\therefore$ Velocity of hammer while striking pile $v$ is given by

$$v^2 - u^2 = 2gh$$
$$v^2 - 0 = 2gh$$
$$v = \sqrt{2gh} \quad \ldots(i)$$

After striking, the hammer and the pile move together. Hence, the momentum is conserved. Applying the principle of conservation of momentum

$$\frac{wv}{g} = \frac{(w+W)}{g}V$$

where, $V$ is the velocity of hammer and the pile immediately after the strike.

$$V = \frac{w}{w+W}\, v \quad \ldots(ii)$$

With this initial velocity, the pile and the hammer start moving downwards and they stop moving after a distance $s$. During this process, work is done against the resistance $R$ of the earth. Applying work-energy equation, we have

$$(w + W)s - R \times s = \frac{1}{2}\frac{w+W}{g}(0 - V^2)$$

$$Rs = (w + W)s + \frac{w+W}{2g}V^2 \quad \ldots(iii)$$

From Eqn. $(ii)$,
$$V = \frac{w}{w+W}\, v$$

$$\therefore \quad Rs = (w + W)s + \frac{w+W}{2g}\frac{w^2}{(w+W)^2}v^2$$

i.e., $$Rs = (w + W)s + \frac{w^2}{2g(w+W)} v^2 \qquad \ldots(iv)$$

From Eqn. (*i*), $v^2 = 2gh$

$$\therefore \quad Rs = (w + W)s + \frac{w^2}{2g(w+W)} 2gh = (w + W)s + \frac{w^2}{w+W} h$$

or $$\boldsymbol{R = w + W + \frac{w^2}{w+W}\frac{h}{s}}$$ **Ans.**

**Loss of Kinetic Energy during the Impact:**

$$= \frac{w}{2g} v^2 - \frac{w+W}{2g} V^2$$

$$= \frac{w}{2g} v^2 - \frac{w+W}{2g}\left(\frac{w}{w+W}\right)^2 v^2 = \frac{v^2}{2g}\left\{w - \frac{w^2}{w+W}\right\}$$

$$= \frac{2gh}{2g} w \left\{1 - \frac{w}{w+W}\right\}$$

$$= wh \frac{W}{w+W}$$

$$\textbf{Loss of K.E.} = \boldsymbol{\frac{w\,W\,h}{w+W}}$$ **Ans.**

**Time during which the pile is in motion:**

Let $t$ be the time during which the pile is in motion. Applying impulse momentum equation

$$[(w + W) - R]t = \frac{(w+W)}{g}(0 - V)$$

$$[R - (w + W)]t = \frac{w+W}{g} V$$

Substituting for $R$ and $V$, we get

$$\frac{w^2}{w+W}\frac{h}{s} t = \frac{w+W}{g}\frac{w}{w+W} v = \frac{w}{g}\sqrt{2gh}$$

$$\boldsymbol{t = \frac{w+W}{w} s\sqrt{2gh}}$$ **Ans.**

**Example 10.19.** A pile hammer weighing 20 kN drops from a height of 750 mm on a pile of 10 kN. The pile penetrates 100 mm per blow. Assuming that the motion of the pile is resisted by a constant force, find the resistance to penetration of the ground.

**Solution.** *Initial velocity of hammer $u = 0$*

Distance moved = $h$ = 750 mm = 0.75 m

Acceleration $g = 9.81$ m/s$^2$

$\therefore$ Velocity at the time of strike = $\sqrt{2gh} = \sqrt{2 \times 9.81 \times 0.75} = 3.836$ m/s

Applying the principle of conservation of momentum of pile and hammer, we get velocity $V$ of the pile and hammer immediately after the impact.

$$\frac{20}{9.81} \times 3.836 = \frac{20 + 10}{9.81} V$$

$$V = \frac{20}{30} \times 3.836 = 2.557 \text{ m/s}$$

Applying work-energy equation to the motion of the hammer and pile, resistance $R$ of the ground can be obtained.

$$(20 + 10 - R)s = \frac{20 + 10}{2g}(0 - V^2)$$

$$(30 - R)\,0.1 = \frac{30}{2 \times 9.81}(-2.557^2)$$

$$\mathbf{R = 130 \text{ kN}}$$ **Ans.**

**Example 10.20.** A pile hammer, weighing 15 kN, drops from a height of 600 mm on a pile of 7.5 kN. How deep does a single blow of hammer drive the pile if the resistance of the ground to pile is 140 kN?

Assume that ground resistance is constant.

**Solution.** $h = 600$ mm $= 0.6$ m

Velocity of hammer at the time of strike

$$v = \sqrt{2gh} = \sqrt{2 \times 9.81 \times 0.6} = 3.431 \text{ m/s}$$

Let $V$ be the velocity of pile and hammer immediately after impact. Applying principle of conservation of momentum to the system of pile and pile hammer, we get

$$\frac{15}{9.81} \times 3.431 + 0 = \frac{15 + 7.5}{9.81} V$$

$$V = \frac{15 \times 3.431}{22.5} = 2.287 \text{ m/s}$$

Now applying work-energy equation to the system, we get

$$(15 + 7.5 - R)s = \frac{15 + 7.5}{2 \times 9.81}(0 - V^2)$$

$$(15 + 7.5 - 140)s = \frac{15 + 7.5}{2 \times 9.81}(-2.287^2)$$

$$s = \frac{22.5}{2 \times 9.81}(+\,2.287^2)\frac{1}{117.5} = 0.051 \text{ m}$$

$$\mathbf{s = 51 \text{ mm}}$$ **Ans.**

**Example 10.21.** A hammer weighing 5 N is used to drive a nail of weight 0.2 N with a velocity of 5 m/s horizontally into a fixed wooden block. If the nail penetrates by 20 mm per blow, calculate the resistance of the block, which may be assumed uniform.

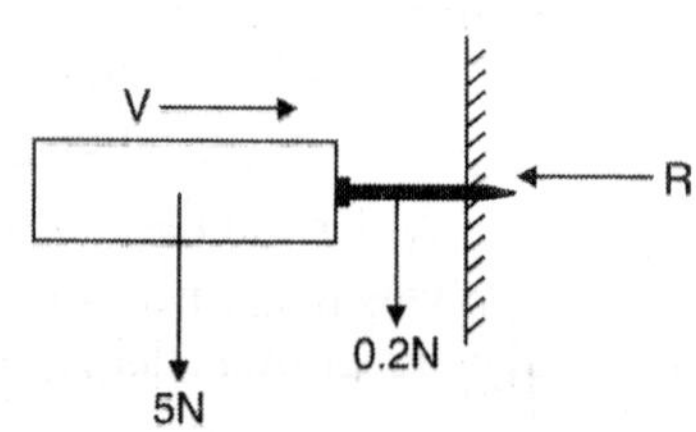

**Fig. 10.17**

**Solution.** Applying impulse momentum equation to the system of hammer and nail, we get

$$\frac{5}{9.81} \times 5 + 0 = \frac{5 + 0.2}{9.81} V$$

$$V = 4.808 \text{ m/s}$$

Applying work-energy equation to the system shown in Fig. 10.17, we get

$$-Rs = \frac{5.2}{2 \times 9.81}(0 - 4.808^2)$$

$$R \times 0.02 = \frac{5.2}{2 \times 9.81} \times 4.808^2$$

$$\mathbf{R = 306.3 \text{ N}}$$ **Ans.**

**Note:** Weights of hammer and nail do not do any work, since there is no displacement in the directions of these forces.

## IMPORTANT DEFINITIONS

1. The product of mass and velocity is called **momentum.**
2. If $R$ is the resultant force acting on the body for an interval zero to $t$, then the term $\int_0^t R\,dt$ is called **impulse.**
3. **Impulse momentum equation** can be stated as 'the component of resultant linear impulse along any direction is equal to change in the component of momentum in that direction'.

## IMPORTANT EQUATIONS

1. $\int_0^t R\,dt = mv - mu.$
2. If momentum is conserved in a system,

$$\frac{W_1 + W_2}{g} v = \frac{W_1}{g} v_1 + \frac{W_2}{g} v_2.$$

## QUESTIONS

**10.1** State impulse momentum equation

**10.2** State principle of conservation of momentum.

## EXERCISES

**10.1** A cricket ball weighing one newton approaches a batsman with a velocity of 18 m/s in the direction shown in Fig. 10.18. After hit by the bat at $B$, it moves out with a velocity of 40 m/s at 45° to horizontal. If the bat and the ball were in contact for 0.02 s, determine the impulsive force exerted by the bat.
[**Ans.** 230.37 N]

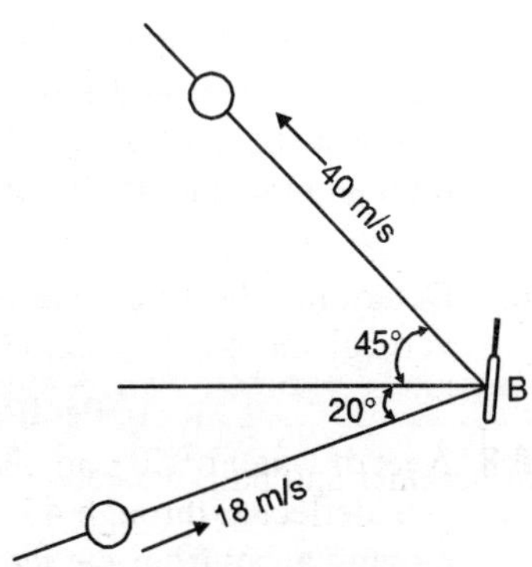

**Fig. 10.18**

**10.2** A block weighing 200 N is pulled up a 30° plane by a force $P$ producing a velocity of 5 m/s in 5 seconds. If the coefficient of friction is 0.2, determine the magnitude of force $P$. At this stage if force $P$ is removed, how much more time will it take to come to rest ? [**Ans.** 155.N; 0.757 s]

**10.3** The initial velocity of 500 N block is 6 m/s towards left. At this stage, a weight of 250 N is applied as shown in Fig. 10.19. Determine the time at which the block has (*a*) no velocity, (*b*) a velocity of 4 m/s. to the right. Take coefficient of friction 0.2 and assume pulley as ideal.

[**Ans.** (*a*) 0.87 s; (*b*) 2.23 s]

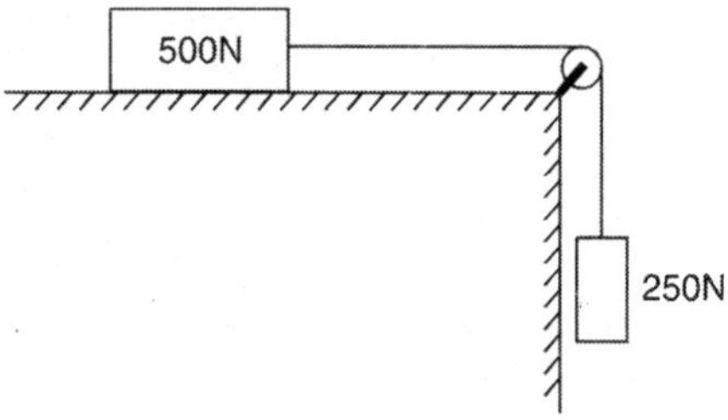


**Fig. 10.19**

**10.4** Determine the tension in the string and the velocity of 2000 N block shown in Fig. 10.20, 6 second after starting with a velocity of 3 m/s.

[**Ans.** $v$ = 25.07 m/s; $T$ = 1280 N]

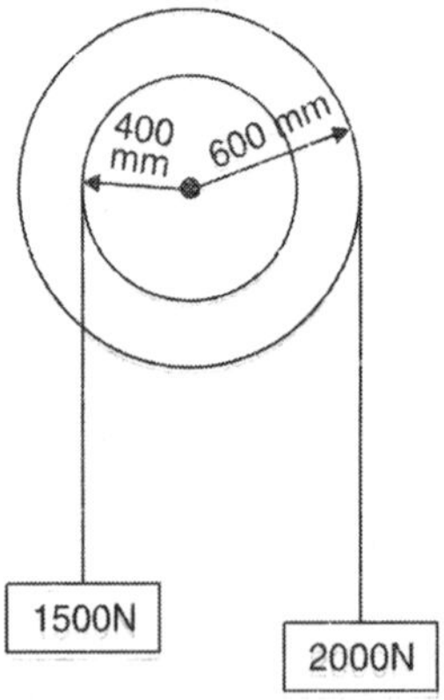


**Fig. 10.20**

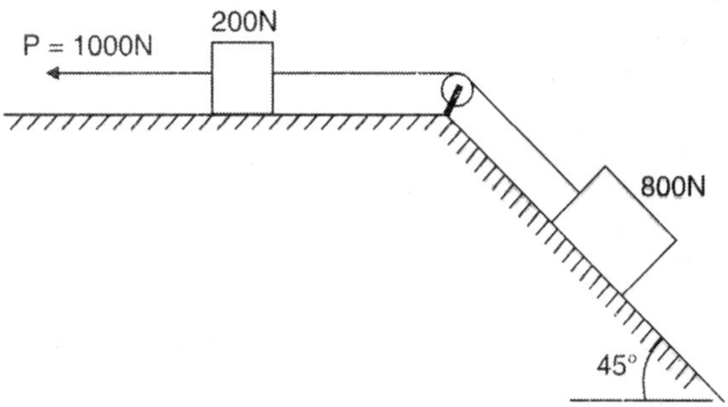


**Fig. 10.21**

**10.5** Force $P$ = 1900 N shown in Fig. 10.21 was applied to 200 N block when the block was moving with rightward velocity of 5 m/s. Determine the time at which the system has (*a*) no velocity, (*b*) a velocity of 3 m/s towards left. Coefficient of friction between blocks and surface = 0.2. Assume pulley to be ideal.

[**Ans.** (*a*) 0.8676 s; (*b*) 1.96 s]

**10.6** An engine of weight 500 kN pulls a train weighing 1500 kN up an incline of 1 in 100. The train starts from rest and moves with a constant acceleration against a resistance of 5 N/kN. It attains a speed of 18 kmph in 60 seconds. Determine the tension in the draw bar connecting train and the engine. What will be its speed 90 second after the start?

[**Ans.** $P$ = 35.242 kN; $v$ = 27 kmph]

**10.7** Determine the force exerted by a 60 mm diameter jet of water flowing at 25 m/s on (*a*) a vertical stationary plate, (*b*) a cup that turns the water through 120°.

[**Ans.** (*a*) $P$ = 1767.146 kN; (*b*) $P$ = 3006.787 kN at θ = 30° to horizontal]

**10.8** A jet of water of 50 mm diameter strikes a series of vanes horizontally at a speed of 36 m/s and gets deflected through 45°. Determine the force exerted by the jet on vanes, if the vanes are moving away from the jet at a velocity of 6 m/s.

[**Ans.** $P$ = 1252.516 kN and θ = 22.5° to horizontal]

**10.9** A shot is fired horizontally from a gun boat towards a target. The total weight of gun boat including men, gun, shells, etc., in it is 15 kN. The weight of shell is 15 N and emerges out at a velocity of 300 m/s. What will be the velocity of the boat when the shell is fired? If the target weighs 1 kN suspended by a rope of length 2 m, by what angle will it swing if the shell gets embedded in it? [**Ans.** $v = 0.3$ m/s; $\theta = 60.06°$]

**10.10** A bullet weighing 0.3 N and moving at 600 m/s penetrates 40 N body shown in Fig. 10.22 and emerges with a velocity of 180 m/s. How far and how long will the block move, if the coefficient of friction between the body and the horizontal floor is 0.3.

[**Ans.** $s = 1.686$ m; $t = 1.07$ s]

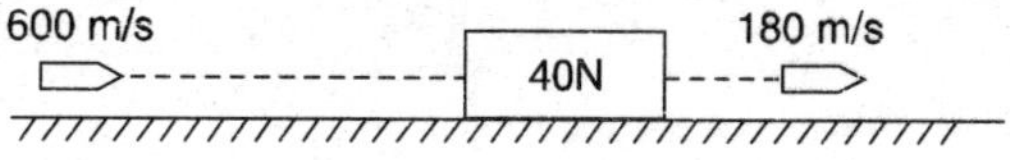


**Fig. 10.22**

**10.11** A cannon weighing 200 kN fires a shell weighing 1 kN with a muzzle velocity of 800 m/s. Calculate the velocity with which the cannon recoils and the uniform force required to stop it within 400 mm distance. In how much time will it stop?

[**Ans.** $v = 4.0$ m/s; $t = 0.20$ s; $F = 407.747$ kN]

**10.12** A man weighing 750 N and a boy weighing 500 N jump from a boat to a pier with a horizontal velocity of 5 m/s relative to the boat. The boat weighs 4000 N and was stationary before they jumped. Determine the velocity of the boat if

(*a*) they jump together; (*b*) boy jumps first and the man later;

(*c*) man jumps first and the boy later.

[**Ans.** (*a*) 1.563 m/s; (*b*) 1.464 m/s and (*c*) 1.458 m/s]

**10.13** A pile hammer weighing 8 kN falls freely from a height of 1.5 m on a pile weighing 5 kN. For each blow, the pile is driven by 80 mm.

Determine:

(*a*) resistance of the ground; (*b*) loss of kinetic energy during the impact;

(*c*) time during which the pile is in motion for each blow.

[**Ans.** $R = 105.31$ kN; Loss of K.E. = 4.615 kN-m, $t = 0.048$ s]

## OBJECTIVE TYPE QUESTIONS

**Select the correct options in the following:**

**10.1** Unit of momentum is

(*a*) N – m (*b*) (N–m)/s

(*c*) N – s (*d*) m – s

**10.2** If a hammer of mass $M$ falls from a height '$h$' on a pile, its velocity at the time of strike is

(*a*) $2gh$ (*b*) $Mh$

(*c*) $\sqrt{2gh}$ (*d*) $M\sqrt{gh}$

CHAPTER

# 11

# IMPACT OF ELASTIC BODIES

A collision between two bodies is said to be impact, if the bodies are in contact for a short interval of time and exert very large force on each other during this short period. On impact, the bodies deform first and then recover due to elastic properties and start moving with different velocities. The velocity with which they separate depends not only on their velocity of approach but also on the shape, size, elastic property and the line of impact. In this book, the velocity of the bodies during the short period of impact is not considered. Only the velocities of the colliding bodies before impact and after impact are considered. Some of the new technical terms used in this chapter are first defined, then the cases of direct impact and oblique impact are analysed. The expression for the loss of kinetic energy on impact is derived.

## 11.1 DEFINITIONS

(*i*) ***Line of Impact:*** Common normal to the colliding surfaces is known as line of impact.

(*ii*) ***Direct Impact:*** If the motion of the two colliding bodies is directed along the line of impact, the impact is said to be direct impact.

(*iii*) ***Oblique Impact:*** If the motion of one or both of the colliding bodies is not directed along the line of impact, the impact is known as oblique impact.

(*iv*) ***Central Impact:*** If the mass centres of colliding bodies are on the line of impact, the impact is called central impact.

(*v*) ***Eccentric Impact:*** Even if mass centre of one of the colliding bodies is not on the line of impact, the impact is called eccentric impact.

These terms are illustrated in Fig. 11.1.

## 11.2 COEFFICIENT OF RESTITUTION

During the collision, the colliding bodies initially undergo a deformation for a small time interval and then recover the deformation in a further small time interval. So the **period of collision** (or time of impact) consists of two time intervals, **Period of Deformation and Period of Restitution.** "Period of

Deformation is the time elapsed between the instant of the initial contact and the instant of maximum deformation of the bodies." Similarly, "Period of Restitution is the time elapsed between the instant of the maximum deformation condition and the instant of separation of the bodies."

Therefore, Impulse during deformation = $F_D\, dt$

where $F_D$ refers to the force that acts during the period of deformation.

The magnitude of $F_D$ varies from zero at the instant contact to the maximum value at the instant of maximum deformation.

Similarly, impulse during restitution = $F_R\, dt$

where $F_R$ refer to the force that acts during period of restitution.

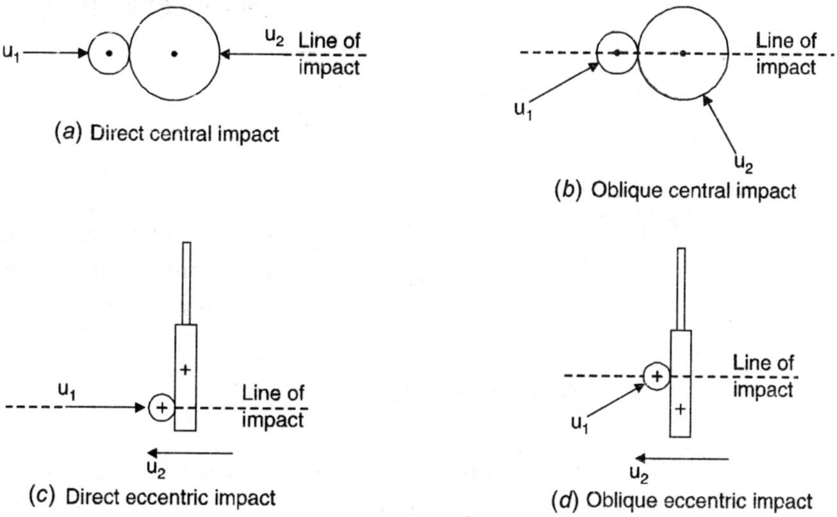


**Fig. 11.1**

The magnitude of $F_R$ varies from a maximum value at the instant of maximum deformation condition to zero at the instant of just separation of the bodies.

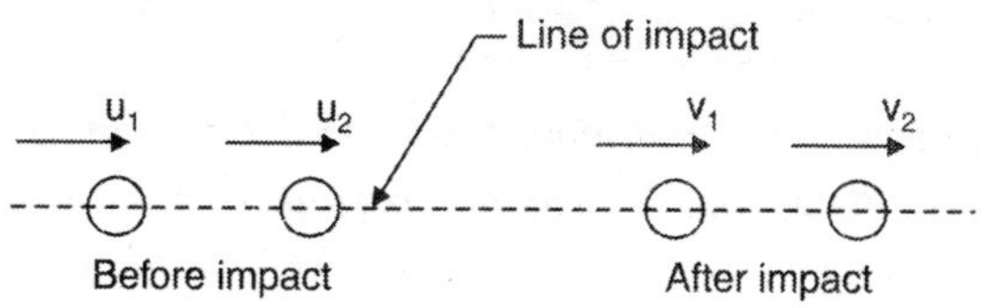


**Fig. 11.2**

Let $m_1$ = Mass of the first body

$m_2$ = Mass of the second body

$u_1$ = Velocity of the first body before impact

$u_2$ = Velocity of the second body before impact

$v_1$ = Velocity of the first body after impact

and $v_2$ = Velocity of the second body after impact.

At the instant of maximum deformation, the colliding bodies will have same velocity. Let the velocity of the bodies at the instant of maximum deformation be $U_{D\max}$.

Applying impulse momentum principle for the first body, we get

$$F_D dt = m_1 U_{D\max} - m_1 u_1 \quad \text{...(11.1)}$$

and
$$F_R dt = m_1 v_1 - m_1 U_{D\max} \quad \text{...(11.2)}$$

Now, dividing Eqn. (11.2) by Eqn. (11.1)

$$\frac{F_R dt}{F_D dt} = \frac{m_1 v_1 - m_1 U_{D\max}}{m_1 U_{D\max} - m_1 u_1}$$

*i.e.*,
$$\frac{F_R dt}{F_D dt} = \frac{v_1 - U_{D\max}}{U_{D\max} - u_1} \quad \text{...(11.3)}$$

Similar analysis for the second body gives,

$$\frac{F_R dt}{F_D dt} = \frac{U_{D\max} - v_2}{u_2 - U_{D\max}} \quad \text{...(11.4)}$$

From Eqns. (11.3) and (11.4), we get

$$\frac{F_R dt}{F_D dt} = \frac{v_1 - U_{D\max}}{U_{D\max} - u_1} = \frac{U_{D\max} - v_2}{u_2 - U_{D\max}}$$

$$= \frac{v_1 - U_{D\max} + U_{D\max} - v_2}{U_{D\max} - u_1 + u_2 - U_{D\max}} = \frac{v_1 - v_2}{u_2 - u_1} = \frac{v_2 - v_1}{u_1 - u_2}$$

$$= \frac{\text{Relative velocity of separation}}{\text{Relative velocity of approach}} \quad \text{...(11.5)}$$

Sir Isaac Newton conducted the experiments and observed that when collision of two bodies takes place, relative velocity of separation bears a constant ratio to the relative velocity of approach, the relative velocities being measured along the line of impact. This constant ratio is called as the coefficient of restitution and is denoted by letter $e$. Hence from (11.5), we have

$$e = \frac{F_R dt}{F_D dt} = \frac{v_2 - v_1}{u_1 - u_2} \quad \text{...(11.6)}$$

*The coefficient of restitution of two colliding bodies may be taken as the ratio of impulse during restitution period to the impulse during the deformation period. This is also equal to the ratio of relative velocity of separation to the relative velocity of approach of the colliding bodies, the relative velocities being measured in the line of impact.*

**Notes:** (*i*) The relative velocities are to be considered only along the line of impact.

(*ii*) Signs of the velocities are to be considered carefully in Eqn. 11.6.

For perfectly elastic bodies, the magnitude of relative velocity after impact will be same as that before impact and hence coefficient of restitution will be 1. Perfectly inelastic bodies cling together and hence the velocity of separation will be zero, *i.e.,* coefficient of restitution will be zero. The coefficient of restitution always lies between 0 and 1.

The value of coefficient of restitution depends not only on the material property, but it also depends on the shape and size of the body. Hence, the coefficient of restitution is the property of two colliding bodies but not merely of material of the colliding bodies.

## 11.3 DIRECT CENTRAL IMPACT

Let $u_1$, $u_2$ be initial velocities and $v_1$ and $v_2$ be the velocities after collision of bodies 1 and 2 respectively.

Let $W_1$ and $W_2$ be the weight of the colliding bodies and '$e$' be coefficient of restitution. $u_1$, $u_2$, $W_1$, $W_2$ and $e$ are known quantities. Velocities $v_1$ and $v_2$ after collision are the unknowns. To find these two unknowns, we need two equations. One equation is obtained by the principle of conservation of momentum as

$$\frac{W_1}{g}u_1 + \frac{W_2}{g}u_2 = \frac{W_1}{g}v_1 + \frac{W_2}{g}v_2$$

*i.e.,* $$W_1u_1 + W_2u_2 = W_1v_1 + W_2v_2 \quad ...(11.7)$$

Another equation based on the definition of coefficient of restitution (Eqn. 11.6), may be written as

$$e(u_1 - u_2) = v_2 - v_1 \quad ...(11.8)$$

From Eqns. 11.7 and 11.8, the unknown velocities $v_1$ and $v_2$ may be found.

**Example 11.1.** Direct central impact occurs between a 300 N body moving to the right with a velocity of 6 m/s and 150 N body moving to the left with a velocity of 10 m/s. Find the velocity of each body after impact if the coefficient of restitution is 0.8.

**Solution.** Referring to Fig. 11.3, initial velocity of 300 N body,

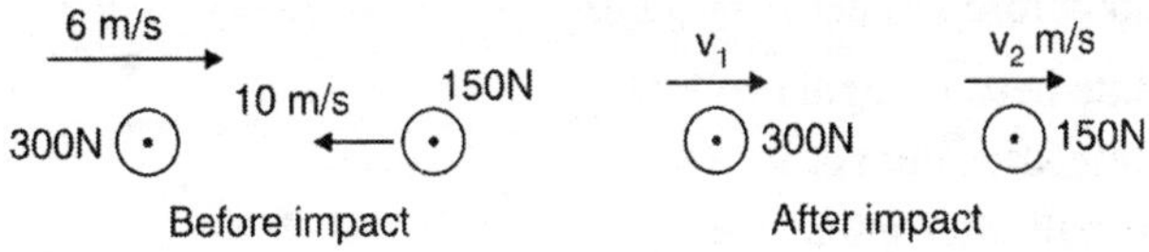


**Fig. 11.3**

$$u_1 = 6 \text{ m/s}$$

Initial velocity of 150 N body, $u_2 = -10$ m/s

Let final velocity of 300 N body, $= v_1$ m/s and

Final velocity of 150 N body, $= v_2$ m/s

From principle of conservation of momentum, we have

$$\frac{300}{g} \times 6 + \frac{150}{g}(-10) = \frac{300}{g}v_1 + \frac{150}{g}v_2$$

*i.e.,* $$2v_1 + v_2 = 2 \quad ...(i)$$

From the definition of coeff. of restitution, we have

$$e(u_1 - u_2) = v_2 - v_1$$

$$0.8(6 + 10) = v_2 - v_1$$

*i.e.,* $$v_2 - v_1 = 12.8 \quad ...(ii)$$

From (*i*) and (*ii*), we get

$$3v_1 = -10.8$$

$$v_1 = \mathbf{-3.6 \text{ m/s}}$$ **Ans.**

Hence $$v_2 = 12.8 - v_1 = 12.8 - 3.6$$

$$v_2 = \mathbf{9.2 \text{ m/s}}$$ **Ans.**

**Example 11.2.** A 80 N body moving to the right at a speed of 3 m/s strikes a 10 N body that is moving to the left at a speed of 10 m/s. The final velocity of 10 N body is 4 m/s to the right. Calculate the coefficient of restitution and the final velocity of the 80 N body.

**Solution.** $v_2 = 3$ m/s, $u_2 = -10$ m/s

$v_1 = ?$, $v_2 = 4$ m/s

Applying the principle of conservation of momentum to the colliding bodies, we get

$$\frac{80}{g} \times 3 + \frac{10}{g}(-10) = \frac{80}{g} v_1 + \frac{10}{g} \times 4$$

*i.e.,*
$$v_1 = \frac{80 \times 3 - 100 - 40}{80} = \mathbf{1.25 \text{ m/s}}$$ **Ans.**

From the definition of coefficient of restitution, we get

$$e(u_1 - u_2) = v_2 - v_1$$
$$e(3 + 10) = 4 - 1.25$$
$$\boldsymbol{e = 0.212}$$ **Ans.**

**Example 11.3.** A golf ball is dropped from a height of 10 m on a fixed steel plate. The coefficient of restitution is 0.894. Find the height to which the ball rebounds on the first, second and third bounces.

**Solution.** Initial height $h_0 = 10$ m

Velocity of golf ball before impact $u_1 = \sqrt{2gh_0}$

Velocity of steel plate before impact $u_2 = 0$

Velocity of steel plate after impact $v_2 = 0$

Let velocity of golf ball after impact be $v_1$.

From the definition of coefficient of restitution, we have

$$e(u_1 - u_2) = v_2 - v_1$$
$$e(\sqrt{2gh_0} - 0) = 0 - v_1$$
$$v_1 = -e\sqrt{2gh_0} = e\sqrt{2gh_0} \quad \text{in upward direction.}$$

From kinematic equation, the height $h_1$ to which the ball will rise is given by

$$v_1^2 - 0 = 2gh_1$$
$$h_1 = \frac{v_1^2}{2g} = \frac{e^2 \times 2gh_0}{2g}$$

*i.e.,*
$$h_1 = e^2 h_0 \qquad ...(i)$$

Now, $e = 0.894$, $h_0 = 10$ m

$\therefore$ $h_1 = 0.894^2 \times 10 = \mathbf{7.992 \text{ m}}$ **Ans.**

Similarly, after second bounce, the height to which the ball will rise is given by

$$h_2 = e^2 h_1 = 0.894^2 \times 7.992$$
$$= \mathbf{6.388 \text{ m}}$$ **Ans.**

And, after third bounce, the height

$$h_3 = e^2 h_2 = 0.894^2 \times 6.388$$
$$\boldsymbol{h_3} = \mathbf{5.105 \text{ m}}$$ **Ans.**

**Example 11.4.** A ball is dropped from a height of 1 m on a smooth floor. The height of first bounce is 0.810 m.

Determine:

(*a*) coefficient of the restitution.

(*b*) expected height of second bounce.

**Solution.** Velocity of ball before striking the floor

$$u_1 = \sqrt{2gh} = \sqrt{2 \times 9.81 \times 1} = 4.429 \text{ m/s}$$

Velocity of ball after striking the floor

$$v_1 = -\sqrt{2gh_1} = -\sqrt{2 \times 9.81 \times 0.810} = -3.987 \text{ m/s}$$

There is no movement of the floor, before and after striking, *i.e.*, $u_2 = v_2 = 0$

From the definition of coefficient of restitution, we have

$$e(u_1 - u_2) = v_2 - v_1$$

$$e \times 4.429 = 0 + 3.987$$

$$e = \frac{3.987}{4.429} = \mathbf{0.9} \qquad \textbf{Ans.}$$

Let velocity of the ball after second bounce be $v_2$.

Velocity of strike in this case, $v_1 = 3.987$ m/s downward.

$$\therefore \qquad e(3.987 - 0) = 0 - v_2$$

$$v_2 = -3.586 \text{ m/s}$$

$$= 3.586 \text{ m/s} \quad \text{upward.}$$

Expected height $h_2$ is given by the kinematic equation as

$$h_2 = \frac{v_2^{\,2}}{2g} = \frac{3.576^2}{2 \times 9.81}$$

$$\mathbf{h_2 = 0.6561 \text{ m}} \qquad \textbf{Ans.}$$

**Example 11.5:** A 10 N ball traverses a frictionless tube, as shown in Fig. 11.4, falling through a height of 2 m. It then strikes a 20 N ball hung from a rope 1.2 m long.

Determine the height to which the hanging ball will rise:

(*i*) if the collision is perfectly elastic.

(*ii*) if the coefficient of restitution is 0.7.

**Solution.** Velocity of 10 N ball after falling through 2 m height

$$u_1 = \sqrt{2gh} = \sqrt{2 \times 9.81 \times 2} = 6.264 \text{ m/s}$$

Velocity of 20 N ball before impact

$$u_2 = 0$$

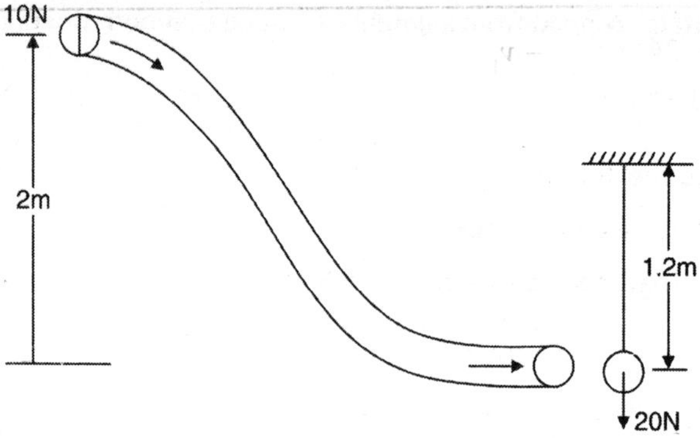


**Fig. 11.4**

Let the velocities of 10 N and 20 N balls after impact be $v_1$ and $v_2$ respectively.

By principle of conservation of momentum

$$m_1u_1 + m_2u_2 = m_1v_1 + m_2v_2$$

$$\frac{10}{g}\,6.264 + 0 = \frac{10}{g}v_1 + \frac{20}{g}v_2$$

i.e., $$v_1 + 2v_2 = 6.264 \qquad ...(i)$$

From the definition of coefficient of restitution, we have

$$e(u_1 - u_2) = v_2 - v_1 \qquad ...(ii)$$

*Case (i)*: $e = 1$

$$\therefore \quad u_1 - u_2 = v_2 - v_1$$

i.e., $$v_2 - v_1 = 6.264 \qquad ...(iii)$$

From Eqns. (*i*) and (*iii*),

$$v_2 = \frac{2 \times 6.264}{3} = 4.176 \text{ m/s}$$

$$\therefore \quad v_1 = 6.264 - 2 \times 4.176$$

$$= -2.088 \text{ m/s}$$

Let $h$ be the height to which hanging ball will rise after impact. Applying work-energy equation

Change in K.E. = Work done

$$0 - \frac{20}{2g}v_2^2 = -20 \times h$$

$$h = \frac{v_2^2}{2g} = \frac{4.176^2}{2 \times 9.81}$$

$$\boldsymbol{h = 0.889 \text{ m}}$$ **Ans.**

*Case (ii)*: $e = 0.7$

From Eqn. (*ii*), we have

$$0.7(6.264) = v_2 - v_1$$

*i.e.*, $$v_2 - v_1 = 4.385 \quad ...(iv)$$

From Eqns. (*i*) and (*iv*), we get

$$v_2 = \frac{6.264 + 4.385}{3} = 3.55 \text{ m/s}$$

∴ The height to which ball will rise

$$h_2 = \frac{v_2^2}{2g} = \frac{4.176^2}{2 \times 9.81}$$

*i.e.*, $$\mathbf{h_2 = 0.642 \text{ m}}$$ **Ans.**

## 11.4 OBLIQUE IMPACT

Referring to Fig. 11.1(*b*) note that, the expression for coefficient of restitution, $e(u_1 - u_2) = v_2 - v_1$, holds good only for the component of velocities in the line of impact. The component of velocities in a direction at right angles to line of impact remain unaltered.

**Example 11.6.** The magnitude and direction of the two identical smooth balls before central oblique impact are as shown in Fig. 11.5. Assuming coefficient of restitution $e = 0.9$, determine the magnitude and direction of the velocity of each ball after the impact.

**Solution.** Component of velocity of *A* before impact:

(*i*) Normal to line of impact = 9 sin 30°

$$u_{AY} = 4.5 \text{ m/s}$$

(*ii*) In the line of impact = 9 cos 30°

$$u_{AX} = 7.794 \text{ m/s}$$

Component of velocity of *B* before impact:

(*i*) Normal to line of impact = 12 sin 60°

$$u_{BY} = 10.392 \text{ m/s}$$

(*ii*) In the line of impact $u_{BX} = -12 \cos 60°$

$$= -6 \text{ m/s}$$

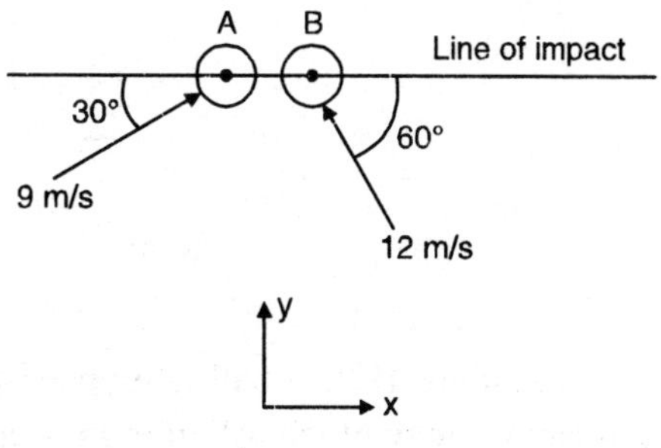


**Fig. 11.5**

Component of velocities only in the line of impact get affected by impact:

$$v_{AY} = u_{AY} = 4.5 \text{ m/s}$$

$$v_{BY} = u_{BY} = 12 \sin 60° = 10.392 \text{ m/s}$$

Let $v_{AX}$ and $v_{BX}$ be the velocities after impact. Applying principles of conservation of momentum in the line of impact, we get

$$mu_{AX} + mu_{BX} = mv_{AX} + mv_{BX}$$

$$7.794 - 6 = v_{AX} + v_{BX}$$

or $$v_{AX} + v_{BX} = 1.794 \quad ...(i)$$

From the definition of coefficient of restitution, we have

$$0.9(u_{AX} - u_{BX}) = v_{BX} - v_{AX}$$

$$v_{BX} - v_{AX} = 0.9(7.794 + 6) = 12.415 \quad ...(ii)$$

From Eqns. (*i*) and (*ii*), we get

$$v_{BX} = \frac{12.415 + 1.794}{2} = 7.104 \text{ m/s}$$

$$\therefore \quad v_{AX} = 1.794 - 7.104 = -5.310 \text{ m/s}$$

$$v_A = \sqrt{5.31^2 + 4.5^2} = \mathbf{6.960 \text{ m/s}}$$ **Ans.**

$$\theta_A = \tan^{-1}\frac{4.5}{5.31} = \mathbf{40.28°} \text{ as shown in Fig. 11.6}(a).$$ **Ans.**

$$v_B = \sqrt{7.104^2 + 10.392^2} = \mathbf{12.588 \text{ m/s}}$$ **Ans.**

and $$\theta_B = \tan^{-1}\left(\frac{10.392}{7.104}\right) = \mathbf{55.643} \text{ as shown in Fig. 11.6 }(b).$$ **Ans.**

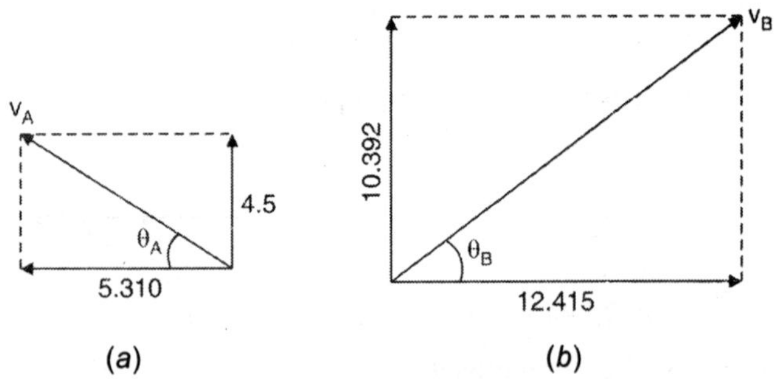


**Fig. 11.6**

**Example 11.7.** A ball is dropped from a height of 3 m upon a 15° incline. If $e = 0.8$, find the resultant velocity of the ball after impact.

**Solution.** The ball falls freely under gravity from a height of $h_0 = 3$ m. Hence, its vertical downward velocity at the instance of striking the incline

$$u_1 = \sqrt{2gh_0} = \sqrt{2 \times 9.81 \times 3}$$

$$= 7.672 \text{ m/s}$$

Line of impact is normal to the plane, *i.e.*, at 15° to vertical. Taking axes *x* and *y* as shown in Fig. 11.7, velocity of the ball before striking the plane:

(*i*) Normal to the line of impact $u_{1x} = u_1 \sin 15° = 1.986$ m/s

(*ii*) In the line of impact $u_{1y} = -u \cos 15° = -7.411$ m/s

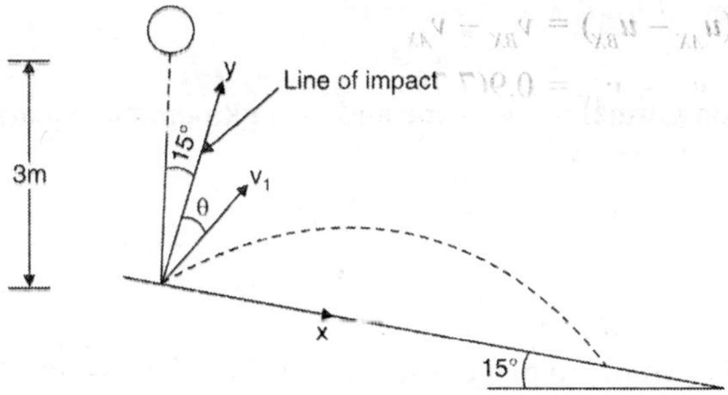


Fig. 11.7

Let the velocity after impact be $v_2$.

In the direction normal to line of impact, the component of velocity is not affected.

$$v_{1x} = u_{1x} = 1.986 \text{ m/s}$$

Initial and final velocities of floor = 0

From the definition of coefficient of restitution, we have

$$0.8(-7.412 - 0) = 0 - v_{1y}$$

$$v_{1y} = 5.929 \text{ m/s}$$

$$v_1 = \sqrt{v_{1x}^2 + v_{1y}^2} = \sqrt{1.986^2 + 5.929^2} = \mathbf{6.253\ m/s}$$

$$\theta = \tan^{-1}\left(\frac{v_{1x}}{v_{1y}}\right) = \tan^{-1}\left(\frac{1.986}{5.929}\right)$$

$$= 18.52° \text{ to the line of impact}$$ **Ans.**

∴ Inclination to the plane

$$= 90° - 18.52° = \mathbf{73.48°}$$ **Ans.**

**Example 11.8.** A ball falls vertically for 3 seconds on a plane inclined at 20° to the horizontal axis. If the coefficient of restitution is 0.8, when and where will the ball strike the plane again?

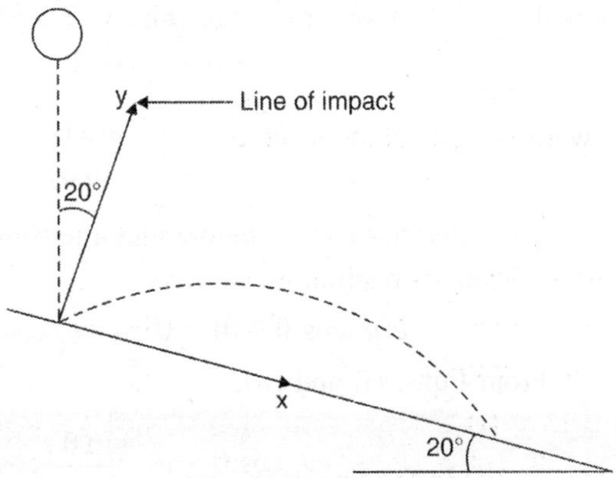


Fig. 11.8

**Solution.** Velocity of ball which striks the plane

$$3 \times g = 3g \text{ downward}$$

Component of velocity down the plane

$$= 3\, g \sin 20°$$

Component of velocity in the line of impact

$$u_y = -3\, g \cos 20°$$

Velocity after the impact

$$v_y = eu_y = 0.8 \times 3g \cos 20° = 2.4g \cos 20°$$

Acceleration in the line of impact $= - g \cos 20°$

Considering the motion normal to the plane and using kinematic equation $s = ut + \frac{1}{2}at^2$, we get

$$0 = 2.4\, g \cos 20°\, t - \left(\frac{1}{2}\right) g \cos 20°\, t^2$$

$$t = 4.8 \text{ s}$$

Component of velocity parallel to plane is not affected by the impact, since it is normal to the line of impact.

$$v_x = 3g \sin 20°$$

Acceleration in this direction $= g \sin 20°$

∴ Distance travelled in 4.8 second is given by

$$s = 3g \sin 20° t - \frac{1}{2} g \sin 20°\, t^2$$

$$= g \sin 20° \left(3 \times 4.8 - \frac{1}{2} \times 4.8^2\right)$$

$$= 9.81 \sin 20° \left(3 \times 4.8 - \frac{1}{2} \times 4.8^2\right)$$

$s =$ **86.967 m.** **Ans.**

**Example 11.9.** A ball is thrown at an angle θ with the normal to a smooth wall. It rebounds at an angle θ′ with the normal. Show that the coefficient of restitution is expressed by

$$e = \frac{\tan \theta}{\tan \theta'}.$$

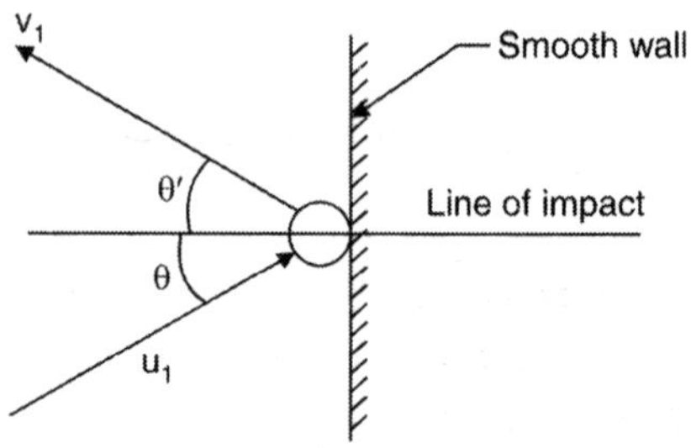

**Fig. 11.9**

**Solution.** Component normal to line of impact is unaltered after impact. If $u_1$ is velocity of approach and $v_1$ is the velocity of separation of the ball (Fig. 11.9), we get

$$u_1 \sin \theta = v_1 \sin \theta'$$

or

$$v_1 = u_1 \frac{\sin \theta}{\sin \theta'} \quad ...(i)$$

The velocities of wall before and after impact, we have *i.e.*, $u_2$, $v_2 = 0$. Hence from the definition of coefficient of restitution, we have

$$e(u_1 \cos \theta - 0) = 0 - (- v_1 \cos \theta') = v_1 \cos \theta' \quad ...(ii)$$

From Eqns. (*i*) and (*ii*),

$$eu_1 \cos \theta = u_1 \frac{\sin \theta}{\sin \theta'} \cos \theta'$$

or

$$e = \frac{\tan \theta}{\tan \theta'}$$

**Example 11.10.** A ball is dropped on an inclined plane and is observed to move horizontally after the impact. If coefficient of restitution is '*e*', determine the inclination of the plane and the velocity after impact.

**Solution.** Let θ be the inclination of the plane to horizontal (Ref. Fig. 11.10). Hence the line of impact is at right angles to the plane. The vertical downward velocity before striking the plane be *u* and the horizontal velocity after impact be *v*. During impact, component of velocity normal to the line of impact is not altered. Hence

$$u \sin\theta = v \cos\theta$$

$$v = u \tan\theta \qquad ...(i)$$

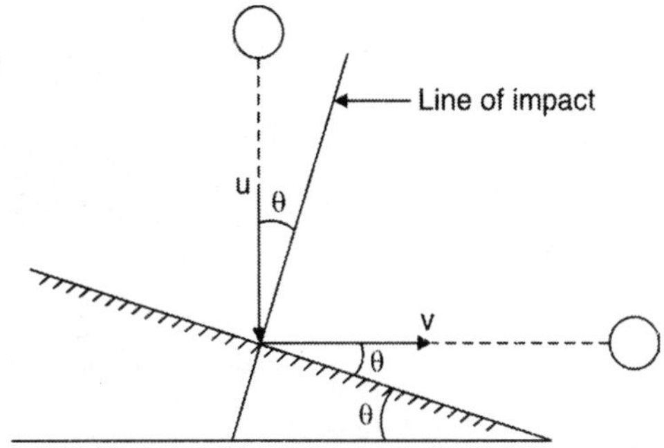


**Fig. 11.10**

Considering the velocities of ball and inclined plane (which is zero), the coefficient of restitution *e* is given by

$$e(u \cos\theta - 0) = [0 - (-v \sin\theta)]$$

$$e\, u \cos\theta = v \sin\theta$$

Substituting the value of *v* from Eqn. (*i*), we get

$$e\, u \cos\theta = u \tan\theta \sin\theta$$

or $$e = \tan^2\theta$$

or $$\tan\theta = \sqrt{e}$$ **Ans.**

From Eqn. (*i*), $$v = u \tan\theta$$

*i.e.*, $$\mathbf{v = u\sqrt{e}}$$ **Ans.**

**Example 11.11.** A ball is dropped from a height $h_0$ = 1.2 m on a smooth floor as shown in Fig. 11.11. Knowing that for the first bounce, $h_1$ = 1 m and $D_1$ = 0.4 m, determine

(*a*) the coefficient of restitution.

(*b*) the height and the range of the second bounce.

**Solution.** Line of impact being vertical, horizontal component of velocity is not affected by impact.

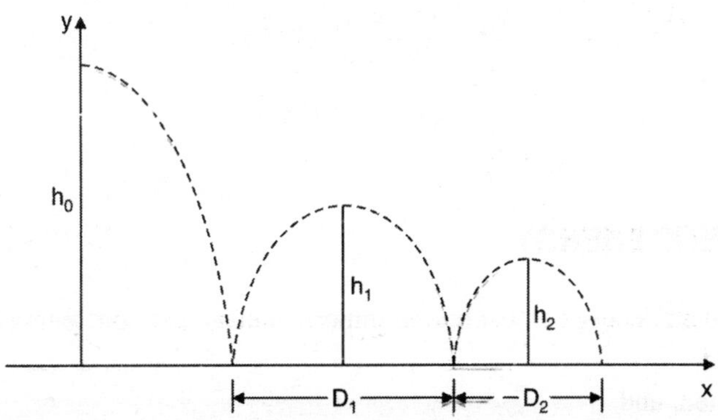


**Fig. 11.11**

Let this value be $u_x$.

The ball is dropped from a height of $h_0$ = 1.2 m.

Hence vertical component of velocity before first impact

$$u_y = \sqrt{2gh_0} \text{ downward}$$

Let the vertical component of velocity after first impact be $v_{1y}$.

Since after first bounce, the ball has raised to

$$h_1 = 1 \text{ m}$$

$$v_{1y} = \sqrt{2gh_1} \text{ upward}$$

$$e = \frac{\text{Relative velocity of separation}}{\text{Relative velocity of approach}}$$

$$= \frac{\sqrt{2gh_1}}{\sqrt{2gh_0}} = \sqrt{\frac{h_1}{h_0}} = \sqrt{\frac{1}{1.2}} = \mathbf{0.913} \qquad \textbf{Ans.}$$

Time of flight $t_1$ in first bounce $= \dfrac{2v_{1y}}{g} = \dfrac{2\sqrt{2gh_1}}{g} = \dfrac{2\sqrt{2 \times 9.81 \times 1}}{9.81}$

$$= 0.903 \text{ s}$$

Range, $D_1 = u_x t$

$$0.4 = u_x \times 0.903$$

$$u_x = 0.443 \text{ m/s}$$

Vertical component of velocity after second bounce

$$v_{2y} = e\sqrt{2gh_1} = 0.903\sqrt{2 \times 9.81 \times 1} = 4.0 \text{ m/s}$$

$$\therefore \qquad h_2 = \frac{v_{2y}^2}{2g} = \mathbf{0.815 \text{ m}} \qquad \textbf{Ans.}$$

Time of flight, $t_2 = 2 \times \dfrac{v_{2y}}{g} = \dfrac{2 \times 4}{9.81} = 0.815 \text{ s}$

Horizontal range $D_2 = u_x \times t_2 = 0.443 \times 0.815$

$$= \mathbf{0.316 \text{ m}} \qquad \textbf{Ans.}$$

## 11.5 LOSS OF KINETIC ENERGY

During collision, the kinetic energy is lost due to imperfect elastic action. Energy is also lost due to

(*a*) heat generated,

(*b*) sound generated, and

(*c*) vibration of colliding bodies.

The loss of kinetic energy can be found by finding kinetic energies before impact and after impact. Let $u_1$, $u_2$ be initial velocities and $v_1$, $v_2$ be final velocities of two bodies colliding in the line of impact, their weights being $W_1$ and $W_2$. Then

$$\text{Initial K.E.} = \frac{W_1}{2g}u_1^2 + \frac{W_2}{2g}u_2^2$$

$$\text{Final K.E.} = \frac{W_1}{2g}v_1^2 + \frac{W_2}{2g}v_2^2$$

$\therefore$ Loss of K.E. = Initial K.E. – Final K.E.

$$= \left(\frac{W_1}{2g}u_1^2 + \frac{W_2}{2g}u_2^2\right) - \left(\frac{W_1}{2g}v_1^2 + \frac{W_2}{2g}v_2^2\right)$$

$$= \frac{W_1}{2g}(u_1^2 - v_1^2) + \frac{W_2}{2g}(u_2^2 - v_2^2)$$

**Example 11.12.** A sphere of weight 10 N moving at 3 m/s collides with another sphere of weight 50 N moving in the same line at 0.6 m/s. Find the loss of kinetic energy during impact and show that the direction of motion of the first sphere is reversed after the impact. Assume coefficient of restitution as 0.75.

**Solution.** Let velocities after impact be $v_1$ and $v_2$.

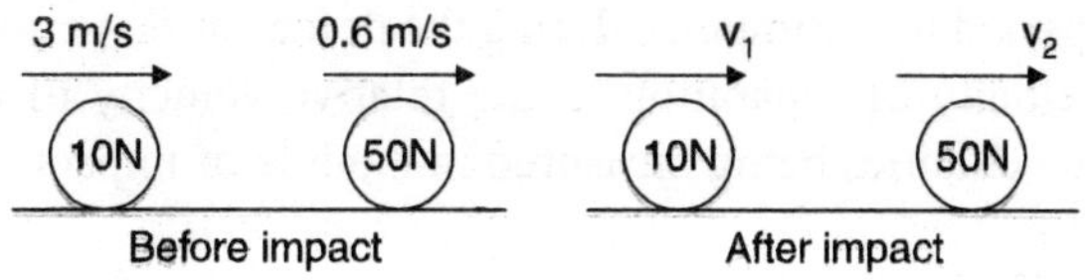


**Fig. 11.12**

$$u_1 = 3 \text{ m/s}, \; u_2 = 0.6 \text{ m/s}$$

From principles of conservation of momentum, we have

$$\frac{10}{g}(3) + \frac{50}{g}(0.6) = \frac{10}{g}v_1 + \frac{50}{g}v_2$$

$$v_1 + 5v_2 = 6 \qquad ...(i)$$

From the definition of coefficient of restitution, we have

$$0.75\,(3 - 0.6) = v_2 - v_1 \quad \text{or} \quad v_2 - v_1 = 1.8 \qquad ...(ii)$$

From Eqns. (*i*) and (*ii*), $v_2 = 1.3$ m/sec

and $v_1 = 6 - 1.3 \times 5 = -0.5$ m/s

Thus, **the velocity of first ball is reversed after impact.**

Loss of K.E. = Initial K.E. – Final K.E.

$$= \frac{10}{2g} \times 3^2 + \frac{50}{2g} \times 0.6^2 - \left\{\frac{10}{2g} \times (0.5)^2 + \frac{50}{2g} \times 1.3^2\right\}$$

$$= \frac{10}{2 \times 9.81}(9 + 1.8 - 0.25 - 8.45)$$

$$= \mathbf{1.07 \text{ joules}}$$ **Ans.**

**Example 11.13.** Find the loss of kinetic energy in Example 11.6 if the weight of each ball is 10 N.

**Solution.** Component of velocities in $y$-direction is not affected. Hence there is no change in K.E. due to the components of velocities in $y$-direction. In $x$-direction,

$u_{AX} = 7.79$ m/s; $u_{BX} = -6$ m/s;

$v_{AX} = -5.31$ m/s; $v_{BX} = 7.104$ m/s.

$$\text{Mass of the ball} = \frac{10}{g} = \frac{10}{9.81}$$

$$\therefore \quad \text{Loss of K.E.} = \frac{10}{2 \times 9.81} [u_{AX}^2 + u_{BX}^2 - v_{AX}^2 - v_{BX}^2]$$

$$= \frac{10}{2 \times 9.81} [7.79^2 + (-6)^2 - (-5.31)^2 - 7.104^2]$$

$$= \mathbf{18.02 \text{ joules}}$$ **Ans.**

## IMPORTANT DEFINITIONS

1. For definitions of the terms 'line of impact', direct impact, oblique impact, central impact, eccentric impact, refer Art 11.1.
2. The **coefficient of restitution** of two colliding bodies may be taken as the ratio of impulse during restitution period to the impulse during the deformation period. This is also equal to the ratio of relative velocity of separation to the relative velocity of approach of the colliding bodies, the relative velocities being measured in the line of impact.

## IMPORTANT FORMULAE

1. $e = \dfrac{v_2 - v_1}{v_1 - v_2}$, for the component of velocities in the line of impact.
2. For a freely falling body from height $h_0$ and having bounce $h_1$,

$$e = \sqrt{h_1 / h_0}\,.$$

## QUESTIONS

**11.1** Define the terms:

(*a*) Line of impact (*b*) Oblique impact

(*c*) Central impact (*d*) Eccentric impact

**11.2** Define the terms:

(*a*) Period of deformation (*b*) Period of restitution

(*c*) Coefficient of restitution

## EXERCISES

**11.1** Two bodies, one of which is 400 N with a velocity of 8 m/s and the other of 250 N with a velocity of 12 m/s, move towards each other along a straight line and impinge centrally. Find the velocity of each body after impact if the coefficient of restitution is 0.8.

[**Ans.** $v_1 = -5.846$ m/s; $v_2 = 10.154$ m/s]

**11.2** A ball is dropped from a height of $h_0$ metres on a floor. Show that after first bounce it will rise to height $h_1$ given by $h_1 = e^2 h_0$, where $e$ is coefficient of restitution. Determine to what height the ball will rise after 3 bounces if dropped from a height of 2 m if the coefficient of restitution is 0.8.

[**Ans.** 0.524 m]

**11.3** 10 N sphere shown in Fig. 11.13 is released from rest when $\theta_A = 90°$. The coefficient of restitution between the sphere and the block is 0.70. If the coefficient of friction between the block and the horizontal surface is 0.3, determine how far the block will move after impact. [**Ans.** $s = 0.934$]

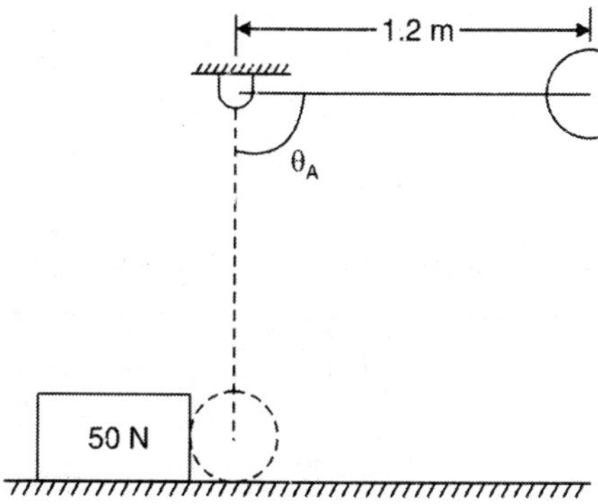


**Fig. 11.13**

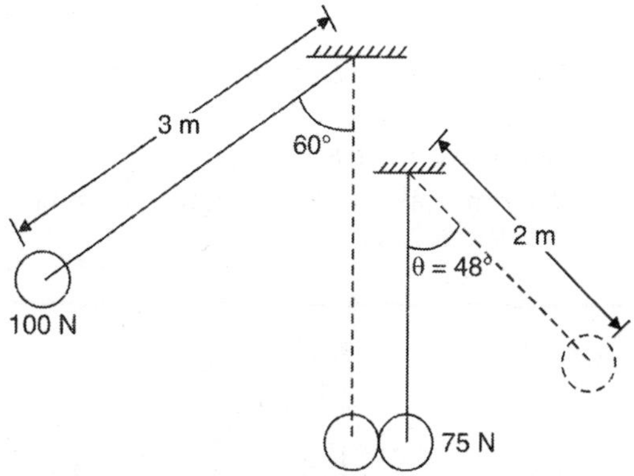


**Fig. 11.14**

**11.4** 100 N ball as shown in Fig. 11.14 is released from the position shown by continuous line. It strikes a freely suspended 75 N ball. After impact, 75 N ball is raised by an angle $\theta = 48°$. Determine the coefficient of restitution. [**Ans.** $e = 0.162$ m]

[**Hint:** $u_1 = \sqrt{2gh_1}$ where, $h_1 = 3(1 - \cos 60°)$

$v_2 = \sqrt{2gh_2}$ where $h_2 = 2(1 - \cos 48°)$]

**11.5** Two identical balls, moving horizontally, collide as shown in Fig. 11.15. Determine their velocities after impact, if the coefficient of restitution is 0.75.

[**Ans.** $v_1 = 8.718$ m/s; $\theta_1 = 23.41°$; $v_2 = 8.718$ m/s; $\theta_2 = 96.587°$]

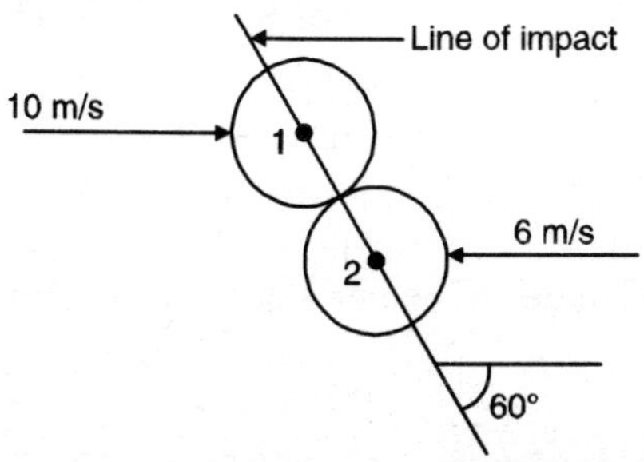


**Fig. 11.15**

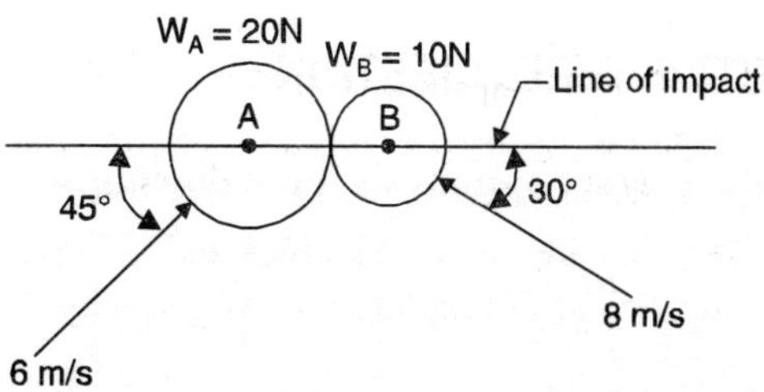


**Fig. 11.16**

**11.6** Find the velocities of the two balls shown in Fig. 11.16 after impact, if the coefficient of restitution = 0.6. [**Ans.** $v_A = 4.576$ m/s; $\theta_A = 67.99°$; $v_B = 6.3932$ m/s; $\theta_B = 38.73°$]

**11.7** Central impact takes place between a 40 N ball and a stationary block of 60 N as shown in Fig. 11.17. The block rests on rollers and move freely on horizontal surface. Find the velocities of the block the and ball after impact. Take coefficient of restitution as 0.8.

[**Ans.** Velocity of block = 5.76 m/s; velocity of ball = 4.04 m/s at 67.88° to horizontal]

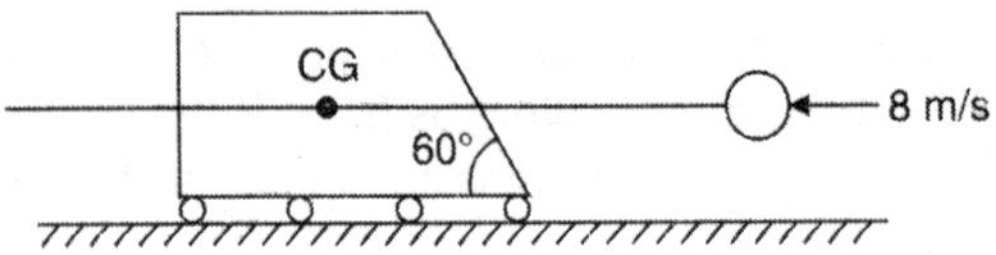


**Fig. 11.17**

**11.8** A ball drops on to a smooth horizontal floor and bounces as shown in Fig. 11.18. Derive expressions for coefficient of restitution in terms of (*a*) two successive heights and (*b*) two successive ranges.

Determine also time of *f* lights in *n*th bounce.

[**Ans.** (*a*) $e = \sqrt{h_n/(h_{n-1})}$ and (*b*) $e = D_n/D_{n-1}$, $t = 2\sqrt{2h_n/g}$]

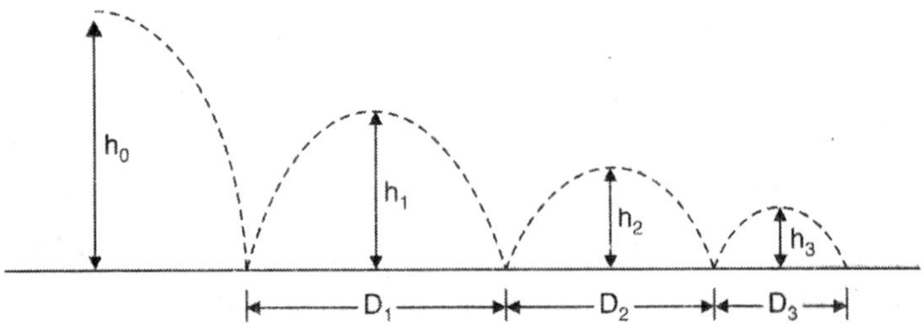


**Fig. 11.18**

**11.9** 80 N and 150 N bodies are approaching each other with a velocity of 20 m/s, and 6 m/s respectively. What will be the velocity of each body after impact ? How much is the loss of kinetic energy? Take coefficient of restitution = 0.6. Assume 80 N block is moving from left to right.

[**Ans.** For 80 N block $v_1 = \overleftarrow{7.13}$ m/s; 150 N block, $v_2 = \overrightarrow{8.47}$ m/s;
Loss of K.E. = 1150.43 joules]

## OBJECTIVE TYPE QUESTION

**Select the correct options in the following:**

**11.1** If $u_1$, $u_2$ are initial velocities and $v_1$, $v_2$ are final velocities of bodies 1 and 2 in an impact, the coefficient of restitution '*e*' is given by

(*a*) $\dfrac{v_1 - v_2}{u_1 - u_2}$ (*b*) $\dfrac{v_2 - v_1}{u_2 - u_1}$

(*c*) $\dfrac{v_2 - v_1}{u_1 - u_2}$ (*d*) $\dfrac{v_1 - v_2}{u_2 - u_1}$